STEAM TABLES

STEAM TABLES

Thermodynamic Properties of Water
Including
Vapor, Liquid, and Solid Phases

(International System of Units—S.I.)

JOSEPH H. KEENAN, B.S., LL.D.

Professor of Mechanical Engineering
Massachusetts Institute of Technology

FREDERICK G. KEYES, B.S. Sc.D., Ph.D.

Professor of Physical Chemistry
Massachusetts Institute of Technology

PHILIP G. HILL, B.Sc., Sc.D.

Professor of Mechanical Engineering
University of British Columbia, Vancouver, Canada

JOAN G. MOORE, B.S.

Member of Research Staff
Massachusetts Institute of Technology

A Wiley-Interscience Publication
John Wiley & Sons
New York · Chichester · Brisbane · Toronto

Library of Congress Cataloging in Publication Data:

Main entry under title:

Steam tables.

   "A Wiley-Interscience publication."
   Introd. in English, French, German, Italian, Japanese,
Russian, and Spanish.
   Includes bibliographical references.
   1.   Thermodynamics—Tables, calculations, etc.
2.   Steam—Tables, calculations, etc.   I.   Keenan,
Joseph Henry, 1900–1977   II.   Title.

QC311.3.S73   1978   621.1′021′2   77-28321
ISBN 0-471-04210-2

10   9   8   7   6   5   4   3

Introduction to the Steam Tables

In the past decade a substantial body of experimental data on thermodynamic and transport properties of water has been produced and published by research groups in the U.S.S.R., Great Britain, Czechoslovakia, Canada, and the United States. This book presents the results of a new and independent correlation of all the new thermodynamic data, and all previously existing data. It constitutes a complete revision of the Keenan and Keyes Tables of 1936 and an extension of range to 100 MPa and about 1300°C.

The tables are based on a new fundamental equation of the form

$$\psi = \psi(v, T)$$

where ψ denotes specific Helmholtz free energy, T temperature on the Kelvin scale and v specific volume. This equation has a relatively small number of coefficients, it is continuous in all derivatives, and it represents a continuum of single phase states from zero pressure to 100 MPa or more, and from 0°C to 1300°C. Within this range it represents virtually all experimental data within reasonable estimates of experimental precision and indicates where some previous estimates may require revision. For the Bridgeman synthesis of observed values of vapor pressure, it yields values of Gibbs free energy for liquid and vapor that are virtually identical, and it is in accord with observed values of saturation properties and the Bridgeman version of the critical state. It yields reasonable numbers for properties of metastable states in both liquid and vapor regions, and it represents faithfully the region of maximum density.

Values of the thermodynamic properties specific volume, internal energy, enthalpy, and entropy are tabulated with pressure and temperature as parameters for vapor-liquid equilibrium, vapor-solid equilibrium, superheated vapor, and compressed liquid. A table of properties of water vapor as a semiperfect gas is provided for analysis of processes at low pressure. A detailed table for the critical region shows volume and temperature as parameters. Intervals are generally small enough for linear interpolation to be adequate.

Mollier and temperature-entropy charts are included along with charts of heat capacity of liquid and vapor, Prandtl number, and isentropic expansion exponent. Extensive tables of viscosity and thermal-conductivity are reproduced from the documents of the Sixth International Conference on the Properties of Steam, along with pertinent conversion factors.

Introduction aux
Tables de la Vapeur d'Eau

Dans le courant des dix dernières années, une quantité appréciable de données expérimentales sur les propriétés thermodynamiques et de transport de l'eau a été produite et publiée par des groupes de recherche en U.R.S.S., Grande Bretagne, Tchécoslovaquie, Canada, et les Etats-Unis d'Amérique. Ce livre présente les résultats d'une corrélation nouvelle et indépendante de toutes ces données et de toutes celles qui existaient préalablement. Il constitue une révision complète des Tables de Keenan et Keyes de 1936, et une extension de valeurs jusqu'à 100 MPa et 1300°C.

Les tables sont basées sur une équation fondamentale nouvelle, de la forme

$$\psi = \psi(v, T)$$

où ψ représente l'énergie libre spécifique de Helmholtz, T la température à l'échelle de Kelvin, et v le volume spécifique. Cette équation a un nombre de coéfficients relativement petit, toutes ses dérivées sont continues, et elle représente un continu d'états d'une seule phase de la pression 0 à 100 MPa ou plus, et de 0°C à 1300°C. Dans ces intervalles elle représente virtuellement toutes les données expérimentales où l'on tient compte des erreurs d'expérience auxquelles il est raisonnable de s'attendre, et elle indique là où les calculs préalables peuvent avoir besoin d'être revus. Pour la synthèse de Bridgeman des valeurs observées de la pression de la vapeur, elle donne des valeurs de l'énergie libre de Gibbs pour le liquide et la vapeur qui sont pratiquement identiques, et elle est en accord avec les valeurs observées des propriétés de saturation et la version de Bridgeman de l'état critique. Elle donne des chiffres raisonnables pour les propriétés d'états métastables aussi bien dans les régions du liquide que dans celles de la vapeur, et elle représente fidèlement la région de densité maximum.

Les valeurs des propriétés thermodynamiques volume spécifique, énergie interne, enthalpie, et entropie sont arrangées en tables avec la pression et la température comme paramètres pour l'équilibre entre la vapeur et le liquide, l'équilibre entre la vapeur et le solide, la vapeur surchauffée, et le liquide comprimé. Une table des propriétés de la vapeur d'eau en tant que gaz presque parfait est fournie pour l'analyse de transformations à basse pression. Une table détaillée pour la région critique utilise le volume et la température comme paramètres. Les intervalles sont généralement suffisamment petits pour permettre l'interpolation linéaire.

Des graphiques de Mollier et de température-entropie sont inclus avec des graphiques de la capacité calorifique du liquide et de la vapeur, du nombre de Prandtl, et de l'exposant d'expansion isentropique. Des tables étendues de viscosité et de conductibilité thermique sont reproduites à partir des documents de la Sixième Conférence Internationale sur les Propriétés de la Vapeur d'Eau, avec les facteurs de changement d'unités pertinents.

Einleitung zu den Dampftafeln

Im letzten Jahrzehnt ist eine Menge experimenteller Daten der thermodynamischen und Transport-Eigenschaften von Wasser von Forschungsgruppen in der UdSSR, Grossbritannien, der Tchechoslowakei, Kanada und den Vereinigten Staaten ermittelt und veröffentlicht worden. Dieses Buch präsentiert alle alten und neuermittelten thermodynamischen Daten als Ergebnis einer neuen und unabhängigen mathematischen Beziehung. Es stellt eine vollständige Überarbeitung der Keenan und Keyes Tafeln von 1936 dar und ist gegenüber diesen auf Bereiche bis zu 100 MPa und ungefähr 1300°C erweitert worden.

Die Tafeln basieren auf einer neuen Grundgleichung der Form

$$\psi - \psi(v, T)$$

worin ψ die spezifische Helmholtzsche Freie Energie, T die Temperatur in Grad Kelvin und v das spezifische Volumen darstellen. Diese Gleichung enthält nur relativ wenige Koeffizienten. Sie ist stetig in allen Ableitungen und stellt ein Kontinuum von Einphasen-Zuständen für Drücke zwischen 0 und 100 MPa oder mehr und Temperaturen von 0°C bis 1300°C dar. In diesem Bereich stimmt sie im Wesentlichen mit allen experimentell gefundenen Werten innerhalb annehmbarer Messgenauigkeiten überein und gibt an, wo eine Überarbeitung früherer Messwerte erforderlich sein könnte. Für die Bridgeman Synthese der beobachteten Werte des Sättigungsdruckes liefert sie Werte der Gibbs'schen Freien Energie für Flüssigkeit und Dampf, die im Wesentlichen gleich sind, und stimmt mit den beobachteten Werten der Sättigungseigenschaften und der Bridgeman Version des kritischen Zustandes überein. Sie gibt vernünftige Werte der Eigenschaften des metastabilen Zustandes im Flüssigkeits- als auch im Dampfgebiet wieder und repräsentiert gut das Gebiet der maximalen Dichte.

Werte der thermodynamischen Eigenschaften, spezifisches Volumen, Innere Energie, Enthalpie und Entropie sind mit Druck und Temperatur als Parameter für Dampf-Flüssigkeits Gleichgewicht, Dampf-Festkörper Gleichgewicht, überhitzten Wasserdampf und comprimierte Flüssigkeit tabelliert. Eine Tafel der Eigenschaften von Wasserdampf als halbideales Gas steht zur Analyse von Niedrigdruck-Prozessen zur Verfügung. Eine ausführliche Tafel mit Volumen und Temperatur als Parametern ist für die kritischen Gebiete beigefügt. Intervalle sind im allgemeinen klein genug, um linear interpolieren zu können.

Mollier- und Temperatur-Entropie-Diagramme sind zusammen mit den Diagrammen der spezifischen Wärme von Flüssigkeit und Dampf, Prandtlzahl und isentropischen Ausdehnungsexponenten eingeschlossen. Ausführliche Tabellen der Viscosität und Wärmeleitfähigkeit sind aus den Dokumenten der Sechsten Internationalen Konferenz über Dampfeigenschaften übernommen worden und die entsprechenden Umrechnungsfaktoren wurden beigefügt.

Introduzione Alle
Tavole Del Vapor D'Acqua

Nell'ultima decade una notevole quantità di dati sperimentali sulle propietà termodinamiche e del trasporto dell'acqua è stata ottenuta e pubblicata da gruppi di ricercatori in Russia, Inghilterra, Ceco-Slovacchia, Canada e Stati Uniti. Questo libro contiene i risultati ottenuti applicando una nuova e originale correlazione agli ultimi e ai già esistenti dati termodinamici. Il libro è una edizione completamente riveduta delle tavole di "Keenan e Keyes" edita nel 1936 con in aggiunta un estensione dei campi di pressione e temperatura rispettivamente a 100 MPa e 1300°C.

Queste tavole si basano su una nuova equazione fondamentale nella forma

$$\psi = \psi(v, T)$$

ψ denota la energia libera di Helmholtz, T la temperatura in scala Kelvin e v il volume specifico. Questa equazione presenta un modesto numero di coefficienti, è continua in tutte le derivate ed esprime con continuità tutti gli stati a fase singola con pressione da 0 a 100 MPa e con temperature da 0°C a 1300°C. In questo campo di valori interpola virtualmente tutti i dati sperimentali con ragionevole stima di precisione sperimentale e indica dove qualche precedente valutazione dovrebbe essere corretta. Per quanto riguarda la sintesi di Bridgeman sui dati esistenti della pressione di vapore, porta a valori di energia libera di Gibbs per liquido e vapore praticamente identici, ed è in accordo con le propietà a saturazione e con la versione di Bridgeman per lo stato critico. L'equazione infine dà ragionevoli risultati per le proprietà degli stati metastabili in entrambi le regioni di liquido e vapore e rappresenta con fedeltà la regione a massima densità.

Valori delle propietà termodinamiche, del volume specifico, dell'energia interna, dell'entalpia, e dell'entropia sono tabulati con pressione e temperatura come parametri per i casi di equilibrio liquido-vapore, solido-vapore, vapore surriscaldato e liquido compresso. Una tavola di propietà del vapor d'acqua considerato gas semiperfetto è riportata per l'analisi dei processi a bassa pressione. Una tavola dettagliata per la regione critica contiene volume e temperatura come parametri. Gli intervalli sono generalmente così piccoli da permettere una adeguata interpolineazione lineare.

Diagramma di Mollier e diagramma temperatura-entropia sono inclusi così come diagrammi del calore specifico dell'acqua e del vapore, del numero di Prandtl e dell'esponente di espansione isoentropica. Dai documenti della Sesta Conferenza Internazionale sulle Proprietà del Vapore sono infine stati riportate tavole dettagliate sulla viscosità e sulla conducibilità termica con relativi fattori di conversione.

新 蒸 気 表 に つ い て

　過去十年間に、水の熱力学的性質並びに輸送現象特性に関する数多くの実験データが、ロシア，英国，チェコスロバキア，カナダ，米国の研究グループによって発表された。本書では、これらの新しい熱力学実験データ全てと、既存の全てのデータとに基づいて得られた独自の新しい関係式を用いて、数表を作製した。本書は1936年出版の "Keenan and Keyes の蒸気表" の全面的改訂版で、圧力は1000バール　温度は約1300℃迄拡張されている。

　表は下記の新基礎方程式に基づいて作製した。

$$\psi = \psi (V, T)$$

ここで、ψ は Helmholtz の自由エネルギ，T は Kelvin の絶対温度，V は比容積である。上式は次の如き利点を含んでいる。即ち、比較的少数の係数を用い、且つ微係数は全て連続である。0バールから1000バール迄の圧力範囲0℃から1300℃迄の温度範囲で連続的な単相をこの　式で表現している。上記の範囲内では、実験精度以内で、殆んど全ての実験データと合致している。その結果、旧蒸気表で改訂を必要とする部分も明らかになった。Bridgeman による蒸気圧関係式を用いれば、上式は気液両相に対し、殆んど等しい Gibbs の自由エネルギを与える。上式は又、飽和特性の実験値ならびに Bridgeman による臨界状態をも満足している。更に上式は、気液両相において、準安定の状態として妥当な特性値を与え、また最大密度の領域をも忠実に表わしている。

　熱力学的諸特性値、即ち、比容積，内部エネルギ，エンタルピ，エントロピが、圧力と温度をパラメーターとして、気液平衡，気固平衡，過熱蒸気，圧縮液の夫々の場合について数表に示されている。低圧におけるプロセスの解析の便の為に、不完全ガスとしての水蒸気の特性表も示されている。臨界領域についての詳細な数表は比容積と温度をパラメータとしている。一般に数値の間隔は、直線的内挿法が適用しうるべく充分細かく選んである。

　気液両相での比熱，プラントル数，等エントロピ膨脹係数についての各表のみならず、モリエ線図と温度－エントロピ線図も含まれている。尚、粘性係数及び熱伝導率の広範な数表と換算係数は、第六回国際蒸気性質会議（The Sixth International Conference on the Properties of Steam）の結果から引用した。

Предисловие к Таблицам Пара

За последнее десятилетие, важные экспериментальные данные о термодинамических и двигающих свойствах воды, были разработаны и опубликованы научно-исследовательскими группами в С.С.С.Р., Великобритании, Чехословакии, Канаде и в С.Ш.А.

Эта книга представляет результат новой и самостоятельной корреляции всех новых термодинамических данных и всех прежде существовавших. Она дает полный пересмотр таблиц Кинан и Кейс 1936 г. и расширяет пределы применения до 1000 бар и примерно до 1300° температуры Цельсия.

Таблицы базируются на новом основном уравнении:

$$\psi = \psi(v, T)$$

где ψ означает удельную свободную энергию Хельмхольца, T — температуру по шкале Кельвина и v — удельный объем. Это уравнение имеет сравнительно мало коэфициентов, все производные непрерывные и оно дает постоянный порядок однофазных состояний, от нуля до 1000 бар давления или больше, и от 0° до 1300° температуры Цельсия.

В этих пределах уравнение дает фактически все экспериментальные данные, умеренно оценивая экспериментальную точность и указывает, где некоторые прежние вычисления должны быть пересмотрены.

Для синтеза Бриджмена, о величинах давления испарений, уравнение дает почти одинаковые значения свободной энергии Гиббса, для жидкостей и для испарений. Оно также соответствует наблюдаемым величинам свойств насыщения и версии критического состояния Бриджмена.

Уравнение дает приемлемые величины свойствам мета-устойчивости, для жидкостей и для испарений и правильно отображает состояние максимальной плотности.

Величины термодинамических свойств удельного объема, внутренней энергии, энтальпии и энтропии, внесены в таблицы давления и температуры как параметры, для состояния равновесия испарений — жидкостей, испарений — плотных веществ, перегретого испарения и сжатых жидкостей.

Таблица свойств водяного испарения, как полуидеального газа, предусмотрена для анализов процессов при низком давлении.

Детальная таблица для критического состояния, показывает объем и температуру как параметры.

Интервалы между величинами в общем небольшие и достаточны для линейной интерполяции.

Схемы Молье и температуры — энтропии помещены вместе с схемами теплоемкости жидкости и испарений, номерами Прандл и показателем степени изоэнтропического расширения.

Екстенсивные таблицы вязкости и удельной теплопроводности скопированы с документов Шестой Международной Конференции по Свойствам Пара, вместе с соответствующими переводными коэфициентами.

Introducción a
las Tablas de Vapor de Agua

Durante la década pasada grupos de investigación de la U.R.S.S., Gran Bretaña, Checoslovaquia, Canadá y los Estados Unidos de Norteamérica produjeron y publicaron un extenso conjunto de datos experimentales sobre las propiedades termodinámicas y de transporte del agua. En este libro se presentan los resultados de una correlación nueva e independiente de todos los nuevos datos termodinámicos junto con todos los ya existentes. Consiste en una revisión completa de las tablas de 1936 de Keenan y Keyes y una extensión hasta 100 MPa y alrededor de 1300°C en el rango cubierto.

Las tablas están basadas en una nueva relación fundamental de la forma

$$\psi = \psi(v, T)$$

donde ψ representa la energía libre específica de Helmholtz, T la temperatura en la escala Kelvin y v el volumen específico. Esta ecuación tiene un número relativamente pequeño de coeficientes, es continua en todas sus derivadas y representa una serie continua de estados compuestos de una sola fase desde una presión nula hasta 100 MPa o más, y desde 0°C hasta 1300°C. En este rango practicamente representa todos los datos experimentales dentro de cálculos razonable de precisión experimental, al mismo tiempo que muestra qué cálculos, de los aceptados previamente, requieren revisión. Con los valores de Bridgeman para datos observados de tensión del vapor, da resultados de la energía libre de Gibbs practicamente iguales para líquido y vapor, lo que concuerda con los valores observados de las propiedades de saturación y la versión de Bridgeman acerca del estado crítico. Da valores razonables para las propiedades de los estados metaestables, tanto en la región del líquido como en la del vapor, y representa con precisión la zona de máxima densidad.

Los valores de las propiedades termodinámicas volumen específico, energía interna, entalpia y entropia, correspondientes a equilibrio entre vapor y líquido, equilibrio entre vapor y sólido, vapor sobrecalentado y líquido comprimido, están tabulados tomando la presión y la temperatura como parámetros. Se incluye una tabla con las propiedades del vapor de agua considerado como gas semiperfecto para análisis de procesos a baja presión. Para una tabla detallada sobre la región crítica se toman como

parámetros la temperatura y el volumen. Los intervalos son en general lo suficiente-
mente pequeños para permitir el uso de la interpolación lineal.

Junto con los diagramas de calor específico del líquido y el vapor, número de Prandtl
y exponente de expansión isentrópica se incluyen diagramas de Mollier y temperatura-
entropia. Se reprodujeron, de la Sexta Conferencia Internacional sobre Propiedades
del Vapor extensas tablas de viscosidad y conductividad térmica, junto con los factores
de conversión pertinentes.

Preface

The tables by Keenan and Keyes of the properties of water, published in 1936, were based on critical correlation and formulation of all the experimental data available at that time. Included were the Keyes and Smith measurements of specific volume of the liquid; the Keyes, Smith, and Gerry measurements of specific volume of vapor; the Osborne, Stimson, Ginnings, and Fiock observations of saturation states; the enthalpy measurements of Callendar and Egerton and of Havlicek and Miskowsky, and the Gordon calculation from spectrographic data of the heat capacity at zero pressure. For specific volumes from 10 cm³/g to infinity the result was the p-v-T relation given by Equation 13 of the Keenan and Keyes tables and the formulation of the spectroscopic observations of specific heat capacity given by Gordon and expressed by Equation 15 in Keenan and Keyes. These two, taken together, were equivalent to a fundamental equation of the form $\zeta = \zeta(p, T)$, where ζ denotes specific Gibbs free energy, p pressure, and T temperature on the Kelvin scale. This fundamental equation can be stated explicitly and completely. It was completed before, and its results made available to, the Third International Conference on the Properties of Steam in 1934.

Values for the compressed liquid region in the Keenan and Keyes tables were taken from a semialgebraic, semigraphic study by Keenan in 1931. These values were also made available to the Third International Conference. Values for the narrow region of states between the critical temperature on the liquid side and a specific volume of 10 cm³/g were the subject of a subsequent algebraic-graphic study based largely on available enthalpy measurements.

Research activity in the U.S.S.R. early in the decade of the 1950's stimulated an international effort to extend experimental knowledge of the properties of water to higher pressures and temperatures than before. The results of this effort make possible a reliable reformulation of the properties of steam which not only extends the range of dependable knowledge but also permits refinement of knowledge of regions previously investigated. Moreover, the modern computer provides a correlation tool that is far more powerful than anything available 30 years ago.

With a background of formulation techniques, discussed by F. G. Keyes from time to time, an extensive examination and correlation of all available experimental data on liquid and vapor water was undertaken at the Massachusetts Institute of Technology about four years ago. Although the International Skeleton Tables of 1963 were then available, it was felt that an independent study was required in order to take full advantage of the new experimental information and computational tools. Consequently the present tables are in no sense a formulation of the International Skeleton Tables. In general they are in good accord with and their values fall within the tolerances of the International Skeleton Tables of 1963. The exceptions to this statement and their implications are discussed in the appendix.

The result of this study was a fundamental equation of the form $\psi = \psi(T, v)$, where ψ denotes specific Helmholtz free energy and v specific volume. This equation represents a continuity of single-phase states which cover both liquid and vapor regions to about 1400°C in temperature and 100 MPa in pressure and provides saturation states that agree in regard to Gibbs free energy. The agreement with experimental observations, which is discussed in detail in the appendix, is generally within the uncertainty in the observations. This fundamental equation is a new formulation. It was used to generate all tabulated values of thermodynamic properties corresponding to liquid and vapor states.

Because the principal objective was a table of thermodynamic properties, no similar independent study of transport properties, namely, viscosity and thermal conductivity, was undertaken. A substantial amount of new experimental observations of these properties has appeared since publication of the Keenan and Keyes tables. Much of it is of high quality. A committee of the Sixth International Conference on the Properties of Steam has studied all the available material and has arrived at reasonable compromises where differences between various sets of observations have seemed irreconcilable. Moreover, the International Skeleton Tables are, in this instance, sufficiently detailed for practical purposes. They were therefore adopted without modification.

The present tables extend the range of the Keenan and Keyes tables to 1300°C in temperature and to 100 MPa in pressure. The changes made in values of properties for states given in the Keenan and Keyes tables are generally small. They are discussed in the appendix.

Extension of the range of temperature makes desirable a change from lines of constant pressure to columns of constant pressure. Internal energies are given for all states in view of their importance in the wider variety of problems encountered in practice and in teaching. The engineering importance of metastable states is recognized by including a substantial band of them on both the liquid side and the vapor side of the two-phase region. A more convenient table of properties for compressed and metastable liquid is provided, along with a separate and detailed table for states near the critical point.

The fundamental equation itself, as distinguished from the tables, will doubtless prove useful in the design of machinery and the analysis of engineering data. The coefficients of the equation and its extension into expressions for the usual thermodynamic properties are exhibited in the appendix.

In order to ensure accuracy, the typesetting of the table of thermodynamic properties has been done by computer; that is, the numbers were generated from the fundamental equation on an IBM 360, converted to Photon control codes, and punched onto paper tape which was fed directly into a Photon typesetter to obtain the tables.

It is a pleasure to acknowledge indebtedness to many colleagues. The experimenters whose data were drawn upon in the development of the fundamental equation devoted untold time, skill, and ingenuity to providing a body of data that is unique in quality and extent. Thanks are gratefully extended to them in general. Thanks are due in particular, however, to those colleagues who have responded generously to requests for information; for example, G. S. Kell and E. Whalley transmitted preliminary data and E. J. LeFevre offered several helpful suggestions. Dr. Oscar Bridgeman of the U.S. National Bureau of Standards sent reports of his work before publication, and J. Kestin and E. A. Bruges reported frequently on their own and other work on transport properties. At one stage in the development of the fundamental equation some assistance was obtained from the Ford Foundation.

Thanks are due in greatest measure, however, to the Massachusetts Institute of Technology for support of three of the authors as members of its faculty and for the use of its extensive computer center. Glenn Peterson wrote the necessary routines for the IBM 360 computer that made the computer typesetting of the tables possible. The M.I.T. Instrumentation Laboratory provided the facilities for producing the paper tapes.

JOSEPH H. KEENAN
FREDERICK G. KEYES
PHILIP G. HILL
JOAN G. MOORE

Cambridge, Massachusetts
January 1969

This edition of the *Steam Tables* incorporates changes to the International Edition — Metric Units in order to provide full conformity with the International System of Units (S.I.).

PHILIP G. HILL

February, 1978
Vancouver, British Columbia

Contents

STEAM TABLES

Symbols Used in Tables

c_{po} specific heat capacity at constant pressure for zero pressure, kilojoules per kilogram kelvin, $kJ/(kg \cdot K)$.*

c_{vo} specific heat capacity at constant volume for zero pressure, kilojoules per kilogram kelvin, $kJ/(kg \cdot K)$.

h specific enthalpy, kilojoules per kilogram, kJ/kg.

h_o specific enthalpy at zero pressure, kilojoules per kilogram, kJ/kg.

k isentropic exponent, $-(\partial \log p/\partial \log v)_s$.

p pressure, megapascals, MPa.

p_r relative pressure, pressure of semiperfect vapor at zero entropy multiplied by 10^{-6}, megapascals, MPa.

s specific entropy, kilojoules per kilogram kelvin, $kJ/(kg \cdot K)$.

s_1 specific entropy of semiperfect vapor at 0.1 MPa; kilojoules per kilogram kelvin, $kJ/(kg \cdot K)$.

t thermodynamic temperature, degrees Celsius, °C.

T thermodynamic temperature, kelvin, K.

u specific internal energy, kilojoules per kilogram, kJ/kg.

u_o specific internal energy at zero pressure, kilojoules per kilogram, kJ/kg.

v specific volume, cubic meters per kilogram, m^3/kg.

v_r relative specific volume, specific volume of semiperfect vapor at zero entropy multiplied by 10^{10}, cubic meters per kilogram, m^3/kg.

ζ_1 specific Gibbs free energy of semiperfect vapor at 0.1 MPa, kilojoules per kilogram, kJ/kg.

ψ_1 specific Helmholtz free energy of semiperfect vapor at 0.1 MPa, kilojoules per kilogram, kJ/kg.

*Units are defined and compared in Table 10.

SUBSCRIPTS

f refers to a property of liquid in equilibrium with vapor

g refers to a property of vapor in equilibrium with liquid

i refers to a property of solid in equilibrium with vapor

fg refers to a change by evaporation

ig refers to a change by sublimation

Table 1. Saturation: Temperatures

Temp. °C	Press. MPa	Specific Volume		Internal Energy			Enthalpy			Entropy		
		Sat. Liquid	Sat. Vapor	Sat. Liquid	Evap.	Sat. Vapor	Sat. Liquid	Evap.	Sat. Vapor	Sat. Liquid	Evap.	Sat. Vapor
t	p	$10^3 v_f$	$10^3 v_g$	u_f	u_{fg}	u_g	h_f	h_{fg}	h_g	s_f	s_{fg}	s_g
0	.0006109	1.0002	206 278	−.03	2375.4	2375.3	−.02	2501.4	2501.3	−.0001	9.1566	9.1565
.01	.0006113	1.0002	206 136	.00	2375.3	2375.3	.01	2501.3	2501.4	.0000	9.1562	9.1562
1	.0006567	1.0002	192 577	4.15	2372.6	2376.7	4.16	2499.0	2503.2	.0152	9.1147	9.1299
2	.0007056	1.0001	179 889	8.36	2369.7	2378.1	8.37	2496.7	2505.0	.0305	9.0730	9.1035
3	.0007577	1.0001	168 132	12.56	2366.9	2379.5	12.57	2494.3	2506.9	.0457	9.0316	9.0773
4	.0008131	1.0001	157 232	16.77	2364.1	2380.9	16.78	2491.9	2508.7	.0610	8.9904	9.0514
5	.0008721	1.0001	147 120	20.97	2361.3	2382.3	20.98	2489.6	2510.6	.0761	8.9496	9.0257
6	.0009349	1.0001	137 734	25.19	2358.4	2383.6	25.20	2487.2	2512.4	.0912	8.9090	9.0003
7	.0010016	1.0002	129 017	29.38	2355.6	2385.0	29.39	2484.8	2514.2	.1062	8.8688	8.9751
8	.0010724	1.0002	120 917	33.59	2352.8	2386.4	33.60	2482.5	2516.1	.1212	8.8289	8.9501
9	.0011477	1.0003	113 386	37.80	2350.0	2387.8	37.80	2480.1	2517.9	.1362	8.7892	8.9253
10	.0012276	1.0004	106 379	42.00	2347.2	2389.2	42.01	2477.7	2519.8	.1510	8.7498	8.9008
11	.0013123	1.0004	99 857	46.20	2344.3	2390.5	46.20	2475.4	2521.6	.1658	8.7107	8.8765
12	.0014022	1.0005	93 784	50.41	2341.5	2391.9	50.41	2473.0	2523.4	.1806	8.6718	8.8524
13	.0014974	1.0007	88 124	54.60	2338.7	2393.3	54.60	2470.7	2525.3	.1953	8.6332	8.8285
14	.0015983	1.0008	82 848	58.79	2335.9	2394.7	58.80	2468.3	2527.1	.2099	8.5949	8.8048
15	.0017051	1.0009	77 926	62.99	2333.1	2396.1	62.99	2465.9	2528.9	.2245	8.5569	8.7814
16	.0018181	1.0011	73 333	67.18	2330.3	2397.4	67.19	2463.6	2530.8	.2390	8.5191	8.7582
17	.0019376	1.0012	69 044	71.38	2327.4	2398.8	71.38	2461.2	2532.6	.2535	8.4816	8.7351
18	.0020640	1.0014	65 038	75.57	2324.6	2400.2	75.58	2458.8	2534.4	.2679	8.4443	8.7123
19	.0021975	1.0016	61 293	79.76	2321.8	2401.6	79.77	2456.5	2536.2	.2823	8.4073	8.6897
20	.002339	1.0018	57 791	83.95	2319.0	2402.9	83.96	2454.1	2538.1	.2966	8.3706	8.6672
21	.002487	1.0020	54 514	88.14	2316.2	2404.3	88.14	2451.8	2539.9	.3109	8.3341	8.6450
22	.002645	1.0022	51 447	92.32	2313.3	2405.7	92.33	2449.4	2541.7	.3251	8.2979	8.6229
23	.002810	1.0024	48 574	96.51	2310.5	2407.0	96.52	2447.0	2543.5	.3393	8.2618	8.6011
24	.002985	1.0027	45 883	100.70	2307.7	2408.4	100.70	2444.7	2545.4	.3534	8.2261	8.5794
25	.003169	1.0029	43 360	104.88	2304.9	2409.8	104.89	2442.3	2547.2	.3674	8.1905	8.5580
26	.003363	1.0032	40 994	109.06	2302.1	2411.1	109.07	2439.9	2549.0	.3814	8.1552	8.5367
27	.003567	1.0035	38 774	113.25	2299.3	2412.5	113.25	2437.6	2550.8	.3954	8.1202	8.5156
28	.003782	1.0037	36 690	117.42	2296.4	2413.9	117.43	2435.2	2552.6	.4093	8.0854	8.4946
29	.004008	1.0040	34 733	121.60	2293.6	2415.2	121.61	2432.8	2554.5	.4231	8.0508	8.4739
30	.004246	1.0043	32 894	125.78	2290.8	2416.6	125.79	2430.5	2556.3	.4369	8.0164	8.4533
31	.004496	1.0046	31 165	129.96	2288.0	2418.0	129.97	2428.1	2558.1	.4507	7.9822	8.4329
32	.004759	1.0050	29 540	134.14	2285.2	2419.3	134.15	2425.7	2559.9	.4644	7.9483	8.4127
33	.005034	1.0053	28 011	138.32	2282.4	2420.7	138.33	2423.4	2561.7	.4781	7.9146	8.3927
34	.005324	1.0056	26 571	142.50	2279.5	2422.0	142.50	2421.0	2563.5	.4917	7.8811	8.3728
35	.005628	1.0060	25 216	146.67	2276.7	2423.4	146.68	2418.6	2565.3	.5053	7.8478	8.3531
36	.005947	1.0063	23 940	150.85	2273.9	2424.7	150.86	2416.2	2567.1	.5188	7.8147	8.3336
37	.006281	1.0067	22 737	155.03	2271.1	2426.1	155.03	2413.9	2568.9	.5323	7.7819	8.3142
38	.006632	1.0071	21 602	159.20	2268.2	2427.4	159.21	2411.5	2570.7	.5458	7.7492	8.2950
39	.006999	1.0074	20 533	163.38	2265.4	2428.8	163.39	2409.1	2572.5	.5592	7.7167	8.2759
40	.007384	1.0078	19 523	167.56	2262.6	2430.1	167.57	2406.7	2574.3	.5725	7.6845	8.2570
41	.007786	1.0082	18 570	171.73	2259.7	2431.5	171.74	2404.3	2576.1	.5858	7.6524	8.2383
42	.008208	1.0086	17 671	175.91	2256.9	2432.8	175.91	2401.9	2577.9	.5991	7.6206	8.2197
43	.008649	1.0090	16 821	180.08	2254.1	2434.2	180.10	2399.5	2579.6	.6123	7.5889	8.2012
44	.009111	1.0095	16 018	184.26	2251.2	2435.5	184.27	2397.2	2581.4	.6255	7.5574	8.1829
45	.009593	1.0099	15 258	188.44	2248.4	2436.8	188.45	2394.8	2583.2	.6387	7.5261	8.1648
46	.010098	1.0103	14 540	192.61	2245.6	2438.2	192.62	2392.4	2585.0	.6518	7.4950	8.1468
47	.010624	1.0108	13 861	196.79	2242.7	2439.5	196.80	2390.0	2586.8	.6648	7.4642	8.1290
48	.011175	1.0112	13 218	200.96	2239.9	2440.8	200.97	2387.6	2588.5	.6779	7.4334	8.1113
49	.011749	1.0117	12 609	205.14	2237.0	2442.2	205.15	2385.2	2590.3	.6908	7.4029	8.0937
50	.012349	1.0121	12 032	209.32	2234.2	2443.5	209.33	2382.7	2592.1	.7038	7.3725	8.0763
51	.012975	1.0126	11 485	213.50	2231.3	2444.8	213.51	2380.3	2593.8	.7167	7.3423	8.0590
52	.013628	1.0131	10 968	217.67	2228.5	2446.1	217.69	2377.9	2595.6	.7296	7.3123	8.0419
53	.014309	1.0136	10 476	221.85	2225.6	2447.5	221.87	2375.5	2597.4	.7424	7.2825	8.0249
54	.015019	1.0141	10 011	226.03	2222.8	2448.8	226.04	2373.1	2599.1	.7552	7.2528	8.0080

Definitions of symbols on page 1.

Temp. °C	Press. MPa	Specific Volume		Internal Energy			Enthalpy			Entropy		
		Sat. Liquid	Sat. Vapor	Sat. Liquid	Evap.	Sat. Vapor	Sat. Liquid	Evap.	Sat. Vapor	Sat. Liquid	Evap.	Sat. Vapor
t	p	$10^3 v_f$	$10^3 v_g$	u_f	u_{fg}	u_g	h_f	h_{fg}	h_g	s_f	s_{fg}	s_g
55	.015758	1.0146	9568.	230.21	2219.9	2450.1	230.23	2370.7	2600.9	.7679	7.2234	7.9913
56	.016529	1.0151	9149.	234.39	2217.0	2451.4	234.41	2368.2	2602.6	.7807	7.1940	7.9747
57	.017331	1.0156	8751.	238.57	2214.2	2452.7	238.59	2365.8	2604.4	.7933	7.1649	7.9582
58	.018166	1.0161	8372.	242.75	2211.3	2454.0	242.77	2363.4	2606.1	.8060	7.1359	7.9419
59	.019036	1.0166	8013.	246.93	2208.4	2455.3	246.95	2360.9	2607.9	.8186	7.1071	7.9257
60	.019940	1.0172	7671.	251.11	2205.5	2456.6	251.13	2358.5	2609.6	.8312	7.0784	7.9096
61	.020881	1.0177	7346.	255.29	2202.7	2458.0	255.31	2356.0	2611.3	.8437	7.0499	7.8936
62	.021860	1.0182	7037.	259.47	2199.8	2459.3	259.49	2353.6	2613.1	.8562	7.0216	7.8778
63	.022877	1.0188	6743.	263.66	2196.9	2460.6	263.68	2351.1	2614.8	.8686	6.9934	7.8621
64	.023934	1.0194	6463.	267.84	2194.0	2461.8	267.86	2348.7	2616.5	.8811	6.9654	7.8465
65	.02503	1.0199	6197.	272.02	2191.1	2463.1	272.06	2346.2	2618.3	.8935	6.9375	7.8310
66	.02617	1.0205	5943.	276.21	2188.2	2464.4	276.23	2343.7	2620.0	.9058	6.9098	7.8156
67	.02736	1.0211	5701.	280.39	2185.3	2465.7	280.42	2341.3	2621.7	.9181	6.8822	7.8004
68	.02859	1.0217	5471.	284.58	2182.4	2467.0	284.61	2338.8	2623.4	.9304	6.8548	7.7852
69	.02986	1.0222	5252.	288.76	2179.5	2468.3	288.80	2336.3	2625.1	.9427	6.8275	7.7702
70	.03119	1.0228	5042.	292.95	2176.6	2469.6	292.98	2333.8	2626.8	.9549	6.8004	7.7553
71	.03256	1.0234	4843.	297.14	2173.7	2470.8	297.17	2331.4	2628.5	.9671	6.7734	7.7405
72	.03399	1.0240	4652.	301.33	2170.8	2472.1	301.36	2328.9	2630.2	.9792	6.7466	7.7258
73	.03546	1.0247	4470.	305.51	2167.9	2473.4	305.56	2326.4	2631.9	.9914	6.7199	7.7113
74	.03699	1.0253	4297.	309.70	2165.0	2474.7	309.74	2323.9	2633.6	1.0034	6.6934	7.6968
75	.03858	1.0259	4131.	313.90	2162.0	2475.9	313.93	2321.4	2635.3	1.0155	6.6669	7.6824
76	.04022	1.0265	3973.	318.00	2159.1	2477.2	318.13	2318.9	2637.0	1.0275	6.6407	7.6682
77	.04192	1.0272	3822.	322.28	2156.2	2478.4	322.32	2316.3	2638.7	1.0395	6.6145	7.6540
78	.04368	1.0278	3677.	326.47	2153.2	2479.7	326.51	2313.8	2640.3	1.0515	6.5885	7.6400
79	.04550	1.0285	3539.	330.66	2150.3	2481.0	330.72	2311.3	2642.0	1.0634	6.5627	7.6260
80	.04739	1.0291	3407.	334.86	2147.4	2482.2	334.91	2308.8	2643.7	1.0753	6.5369	7.6122
81	.04934	1.0298	3281.	339.05	2144.4	2483.5	339.10	2306.2	2645.3	1.0871	6.5113	7.5985
82	.05136	1.0305	3160.	343.25	2141.5	2484.7	343.30	2303.7	2647.0	1.0990	6.4858	7.5848
83	.05345	1.0311	3044.	347.45	2138.5	2485.9	347.50	2301.1	2648.6	1.1108	6.4605	7.5713
84	.05560	1.0318	2934.	351.64	2135.5	2487.2	351.70	2298.6	2650.3	1.1225	6.4353	7.5578
85	.05783	1.0325	2828.	355.84	2132.6	2488.4	355.90	2296.0	2651.9	1.1343	6.4102	7.5445
86	.06014	1.0332	2726.	360.04	2129.6	2489.6	360.10	2293.5	2653.6	1.1460	6.3852	7.5312
87	.06252	1.0339	2629.	364.24	2126.6	2490.9	364.30	2290.9	2655.2	1.1577	6.3604	7.5180
88	.06498	1.0346	2536.	368.44	2123.7	2492.1	368.51	2288.3	2656.9	1.1693	6.3356	7.5050
89	.06752	1.0353	2446.	372.64	2120.7	2493.3	372.71	2285.8	2658.5	1.1809	6.3110	7.4920
90	.07014	1.0360	2361.	376.85	2117.7	2494.5	376.92	2283.2	2660.1	1.1925	6.2866	7.4791
91	.07284	1.0367	2278.	381.05	2114.7	2495.8	381.12	2280.6	2661.7	1.2041	6.2622	7.4663
92	.07564	1.0375	2200.	385.26	2111.7	2497.0	385.33	2278.0	2663.3	1.2156	6.2379	7.4536
93	.07852	1.0382	2124.	389.46	2108.7	2498.2	389.54	2275.4	2664.9	1.2271	6.2138	7.4409
94	.08149	1.0389	2052.	393.67	2105.7	2499.4	393.75	2272.8	2666.5	1.2386	6.1898	7.4284
95	.08455	1.0397	1981.9	397.88	2102.7	2500.6	397.96	2270.2	2668.1	1.2500	6.1659	7.4159
96	.08771	1.0404	1915.0	402.09	2099.7	2501.8	402.17	2267.6	2669.7	1.2615	6.1421	7.4036
97	.09097	1.0412	1850.8	406.30	2096.7	2503.0	406.39	2264.9	2671.3	1.2728	6.1184	7.3913
98	.09433	1.0420	1789.1	410.51	2093.6	2504.1	410.61	2262.3	2672.9	1.2842	6.0948	7.3791
99	.09778	1.0427	1729.9	414.72	2090.6	2505.3	414.83	2259.7	2674.5	1.2956	6.0714	7.3669
100	.10135	1.0435	1672.9	418.94	2087.6	2506.5	419.04	2257.0	2676.1	1.3069	6.0480	7.3549
101	.10502	1.0443	1618.2	423.15	2084.5	2507.7	423.26	2254.4	2677.6	1.3181	6.0248	7.3429
102	.10880	1.0451	1565.5	427.37	2081.5	2508.9	427.48	2251.7	2679.2	1.3294	6.0016	7.3310
103	.11269	1.0459	1514.9	431.58	2078.5	2510.0	431.71	2249.0	2680.7	1.3406	5.9786	7.3192
104	.11669	1.0467	1466.2	435.80	2075.4	2511.2	435.92	2246.4	2682.3	1.3518	5.9557	7.3075
105	.12082	1.0475	1419.4	440.02	2072.3	2512.4	440.15	2243.7	2683.8	1.3630	5.9328	7.2958
106	.12506	1.0483	1374.3	444.24	2069.3	2513.5	444.37	2241.0	2685.4	1.3741	5.9101	7.2843
107	.12942	1.0491	1330.9	448.47	2066.2	2514.7	448.60	2238.3	2686.9	1.3853	5.8875	7.2728
108	.13391	1.0499	1289.1	452.69	2063.1	2515.8	452.83	2235.6	2688.4	1.3964	5.8650	7.2613
109	.13853	1.0508	1248.9	456.92	2060.0	2517.0	457.06	2232.9	2690.0	1.4074	5.8425	7.2500
110	.14327	1.0516	1210.2	461.14	2057.0	2518.1	461.30	2230.2	2691.5	1.4185	5.8202	7.2387
111	.14815	1.0525	1172.8	465.37	2053.9	2519.2	465.53	2227.5	2693.0	1.4295	5.7980	7.2275
112	.15317	1.0533	1136.9	469.60	2050.8	2520.4	469.76	2224.7	2694.5	1.4405	5.7758	7.2163
113	.15832	1.0542	1102.2	473.84	2047.7	2521.5	474.01	2222.0	2696.0	1.4515	5.7538	7.2052
114	.16362	1.0550	1068.8	478.07	2044.5	2522.6	478.24	2219.2	2697.5	1.4624	5.7318	7.1942

Definitions of symbols on page 1.

Table 1. Saturation: Temperatures

Temp. °C	Press. MPa	Specific Volume		Internal Energy			Enthalpy			Entropy		
		Sat. Liquid	Sat. Vapor	Sat. Liquid	Evap.	Sat. Vapor	Sat. Liquid	Evap.	Sat. Vapor	Sat. Liquid	Evap.	Sat. Vapor
t	p	$10^3 v_f$	$10^3 v_g$	u_f	u_{fg}	u_g	h_f	h_{fg}	h_g	s_f	s_{fg}	s_g
115	.16906	1.0559	1036.6	482.30	2041.4	2523.7	482.48	2216.5	2699.0	1.4734	5.7100	7.1833
116	.17465	1.0568	1005.5	486.54	2038.3	2524.8	486.72	2213.7	2700.5	1.4842	5.6882	7.1724
117	.18038	1.0576	975.6	490.77	2035.2	2526.0	490.97	2211.0	2701.9	1.4951	5.6665	7.1616
118	.18627	1.0585	946.7	495.01	2032.0	2527.1	495.21	2208.2	2703.4	1.5060	5.6449	7.1509
119	.19232	1.0594	918.8	499.26	2028.9	2528.2	499.46	2205.4	2704.9	1.5168	5.6234	7.1402
120	.19853	1.0603	891.9	503.50	2025.8	2529.3	503.71	2202.6	2706.3	1.5276	5.6020	7.1296
121	.20490	1.0612	865.9	507.74	2022.6	2530.3	507.96	2199.8	2707.8	1.5384	5.5807	7.1191
122	.21143	1.0622	840.8	511.99	2019.4	2531.4	512.21	2197.0	2709.2	1.5492	5.5594	7.1086
123	.21814	1.0631	816.6	516.24	2016.3	2532.5	516.47	2194.2	2710.6	1.5599	5.5383	7.0982
124	.22502	1.0640	793.2	520.49	2013.1	2533.6	520.73	2191.3	2712.1	1.5706	5.5172	7.0878
125	.2321	1.0649	770.6	524.74	2009.9	2534.6	524.99	2188.5	2713.5	1.5813	5.4962	7.0775
126	.2393	1.0659	748.8	528.99	2006.7	2535.7	529.24	2185.6	2714.9	1.5920	5.4753	7.0673
127	.2467	1.0668	727.7	533.24	2003.5	2536.8	533.51	2182.8	2716.3	1.6026	5.4545	7.0571
128	.2543	1.0678	707.3	537.50	2000.3	2537.8	537.77	2179.9	2717.7	1.6132	5.4338	7.0470
129	.2621	1.0687	687.6	541.76	1997.1	2538.9	542.04	2177.0	2719.1	1.6238	5.4131	7.0369
130	.2701	1.0697	668.5	546.02	1993.9	2539.9	546.31	2174.2	2720.5	1.6344	5.3925	7.0269
131	.2783	1.0707	650.1	550.28	1990.7	2541.0	550.58	2171.3	2721.8	1.6450	5.3720	7.0170
132	.2866	1.0717	632.3	554.55	1987.4	2542.0	554.86	2168.4	2723.2	1.6555	5.3516	7.0071
133	.2952	1.0726	615.0	558.81	1984.2	2543.0	559.13	2165.4	2724.6	1.6660	5.3312	6.9973
134	.3040	1.0736	598.3	563.08	1980.9	2544.0	563.41	2162.5	2725.9	1.6765	5.3109	6.9875
135	.3130	1.0746	582.2	567.35	1977.7	2545.0	567.69	2159.6	2727.3	1.6870	5.2907	6.9777
136	.3222	1.0756	566.6	571.63	1974.4	2546.1	571.97	2156.6	2728.6	1.6975	5.2706	6.9681
137	.3317	1.0767	551.4	575.90	1971.2	2547.1	576.26	2153.7	2729.9	1.7079	5.2505	6.9584
138	.3413	1.0777	536.8	580.18	1967.9	2548.1	580.54	2150.7	2731.3	1.7183	5.2306	6.9489
139	.3512	1.0787	522.6	584.46	1964.6	2549.0	584.83	2147.7	2732.6	1.7287	5.2106	6.9394
140	.3613	1.0797	508.9	588.74	1961.3	2550.0	589.13	2144.7	2733.9	1.7391	5.1908	6.9299
141	.3716	1.0808	495.6	593.02	1958.0	2551.0	593.42	2141.8	2735.2	1.7495	5.1710	6.9205
142	.3822	1.0818	482.7	597.31	1954.7	2552.0	597.72	2138.7	2736.5	1.7598	5.1513	6.9111
143	.3930	1.0829	470.2	601.60	1951.4	2553.0	602.02	2135.7	2737.7	1.7701	5.1317	6.9018
144	.4041	1.0840	458.1	605.89	1948.0	2553.9	606.33	2132.7	2739.0	1.7804	5.1121	6.8925
145	.4154	1.0850	446.3	610.18	1944.7	2554.9	610.63	2129.6	2740.3	1.7907	5.0926	6.8833
146	.4270	1.0861	435.0	614.48	1941.3	2555.8	614.94	2126.6	2741.5	1.8009	5.0732	6.8741
147	.4388	1.0872	423.9	618.77	1938.0	2556.8	619.25	2123.5	2742.8	1.8112	5.0538	6.8650
148	.4509	1.0883	413.2	623.07	1934.6	2557.7	623.57	2120.4	2744.0	1.8214	5.0345	6.8559
149	.4632	1.0894	402.9	627.38	1931.3	2558.6	627.88	2117.4	2745.2	1.8316	5.0152	6.8468
150	.4758	1.0905	392.8	631.68	1927.9	2559.5	632.20	2114.3	2746.5	1.8418	4.9960	6.8379
151	.4887	1.0916	383.0	635.99	1924.5	2560.5	636.52	2111.1	2747.7	1.8520	4.9769	6.8289
152	.5019	1.0927	373.5	640.30	1921.1	2561.4	640.85	2108.0	2748.9	1.8621	4.9579	6.8200
153	.5154	1.0939	364.4	644.61	1917.7	2562.3	645.17	2104.9	2750.1	1.8723	4.9389	6.8111
154	.5291	1.0950	355.4	648.92	1914.2	2563.2	649.50	2101.7	2751.2	1.8824	4.9199	6.8023
155	.5431	1.0961	346.8	653.24	1910.8	2564.1	653.84	2098.6	2752.4	1.8925	4.9010	6.7935
156	.5575	1.0973	338.4	657.56	1907.4	2564.9	658.17	2095.4	2753.6	1.9025	4.8822	6.7848
157	.5721	1.0985	330.2	661.88	1903.9	2565.8	662.51	2092.2	2754.7	1.9126	4.8635	6.7761
158	.5870	1.0996	322.3	666.21	1900.5	2566.7	666.85	2089.0	2755.8	1.9227	4.8447	6.7674
159	.6023	1.1008	314.5	670.54	1897.0	2567.5	671.20	2085.8	2757.0	1.9327	4.8261	6.7588
160	.6178	1.1020	307.1	674.87	1893.5	2568.4	675.55	2082.6	2758.1	1.9427	4.8075	6.7502
161	.6337	1.1032	299.8	679.20	1890.0	2569.2	679.90	2079.3	2759.2	1.9527	4.7890	6.7416
162	.6499	1.1044	292.7	683.54	1886.5	2570.1	684.25	2076.1	2760.3	1.9627	4.7705	6.7331
163	.6664	1.1056	285.9	687.88	1883.0	2570.9	688.61	2072.8	2761.4	1.9726	4.7520	6.7247
164	.6833	1.1068	279.2	692.22	1879.5	2571.7	692.98	2069.5	2762.5	1.9826	4.7336	6.7162
165	.7005	1.1080	272.7	696.56	1876.0	2572.5	697.34	2066.2	2763.5	1.9925	4.7153	6.7078
166	.7180	1.1093	266.4	700.91	1872.4	2573.3	701.71	2062.9	2764.6	2.0024	4.6970	6.6995
167	.7359	1.1105	260.2	705.26	1868.9	2574.1	706.08	2059.6	2765.6	2.0123	4.6788	6.6911
168	.7541	1.1117	254.3	709.61	1865.3	2574.9	710.45	2056.2	2766.7	2.0222	4.6606	6.6828
169	.7727	1.1130	248.5	713.97	1861.7	2575.7	714.83	2052.9	2767.7	2.0321	4.6425	6.6746
170	.7917	1.1143	242.8	718.33	1858.1	2576.5	719.21	2049.5	2768.7	2.0419	4.6244	6.6663
171	.8110	1.1155	237.3	722.69	1854.5	2577.2	723.59	2046.1	2769.7	2.0517	4.6064	6.6581
172	.8307	1.1168	232.0	727.06	1850.9	2578.0	727.99	2042.7	2770.7	2.0616	4.5884	6.6500
173	.8507	1.1181	226.8	731.43	1847.3	2578.7	732.37	2039.3	2771.7	2.0714	4.5705	6.6418
174	.8712	1.1194	221.7	735.80	1843.7	2579.5	736.77	2035.9	2772.6	2.0812	4.5526	6.6337

Definitions of symbols on page 1.

Temp. °C t	Press. MPa p	Specific Volume		Internal Energy			Enthalpy			Entropy		
		Sat. Liquid $10^3 v_f$	Sat. Vapor $10^3 v_g$	Sat. Liquid u_f	Evap. u_{fg}	Sat. Vapor u_g	Sat. Liquid h_f	Evap. h_{fg}	Sat. Vapor h_g	Sat. Liquid s_f	Evap. s_{fg}	Sat. Vapor s_g
175	.8920	1.1207	216.8	740.17	1840.0	2580.2	741.17	2032.4	2773.6	2.0909	4.5347	6.6256
176	.9132	1.1220	212.0	744.55	1836.4	2580.9	745.57	2029.0	2774.5	2.1007	4.5169	6.6176
177	.9348	1.1234	207.3	748.93	1832.7	2581.6	749.98	2025.5	2775.5	2.1104	4.4992	6.6096
178	.9569	1.1247	202.8	753.31	1829.0	2582.3	754.39	2022.0	2776.4	2.1202	4.4814	6.6016
179	.9793	1.1260	198.4	757.70	1825.3	2583.0	758.80	2018.5	2777.3	2.1299	4.4638	6.5936
180	1.0021	1.1274	194.05	762.09	1821.6	2583.7	763.22	2015.0	2778.2	2.1396	4.4461	6.5857
181	1.0254	1.1288	189.85	766.49	1817.9	2584.4	767.64	2011.4	2779.1	2.1493	4.4285	6.5778
182	1.0491	1.1301	185.75	770.88	1814.2	2585.1	772.07	2007.9	2779.9	2.1590	4.4110	6.5699
183	1.0732	1.1315	181.77	775.28	1810.4	2585.7	776.50	2004.3	2780.8	2.1686	4.3935	6.5621
184	1.0977	1.1329	177.88	779.69	1806.7	2586.4	780.93	2000.7	2781.6	2.1783	4.3760	6.5543
185	1.1227	1.1343	174.09	784.10	1802.9	2587.0	785.37	1997.1	2782.4	2.1879	4.3586	6.5465
186	1.1481	1.1357	170.40	788.51	1799.1	2587.6	789.81	1993.4	2783.3	2.1975	4.3412	6.5387
187	1.1740	1.1371	166.80	792.92	1795.3	2588.2	794.26	1989.8	2784.1	2.2071	4.3238	6.5310
188	1.2003	1.1386	163.29	797.34	1791.5	2588.8	798.71	1986.1	2784.8	2.2167	4.3065	6.5232
189	1.2271	1.1400	159.87	801.76	1787.7	2589.4	803.16	1982.5	2785.6	2.2263	4.2892	6.5155
190	1.2544	1.1414	156.54	806.19	1783.8	2590.0	807.62	1978.8	2786.4	2.2359	4.2720	6.5079
191	1.2821	1.1429	153.28	810.62	1780.0	2590.6	812.08	1975.0	2787.1	2.2454	4.2548	6.5002
192	1.3103	1.1444	150.11	815.05	1776.1	2591.2	816.55	1971.3	2787.9	2.2550	4.2376	6.4926
193	1.3390	1.1458	147.02	819.49	1772.2	2591.7	821.02	1967.6	2788.6	2.2645	4.2205	6.4850
194	1.3682	1.1473	144.00	823.93	1768.3	2592.3	825.50	1963.8	2789.3	2.2740	4.2034	6.4774
195	1.3978	1.1488	141.05	828.37	1764.4	2592.8	829.98	1960.0	2790.0	2.2835	4.1863	6.4698
196	1.4280	1.1503	138.18	832.82	1760.5	2593.3	834.46	1956.2	2790.6	2.2930	4.1692	6.4623
197	1.4587	1.1519	135.38	837.27	1756.6	2593.8	838.95	1952.4	2791.3	2.3025	4.1522	6.4547
198	1.4899	1.1534	132.64	841.73	1752.6	2594.3	843.45	1948.5	2791.9	2.3120	4.1352	6.4472
199	1.5216	1.1549	129.97	846.19	1748.6	2594.8	847.94	1944.6	2792.6	2.3214	4.1183	6.4397
200	1.5538	1.1565	127.36	850.65	1744.7	2595.3	852.45	1940.7	2793.2	2.3309	4.1014	6.4323
201	1.5866	1.1581	124.81	855.12	1740.7	2595.8	856.96	1936.8	2793.8	2.3403	4.0845	6.4248
202	1.6198	1.1596	122.33	859.59	1736.6	2596.2	861.47	1932.9	2794.4	2.3498	4.0676	6.4174
203	1.6537	1.1612	119.90	864.07	1732.6	2596.7	865.99	1929.0	2794.9	2.3592	4.0508	6.4100
204	1.6881	1.1628	117.53	868.55	1728.6	2597.1	870.51	1925.0	2795.5	2.3686	4.0340	6.4026
205	1.7230	1.1644	115.21	873.04	1724.5	2597.5	875.04	1921.0	2796.0	2.3780	4.0172	6.3952
206	1.7585	1.1660	112.95	877.53	1720.4	2597.9	879.57	1917.0	2796.6	2.3874	4.0004	6.3878
207	1.7946	1.1677	110.74	882.02	1716.3	2598.3	884.11	1913.0	2797.1	2.3967	3.9837	6.3805
208	1.8312	1.1693	108.58	886.52	1712.2	2598.7	888.66	1908.9	2797.6	2.4061	3.9670	6.3731
209	1.8684	1.1710	106.48	891.02	1708.1	2599.1	893.21	1904.8	2798.0	2.4155	3.9503	6.3658
210	1.9062	1.1726	104.41	895.53	1703.9	2599.5	897.76	1900.7	2798.5	2.4248	3.9337	6.3585
211	1.9446	1.1743	102.40	900.04	1699.8	2599.8	902.32	1896.6	2798.9	2.4341	3.9170	6.3512
212	1.9836	1.1760	100.43	904.56	1695.6	2600.1	906.89	1892.5	2799.4	2.4435	3.9004	6.3439
213	2.0232	1.1777	98.51	909.08	1691.4	2600.5	911.46	1888.3	2799.8	2.4528	3.8838	6.3366
214	2.0634	1.1794	96.63	913.61	1687.2	2600.8	916.04	1884.1	2800.2	2.4621	3.8673	6.3294
215	2.104	1.1812	94.79	918.14	1682.9	2601.1	920.62	1879.9	2800.5	2.4714	3.8507	6.3221
216	2.146	1.1829	92.99	922.67	1678.7	2601.4	925.21	1875.7	2800.9	2.4807	3.8342	6.3149
217	2.188	1.1847	91.23	927.21	1674.4	2601.6	929.81	1871.4	2801.2	2.4900	3.8177	6.3077
218	2.230	1.1864	89.52	931.76	1670.1	2601.9	934.40	1867.1	2801.5	2.4992	3.8012	6.3005
219	2.274	1.1882	87.84	936.31	1665.8	2602.1	939.01	1862.8	2801.8	2.5085	3.7848	6.2933
220	2.318	1.1900	86.19	940.87	1661.5	2602.4	943.62	1858.5	2802.1	2.5178	3.7683	6.2861
221	2.362	1.1918	84.58	945.43	1657.1	2602.6	948.24	1854.2	2802.4	2.5270	3.7519	6.2789
222	2.408	1.1936	83.01	950.00	1652.8	2602.8	952.87	1849.8	2802.6	2.5363	3.7355	6.2717
223	2.454	1.1955	81.47	954.57	1648.4	2603.0	957.50	1845.4	2802.9	2.5455	3.7191	6.2646
224	2.500	1.1973	79.97	959.15	1644.0	2603.2	962.14	1840.9	2803.1	2.5547	3.7027	6.2574
225	2.548	1.1992	78.49	963.73	1639.6	2603.3	966.78	1836.5	2803.3	2.5639	3.6863	6.2503
226	2.596	1.2011	77.05	968.32	1635.1	2603.5	971.44	1832.0	2803.5	2.5732	3.6700	6.2431
227	2.644	1.2030	75.64	972.92	1630.7	2603.6	976.10	1827.5	2803.6	2.5824	3.6536	6.2360
228	2.694	1.2049	74.26	977.52	1626.2	2603.7	980.76	1823.0	2803.7	2.5916	3.6373	6.2289
229	2.744	1.2068	72.90	982.12	1621.7	2603.8	985.43	1818.4	2803.9	2.6008	3.6210	6.2217
230	2.795	1.2088	71.58	986.74	1617.2	2603.9	990.12	1813.8	2804.0	2.6099	3.6047	6.2146
231	2.846	1.2107	70.29	991.36	1612.6	2604.0	994.80	1809.2	2804.0	2.6191	3.5884	6.2075
232	2.899	1.2127	69.02	995.98	1608.1	2604.0	999.49	1804.6	2804.1	2.6283	3.5721	6.2004
233	2.952	1.2147	67.77	1000.61	1603.5	2604.1	1004.19	1799.9	2804.1	2.6375	3.5558	6.1933
234	3.006	1.2167	66.56	1005.25	1598.9	2604.1	1008.90	1795.3	2804.2	2.6466	3.5396	6.1862

Definitions of symbols on page 1.

(5)

Table 1. Saturation: Temperatures

Temp. °C	Press. MPa	Specific Volume		Internal Energy			Enthalpy			Entropy		
		Sat. Liquid $10^3 v_f$	Sat. Vapor $10^3 v_g$	Sat. Liquid u_f	Evap. u_{fg}	Sat. Vapor u_g	Sat. Liquid h_f	Evap. h_{fg}	Sat. Vapor h_g	Sat. Liquid s_f	Evap. s_{fg}	Sat. Vapor s_g
t	p											
235	3.060	1.2187	65.37	1009.89	1594.2	2604.1	1013.62	1790.5	2804.2	2.6558	3.5233	6.1791
236	3.115	1.2208	64.20	1014.54	1589.6	2604.1	1018.35	1785.8	2804.1	2.6650	3.5071	6.1720
237	3.171	1.2228	63.06	1019.20	1584.9	2604.1	1023.08	1781.0	2804.1	2.6741	3.4909	6.1650
238	3.228	1.2249	61.94	1023.86	1580.2	2604.1	1027.82	1776.2	2804.0	2.6832	3.4746	6.1579
239	3.286	1.2270	60.84	1028.54	1575.5	2604.0	1032.57	1771.4	2803.9	2.6924	3.4584	6.1508
240	3.344	1.2291	59.76	1033.21	1570.8	2604.0	1037.32	1766.5	2803.8	2.7015	3.4422	6.1437
241	3.403	1.2312	58.71	1037.90	1566.0	2603.9	1042.09	1761.6	2803.7	2.7107	3.4260	6.1366
242	3.463	1.2334	57.68	1042.59	1561.2	2603.8	1046.86	1756.7	2803.5	2.7198	3.4098	6.1296
243	3.524	1.2355	56.67	1047.29	1556.4	2603.7	1051.64	1751.7	2803.4	2.7289	3.3936	6.1225
244	3.586	1.2377	55.68	1051.99	1551.5	2603.5	1056.43	1746.8	2803.2	2.7380	3.3774	6.1154
245	3.648	1.2399	54.71	1056.71	1546.7	2603.4	1061.23	1741.7	2803.0	2.7472	3.3612	6.1083
246	3.711	1.2422	53.75	1061.43	1541.8	2603.2	1066.03	1736.7	2802.7	2.7563	3.3450	6.1013
247	3.776	1.2444	52.82	1066.16	1536.9	2603.0	1070.85	1731.6	2802.5	2.7654	3.3288	6.0942
248	3.840	1.2467	51.90	1070.89	1531.9	2602.8	1075.68	1726.5	2802.2	2.7745	3.3126	6.0871
249	3.906	1.2489	51.01	1075.64	1527.0	2602.6	1080.51	1721.4	2801.9	2.7836	3.2964	6.0800
250	3.973	1.2512	50.13	1080.39	1522.0	2602.4	1085.36	1716.2	2801.5	2.7927	3.2802	6.0730
251	4.040	1.2536	49.26	1085.15	1517.0	2602.1	1090.21	1711.0	2801.2	2.8018	3.2640	6.0659
252	4.109	1.2559	48.42	1089.92	1511.9	2601.9	1095.08	1705.7	2800.8	2.8110	3.2478	6.0588
253	4.178	1.2583	47.59	1094.70	1506.9	2601.6	1099.96	1700.5	2800.4	2.8201	3.2316	6.0517
254	4.248	1.2607	46.77	1099.49	1501.8	2601.3	1104.84	1695.1	2800.0	2.8292	3.2154	6.0446
255	4.319	1.2631	45.98	1104.28	1496.7	2600.9	1109.73	1689.8	2799.5	2.8383	3.1992	6.0375
256	4.391	1.2655	45.19	1109.08	1491.5	2600.6	1114.64	1684.4	2799.1	2.8474	3.1830	6.0304
257	4.464	1.2680	44.42	1113.90	1486.3	2600.2	1119.55	1679.0	2798.6	2.8565	3.1668	6.0233
258	4.538	1.2705	43.67	1118.72	1481.1	2599.9	1124.48	1673.6	2798.0	2.8656	3.1506	6.0162
259	4.613	1.2730	42.93	1123.55	1475.9	2599.5	1129.42	1668.1	2797.5	2.8747	3.1343	6.0091
260	4.688	1.2755	42.21	1128.39	1470.6	2599.0	1134.37	1662.5	2796.9	2.8838	3.1181	6.0019
261	4.765	1.2781	41.49	1133.24	1465.3	2598.6	1139.33	1657.0	2796.3	2.8929	3.1019	5.9948
262	4.843	1.2807	40.79	1138.10	1460.0	2598.1	1144.30	1651.4	2795.7	2.9020	3.0856	5.9876
263	4.921	1.2833	40.11	1142.97	1454.7	2597.6	1149.28	1645.7	2795.0	2.9112	3.0693	5.9805
264	5.001	1.2859	39.43	1147.85	1449.3	2597.1	1154.28	1640.1	2794.3	2.9203	3.0530	5.9733
265	5.081	1.2886	38.77	1152.74	1443.9	2596.6	1159.28	1634.4	2793.6	2.9294	3.0368	5.9662
266	5.163	1.2913	38.12	1157.64	1438.4	2596.1	1164.30	1628.6	2792.9	2.9385	3.0205	5.9590
267	5.245	1.2940	37.49	1162.56	1433.0	2595.5	1169.34	1622.8	2792.1	2.9477	3.0041	5.9518
268	5.329	1.2967	36.86	1167.48	1427.4	2594.9	1174.38	1617.0	2791.3	2.9568	2.9878	5.9446
269	5.413	1.2995	36.25	1172.41	1421.9	2594.3	1179.44	1611.1	2790.5	2.9659	2.9714	5.9374
270	5.499	1.3023	35.64	1177.36	1416.3	2593.7	1184.51	1605.2	2789.7	2.9751	2.9551	5.9301
271	5.585	1.3051	35.05	1182.31	1410.7	2593.0	1189.60	1599.2	2788.8	2.9842	2.9387	5.9229
272	5.673	1.3080	34.47	1187.28	1405.1	2592.4	1194.69	1593.2	2787.9	2.9934	2.9223	5.9157
273	5.761	1.3109	33.90	1192.26	1399.4	2591.7	1199.81	1587.2	2787.0	3.0025	2.9059	5.9084
274	5.851	1.3138	33.34	1197.25	1393.7	2590.9	1204.93	1581.1	2786.0	3.0117	2.8894	5.9011
275	5.942	1.3168	32.79	1202.25	1387.9	2590.2	1210.07	1574.9	2785.0	3.0208	2.8730	5.8938
276	6.034	1.3198	32.24	1207.27	1382.2	2589.4	1215.22	1568.7	2784.0	3.0300	2.8565	5.8865
277	6.126	1.3228	31.71	1212.30	1376.3	2588.6	1220.40	1562.5	2782.9	3.0392	2.8400	5.8792
278	6.220	1.3259	31.19	1217.34	1370.5	2587.8	1225.58	1556.3	2781.8	3.0484	2.8235	5.8718
279	6.315	1.3290	30.68	1222.39	1364.6	2587.0	1230.78	1549.9	2780.7	3.0576	2.8069	5.8645
280	6.412	1.3321	30.17	1227.46	1358.7	2586.1	1235.99	1543.6	2779.6	3.0668	2.7903	5.8571
281	6.509	1.3353	29.67	1232.54	1352.7	2585.2	1241.22	1537.1	2778.4	3.0760	2.7737	5.8497
282	6.607	1.3385	29.19	1237.64	1346.7	2584.3	1246.47	1530.7	2777.2	3.0852	2.7571	5.8423
283	6.707	1.3417	28.71	1242.74	1340.6	2583.4	1251.73	1524.2	2775.9	3.0945	2.7404	5.8348
284	6.807	1.3450	28.24	1247.87	1334.5	2582.4	1257.02	1517.6	2774.6	3.1037	2.7237	5.8274
285	6.909	1.3483	27.77	1253.00	1328.4	2581.4	1262.31	1511.0	2773.3	3.1130	2.7070	5.8199
286	7.012	1.3517	27.32	1258.16	1322.2	2580.4	1267.63	1504.3	2771.9	3.1222	2.6902	5.8124
287	7.116	1.3551	26.87	1263.32	1316.0	2579.3	1272.96	1497.6	2770.5	3.1315	2.6734	5.8049
288	7.222	1.3585	26.43	1268.51	1309.8	2578.3	1278.31	1490.8	2769.1	3.1408	2.6565	5.7973
289	7.328	1.3620	26.00	1273.70	1303.4	2577.2	1283.68	1484.0	2767.7	3.1501	2.6397	5.7897
290	7.436	1.3656	25.57	1278.92	1297.1	2576.0	1289.07	1477.1	2766.2	3.1594	2.6227	5.7821
291	7.545	1.3691	25.15	1284.15	1290.7	2574.9	1294.47	1470.1	2764.6	3.1687	2.6058	5.7745
292	7.655	1.3728	24.74	1289.40	1284.3	2573.7	1299.90	1463.1	2763.0	3.1781	2.5888	5.7669
293	7.766	1.3764	24.33	1294.66	1277.8	2572.4	1305.34	1456.1	2761.4	3.1874	2.5718	5.7592
294	7.879	1.3802	23.94	1299.94	1271.2	2571.2	1310.81	1449.0	2759.8	3.1968	2.5547	5.7515

Definitions of symbols on page 1.

Temp. °C t	Press. MPa p	Specific Volume		Internal Energy			Enthalpy			Entropy		
		Sat. Liquid $10^3 v_f$	Sat. Vapor $10^3 v_g$	Sat. Liquid u_f	Evap. u_{fg}	Sat. Vapor u_g	Sat. Liquid h_f	Evap. h_{fg}	Sat. Vapor h_g	Sat. Liquid s_f	Evap. s_{fg}	Sat. Vapor s_g
295	7.993	1.3839	23.54	1305.2	1264.7	2569.9	1316.3	1441.8	2758.1	3.2062	2.5375	5.7437
296	8.108	1.3878	23.16	1310.6	1258.0	2568.6	1321.8	1434.5	2756.3	3.2156	2.5204	5.7359
297	8.224	1.3916	22.78	1315.9	1251.3	2567.2	1327.3	1427.2	2754.6	3.2250	2.5031	5.7281
298	8.342	1.3956	22.40	1321.2	1244.6	2565.9	1332.9	1419.9	2752.7	3.2344	2.4858	5.7203
299	8.461	1.3995	22.04	1326.6	1237.8	2564.4	1338.4	1412.4	2750.9	3.2439	2.4685	5.7124
300	8.581	1.4036	21.67	1332.0	1231.0	2563.0	1344.0	1404.9	2749.0	3.2534	2.4511	5.7045
301	8.702	1.4077	21.32	1337.4	1224.1	2561.5	1349.7	1397.4	2747.0	3.2629	2.4337	5.6965
302	8.825	1.4118	20.97	1342.8	1217.1	2560.0	1355.3	1389.7	2745.0	3.2724	2.4162	5.6885
303	8.949	1.4160	20.62	1348.3	1210.1	2558.4	1361.0	1382.0	2743.0	3.2819	2.3986	5.6805
304	9.075	1.4203	20.28	1353.8	1203.1	2556.8	1366.6	1374.2	2740.9	3.2915	2.3810	5.6724
305	9.202	1.4247	19.948	1359.3	1195.9	2555.2	1372.4	1366.4	2738.7	3.3010	2.3633	5.6643
306	9.330	1.4291	19.619	1364.8	1188.7	2553.5	1378.1	1358.5	2736.6	3.3106	2.3455	5.6561
307	9.459	1.4335	19.294	1370.3	1181.5	2551.8	1383.9	1350.5	2734.3	3.3203	2.3277	5.6479
308	9.590	1.4381	18.975	1375.9	1174.2	2550.1	1389.7	1342.4	2732.0	3.3299	2.3098	5.6397
309	9.723	1.4427	18.660	1381.5	1166.8	2548.3	1395.5	1334.2	2729.7	3.3396	2.2918	5.6314
310	9.856	1.4474	18.350	1387.1	1159.4	2546.4	1401.3	1326.0	2727.3	3.3493	2.2737	5.6230
311	9.992	1.4522	18.045	1392.7	1151.9	2544.6	1407.2	1317.6	2724.8	3.3590	2.2556	5.6146
312	10.128	1.4570	17.744	1398.4	1144.3	2542.6	1413.1	1309.2	2722.3	3.3688	2.2373	5.6061
313	10.266	1.4619	17.447	1404.0	1136.6	2540.7	1419.0	1300.7	2719.8	3.3786	2.2190	5.5976
314	10.406	1.4669	17.155	1409.7	1128.9	2538.6	1425.0	1292.2	2717.2	3.3884	2.2006	5.5890
315	10.547	1.4720	16.867	1415.5	1121.1	2536.6	1431.0	1283.5	2714.5	3.3982	2.1821	5.5804
316	10.689	1.4772	16.583	1421.2	1113.2	2534.5	1437.0	1274.7	2711.7	3.4081	2.1635	5.5717
317	10.833	1.4824	16.303	1427.0	1105.3	2532.3	1443.1	1265.8	2708.9	3.4180	2.1449	5.5629
318	10.978	1.4878	16.028	1432.9	1097.2	2530.1	1449.2	1256.9	2706.1	3.4280	2.1261	5.5541
319	11.125	1.4933	15.756	1438.7	1089.1	2527.8	1455.3	1247.8	2703.1	3.4380	2.1072	5.5452
320	11.274	1.4988	15.488	1444.6	1080.9	2525.5	1461.5	1238.6	2700.1	3.4480	2.0882	5.5362
322	11.575	1.5102	14.962	1456.4	1064.3	2520.7	1473.9	1220.0	2693.9	3.4682	2.0498	5.5180
324	11.883	1.5221	14.451	1468.4	1047.2	2515.7	1486.5	1200.9	2687.4	3.4885	2.0109	5.4995
326	12.197	1.5344	13.953	1480.6	1029.8	2510.4	1499.3	1181.3	2680.6	3.5091	1.9715	5.4806
328	12.518	1.5473	13.469	1492.8	1011.9	2504.8	1512.2	1161.2	2673.4	3.5298	1.9315	5.4613
330	12.845	1.5607	12.996	1505.3	993.7	2498.9	1525.3	1140.6	2665.9	3.5507	1.8909	5.4417
332	13.179	1.5747	12.535	1517.9	974.9	2492.8	1538.6	1119.3	2658.0	3.5719	1.8496	5.4215
334	13.520	1.5893	12.086	1530.7	955.6	2486.3	1552.2	1097.5	2649.7	3.5933	1.8076	5.4009
336	13.868	1.6047	11.647	1543.7	935.8	2479.4	1565.9	1075.0	2641.0	3.6150	1.7647	5.3798
338	14.223	1.6209	11.217	1556.9	915.3	2472.2	1579.9	1051.8	2631.8	3.6370	1.7210	5.3580
340	14.586	1.6379	10.797	1570.3	894.3	2464.6	1594.2	1027.9	2622.0	3.6594	1.6763	5.3357
342	14.956	1.6559	10.386	1584.0	872.5	2456.5	1608.8	1003.0	2611.8	3.6821	1.6305	5.3126
344	15.333	1.6750	9.982	1598.0	849.9	2447.8	1623.6	977.3	2600.9	3.7052	1.5834	5.2887
346	15.718	1.6953	9.586	1612.2	826.4	2438.7	1638.9	950.5	2589.4	3.7288	1.5351	5.2639
348	16.112	1.7170	9.197	1626.9	802.0	2428.9	1654.5	922.6	2577.1	3.7530	1.4852	5.2381
350	16.513	1.7403	8.813	1641.9	776.6	2418.4	1670.6	893.4	2563.9	3.7777	1.4335	5.2112
352	16.923	1.7654	8.434	1657.3	749.9	2407.2	1687.2	862.7	2549.9	3.8031	1.3799	5.1830
354	17.342	1.7927	8.059	1673.3	721.7	2395.0	1704.4	830.4	2534.8	3.8293	1.3240	5.1534
356	17.769	1.8225	7.687	1689.8	692.0	2381.8	1722.2	796.2	2518.4	3.8565	1.2654	5.1220
358	18.205	1.8556	7.316	1707.1	660.3	2367.4	1740.9	759.7	2500.6	3.8849	1.2036	5.0885
360	18.651	1.8925	6.945	1725.2	626.3	2351.5	1760.5	720.5	2481.0	3.9147	1.1379	5.0526
362	19.106	1.9345	6.570	1744.5	589.3	2333.8	1781.5	677.9	2459.3	3.9463	1.0672	5.0135
364	19.571	1.9833	6.189	1765.2	548.6	2313.8	1804.0	630.9	2434.9	3.9803	.9901	4.9704
366	20.046	2.0416	5.795	1787.8	502.8	2290.7	1828.8	578.1	2406.8	4.0176	.9044	4.9220
368	20.531	2.1145	5.380	1813.4	449.7	2263.1	1856.8	516.8	2373.6	4.0599	.8059	4.8658
370	21.03	2.213	4.925	1844.0	384.5	2228.5	1890.5	441.6	2332.1	4.1106	.6865	4.7971
371	21.28	2.280	4.671	1862.5	344.3	2206.7	1911.0	395.1	2306.1	4.1416	.6134	4.7550
372	21.53	2.369	4.380	1885.2	294.3	2179.5	1936.2	337.6	2273.8	4.1798	.5233	4.7031
373	21.79	2.509	4.019	1916.6	225.0	2141.6	1971.3	258.0	2229.2	4.2331	.3992	4.6324
374	22.05	2.880	3.404	1985.9	79.5	2065.4	2049.4	91.0	2140.4	4.3529	.1406	4.4935
374.136	22.09	3.155	3.155	2029.6	0	2029.6	2099.3	0	2099.3	4.4298	0	4.4298

Definitions of symbols on page 1.

Table 2. Saturation: Pressures

Press. MPa	Temp °C	Specific Volume		Internal Energy			Enthalpy			Entropy		
		Sat. Liquid $10^3 v_f$	Sat. Vapor $10^3 v_g$	Sat. Liquid u_f	Evap. u_{fg}	Sat. Vapor u_g	Sat. Liquid h_f	Evap. h_{fg}	Sat. Vapor h_g	Sat. Liquid s_f	Evap. s_{fg}	Sat. Vapor s_g
p	t											
.0006113	.01	1.0002	206 136	.00	2375.3	2375.3	.01	2501.3	2501.4	.0000	9.1562	9.1562
.0007	1.89	1.0001	181 255	7.90	2370.0	2377.9	7.91	2496.9	2504.8	.0288	9.0775	9.1064
.0008	3.77	1.0001	159 675	15.81	2364.7	2380.6	15.81	2492.5	2508.3	.0575	8.9999	9.0573
.0009	5.45	1.0001	142 789	22.88	2360.0	2382.9	22.89	2488.5	2511.4	.0829	8.9312	9.0142
.0010	6.98	1.0002	129 208	29.30	2355.7	2385.0	29.30	2484.9	2514.2	.1059	8.8697	8.9756
.0011	8.37	1.0002	118 042	35.17	2351.8	2386.9	35.17	2481.6	2516.8	.1268	8.8140	8.9408
.0012	9.66	1.0003	108 696	40.58	2348.1	2388.7	40.58	2478.6	2519.1	.1460	8.7631	8.9091
.0013	10.86	1.0004	100 755	45.60	2344.7	2390.3	45.60	2475.7	2521.3	.1637	8.7162	8.8799
.0014	11.98	1.0005	93 922	50.31	2341.6	2391.9	50.31	2473.1	2523.4	.1802	8.6727	8.8529
.0015	13.03	1.0007	87 980	54.71	2338.6	2393.3	54.71	2470.6	2525.3	.1957	8.6322	8.8279
.0016	14.02	1.0008	82 763	58.87	2335.8	2394.7	58.87	2468.3	2527.1	.2102	8.5943	8.8044
.0017	14.95	1.0009	78 146	62.80	2333.2	2396.0	62.80	2466.0	2528.8	.2238	8.5586	8.7825
.0018	15.84	1.0010	74 030	66.53	2330.7	2397.2	66.54	2463.9	2530.5	.2368	8.5250	8.7618
.0019	16.69	1.0012	70 337	70.09	2328.3	2398.4	70.10	2461.9	2532.0	.2491	8.4931	8.7422
.0020	17.50	1.0013	67 004	73.48	2326.0	2399.5	73.48	2460.0	2533.5	.2607	8.4629	8.7237
.0021	18.28	1.0014	63 981	76.73	2323.8	2400.6	76.74	2458.2	2534.9	.2719	8.4341	8.7060
.0022	19.02	1.0016	61 226	79.84	2321.7	2401.6	79.85	2456.4	2536.3	.2826	8.4067	8.6892
.0023	19.73	1.0017	58 705	82.83	2319.7	2402.6	82.83	2454.8	2537.6	.2928	8.3804	8.6732
.0024	20.42	1.0019	56 389	85.71	2317.8	2403.5	85.72	2453.1	2538.8	.3026	8.3552	8.6579
.0025	21.08	1.0020	54 254	88.48	2315.9	2404.4	88.49	2451.6	2540.0	.3120	8.3311	8.6432
.0026	21.72	1.0021	52 279	91.16	2314.1	2405.3	91.17	2450.1	2541.2	.3211	8.3079	8.6290
.0027	22.34	1.0023	50 446	93.75	2312.4	2406.1	93.75	2448.6	2542.3	.3299	8.2856	8.6155
.0028	22.94	1.0024	48 742	96.26	2310.7	2407.0	96.27	2447.2	2543.4	.3384	8.2640	8.6024
.0029	23.52	1.0026	47 152	98.69	2309.1	2407.8	98.70	2445.8	2544.5	.3466	8.2432	8.5898
.0030	24.08	1.0027	45 665	101.04	2307.5	2408.5	101.05	2444.5	2545.5	.3545	8.2231	8.5776
.0032	25.16	1.0030	42 964	105.56	2304.4	2410.0	105.57	2441.9	2547.5	.3697	8.1848	8.5545
.0034	26.19	1.0032	40 572	109.83	2301.6	2411.4	109.84	2439.5	2549.3	.3840	8.1488	8.5327
.0036	27.16	1.0035	38 440	113.89	2298.8	2412.7	113.90	2437.2	2551.1	.3975	8.1148	8.5123
.0038	28.08	1.0038	36 527	117.76	2296.2	2414.0	117.77	2435.0	2552.8	.4104	8.0826	8.4930
.0040	28.96	1.0040	34 800	121.45	2293.7	2415.2	121.46	2432.9	2554.4	.4226	8.0520	8.4746
.0042	29.81	1.0043	33 234	124.99	2291.3	2416.3	125.00	2430.9	2555.9	.4343	8.0229	8.4572
.0044	30.62	1.0045	31 806	128.39	2289.1	2417.4	128.39	2429.0	2557.4	.4455	7.9951	8.4406
.0046	31.40	1.0048	30 500	131.64	2286.9	2418.5	131.65	2427.2	2558.8	.4562	7.9686	8.4248
.0048	32.15	1.0050	29 299	134.78	2284.7	2419.5	134.79	2425.4	2560.2	.4665	7.9431	8.4096
.0050	32.88	1.0053	28 192	137.81	2282.7	2420.5	137.82	2423.7	2561.5	.4764	7.9187	8.3951
.0055	34.58	1.0058	25 769	144.94	2277.9	2422.8	144.95	2419.6	2564.5	.4997	7.8616	8.3613
.0060	36.16	1.0064	23 739	151.53	2273.4	2425.0	151.53	2415.9	2567.4	.5210	7.8094	8.3304
.0065	37.63	1.0069	22 014	157.66	2269.3	2426.9	157.67	2412.4	2570.0	.5408	7.7613	8.3020
.0070	39.00	1.0074	20 530	163.39	2265.4	2428.8	163.40	2409.1	2572.5	.5592	7.7167	8.2758
.0075	40.29	1.0079	19 238	168.78	2261.7	2430.5	168.79	2406.0	2574.8	.5764	7.6750	8.2515
.0080	41.51	1.0084	18 103	173.87	2258.3	2432.2	173.88	2403.1	2577.0	.5926	7.6361	8.2287
.0085	42.67	1.0089	17 099	178.69	2255.0	2433.7	178.70	2400.3	2579.0	.6079	7.5994	8.2073
.0090	43.76	1.0094	16 203	183.27	2251.9	2435.2	183.29	2397.7	2581.0	.6224	7.5648	8.1872
.0095	44.81	1.0098	15 399	187.64	2248.9	2436.6	187.65	2395.2	2582.9	.6362	7.5321	8.1682
.010	45.81	1.0102	14 674	191.82	2246.1	2437.9	191.83	2392.8	2584.7	.6493	7.5009	8.1502
.011	47.69	1.0111	13 415	199.66	2240.8	2440.4	199.67	2388.3	2588.0	.6738	7.4430	8.1168
.012	49.42	1.0119	12 361	206.91	2235.8	2442.7	206.92	2384.1	2591.1	.6963	7.3900	8.0863
.013	51.04	1.0126	11 465	213.66	2231.2	2444.9	213.67	2380.2	2593.9	.7172	7.3412	8.0584
.014	52.55	1.0134	10 693	219.98	2226.9	2446.9	219.99	2376.6	2596.6	.7366	7.2959	8.0325
.015	53.97	1.0141	10 022	225.92	2222.8	2448.7	225.94	2373.1	2599.1	.7549	7.2536	8.0085
.016	55.32	1.0147	9433	231.54	2219.0	2450.5	231.56	2369.9	2601.4	.7720	7.2140	7.9860
.017	56.59	1.0154	8910	236.86	2215.3	2452.2	236.89	2366.8	2603.7	.7882	7.1767	7.9649
.018	57.80	1.0160	8445	241.93	2211.8	2453.8	241.95	2363.8	2605.8	.8035	7.1416	7.9451
.019	58.96	1.0166	8027	246.76	2208.5	2455.3	246.78	2361.0	2607.8	.8181	7.1082	7.9263
.020	60.06	1.0172	7649	251.38	2205.4	2456.7	251.40	2358.3	2609.7	.8320	7.0766	7.9085
.021	61.12	1.0178	7307	255.81	2202.3	2458.1	255.83	2355.7	2611.6	.8452	7.0464	7.8916
.022	62.14	1.0183	6995	260.06	2199.4	2459.4	260.08	2353.2	2613.3	.8579	7.0176	7.8756
.023	63.12	1.0189	6709	264.15	2196.6	2460.7	264.18	2350.8	2615.0	.8701	6.9901	7.8602
.024	64.06	1.0194	6446	268.09	2193.8	2461.9	268.12	2348.5	2616.6	.8818	6.9637	7.8455

Definitions of symbols on page 1.

Press. MPa	Temp °C	Specific Volume		Internal Energy			Enthalpy			Entropy		
		Sat. Liquid	Sat. Vapor	Sat. Liquid	Evap.	Sat. Vapor	Sat. Liquid	Evap.	Sat. Vapor	Sat. Liquid	Evap.	Sat. Vapor
p	t	$10^3 v_f$	$10^3 v_g$	u_f	u_{fg}	u_g	h_f	h_{fg}	h_g	s_f	s_{fg}	s_g
.025	64.97	1.0199	6204.	271.90	2191.2	2463.1	271.93	2346.3	2618.2	.8931	6.9383	7.8314
.026	65.85	1.0204	5980.	275.58	2188.7	2464.2	275.61	2344.1	2619.7	.9040	6.9139	7.8179
.027	66.70	1.0209	5772.	279.14	2186.2	2465.3	279.17	2342.0	2621.2	.9145	6.8904	7.8049
.028	67.53	1.0214	5579.	282.60	2183.8	2466.4	282.62	2340.0	2622.6	.9246	6.8678	7.7924
.029	68.33	1.0218	5398.	285.95	2181.5	2467.4	285.98	2338.0	2624.0	.9344	6.8459	7.7803
.030	69.10	1.0223	5229.	289.20	2179.2	2468.4	289.23	2336.1	2625.3	.9439	6.8247	7.7686
.032	70.60	1.0232	4922.	295.45	2174.9	2470.3	295.48	2332.4	2627.8	.9622	6.7843	7.7465
.034	72.01	1.0241	4650.	301.37	2170.8	2472.1	301.40	2328.8	2630.2	.9793	6.7463	7.7257
.036	73.36	1.0249	4408.	307.01	2166.8	2473.8	307.05	2325.5	2632.5	.9956	6.7104	7.7061
.038	74.64	1.0257	4190.	312.39	2163.1	2475.5	312.43	2322.3	2634.7	1.0111	6.6764	7.6876
.040	75.87	1.0265	3993.	317.53	2159.5	2477.0	317.58	2319.2	2636.8	1.0259	6.6441	7.6700
.042	77.05	1.0272	3815.	322.47	2156.0	2478.5	322.51	2316.2	2638.7	1.0400	6.6133	7.6534
.044	78.18	1.0279	3652.	327.22	2152.7	2479.9	327.26	2313.4	2640.6	1.0536	6.5839	7.6375
.046	79.27	1.0286	3503.	331.78	2149.5	2481.3	331.83	2310.6	2642.5	1.0666	6.5558	7.6223
.048	80.32	1.0293	3367.	336.19	2146.4	2482.6	336.23	2308.0	2644.2	1.0790	6.5288	7.6078
.050	81.33	1.0300	3240.	340.44	2143.4	2483.9	340.49	2305.4	2645.9	1.0910	6.5029	7.5939
.055	83.72	1.0316	2964.	350.48	2136.4	2486.8	350.54	2299.3	2649.8	1.1193	6.4422	7.5615
.060	85.94	1.0331	2732.	359.79	2129.8	2489.6	359.86	2293.6	2653.5	1.1453	6.3867	7.5320
.065	88.01	1.0346	2535.	368.48	2123.6	2492.1	368.54	2288.3	2656.9	1.1694	6.3354	7.5048
.070	89.95	1.0360	2365.	376.63	2117.8	2494.5	376.70	2283.3	2660.0	1.1919	6.2878	7.4797
.075	91.78	1.0373	2217.	384.31	2112.4	2496.7	384.39	2278.6	2663.0	1.2130	6.2434	7.4564
.080	93.50	1.0386	2087.	391.58	2107.2	2498.8	391.66	2274.1	2665.8	1.2329	6.2017	7.4346
.085	95.14	1.0398	1972.	398.48	2102.3	2500.8	398.57	2269.8	2668.4	1.2517	6.1625	7.4141
.090	96.71	1.0410	1869.	405.06	2097.6	2502.6	405.15	2265.7	2670.9	1.2695	6.1254	7.3949
.095	98.20	1.0421	1777.	411.34	2093.0	2504.4	411.43	2261.8	2673.2	1.2864	6.0902	7.3766
.100	99.63	1.0432	1694.0	417.36	2088.7	2506.1	417.46	2258.0	2675.5	1.3026	6.0568	7.3594
.105	101.00	1.0443	1618.4	423.13	2084.6	2507.7	423.24	2254.4	2677.6	1.3181	6.0249	7.3430
.110	102.31	1.0453	1549.5	428.68	2080.5	2509.2	428.79	2250.9	2679.7	1.3329	5.9944	7.3273
.115	103.58	1.0463	1486.4	434.03	2076.7	2510.7	434.15	2247.5	2681.7	1.3471	5.9653	7.3124
.120	104.80	1.0473	1428.4	439.20	2072.9	2512.1	439.32	2244.2	2683.5	1.3608	5.9373	7.2981
.125	105.99	1.0483	1374.9	444.19	2069.3	2513.5	444.32	2241.0	2685.4	1.3740	5.9104	7.2844
.130	107.13	1.0492	1325.4	449.02	2065.8	2514.8	449.15	2238.0	2687.1	1.3867	5.8845	7.2712
.135	108.24	1.0501	1279.4	453.70	2062.4	2516.1	453.83	2235.0	2688.8	1.3990	5.8596	7.2586
.140	109.31	1.0510	1236.6	458.24	2059.1	2517.3	458.39	2232.1	2690.4	1.4109	5.8355	7.2464
.145	110.36	1.0519	1196.7	462.65	2055.9	2518.5	462.80	2229.2	2692.0	1.4224	5.8123	7.2347
.150	111.37	1.0528	1159.3	466.94	2052.7	2519.7	467.11	2226.5	2693.6	1.4336	5.7897	7.2233
.155	112.36	1.0536	1124.3	471.12	2049.7	2520.8	471.28	2223.8	2695.0	1.4444	5.7679	7.2123
.160	113.32	1.0544	1091.4	475.19	2046.7	2521.9	475.36	2221.1	2696.5	1.4550	5.7467	7.2017
.165	114.26	1.0552	1060.4	479.15	2043.8	2522.9	479.33	2218.5	2697.9	1.4652	5.7262	7.1914
.170	115.17	1.0560	1031.2	483.02	2040.9	2523.9	483.20	2216.0	2699.2	1.4752	5.7062	7.1814
.175	116.06	1.0568	1003.6	486.80	2038.1	2524.9	486.99	2213.6	2700.6	1.4849	5.6868	7.1717
.180	116.93	1.0576	977.5	490.49	2035.4	2525.9	490.68	2211.2	2701.8	1.4944	5.6679	7.1623
.185	117.79	1.0583	952.8	494.11	2032.7	2526.8	494.30	2208.8	2703.1	1.5036	5.6495	7.1532
.190	118.62	1.0591	929.3	497.64	2030.1	2527.7	497.84	2206.5	2704.3	1.5127	5.6316	7.1443
.195	119.43	1.0598	907.0	501.10	2027.5	2528.6	501.31	2204.2	2705.5	1.5215	5.6141	7.1356
.200	120.23	1.0605	885.7	504.49	2025.0	2529.5	504.70	2201.9	2706.7	1.5301	5.5970	7.1271
.205	121.02	1.0613	865.5	507.81	2022.5	2530.4	508.03	2199.8	2707.8	1.5386	5.5803	7.1189
.210	121.78	1.0620	846.2	511.06	2020.1	2531.2	511.29	2197.6	2708.9	1.5468	5.5640	7.1109
.215	122.53	1.0626	827.7	514.26	2017.7	2532.0	514.48	2195.5	2710.0	1.5549	5.5481	7.1030
.220	123.27	1.0633	810.1	517.40	2015.4	2532.8	517.63	2193.4	2711.0	1.5628	5.5325	7.0953
.225	124.00	1.0640	793.3	520.47	2013.1	2533.6	520.72	2191.3	2712.1	1.5706	5.5173	7.0878
.230	124.71	1.0647	777.1	523.50	2010.8	2534.3	523.74	2189.3	2713.1	1.5782	5.5023	7.0805
.235	125.41	1.0653	761.6	526.47	2008.6	2535.1	526.72	2187.3	2714.1	1.5856	5.4877	7.0733
.240	126.10	1.0660	746.7	529.39	2006.4	2535.8	529.65	2185.4	2715.0	1.5930	5.4733	7.0663
.245	126.77	1.0666	732.4	532.27	2004.3	2536.5	532.53	2183.5	2716.0	1.6002	5.4593	7.0594
.250	127.44	1.0672	718.7	535.10	2002.1	2537.2	535.37	2181.5	2716.9	1.6072	5.4455	7.0527
.255	128.09	1.0679	705.5	537.88	2000.0	2537.9	538.15	2179.7	2717.8	1.6142	5.4319	7.0461
.260	128.73	1.0685	692.8	540.62	1998.0	2538.6	540.90	2177.8	2718.7	1.6210	5.4186	7.0396
.265	129.37	1.0691	680.5	543.32	1995.9	2539.3	543.60	2176.0	2719.6	1.6277	5.4056	7.0333
.270	129.99	1.0697	668.7	545.97	1993.9	2539.9	546.27	2174.2	2720.5	1.6343	5.3927	7.0270

Definitions of symbols on page 1.

Table 2. Saturation: Pressures

Press. MPa	Temp °C	Specific Volume		Internal Energy			Enthalpy			Entropy		
		Sat. Liquid	Sat. Vapor	Sat. Liquid	Evap.	Sat. Vapor	Sat. Liquid	Evap.	Sat. Vapor	Sat. Liquid	Evap.	Sat. Vapor
p	t	$10^3 v_f$	$10^3 v_g$	u_f	u_{fg}	u_g	h_f	h_{fg}	h_g	s_f	s_{fg}	s_g
.275	130.60	1.0703	657.3	548.59	1991.9	2540.5	548.89	2172.4	2721.3	1.6408	5.3801	7.0209
.280	131.21	1.0709	646.3	551.18	1990.0	2541.2	551.48	2170.7	2722.1	1.6472	5.3677	7.0149
.285	131.81	1.0715	635.7	553.72	1988.1	2541.8	554.02	2168.9	2723.0	1.6535	5.3555	7.0090
.290	132.39	1.0720	625.4	556.23	1986.2	2542.4	556.54	2167.2	2723.8	1.6597	5.3435	7.0032
.295	132.97	1.0726	615.4	558.71	1984.3	2543.0	559.02	2165.5	2724.5	1.6658	5.3317	6.9975
.300	133.55	1.0732	605.8	561.15	1982.4	2543.6	561.47	2163.8	2725.3	1.6718	5.3201	6.9919
.305	134.11	1.0737	596.5	563.56	1980.6	2544.1	563.88	2162.2	2726.1	1.6777	5.3087	6.9864
.310	134.67	1.0743	587.5	565.94	1978.8	2544.7	566.27	2160.6	2726.8	1.6835	5.2974	6.9810
.315	135.22	1.0749	578.7	568.29	1977.0	2545.3	568.62	2158.9	2727.6	1.6893	5.2863	6.9756
.320	135.76	1.0754	570.2	570.61	1975.2	2545.8	570.95	2157.3	2728.3	1.6950	5.2754	6.9704
.325	136.30	1.0759	562.0	572.90	1973.5	2546.4	573.25	2155.8	2729.0	1.7006	5.2646	6.9652
.330	136.83	1.0765	554.0	575.16	1971.7	2546.9	575.52	2154.2	2729.7	1.7061	5.2540	6.9601
.335	137.35	1.0770	546.3	577.40	1970.0	2547.4	577.76	2152.6	2730.4	1.7116	5.2435	6.9551
.340	137.87	1.0775	538.7	579.61	1968.3	2547.9	579.97	2151.1	2731.1	1.7169	5.2332	6.9502
.345	138.38	1.0781	531.4	581.79	1966.6	2548.4	582.16	2149.6	2731.8	1.7222	5.2230	6.9453
.350	138.88	1.0786	524.3	583.95	1965.0	2548.9	584.33	2148.1	2732.4	1.7275	5.2130	6.9405
.355	139.38	1.0791	517.3	586.09	1963.3	2549.4	586.47	2146.6	2733.1	1.7327	5.2031	6.9358
.360	139.87	1.0796	510.6	588.20	1961.7	2549.9	588.59	2145.1	2733.7	1.7378	5.1933	6.9311
.365	140.36	1.0801	504.0	590.29	1960.1	2550.4	590.68	2143.7	2734.4	1.7428	5.1836	6.9265
.370	140.84	1.0806	497.6	592.36	1958.5	2550.9	592.75	2142.2	2735.0	1.7478	5.1741	6.9219
.375	141.32	1.0811	491.4	594.40	1956.9	2551.3	594.81	2140.8	2735.6	1.7528	5.1647	6.9175
.380	141.79	1.0816	485.3	596.42	1955.4	2551.8	596.83	2139.4	2736.2	1.7577	5.1554	6.9130
.385	142.26	1.0821	479.4	598.43	1953.8	2552.2	598.84	2138.0	2736.8	1.7625	5.1462	6.9087
.390	142.72	1.0826	473.6	600.41	1952.3	2552.7	600.83	2136.6	2737.4	1.7673	5.1371	6.9044
.395	143.18	1.0831	468.0	602.37	1950.8	2553.1	602.80	2135.2	2738.0	1.7720	5.1281	6.9001
.40	143.63	1.0836	462.5	604.31	1949.3	2553.6	604.74	2133.8	2738.6	1.7766	5.1193	6.8959
.41	144.53	1.0845	451.9	608.14	1946.3	2554.4	608.59	2131.1	2739.7	1.7858	5.1018	6.8877
.42	145.40	1.0855	441.7	611.90	1943.4	2555.3	612.36	2128.4	2740.8	1.7948	5.0848	6.8796
.43	146.26	1.0864	432.1	615.59	1940.5	2556.1	616.06	2125.8	2741.9	1.8036	5.0681	6.8717
.44	147.10	1.0873	422.8	619.21	1937.6	2556.9	619.68	2123.2	2742.9	1.8122	5.0518	6.8641
.45	147.93	1.0882	414.0	622.77	1934.9	2557.6	623.25	2120.7	2743.9	1.8207	5.0359	6.8565
.46	148.74	1.0891	405.5	626.26	1932.1	2558.4	626.76	2118.2	2744.9	1.8290	5.0202	6.8492
.47	149.54	1.0900	397.4	629.70	1929.4	2559.1	630.21	2115.7	2745.9	1.8371	5.0049	6.8420
.48	150.32	1.0909	389.6	633.08	1926.8	2559.8	633.60	2113.2	2746.8	1.8451	4.9898	6.8349
.49	151.10	1.0917	382.1	636.40	1924.1	2560.6	636.94	2110.8	2747.8	1.8530	4.9751	6.8280
.50	151.86	1.0926	374.9	639.68	1921.6	2561.2	640.23	2108.5	2748.7	1.8607	4.9606	6.8213
.51	152.60	1.0934	368.0	642.90	1919.0	2561.9	643.46	2106.1	2749.6	1.8682	4.9464	6.8146
.52	153.34	1.0942	361.3	646.08	1916.5	2562.6	646.65	2103.8	2750.5	1.8757	4.9324	6.8081
.53	154.06	1.0951	354.9	649.20	1914.0	2563.2	649.78	2101.5	2751.3	1.8830	4.9187	6.8017
.54	154.78	1.0959	348.7	652.28	1911.6	2563.9	652.87	2099.3	2752.1	1.8902	4.9052	6.7955
.55	155.48	1.0967	342.7	655.32	1909.2	2564.5	655.93	2097.0	2753.0	1.8973	4.8920	6.7893
.56	156.17	1.0975	336.9	658.32	1906.8	2565.1	658.93	2094.8	2753.8	1.9043	4.8789	6.7832
.57	156.86	1.0983	331.3	661.27	1904.4	2565.7	661.90	2092.7	2754.5	1.9112	4.8661	6.7773
.58	157.53	1.0991	325.9	664.18	1902.1	2566.3	664.83	2090.5	2755.3	1.9180	4.8535	6.7714
.59	158.20	1.0999	320.7	667.06	1899.8	2566.8	667.71	2088.4	2756.1	1.9246	4.8411	6.7657
.60	158.85	1.1006	315.7	669.90	1897.5	2567.4	670.56	2086.3	2756.8	1.9312	4.8288	6.7600
.61	159.50	1.1014	310.8	672.70	1895.3	2568.0	673.37	2084.2	2757.5	1.9377	4.8168	6.7545
.62	160.14	1.1021	306.0	675.47	1893.0	2568.5	676.15	2082.1	2758.3	1.9441	4.8049	6.7490
.63	160.77	1.1029	301.5	678.20	1890.8	2569.0	678.89	2080.1	2759.0	1.9504	4.7932	6.7436
.64	161.39	1.1036	297.0	680.90	1888.7	2569.6	681.60	2078.0	2759.6	1.9566	4.7817	6.7383
.65	162.01	1.1044	292.7	683.56	1886.5	2570.1	684.28	2076.0	2760.3	1.9627	4.7703	6.7331
.66	162.61	1.1051	288.5	686.20	1884.4	2570.6	686.93	2074.0	2761.0	1.9688	4.7591	6.7279
.67	163.21	1.1058	284.4	688.80	1882.3	2571.1	689.55	2072.1	2761.6	1.9748	4.7481	6.7228
.68	163.81	1.1066	280.5	691.38	1880.2	2571.6	692.13	2070.1	2762.3	1.9807	4.7372	6.7178
.69	164.39	1.1073	276.6	693.93	1878.1	2572.0	694.69	2068.2	2762.9	1.9865	4.7264	6.7129
.70	164.97	1.1080	272.9	696.44	1876.1	2572.5	697.22	2066.3	2763.5	1.9922	4.7158	6.7080
.71	165.55	1.1087	269.2	698.93	1874.0	2573.0	699.72	2064.4	2764.1	1.9979	4.7053	6.7032
.72	166.11	1.1094	265.7	701.40	1872.0	2573.4	702.20	2062.5	2764.7	2.0035	4.6950	6.6985
.73	166.67	1.1101	262.2	703.84	1870.0	2573.9	704.64	2060.7	2765.3	2.0091	4.6848	6.6938
.74	167.23	1.1108	258.9	706.25	1868.1	2574.3	707.07	2058.8	2765.9	2.0146	4.6747	6.6892

Press. MPa	Temp °C	Specific Volume		Internal Energy			Enthalpy			Entropy		
	t	Sat. Liquid $10^3 v_f$	Sat. Vapor $10^3 v_g$	Sat. Liquid u_f	Evap. u_{fg}	Sat. Vapor u_g	Sat. Liquid h_f	Evap. h_{fg}	Sat. Vapor h_g	Sat. Liquid s_f	Evap. s_{fg}	Sat. Vapor s_g
.75	167.78	1.1115	255.6	708.64	1866.1	2574.7	709.47	2057.0	2766.4	2.0200	4.6647	6.6847
.76	168.32	1.1121	252.4	711.00	1864.2	2575.2	711.85	2055.2	2767.0	2.0253	4.6549	6.6802
.77	168.86	1.1128	249.3	713.34	1862.2	2575.5	714.20	2053.4	2767.5	2.0306	4.6451	6.6758
.78	169.39	1.1135	246.3	715.66	1860.3	2576.0	716.52	2051.6	2768.1	2.0359	4.6355	6.6714
.79	169.91	1.1142	243.3	717.95	1858.5	2576.4	718.83	2049.8	2768.6	2.0411	4.6260	6.6670
.80	170.43	1.1148	240.4	720.22	1856.6	2576.8	721.11	2048.0	2769.1	2.0462	4.6166	6.6628
.81	170.95	1.1155	237.6	722.47	1854.7	2577.2	723.38	2046.3	2769.7	2.0513	4.6073	6.6585
.82	171.46	1.1161	234.9	724.70	1852.9	2577.6	725.62	2044.5	2770.2	2.0563	4.5981	6.6544
.83	171.97	1.1168	232.2	726.91	1851.1	2578.0	727.83	2042.8	2770.7	2.0612	4.5890	6.6502
.84	172.47	1.1174	229.5	729.10	1849.2	2578.3	730.04	2041.1	2771.2	2.0662	4.5800	6.6462
.85	172.96	1.1181	227.0	731.27	1847.4	2578.7	732.22	2039.4	2771.6	2.0710	4.5711	6.6421
.86	173.46	1.1187	224.5	733.42	1845.7	2579.1	734.38	2037.7	2772.1	2.0758	4.5623	6.6381
.87	173.94	1.1193	222.0	735.55	1843.9	2579.4	736.52	2036.1	2772.6	2.0806	4.5536	6.6342
.88	174.43	1.1200	219.6	737.66	1842.1	2579.8	738.64	2034.4	2773.0	2.0853	4.5450	6.6303
.89	174.90	1.1206	217.3	739.75	1840.4	2580.1	740.75	2032.8	2773.5	2.0900	4.5364	6.6264
.90	175.38	1.1212	215.0	741.83	1838.6	2580.5	742.83	2031.1	2773.9	2.0946	4.5280	6.6226
.91	175.85	1.1218	212.7	743.89	1836.9	2580.8	744.91	2029.5	2774.4	2.0992	4.5196	6.6188
.92	176.31	1.1225	210.5	745.93	1835.2	2581.1	746.96	2027.9	2774.8	2.1038	4.5113	6.6151
.93	176.78	1.1231	208.4	747.95	1833.5	2581.5	749.00	2026.3	2775.3	2.1083	4.5031	6.6114
.94	177.24	1.1237	206.3	749.96	1831.8	2581.8	751.02	2024.7	2775.7	2.1127	4.4950	6.6077
.95	177.69	1.1243	204.2	751.95	1830.2	2582.1	753.02	2023.1	2776.1	2.1172	4.4869	6.6041
.96	178.14	1.1249	202.2	753.93	1828.5	2582.4	755.01	2021.5	2776.5	2.1215	4.4789	6.6005
.97	178.59	1.1255	200.2	755.89	1826.9	2582.7	756.98	2019.9	2776.9	2.1259	4.4710	6.5969
.98	179.03	1.1261	198.2	757.84	1825.2	2583.1	758.94	2018.4	2777.3	2.1302	4.4632	6.5934
.99	179.47	1.1267	196.3	759.77	1823.6	2583.4	760.88	2016.8	2777.7	2.1345	4.4555	6.5899
1.00	179.91	1.1273	194.44	761.68	1822.0	2583.6	762.81	2015.3	2778.1	2.1387	4.4478	6.5865
1.02	180.77	1.1284	190.80	765.47	1818.8	2584.2	766.63	2012.2	2778.9	2.1471	4.4326	6.5796
1.04	181.62	1.1296	187.30	769.21	1815.6	2584.8	770.38	2009.2	2779.6	2.1553	4.4177	6.5729
1.06	182.46	1.1308	183.92	772.89	1812.5	2585.4	774.08	2006.2	2780.3	2.1634	4.4030	6.5664
1.08	183.28	1.1319	180.67	776.52	1809.4	2585.9	777.74	2003.3	2781.0	2.1713	4.3886	6.5599
1.10	184.09	1.1330	177.53	780.09	1806.3	2586.4	781.34	2000.4	2781.7	2.1792	4.3744	6.5536
1.12	184.89	1.1342	174.49	783.62	1803.3	2586.9	784.89	1997.5	2782.4	2.1869	4.3605	6.5473
1.14	185.68	1.1353	171.56	787.11	1800.3	2587.4	788.40	1994.6	2783.0	2.1945	4.3467	6.5412
1.16	186.46	1.1364	168.73	790.54	1797.4	2587.9	791.86	1991.8	2783.6	2.2020	4.3332	6.5351
1.18	187.23	1.1375	165.99	793.94	1794.4	2588.4	795.28	1989.0	2784.2	2.2093	4.3199	6.5292
1.20	187.99	1.1385	163.33	797.29	1791.5	2588.8	798.65	1986.2	2784.8	2.2166	4.3067	6.5233
1.22	188.74	1.1396	160.77	800.60	1788.7	2589.3	801.98	1983.4	2785.4	2.2238	4.2938	6.5176
1.24	189.48	1.1407	158.28	803.87	1785.9	2589.7	805.28	1980.7	2786.0	2.2309	4.2810	6.5119
1.26	190.20	1.1417	155.86	807.09	1783.0	2590.1	808.53	1978.0	2786.5	2.2378	4.2685	6.5063
1.28	190.93	1.1428	153.53	810.29	1780.3	2590.6	811.75	1975.3	2787.1	2.2447	4.2561	6.5008
1.30	191.64	1.1438	151.25	813.44	1777.5	2591.0	814.93	1972.7	2787.6	2.2515	4.2438	6.4953
1.32	192.34	1.1449	149.05	816.56	1774.8	2591.4	818.07	1970.0	2788.1	2.2582	4.2318	6.4900
1.34	193.04	1.1459	146.91	819.64	1772.1	2591.7	821.18	1967.4	2788.6	2.2648	4.2199	6.4847
1.36	193.72	1.1469	144.83	822.69	1769.4	2592.1	824.25	1964.8	2789.1	2.2714	4.2081	6.4795
1.38	194.40	1.1479	142.81	825.71	1766.8	2592.5	827.29	1962.3	2789.6	2.2778	4.1965	6.4743
1.40	195.07	1.1489	140.84	828.70	1764.1	2592.8	830.30	1959.7	2790.0	2.2842	4.1850	6.4693
1.42	195.74	1.1499	138.93	831.65	1761.5	2593.2	833.28	1957.2	2790.5	2.2905	4.1737	6.4643
1.44	196.39	1.1509	137.07	834.57	1759.0	2593.5	836.23	1954.7	2790.9	2.2968	4.1625	6.4593
1.46	197.04	1.1519	135.25	837.46	1756.4	2593.9	839.14	1952.2	2791.3	2.3029	4.1515	6.4544
1.48	197.69	1.1529	133.49	840.33	1753.9	2594.2	842.03	1949.7	2791.7	2.3090	4.1406	6.4496
1.50	198.32	1.1539	131.77	843.16	1751.3	2594.5	844.89	1947.3	2792.2	2.3150	4.1298	6.4448
1.55	199.88	1.1563	127.66	850.13	1745.1	2595.3	851.92	1941.2	2793.1	2.3298	4.1033	6.4331
1.60	201.41	1.1587	123.80	856.94	1739.0	2596.0	858.79	1935.2	2794.0	2.3442	4.0776	6.4218
1.65	202.89	1.1610	120.16	863.59	1733.0	2596.6	865.50	1929.4	2794.9	2.3582	4.0526	6.4108
1.70	204.34	1.1634	116.73	870.09	1727.2	2597.3	872.06	1923.6	2795.7	2.3718	4.0282	6.4000
1.75	205.76	1.1656	113.49	876.46	1721.4	2597.8	878.50	1917.9	2796.4	2.3851	4.0044	6.3896
1.80	207.15	1.1679	110.42	882.69	1715.7	2598.4	884.79	1912.4	2797.1	2.3981	3.9812	6.3794
1.85	208.51	1.1701	107.51	888.80	1710.1	2598.9	890.96	1906.8	2797.8	2.4109	3.9585	6.3694
1.90	209.84	1.1724	104.75	894.79	1704.6	2599.4	897.02	1901.4	2798.4	2.4233	3.9364	6.3597
1.95	211.14	1.1745	102.12	900.67	1699.2	2599.9	902.96	1896.0	2799.0	2.4354	3.9147	6.3502

Definitions of symbols on page 1.

Table 2. Saturation: Pressures

Press. MPa	Temp °C	Specific Volume		Internal Energy			Enthalpy			Entropy		
		Sat. Liquid	Sat. Vapor	Sat. Liquid	Evap.	Sat. Vapor	Sat. Liquid	Evap.	Sat. Vapor	Sat. Liquid	Evap.	Sat. Vapor
p	t	$10^3 v_f$	$10^3 v_g$	u_f	u_{fg}	u_g	h_f	h_{fg}	h_g	s_f	s_{fg}	s_g
2.00	212.42	1.1767	99.63	906.44	1693.8	2600.3	908.79	1890.7	2799.5	2.4474	3.8935	6.3409
2.05	213.67	1.1789	97.25	912.11	1688.6	2600.7	914.52	1885.5	2800.0	2.4590	3.8728	6.3318
2.10	214.90	1.1810	94.98	917.67	1683.4	2601.0	920.15	1880.3	2800.5	2.4704	3.8524	6.3229
2.15	216.10	1.1831	92.81	923.15	1678.2	2601.4	925.69	1875.2	2800.9	2.4817	3.8325	6.3141
2.20	217.29	1.1852	90.73	928.53	1673.2	2601.7	931.14	1870.2	2801.3	2.4927	3.8129	6.3056
2.25	218.45	1.1872	88.75	933.83	1668.2	2602.0	936.49	1865.2	2801.7	2.5035	3.7937	6.2972
2.30	219.60	1.1893	86.85	939.04	1663.2	2602.3	941.77	1860.2	2802.0	2.5141	3.7749	6.2890
2.35	220.72	1.1913	85.02	944.17	1658.4	2602.5	946.97	1855.4	2802.3	2.5245	3.7564	6.2809
2.40	221.83	1.1933	83.27	949.22	1653.5	2602.8	952.09	1850.5	2802.6	2.5347	3.7382	6.2729
2.45	222.92	1.1953	81.59	954.21	1648.8	2603.0	957.13	1845.7	2802.9	2.5448	3.7204	6.2651
2.5	223.99	1.1973	79.98	959.11	1644.0	2603.1	962.11	1841.0	2803.1	2.5547	3.7028	6.2575
2.6	226.09	1.2013	76.92	968.73	1634.7	2603.5	971.85	1831.6	2803.5	2.5740	3.6685	6.2425
2.7	228.12	1.2051	74.09	978.09	1625.6	2603.7	981.34	1822.4	2803.8	2.5927	3.6353	6.2280
2.8	230.10	1.2090	71.45	987.20	1616.7	2603.9	990.59	1813.4	2804.0	2.6109	3.6030	6.2139
2.9	232.02	1.2127	68.99	996.10	1607.9	2604.0	999.61	1804.5	2804.1	2.6285	3.5717	6.2002
3.0	233.90	1.2165	66.68	1004.78	1599.3	2604.1	1008.42	1795.7	2804.2	2.6457	3.5412	6.1869
3.1	235.72	1.2202	64.52	1013.26	1590.9	2604.1	1017.04	1787.1	2804.1	2.6624	3.5116	6.1740
3.2	237.51	1.2239	62.49	1021.56	1582.5	2604.1	1025.47	1778.6	2804.1	2.6787	3.4827	6.1614
3.3	239.24	1.2275	60.57	1029.68	1574.3	2604.0	1033.72	1770.2	2803.9	2.6946	3.4544	6.1491
3.4	240.94	1.2311	58.77	1037.63	1566.3	2603.9	1041.82	1761.9	2803.7	2.7101	3.4269	6.1370
3.5	242.60	1.2347	57.07	1045.43	1558.3	2603.7	1049.75	1753.7	2803.4	2.7253	3.4000	6.1253
3.6	244.23	1.2382	55.45	1053.07	1550.4	2603.5	1057.53	1745.6	2803.1	2.7401	3.3737	6.1138
3.7	245.82	1.2418	53.92	1060.58	1542.7	2603.3	1065.17	1737.6	2802.8	2.7546	3.3479	6.1025
3.8	247.38	1.2453	52.47	1067.95	1535.0	2603.0	1072.68	1729.7	2802.4	2.7688	3.3227	6.0915
3.9	248.91	1.2487	51.09	1075.19	1527.5	2602.6	1080.05	1721.8	2801.9	2.7828	3.2980	6.0807
4.0	250.40	1.2522	49.78	1082.31	1520.0	2602.3	1087.31	1714.1	2801.4	2.7964	3.2737	6.0701
4.2	253.31	1.2590	47.33	1096.20	1505.3	2601.5	1101.48	1698.8	2800.3	2.8229	3.2266	6.0495
4.4	256.12	1.2658	45.10	1109.66	1490.9	2600.6	1115.22	1683.8	2799.0	2.8485	3.1811	6.0296
4.6	258.83	1.2726	43.06	1122.73	1476.8	2599.5	1128.58	1669.0	2797.6	2.8732	3.1371	6.0103
4.8	261.45	1.2792	41.18	1135.43	1462.9	2598.4	1141.57	1654.5	2796.0	2.8970	3.0945	5.9916
5.0	263.99	1.2859	39.44	1147.81	1449.3	2597.1	1154.23	1640.1	2794.3	2.9202	3.0532	5.9734
5.2	266.45	1.2925	37.83	1159.87	1435.9	2595.8	1166.58	1626.0	2792.6	2.9427	3.0131	5.9557
5.4	268.84	1.2991	36.34	1171.65	1422.8	2594.4	1178.66	1612.0	2790.7	2.9645	2.9740	5.9385
5.6	271.17	1.3056	34.95	1183.16	1409.8	2592.9	1190.46	1598.2	2788.6	2.9858	2.9359	5.9217
5.8	273.43	1.3122	33.65	1194.41	1396.9	2591.3	1202.02	1584.5	2786.5	3.0065	2.8988	5.9052
6.0	275.64	1.3187	32.44	1205.44	1384.3	2589.7	1213.35	1571.0	2784.3	3.0267	2.8625	5.8892
6.2	277.78	1.3252	31.30	1216.25	1371.7	2588.0	1224.46	1557.6	2782.1	3.0464	2.8270	5.8734
6.4	279.88	1.3317	30.23	1226.85	1359.4	2586.2	1235.37	1544.3	2779.7	3.0657	2.7923	5.8580
6.6	281.93	1.3382	29.22	1237.26	1347.1	2584.4	1246.09	1531.2	2777.2	3.0845	2.7583	5.8428
6.8	283.93	1.3448	28.27	1247.49	1335.0	2582.5	1256.63	1518.1	2774.7	3.1030	2.7249	5.8279
7.0	285.88	1.3513	27.37	1257.55	1323.0	2580.5	1267.00	1505.1	2772.1	3.1211	2.6922	5.8133
7.2	287.79	1.3578	26.52	1267.44	1311.0	2578.5	1277.21	1492.2	2769.4	3.1389	2.6600	5.7989
7.4	289.67	1.3644	25.71	1277.18	1299.2	2576.4	1287.28	1479.4	2766.7	3.1563	2.6284	5.7847
7.6	291.50	1.3710	24.94	1286.78	1287.5	2574.3	1297.19	1466.6	2763.8	3.1734	2.5973	5.7707
7.8	293.30	1.3776	24.21	1296.24	1275.8	2572.1	1306.98	1453.9	2760.9	3.1902	2.5666	5.7569
8.0	295.06	1.3842	23.52	1305.57	1264.2	2569.8	1316.64	1441.3	2758.0	3.2068	2.5364	5.7432
8.2	296.79	1.3908	22.86	1314.78	1252.7	2567.5	1326.18	1428.7	2754.9	3.2230	2.5067	5.7297
8.4	298.49	1.3975	22.22	1323.87	1241.3	2565.2	1335.61	1416.2	2751.8	3.2391	2.4773	5.7164
8.6	300.16	1.4042	21.62	1332.86	1229.9	2562.8	1344.93	1403.7	2748.7	3.2549	2.4484	5.7032
8.8	301.80	1.4110	21.04	1341.73	1218.6	2560.3	1354.14	1391.3	2745.4	3.2704	2.4197	5.6902
9.0	303.40	1.4178	20.48	1350.51	1207.3	2557.8	1363.26	1378.9	2742.1	3.2858	2.3915	5.6772
9.2	304.99	1.4246	19.95	1359.19	1196.0	2555.2	1372.29	1366.5	2738.8	3.3009	2.3635	5.6644
9.4	306.54	1.4315	19.44	1367.78	1184.8	2552.6	1381.23	1354.1	2735.4	3.3159	2.3358	5.6517
9.6	308.07	1.4384	18.95	1376.28	1173.6	2549.9	1390.08	1341.8	2731.9	3.3306	2.3084	5.6391
9.8	309.58	1.4454	18.48	1384.70	1162.5	2547.2	1398.86	1329.5	2728.3	3.3452	2.2813	5.6265
10.0	311.06	1.4524	18.026	1393.04	1151.4	2544.4	1407.56	1317.1	2724.7	3.3596	2.2544	5.6141
10.2	312.52	1.4595	17.588	1401.31	1140.3	2541.6	1416.19	1304.8	2721.0	3.3739	2.2278	5.6017
10.4	313.96	1.4667	17.167	1409.51	1129.2	2538.7	1424.76	1292.5	2717.3	3.3880	2.2014	5.5894
10.6	315.38	1.4739	16.760	1417.64	1118.2	2535.8	1433.26	1280.2	2713.5	3.4020	2.1752	5.5771
10.8	316.77	1.4812	16.367	1425.71	1107.1	2532.8	1441.70	1267.9	2709.6	3.4158	2.1491	5.5649

Definitions of symbols on page 1.

Press. MPa	Temp °C	Specific Volume		Internal Energy			Enthalpy			Entropy		
		Sat. Liquid $10^3 v_f$	Sat. Vapor $10^3 v_g$	Sat. Liquid u_f	Evap. u_{fg}	Sat. Vapor u_g	Sat. Liquid h_f	Evap. h_{fg}	Sat. Vapor h_g	Sat. Liquid s_f	Evap. s_{fg}	Sat. Vapor s_g
p	t											
11.0	318.15	1.4886	15.987	1433.7	1096.0	2529.8	1450.1	1255.5	2705.6	3.4295	2.1233	5.5527
11.2	319.50	1.4960	15.620	1441.7	1085.0	2526.7	1458.4	1243.2	2701.6	3.4430	2.0976	5.5406
11.4	320.84	1.5036	15.264	1449.6	1073.9	2523.5	1466.7	1230.8	2697.5	3.4565	2.0720	5.5285
11.6	322.16	1.5112	14.920	1457.4	1062.9	2520.3	1474.9	1218.4	2693.4	3.4698	2.0466	5.5165
11.8	323.46	1.5189	14.587	1465.2	1051.8	2517.0	1483.1	1206.0	2689.2	3.4831	2.0214	5.5044
12.0	324.75	1.5267	14.263	1473.0	1040.7	2513.7	1491.3	1193.6	2684.9	3.4962	1.9962	5.4924
12.2	326.02	1.5345	13.949	1480.7	1029.6	2510.3	1499.4	1181.1	2680.5	3.5092	1.9712	5.4804
12.4	327.27	1.5425	13.644	1488.3	1018.5	2506.9	1507.5	1168.6	2676.0	3.5222	1.9462	5.4684
12.6	328.51	1.5506	13.348	1496.0	1007.4	2503.3	1515.5	1156.0	2671.5	3.5351	1.9213	5.4564
12.8	329.73	1.5588	13.060	1503.6	996.2	2499.8	1523.5	1143.4	2666.9	3.5478	1.8965	5.4444
13.0	330.93	1.5671	12.780	1511.1	985.0	2496.1	1531.5	1130.7	2662.2	3.5606	1.8718	5.4323
13.2	332.12	1.5756	12.508	1518.7	973.7	2492.4	1539.5	1118.0	2657.5	3.5732	1.8471	5.4203
13.4	333.30	1.5841	12.242	1526.2	962.4	2488.6	1547.4	1105.2	2652.6	3.5858	1.8224	5.4082
13.6	334.46	1.5928	11.983	1533.7	951.1	2484.7	1555.3	1092.4	2647.7	3.5983	1.7978	5.3961
13.8	335.61	1.6017	11.731	1541.1	939.7	2480.8	1563.2	1079.5	2642.7	3.6108	1.7731	5.3839
14.0	336.75	1.6107	11.485	1548.6	928.2	2476.8	1571.1	1066.5	2637.6	3.6232	1.7485	5.3717
14.2	337.87	1.6198	11.245	1556.0	916.7	2472.7	1579.0	1053.4	2632.4	3.6356	1.7239	5.3595
14.4	338.98	1.6291	11.010	1563.4	905.1	2468.5	1586.9	1040.2	2627.1	3.6479	1.6992	5.3471
14.6	340.08	1.6386	10.781	1570.8	893.4	2464.3	1594.8	1026.9	2621.7	3.6603	1.6745	5.3348
14.8	341.16	1.6483	10.557	1578.2	881.7	2459.9	1602.6	1013.5	2616.1	3.6726	1.6498	5.3223
15.0	342.24	1.6581	10.337	1585.6	869.8	2455.5	1610.5	1000.0	2610.5	3.6848	1.6249	5.3098
15.2	343.30	1.6682	10.123	1593.0	857.9	2450.9	1618.4	986.4	2604.8	3.6971	1.6001	5.2971
15.4	344.35	1.6785	9.912	1600.4	845.8	2446.3	1626.3	972.7	2598.9	3.7093	1.5751	5.2844
15.6	345.39	1.6890	9.706	1607.8	833.7	2441.5	1634.2	958.8	2593.0	3.7216	1.5500	5.2716
15.8	346.42	1.6997	9.504	1615.3	821.4	2436.7	1642.1	944.8	2586.9	3.7338	1.5248	5.2586
16.0	347.44	1.7107	9.306	1622.7	809.0	2431.7	1650.1	930.6	2580.6	3.7461	1.4994	5.2455
16.2	348.44	1.7220	9.111	1630.1	796.5	2426.6	1658.0	916.2	2574.3	3.7584	1.4739	5.2323
16.4	349.44	1.7336	8.920	1637.6	783.8	2421.4	1666.0	901.7	2567.7	3.7707	1.4482	5.2189
16.6	350.43	1.7454	8.732	1645.1	771.0	2416.1	1674.1	886.9	2561.0	3.7830	1.4223	5.2053
16.8	351.40	1.7577	8.547	1652.6	758.0	2410.6	1682.2	872.0	2554.2	3.7954	1.3961	5.1916
17.0	352.37	1.7702	8.364	1660.2	744.8	2405.0	1690.3	856.9	2547.2	3.8079	1.3698	5.1777
17.2	353.33	1.7832	8.185	1667.8	731.4	2399.2	1698.5	841.5	2540.0	3.8204	1.3431	5.1635
17.4	354.28	1.7966	8.008	1675.5	717.7	2393.2	1706.8	825.8	2532.6	3.8330	1.3161	5.1491
17.6	355.21	1.8104	7.833	1683.2	703.9	2387.1	1715.1	809.9	2525.0	3.8457	1.2888	5.1345
17.8	356.14	1.8248	7.660	1691.0	689.8	2380.8	1723.5	793.6	2517.2	3.8586	1.2611	5.1196
18.0	357.06	1.8397	7.489	1698.9	675.4	2374.3	1732.0	777.1	2509.1	3.8715	1.2329	5.1044
18.2	357.98	1.8551	7.320	1706.9	660.7	2367.6	1740.7	760.1	2500.8	3.8846	1.2043	5.0889
18.4	358.88	1.8713	7.153	1715.0	645.6	2360.6	1749.4	742.8	2492.2	3.8978	1.1752	5.0730
18.6	359.77	1.8881	6.987	1723.2	630.2	2353.4	1758.3	725.1	2483.3	3.9112	1.1455	5.0567
18.8	360.66	1.9058	6.822	1731.5	614.4	2345.9	1767.3	706.8	2474.1	3.9249	1.1151	5.0400
19.0	361.54	1.9243	6.657	1739.9	598.1	2338.1	1776.5	688.0	2464.5	3.9388	1.0839	5.0228
19.2	362.41	1.9439	6.493	1748.6	581.3	2329.9	1785.9	668.7	2454.6	3.9530	1.0520	5.0050
19.4	363.27	1.9646	6.329	1757.4	563.9	2321.4	1795.6	648.6	2444.2	3.9676	1.0191	4.9866
19.6	364.12	1.9866	6.165	1766.5	545.9	2312.4	1805.5	627.8	2433.3	3.9825	.9851	4.9676
19.8	364.97	2.0102	6.000	1775.9	527.1	2303.0	1815.7	606.1	2421.8	3.9979	.9498	4.9477
20.0	365.81	2.036	5.834	1785.6	507.5	2293.0	1826.3	583.4	2409.7	4.0139	.9130	4.9269
20.2	366.64	2.063	5.665	1795.6	486.8	2282.4	1837.3	559.6	2396.9	4.0305	.8745	4.9050
20.4	367.46	2.093	5.495	1806.2	464.9	2271.1	1848.9	534.3	2383.2	4.0479	.8340	4.8819
20.6	368.28	2.126	5.320	1817.3	441.5	2258.8	1861.1	507.3	2368.4	4.0663	.7908	4.8571
20.8	369.09	2.164	5.139	1829.2	416.2	2245.4	1874.2	478.1	2352.3	4.0861	.7444	4.8304
21.0	369.89	2.207	4.952	1842.1	388.5	2230.6	1888.4	446.2	2334.6	4.1075	.6938	4.8013
21.2	370.69	2.257	4.754	1856.3	357.7	2214.0	1904.2	410.6	2314.8	4.1313	.6377	4.7690
21.4	371.47	2.318	4.538	1872.6	322.1	2194.6	1922.2	369.6	2291.8	4.1585	.5733	4.7318
21.6	372.25	2.398	4.298	1892.0	279.2	2171.3	1943.8	320.3	2264.1	4.1914	.4962	4.6876
21.8	373.03	2.514	4.007	1917.8	222.5	2140.3	1972.6	255.1	2227.6	4.2351	.3947	4.6298
22.0	373.80	2.756	3.568	1964.4	122.7	2087.1	2025.0	140.6	2165.6	4.3154	.2173	4.5327
22.09	374.14	3.155	3.155	2029.6	0	2029.6	2099.3	0	2099.3	4.4298	0	4.4298

Definitions of symbols on page 1.

Table 3. Vapor

Water Vapor at Low Pressures $\left(\dfrac{pv}{RT} = 1, \; R = 0.46151 \text{ kJ/kg·K} \right)$

t	T	pv kJ/kg	u_o	h_o	s_1	ψ_1	ζ_1	p_r	v_r	c_{po}	c_{vo}	k
0	273.2	126.06	2375.5	2501.5	6.8042	516.9	643.0	.2529	4984.	1.8584	1.3969	1.3304
5	278.2	128.37	2382.5	2510.8	6.8379	480.5	608.8	.2721	4718.	1.8592	1.3977	1.3302
10	283.2	130.68	2389.4	2520.1	6.8711	443.9	574.6	.2923	4470.	1.8601	1.3986	1.3300
15	288.2	132.98	2396.4	2529.4	6.9036	407.2	540.1	.3137	4239.	1.8611	1.3996	1.3297
20	293.2	135.29	2403.4	2538.7	6.9357	370.2	505.5	.3363	4024.	1.8622	1.4007	1.3295
25	298.2	137.60	2410.4	2548.0	6.9672	333.2	470.8	.3600	3822.	1.8634	1.4019	1.3292
30	303.2	139.91	2417.5	2557.4	6.9982	296.0	435.9	.3850	3634.	1.8647	1.4031	1.3289
35	308.2	142.21	2424.5	2566.7	7.0287	258.6	400.8	.4113	3457.	1.8660	1.4045	1.3286
40	313.2	144.52	2431.5	2576.0	7.0587	221.1	365.6	.4390	3292.	1.8674	1.4059	1.3283
45	318.2	146.83	2438.5	2585.4	7.0883	183.4	330.2	.4681	3137.	1.8689	1.4074	1.3279
50	323.2	149.14	2445.6	2594.7	7.1175	145.6	294.7	.4986	2991.	1.8705	1.4090	1.3275
55	328.2	151.44	2452.6	2604.1	7.1462	107.6	259.0	.5306	2854.	1.8721	1.4106	1.3272
60	333.2	153.75	2459.7	2613.4	7.1745	69.5	223.2	.5642	2725.	1.8738	1.4123	1.3268
65	338.2	156.06	2466.7	2622.8	7.2025	31.2	187.3	.5994	2604.	1.8756	1.4141	1.3264
70	343.2	158.37	2473.8	2632.2	7.2300	-7.2	151.2	.6363	2489.	1.8774	1.4159	1.3259
75	348.2	160 67	2480.9	2641.6	7.2572	-45.7	115.0	.6749	2381.	1.8793	1.4178	1.3255
80	353.2	162.98	2488.0	2651.0	7.2840	-84.3	78.6	.7152	2279.	1.8812	1.4197	1.3251
85	358.2	165.29	2495.1	2660.4	7.3104	-123.1	42.2	.7574	2182.	1.8832	1.4217	1.3246
90	363.2	167.60	2502.2	2669.8	7.3366	-162.1	5.5	.8015	2091.	1.8852	1.4237	1.3242
95	368.2	169.90	2509.3	2679.2	7.3624	-201.1	-31.2	.8476	2004.	1.8873	1.4258	1.3237
100	373.2	172.21	2516.5	2688.7	7.3878	-240.3	-68.1	.8957	1922.6	1.8894	1.4279	1.3232
110	383.2	176.83	2530.8	2707.6	7.4379	-319.0	-142.2	.9983	1771.3	1.8937	1.4322	1.3222
120	393.2	181.44	2545.1	2726.6	7.4867	-398.3	-216.8	1.1097	1635.0	1.8983	1.4367	1.3212
130	403.2	186.06	2559.5	2745.6	7.5344	-478.0	-291.9	1.2307	1511.9	1.9029	1.4414	1.3202
140	413.2	190.67	2573.9	2764.6	7.5811	-558.2	-367.5	1.3617	1400.3	1.9077	1.4462	1.3191
150	423.2	195.29	2588.4	2783.7	7.6268	-638.9	-443.6	1.5033	1299.0	1.9126	1.4511	1.3180
160	433.2	199.90	2603.0	2802.9	7.6715	-720.0	-520.1	1.6564	1206.9	1.9177	1.4562	1.3169
170	443.2	204.52	2617.6	2822.1	7.7154	-801.5	-597.0	1.8214	1122.9	1.9229	1.4613	1.3158
180	453.2	209.13	2632.2	2841.3	7.7583	-883.5	-674.4	1.9991	1046.1	1.9281	1.4666	1.3147
190	463.2	213.75	2646.9	2860.6	7.8005	-965.9	-752.2	2.1902	975.9	1.9335	1.4720	1.3135
200	473.2	218.4	2661.6	2880.0	7.8418	-1048.7	-830.4	2.396	911.5	1.9389	1.4774	1.3124
220	493.2	227.6	2691.3	2918.9	7.9223	-1215.6	-988.0	2.852	798.0	1.9501	1.4886	1.3100
240	513.2	236.8	2721.2	2958.0	8.0001	-1384.1	-1147.2	3.375	701.6	1.9616	1.5001	1.3077
260	533.2	246.0	2751.3	2997.4	8.0753	-1554.1	-1308.0	3.973	619.3	1.9733	1.5118	1.3053
280	553.2	255.3	2781.7	3036.9	8.1482	-1725.5	-1470.2	4.653	548.7	1.9853	1.5238	1.3029
300	573.2	264.5	2812.3	3076.8	8.2189	-1898.4	-1633.9	5.423	487.7	1.9975	1.5360	1.3005
320	593.2	273.7	2843.1	3116.8	8.2877	-2072.7	-1799.0	6.294	434.9	2.0100	1.5485	1.2980
340	613.2	283.0	2874.2	3157.2	8.3545	-2248.4	-1965.4	7.275	388.9	2.0226	1.5611	1.2956
360	633.2	292.2	2905.5	3197.8	8.4197	-2425.4	-2133.2	8.378	348.8	2.0354	1.5739	1.2932
380	653.2	301.4	2937.2	3238.6	8.4832	-2603.6	-2302.2	9.614	313.5	2.0483	1.5868	1.2908
400	673.2	310.7	2969.0	3279.7	8.5451	-2783.1	-2472.5	10.996	282.5	2.0614	1.5999	1.2885
420	693.2	319.9	3001.2	3321.0	8.6057	-2963.9	-2644.0	12.537	255.2	2.0747	1.6132	1.2861
440	713.2	329.1	3033.5	3362.7	8.6649	-3145.8	-2816.7	14.253	230.9	2.0880	1.6265	1.2837
460	733.2	338.4	3066.2	3404.6	8.7228	-3328.9	-2990.6	16.160	209.4	2.1015	1.6400	1.2814
480	753.2	347.6	3099.1	3446.7	8.7796	-3513.2	-3165.6	18.274	190.2	2.1150	1.6535	1.2791
500	773.2	356.8	3132.4	3489.2	8.8352	-3699	-3342	20.61	173.10	2.1287	1.6672	1.2768
550	823.2	379.9	3216.6	3596.5	8.9696	-4167	-3787	27.59	137.71	2.1631	1.7016	1.2712
600	873.2	403.0	3302.5	3705.5	9.0982	-4642	-4239	36.45	110.56	2.1980	1.7365	1.2658
650	923.2	426.0	3390.2	3816.3	9.2216	-5123	-4697	47.62	89.47	2.2331	1.7716	1.2605
700	973.2	449.1	3479.7	3928.8	9.3403	-5610	-5161	61.58	72.93	2.2683	1.8068	1.2554
750	1023.2	472.2	3570.9	4043.1	9.4548	-6103	-5631	78.93	59.82	2.3036	1.8421	1.2505
800	1073.2	495.3	3663.9	4159.2	9.5655	-6601	-6106	100.34	49.36	2.3387	1.8771	1.2459
850	1123.2	518.3	3758.6	4277.0	9.6728	-7105	-6587	126.60	40.95	2.3734	1.9119	1.2414
900	1173.2	541.4	3855.1	4396.5	9.7769	-7615	-7073	158.63	34.13	2.4078	1.9462	1.2371
950	1223.2	564.5	3953.2	4517.7	9.8781	-8129	-7565	197.53	28.58	2.4415	1.9799	1.2331
1000	1273.2	587.6	4053.1	4640.6	9.9766	-8649	-8061	244.5	24.03	2.4744	2.0128	1.2293
1100	1373.2	633.7	4257.5	4891.2	10.1661	-9702	-9068	368.6	17.19	2.5369	2.0754	1.2224
1200	1473.2	679.9	4467.9	5147.8	10.3464	-10 774	-10 094	544.9	12.48	2.5938	2.1323	1.2164
1300	1573.2	726.0	4683.7	5409.7	10.5184	-11 863	-11 137	791.0	9.18	2.6431	2.1816	1.2115

Definitions of symbols on page 1.

	.001 (6.98)				.002 (17.50)				.004 (28.96)			p (t Sat.)
10³ v	u	h	s	10³ v	u	h	s	10³ v	u	h	s	t
129 208	2385.0	2514.2	8.9756	67 004	2399.5	2533.5	8.7237	34 800	2415.2	2554.4	8.4746	Sat.
125 982	*2375.2*	*2501.2*	*8.9287*	*62 952*	*2375.0*	*2500.9*	*8.6080*	*31 436*	*2374.6*	*2500.3*	*8.2865*	**0**
128 294	*2382.2*	*2510.5*	*8.9624*	*64 109*	*2382.0*	*2510.2*	*8.6418*	*32 017*	*2381.6*	*2509.6*	*8.3202*	**5**
130 605	2389.2	2519.8	8.9956	65 267	2389.0	2519.5	8.6749	*32 598*	*2388.5*	*2518.9*	*8.3534*	**10**
132 917	2396.2	2529.1	9.0282	66 425	2396.0	2528.8	8.7075	*33 179*	*2395.5*	*2528.3*	*8.3861*	**15**
135 228	2403.2	2538.5	9.0603	67 582	2403.0	2538.2	8.7396	*33 759*	*2402.6*	*2537.6*	*8.4182*	**20**
137 539	2410.2	2547.8	9.0918	68 740	2410.0	2547.5	8.7712	*34 340*	*2409.6*	*2547.0*	*8.4499*	**25**
139 850	2417.3	2557.1	9.1228	69 897	2417.1	2556.8	8.8023	34 920	2416.6	2556.3	8.4810	**30**
142 161	2424.3	2566.4	9.1534	71 054	2424.1	2566.2	8.8329	35 500	2423.7	2565.7	8.5117	**35**
144 472	2431.3	2575.8	9.1835	72 211	2431.1	2575.6	8.8630	36 080	2430.8	2575.1	8.5419	**40**
146 782	2438.4	2585.1	9.2131	73 368	2438.2	2584.9	8.8927	36 660	2437.8	2584.5	8.5716	**45**
149 093	2445.4	2594.5	9.2423	74 524	2445.2	2594.3	8.9219	37 240	2444.9	2593.9	8.6009	**50**
151 403	2452.5	2603.9	9.2711	75 680	2452.3	2603.7	8.9507	37 819	2452.0	2603.3	8.6298	**55**
153 713	2459.5	2613.2	9.2994	76 836	2459.4	2613.1	8.9791	38 398	2459.1	2612.7	8.6583	**60**
156 022	2466.6	2622.6	9.3274	77 992	2466.5	2622.5	9.0071	38 978	2466.2	2622.1	8.6863	**65**
158 332	2473.7	2632.0	9.3550	79 148	2473.6	2631.9	9.0347	39 557	2473.3	2631.5	8.7140	**70**
160 641	2480.8	2641.4	9.3822	80 304	2480.7	2641.3	9.0619	40 135	2480.4	2640.9	8.7413	**75**
162 951	2487.9	2650.8	9.4090	81 460	2487.8	2650.7	9.0888	40 714	2487.5	2650.4	8.7682	**80**
165 260	2495.0	2660.3	9.4355	82 615	2494.9	2660.1	9.1153	41 293	2494.7	2659.8	8.7947	**85**
167 569	2502.1	2669.7	9.4616	83 770	2502.0	2669.5	9.1414	41 871	2501.8	2669.3	8.8209	**90**
169 878	2509.2	2679.1	9.4874	84 925	2509.1	2679.0	9.1673	42 449	2508.9	2678.7	8.8468	**95**
172 187	2516.4	2688.6	9.5129	86 081	2516.3	2688.4	9.1928	43 028	2516.1	2688.2	8.8724	**100**
176 804	2530.7	2707.5	9.5630	88 391	2530.6	2707.4	9.2429	44 184	2530.4	2707.2	8.9225	**110**
181 422	2545.0	2726.5	9.6119	90 700	2545.0	2726.4	9.2918	45 340	2544.8	2726.2	8.9715	**120**
186 038	2559.4	2745.5	9.6596	93 010	2559.4	2745.4	9.3395	46 495	2559.2	2745.2	9.0193	**130**
190 655	2573.9	2764.5	9.7063	95 319	2573.8	2764.5	9.3863	47 651	2573.7	2764.3	9.0661	**140**
195 272	2588.4	2783.6	9.7520	97 628	2588.3	2783.6	9.4320	48 806	2588.2	2783.4	9.1118	**150**
199 888	2602.9	2802.8	9.7968	99 936	2602.9	2802.7	9.4767	49 961	2602.8	2802.6	9.1566	**160**
204 504	2617.5	2822.0	9.8406	102 245	2617.5	2822.0	9.5206	51 116	2617.4	2821.8	9.2005	**170**
209 120	2632.2	2841.3	9.8836	104 554	2632.1	2841.2	9.5636	52 270	2632.0	2841.1	9.2435	**180**
213 736	2646.8	2860.6	9.9257	106 862	2646.8	2860.5	9.6058	53 425	2646.7	2860.4	9.2857	**190**
218 352	2661.6	2880.0	9.9671	109 170	2661.6	2879.9	9.6471	54 580	2661.5	2879.8	9.3271	**200**
227 584	2691.3	2918.8	10.0476	113 787	2691.2	2918.8	9.7277	56 888	2691.2	2918.7	9.4076	**220**
236 815	2721.2	2958.0	10.1254	118 403	2721.1	2957.9	9.8054	59 197	2721.1	2957.9	9.4854	**240**
246 046	2751.3	2997.3	10.2006	123 019	2751.3	2997.3	9.8807	61 506	2751.2	2997.2	9.5607	**260**
255 277	2781.6	3036.9	10.2735	127 635	2781.6	3036.9	9.9536	63 814	2781.6	3036.8	9.6336	**280**
264 508	2812.2	3076.8	10.3443	132 251	2812.2	3076.7	10.0243	66 122	2812.2	3076.7	9.7044	**300**
273 739	2843.1	3116.8	10.4130	136 866	2843.1	3116.8	10.0931	68 430	2843.0	3116.8	9.7731	**320**
282 970	2874.2	3157.2	10.4799	141 482	2874.2	3157.1	10.1599	70 738	2874.2	3157.1	9.8400	**340**
292 200	2905.6	3197.8	10.5450	146 098	2905.5	3197.7	10.2251	73 046	2905.5	3197.7	9.9051	**360**
301 431	2937.2	3238.6	10.6085	150 713	2937.1	3238.6	10.2886	75 354	2937.1	3238.5	9.9687	**380**
310 661	2969.0	3279.7	10.6705	155 329	2969.0	3279.7	10.3506	77 662	2969.0	3279.6	10.0307	**400**
319 892	3001.2	3321.1	10.7310	159 944	3001.2	3321.0	10.4111	79 970	3001.1	3321.0	10.0912	**420**
329 122	3033.6	3362.7	10.7902	164 559	3033.6	3362.7	10.4703	82 278	3033.5	3362.6	10.1504	**440**
338 353	3066.2	3404.6	10.8482	169 175	3066.2	3404.6	10.5283	84 586	3066.2	3404.5	10.2084	**460**
347 583	3099.2	3446.7	10.9049	173 790	3099.2	3446.7	10.5850	86 894	3099.1	3446.7	10.2651	**480**
356 814	3132.4	3489.2	10.9605	178 405	3132.4	3489.2	10.6406	89 201	3132.3	3489.2	10.3207	**500**
379 890	3216.6	3596.5	11.0950	189 944	3216.6	3596.5	10.7751	94 971	3216.6	3596.4	10.4552	**550**
402 966	3302.5	3705.5	11.2235	201 482	3302.5	3705.5	10.9036	100 740	3302.5	3705.4	10.5837	**600**
426 041	3390.2	3816.2	11.3469	213 020	3390.2	3816.2	11.0270	106 509	3390.2	3816.2	10.7070	**650**
449 117	3479.6	3928.7	11.4655	224 558	3479.6	3928.7	11.1456	112 278	3479.6	3928.7	10.8257	**700**
472 193	3570.8	4043.0	11.5800	236 096	3570.8	4043.0	11.2601	118 047	3570.8	4043.0	10.9402	**750**
495 268	3663.8	4159.1	11.6908	247 634	3663.8	4159.1	11.3709	123 816	3663.8	4159.1	11.0510	**800**
518 344	3758.6	4276.9	11.7981	259 171	3758.6	4276.9	11.4782	129 585	3758.6	4276.9	11.1583	**850**
541 420	3855.1	4396.5	11.9023	270 709	3855.1	4396.5	11.5824	135 354	3855.1	4396.5	11.2625	**900**
564 495	3953.2	4517.7	12.0035	282 247	3953.2	4517.7	11.6836	141 123	3953.2	4517.7	11.3637	**950**
587 571	4053.0	4640.6	12.1019	293 785	4053.0	4640.6	11.7820	146 892	4053.0	4640.6	11.4621	**1000**
633 722	4257.5	4891.2	12.2914	316 861	4257.5	4891.2	11.9715	158 430	4257.5	4891.2	11.6516	**1100**
679 873	4467.9	5147.8	12.4718	339 936	4467.9	5147.8	12.1519	169 968	4467.9	5147.8	11.8320	**1200**
726 025	4683.7	5409.7	12.6438	363 012	4683.7	5409.7	12.3239	181 506	4683.7	5409.7	12.0040	**1300**

Definitions of symbols on page 1.

Table 3. Vapor

p (t Sat.)		.006 (36.16)				.008 (41.51)				.010 (45.81)		
t	$10^3 v$	u	h	s	$10^3 v$	u	h	s	$10^3 v$	u	h	s
Sat.	23 739	2425.0	2567.4	8.3304	18 103.	2432.2	2577.0	8.2287	14 674.	2437.9	2584.7	8.1502
25	*22 873*	*2409.2*	*2546.4*	*8.2614*	*17 140.*	*2408.8*	*2545.9*	*8.1272*	*13 700.*	*2408.3*	*2545.3*	*8.0228*
30	*23 261*	*2416.2*	*2555.8*	*8.2926*	*17 432.*	*2415.8*	*2555.3*	*8.1585*	*13 934.*	*2415.4*	*2554.8*	*8.0541*
35	*23 649*	*2423.3*	*2565.2*	*8.3233*	*17 724.*	*2422.9*	*2564.7*	*8.1893*	*14 168.*	*2422.5*	*2564.2*	*8.0850*
40	24 037	2430.4	2574.6	8.3536	*18 015.*	*2430.0*	*2574.1*	*8.2196*	*14 402.*	*2429.6*	*2573.7*	*8.1154*
45	24 424	2437.5	2584.0	8.3834	18 306.	2437.1	2583.6	8.2495	*14 636.*	*2436.8*	*2583.1*	*8.1454*
50	24 812	2444.6	2593.4	8.4128	18 598.	2444.2	2593.0	8.2790	14 869.	2443.9	2592.6	8.1749
55	25 199	2451.7	2602.9	8.4417	18 889.	2451.3	2602.5	8.3080	15 102.	2451.0	2602.1	8.2040
60	25 586	2458.8	2612.3	8.4702	19 179.	2458.5	2611.9	8.3366	15 336.	2458.2	2611.5	8.2327
65	25 973	2465.9	2621.7	8.4984	19 470.	2465.6	2621.4	8.3647	15 569.	2465.3	2621.0	8.2609
70	26 359	2473.0	2631.2	8.5261	19 761.	2472.7	2630.8	8.3925	15 801.	2472.5	2630.5	8.2887
75	26 746	2480.1	2640.6	8.5534	20 051	2479.9	2640.3	8.4199	16 034.	2479.6	2640.0	8.3162
80	27 132	2487.3	2650.1	8.5804	20 341	2487.0	2649.8	8.4469	16 267.	2486.8	2649.5	8.3432
85	27 518	2494.4	2659.5	8.6070	20 631	2494.2	2659.2	8.4736	16 499.	2494.0	2659.0	8.3699
90	27 904	2501.6	2669.0	8.6332	20 921	2501.4	2668.7	8.4999	16 731.	2501.1	2668.5	8.3963
95	28 291	2508.7	2678.5	8.6591	21 211	2508.5	2678.2	8.5258	16 964.	2508.3	2678.0	8.4223
100	28 676	2515.9	2688.0	8.6847	21 501	2515.7	2687.7	8.5514	17 196.	2515.5	2687.5	8.4479
110	29 448	2530.3	2707.0	8.7350	22 080	2530.1	2706.7	8.6017	17 660.	2529.9	2706.5	8.4983
120	30 219	2544.7	2726.0	8.7840	22 659	2544.5	2725.8	8.6508	18 123.	2544.4	2725.6	8.5474
130	30 990	2559.1	2745.0	8.8318	23 238	2559.0	2744.9	8.6987	18 586.	2558.8	2744.7	8.5954
140	31 761	2573.6	2764.1	8.8787	23 816	2573.5	2764.0	8.7456	19 050.	2573.3	2763.8	8.6423
150	32 532	2588.1	2783.3	8.9244	24 395	2588.0	2783.1	8.7914	19 512.	2587.9	2783.0	8.6882
160	33 302	2602.7	2802.5	8.9693	24 973	2602.6	2802.3	8.8363	19 975.	2602.5	2802.2	8.7331
170	34 072	2617.3	2821.7	9.0132	25 551	2617.2	2821.6	8.8802	20 438.	2617.1	2821.5	8.7770
180	34 843	2631.9	2841.0	9.0562	26 129	2631.9	2840.9	8.9232	20 900.	2631.8	2840.8	8.8201
190	35 613	2646.7	2860.3	9.0984	26 706	2646.6	2860.2	8.9655	21 363.	2646.5	2860.1	8.8623
200	36 383	2661.4	2879.7	9.1398	27 284	2661.4	2879.6	9.0069	21 825	2661.3	2879.5	8.9038
210	37 153	2676.2	2899.2	9.1805	27 862	2676.2	2899.1	9.0476	22 287	2676.1	2899.0	8.9444
220	37 922	2691.1	2918.6	9.2204	28 439	2691.1	2918.6	9.0875	22 749	2691.0	2918.5	8.9844
230	38 692	2706.0	2938.2	9.2596	29 017	2706.0	2938.1	9.1268	23 212	2705.9	2938.0	9.0237
240	39 462	2721.0	2957.8	9.2982	29 594	2721.0	2957.7	9.1653	23 674	2720.9	2957.7	9.0623
250	40 232	2736.1	2977.5	9.3362	30 172	2736.0	2977.4	9.2033	24 136	2736.0	2977.3	9.1002
260	41 001	2751.2	2997.2	9.3735	30 749	2751.1	2997.1	9.2406	24 598	2751.1	2997.0	9.1376
270	41 771	2766.3	3016.9	9.4102	31 326	2766.3	3016.9	9.2774	25 059	2766.2	3016.8	9.1743
280	42 540	2781.5	3036.8	9.4464	31 903	2781.5	3036.7	9.3136	25 521	2781.5	3036.7	9.2105
290	43 310	2796.8	3056.7	9.4821	32 481	2796.8	3056.6	9.3492	25 983	2796.7	3056.6	9.2462
300	44 079	2812.2	3076.6	9.5172	33 058	2812.1	3076.6	9.3844	26 445	2812.1	3076.5	9.2813
320	45 618	2843.0	3116.7	9.5859	34 212	2843.0	3116.7	9.4531	27 369	2842.9	3116.6	9.3501
340	47 157	2874.1	3157.1	9.6528	35 367	2874.1	3157.0	9.5200	28 292	2874.1	3157.0	9.4170
360	48 696	2905.5	3197.7	9.7180	36 521	2905.5	3197.6	9.5852	29 216	2905.4	3197.6	9.4821
380	50 235	2937.1	3238.5	9.7815	37 675	2937.1	3238.5	9.6487	30 139	2937.0	3238.4	9.5457
400	51 774	2969.0	3279.6	9.8435	38 829	2969.0	3279.6	9.7107	31 063	2968.9	3279.6	9.6077
420	53 312	3001.1	3321.0	9.9041	39 983	3001.1	3321.0	9.7713	31 986	3001.1	3320.9	9.6682
440	54 851	3033.5	3362.6	9.9633	41 137	3033.5	3362.6	9.8305	32 909	3033.5	3362.6	9.7275
460	56 390	3066.2	3404.5	10.0212	42 291	3066.2	3404.5	9.8884	33 832	3066.1	3404.5	9.7854
480	57 928	3099.1	3446.7	10.0780	43 445	3099.1	3446.7	9.9452	34 756	3099.1	3446.6	9.8422
500	59 467	3132.3	3489.1	10.1336	44 599	3132.3	3489.1	10.0008	35 679	3132.3	3489.1	9.8978
550	63 313	3216.5	3596.4	10.2680	47 484	3216.5	3596.4	10.1352	37 987	3216.5	3596.4	10.0323
600	67 159	3302.5	3705.4	10.3966	50 369	3302.5	3705.4	10.2638	40 295	3302.5	3705.4	10.1608
650	71 005	3390.2	3816.2	10.5199	53 254	3390.1	3816.2	10.3871	42 603	3390.1	3816.2	10.2842
700	74 852	3479.6	3928.7	10.6386	56 138	3479.6	3928.7	10.5058	44 911	3479.6	3928.7	10.4028
750	78 698	3570.8	4043.0	10.7531	59 023	3570.8	4043.0	10.6203	47 218	3570.8	4043.0	10.5174
800	82 544	3663.8	4159.0	10.8639	61 908	3663.8	4159.0	10.7311	49 526	3663.8	4159.0	10.6281
850	86 390	3758.6	4276.9	10.9712	64 792	3758.6	4276.9	10.8384	51 834	3758.6	4276.9	10.7354
900	90 236	3855.1	4396.5	11.0753	67 677	3855.0	4396.5	10.9426	54 141	3855.0	4396.4	10.8396
950	94 082	3953.2	4517.7	11.1765	70 561	3953.2	4517.7	11.0438	56 449	3953.2	4517.7	10.9408
1000	97 928	4053.0	4640.6	11.2750	73 446	4053.0	4640.6	11.1422	58 757	4053.0	4640.6	11.0393
1100	105 620	4257.5	4891.2	11.4645	79 215	4257.5	4891.2	11.3317	63 372	4257.5	4891.2	11.2287
1200	113 312	4467.9	5147.8	11.6448	84 984	4467.9	5147.8	11.5121	67 987	4467.9	5147.8	11.4091
1300	121 004	4683.7	5409.7	11.8168	90 753	4683.7	5409.7	11.6841	72 602	4683.7	5409.7	11.5811

Definitions of symbols on page 1.

	.015 (53.97)				.020 (60.06)				.025 (64.97)			p (t Sat.)
10^3 v	u	h	s	10^3 v	u	h	s	10^3 v	u	n	s	t
10 022.	2448.7	2599.1	8.0085	7649.	2456.7	2609.7	7.9085	6204.	2463.1	2618.2	7.8314	Sat.
9113.	2407.3	2544.0	7.8321	6820.	2406.2	2542.6	7.6957	5444.	2405.1	2541.2	7.5891	25
9270.	2414.4	2553.5	7.8636	6939.	2413.4	2552.1	7.7275	5539.	2412.3	2550.8	7.6211	30
9428.	2421.5	2563.0	7.8947	7057.	2420.6	2561.7	7.7587	5635.	2419.6	2560.4	7.6525	35
9585.	2428.7	2572.5	7.9253	7176.	2427.8	2571.3	7.7895	5730.	2426.8	2570.1	7.6835	40
9741.	2435.9	2582.0	7.9555	7294.	2435.0	2580.8	7.8199	5826.	2434.1	2579.7	7.7141	45
9898.	2443.0	2591.5	7.9852	7412.	2442.2	2590.4	7.8498	5921.	2441.3	2589.4	7.7441	50
10 054.	2450.2	2601.0	8.0144	7530.	2449.4	2600.0	7.8792	6016.	2448.6	2599.0	7.7737	55
10 210.	2457.4	2610.6	8.0432	7648.	2456.6	2609.6	7.9082	6110.	2455.9	2608.6	7.8029	60
10 367.	2464.6	2620.1	8.0716	7765.	2463.9	2619.2	7.9367	6205.	2463.1	2618.3	7.8316	65
10 522.	2471.8	2629.6	8.0996	7883.	2471.1	2628.8	7.9649	6299.	2470.4	2627.9	7.8599	70
10 678.	2479.0	2639.2	8.1272	8000.	2478.3	2638.3	7.9926	6393.	2477.7	2637.5	7.8877	75
10 834.	2486.2	2648.7	8.1544	8117.	2485.6	2647.9	8.0199	6488.	2485.0	2647.2	7.9152	80
10 989.	2493.4	2658.2	8.1812	8234.	2492.8	2657.5	8.0468	6581.	2492.2	2656.8	7.9422	85
11 145.	2500.6	2667.8	8.2077	8351.	2500.1	2667.1	8.0734	6675.	2499.5	2666.4	7.9689	90
11 300.	2507.8	2677.3	8.2338	8468.	2507.3	2676.7	8.0996	6769.	2506.8	2676.0	7.9952	95
11 455.	2515.0	2686.9	8.2595	8585.	2514.6	2686.2	8.1255	6863.	2514.1	2685.6	8.0212	100
11 765.	2529.5	2706.0	8.3101	8818.	2529.1	2705.4	8.1762	7050.	2528.6	2704.9	8.0720	110
12 075.	2544.0	2725.1	8.3593	9051.	2543.6	2724.6	8.2256	7237.	2543.2	2724.1	8.1216	120
12 385.	2558.5	2744.2	8.4074	9284.	2558.1	2743.8	8.2738	7423.	2557.8	2743.4	8.1700	130
12 694.	2573.0	2763.4	8.4544	9516.	2572.7	2763.0	8.3209	7609.	2572.4	2762.6	8.2172	140
13 003.	2587.6	2782.6	8.5004	9748.	2587.3	2782.3	8.3669	7795.	2587.0	2781.9	8.2633	150
13 312.	2602.2	2801.9	8.5453	9980.	2602.0	2801.6	8.4120	7981.	2601.7	2801.2	8.3084	160
13 621.	2616.9	2821.2	8.5894	10 212.	2616.6	2820.9	8.4561	8167.	2616.4	2820.6	8.3526	170
13 929.	2631.6	2840.5	8.6325	10 444.	2631.4	2840.2	8.4993	8352.	2631.1	2840.0	8.3958	180
14 238.	2646.3	2859.9	8.6748	10 675.	2646.1	2859.6	8.5416	8538.	2645.9	2859.4	8.4382	190
14 546.	2661.1	2879.3	8.7163	10 907.	2660.9	2879.1	8.5831	8723.	2660.7	2878.8	8.4798	200
14 855.	2675.9	2898.8	8.7570	11 138.	2675.8	2898.6	8.6239	8909.	2675.6	2898.3	8.5206	210
15 163.	2690.8	2918.3	8.7970	11 370.	2690.7	2918.1	8.6639	9094.	2690.5	2917.9	8.5606	220
15 471.	2705.8	2937.9	8.8363	11 601.	2705.6	2937.7	8.7032	9279.	2705.5	2937.5	8.5999	230
15 779.	2720.8	2957.5	8.8749	11 832.	2720.7	2957.3	8.7418	9464.	2720.5	2957.1	8.6386	240
16 088.	2735.8	2977.2	8.9129	12 064.	2735.7	2977.0	8.7798	9649.	2735.6	2976.8	8.6766	250
16 396.	2751.0	2996.9	8.9502	12 295.	2750.8	2996.7	8.8172	9834.	2750.7	2996.6	8.7140	260
16 704.	2766.1	3016.7	8.9870	12 526.	2766.0	3016.5	8.8540	10 019.	2765.9	3016.4	8.7508	270
17 012.	2781.4	3036.5	9.0232	12 757.	2781.3	3036.4	8.8903	10 204.	2781.2	3036.3	8.7871	280
17 320.	2796.6	3056.4	9.0589	12 988.	2796.5	3056.3	8.9259	10 389.	2796.4	3056.2	8.8228	290
17 628.	2812.0	3076.4	9.0940	13 219.	2811.9	3076.3	8.9611	10 574.	2811.8	3076.2	8.8579	300
18 244.	2842.9	3116.5	9.1628	13 681.	2842.8	3116.4	9.0299	10 944.	2842.7	3116.3	8.9268	320
18 860.	2874.0	3156.9	9.2297	14 143.	2873.9	3156.8	9.0968	11 314.	2873.8	3156.7	8.9937	340
19 476.	2905.4	3197.5	9.2949	14 605.	2905.3	3197.4	9.1620	11 683.	2905.2	3197.3	9.0589	360
20 091.	2937.0	3238.4	9.3585	15 067.	2936.9	3238.3	9.2256	12 053.	2936.9	3238.2	9.1225	380
20 707	2968.9	3279.5	9.4205	15 529.	2968.8	3279.4	9.2876	12 423.	2968.8	3279.3	9.1845	400
21 323	3001.0	3320.9	9.4810	15 991.	3001.0	3320.8	9.3482	12 792.	3000.9	3320 7	9.2451	420
21 938	3033.4	3362.5	9.5403	16 453.	3033.4	3362.4	9.4074	13 162.	3033.3	3362.4	9.3044	440
22 554	3066.1	3404.4	9.5982	16 915.	3066.1	3404.3	9.4654	13 531.	3066.0	3404.3	9.3623	460
23 170	3099.0	3446.6	9.6550	17 376.	3099.0	3446.5	9.5222	13 901.	3099.0	3446.5	9.4191	480
23 785	3132.3	3489.0	9.7106	17 838.	3132.2	3489.0	9.5778	14 270.	3132.2	3488.9	9.4748	500
25 324	3216.5	3596.3	9.8451	18 992.	3216.4	3596.3	9.7123	15 193.	3216.4	3596.2	9.6092	550
26 863	3302.4	3705.4	9.9737	20 147.	3302.4	3705.3	9.8409	16 117.	3302.4	3705.3	9.7378	600
28 401	3390.1	3816.1	10.0970	21 301.	3390.1	3816.1	9.9642	17 040.	3390.1	3816.1	9.8612	650
29 940	3479.6	3928.6	10.2157	22 455.	3479.5	3928.6	10.0829	17 963.	3479.5	3928.6	9.9799	700
31 478	3570.8	4043.0	10.3303	23 609.	3570.8	4043.0	10.1975	18 887.	3570.8	4043.0	10.0945	750
33 017	3663.8	4159.1	10.4410	24 763.	3663.8	4159.1	10.3082	19 810.	3663.8	4159.0	10.2052	800
34 555	3758.6	4276.9	10.5483	25 916.	3758.5	4276.9	10.4155	20 733.	3758.5	4276.9	10.3125	850
36 094	3855.0	4396.4	10.6524	27 070.	3855.0	4396.4	10.5197	21 656.	3855.0	4396.4	10.4167	900
37 632	3953.2	4517.7	10.7536	28 224.	3953.2	4517.7	10.6209	22 579.	3953.2	4517.6	10.5179	950
39 171	4053.0	4640.6	10.8521	29 378	4053.0	4640.6	10.7193	23 502.	4053.0	4640.5	10.6163	1000
42 248	4257.5	4891.2	11.0416	31 686.	4257.5	4891.2	10.9088	25 349.	4257.4	4891.2	10.8058	1100
45 325	4467.9	5147.8	11.2219	33 994.	4467.9	5147.8	11.0892	27 195.	4467.9	5147.8	10.9862	1200
48 402	4683.7	5409.7	11.3939	36 301.	4683.7	5409.7	11.2612	29 041.	4683.7	5409.7	11.1582	1300

Definitions of symbols on page 1.

Table 3. Vapor

p (t Sat.)	.030 (69.10)				.035 (72.69)				.040 (75.87)			
t	10^3 v	u	h	s	10^3 v	u	h	s	10^3 v	u	h	s
Sat.	5229.	2468.4	2625.3	7.7687	4526.	2473.0	2631.4	7.7158	3993.	2477.0	2636.8	7.6701
25	4526.	2404.1	2539.9	7.5014	3871.	2403.0	2538.5	7.4266	3379.	2401.9	2537.1	7.3614
30	4606.	2411.3	2549.5	7.5335	3940.	2410.3	2548.2	7.4589	3440.	2409.2	2546.8	7.3939
35	4687.	2418.6	2559.2	7.5652	4009.	2417.6	2557.9	7.4908	3501.	2416.6	2556.6	7.4259
40	4767.	2425.9	2568.9	7.5963	4078.	2424.9	2567.7	7.5222	3562.	2424.0	2566.4	7.4575
45	4847.	2433.2	2578.6	7.6271	4147.	2432.3	2577.4	7.5531	3623.	2431.4	2576.3	7.4886
50	4926.	2440.5	2588.3	7.6573	4216.	2439.6	2587.2	7.5835	3683.	2438.8	2586.1	7.5192
55	5006.	2447.8	2598.0	7.6871	4285.	2447.0	2596.9	7.6135	3744.	2446.2	2595.9	7.5493
60	5085.	2455.1	2607.7	7.7164	4353.	2454.3	2606.7	7.6429	3804.	2453.6	2605.7	7.5790
65	5164.	2462.4	2617.3	7.7453	4421.	2461.7	2616.4	7.6720	3864.	2461.0	2615.5	7.6082
70	5243.	2469.7	2627.0	7.7737	4489.	2469.0	2626.2	7.7006	3923.	2468.4	2625.3	7.6369
75	5322.	2477.0	2636.7	7.8017	4557.	2476.4	2635.9	7.7287	3983.	2475.7	2635.1	7.6652
80	5401.	2484.4	2646.4	7.8293	4625.	2483.7	2645.6	7.7564	4043.	2483.1	2644.8	7.6930
85	5480.	2491.7	2656.0	7.8565	4692.	2491.1	2655.3	7.7837	4102.	2490.5	2654.6	7.7205
90	5558.	2499.0	2665.7	7.8833	4760.	2498.4	2665.0	7.8106	4161.	2497.9	2664.3	7.7475
95	5636.	2506.3	2675.4	7.9097	4827.	2505.8	2674.7	7.8371	4220.	2505.2	2674.1	7.7741
100	5715.	2513.6	2685.0	7.9357	4895.	2513.1	2684.4	7.8633	4279.	2512.6	2683.8	7.8003
110	5871.	2528.2	2704.3	7.9868	5029.	2527.8	2703.8	7.9145	4397.	2527.3	2703.2	7.8517
120	6027.	2542.8	2723.6	8.0365	5163.	2542.4	2723.1	7.9644	4515.	2542.0	2722.6	7.9017
130	6183.	2557.4	2742.9	8.0850	5297.	2557.1	2742.5	8.0130	4632.	2556.7	2742.0	7.9505
140	6338.	2572.1	2762.2	8.1323	5430.	2571.8	2761.8	8.0604	4749.	2571.5	2761.4	7.9980
150	6493.	2586.8	2781.5	8.1785	5563.	2586.5	2781.2	8.1067	4866.	2586.2	2780.8	8.0444
160	6648.	2601.4	2800.9	8.2237	5696.	2601.2	2800.6	8.1519	4982.	2600.9	2800.2	8.0897
170	6803.	2616.2	2820.3	8.2679	5829.	2615.9	2820.0	8.1962	5099.	2615.7	2819.7	8.1341
180	6958.	2630.9	2839.7	8.3112	5962.	2630.7	2839.4	8.2396	5215.	2630.5	2839.1	8.1775
190	7113.	2645.7	2859.1	8.3536	6095.	2645.5	2858.9	8.2821	5332.	2645.3	2858.6	8.2200
200	7267.	2660.6	2878.6	8.3952	6228.	2660.4	2878.4	8.3237	5448.	2660.2	2878.1	8.2617
210	7422.	2675.5	2898.1	8.4361	6360.	2675.3	2897.9	8.3646	5564.	2675.1	2897.7	8.3026
220	7577.	2690.4	2917.7	8.4762	6493.	2690.2	2917.5	8.4047	5680.	2690.1	2917.3	8.3428
230	7731.	2705.4	2937.3	8.5155	6625.	2705.2	2937.1	8.4441	5796.	2705.1	2936.9	8.3822
240	7885.	2720.4	2957.0	8.5542	6758.	2720.3	2956.8	8.4828	5912.	2720.1	2956.6	8.4209
250	8040.	2735.5	2976.7	8.5923	6890.	2735.4	2976.5	8.5209	6028.	2735.2	2976.3	8.4590
260	8194.	2750.6	2996.4	8.6297	7022.	2750.5	2996.3	8.5583	6143.	2750.4	2996.1	8.4965
270	8348.	2765.8	3016.2	8.6665	7155.	2765.7	3016.1	8.5952	6259.	2765.6	3016.0	8.5333
280	8502.	2781.0	3036.1	8.7028	7287.	2780.9	3036.0	8.6314	6375.	2780.8	3035.8	8.5696
290	8657.	2796.4	3056.1	8.7385	7419.	2796.3	3055.9	8.6672	6491.	2796.2	3055.8	8.6054
300	8811.	2811.7	3076.0	8.7736	7551.	2811.6	3075.9	8.7024	6606.	2811.5	3075.8	8.6406
320	9119.	2842.6	3116.2	8.8425	7815.	2842.5	3116.1	8.7712	6838.	2842.5	3116.0	8.7095
340	9427.	2873.8	3156.6	8.9095	8080.	2873.7	3156.5	8.8382	7069.	2873.6	3156.4	8.7765
360	9735.	2905.2	3197.2	8.9747	8344.	2905.1	3197.1	8.9034	7300.	2905.0	3197.0	8.8417
380	10 043.	2936.8	3238.1	9.0383	8608.	2936.7	3238.0	8.9670	7531.	2936.7	3237.9	8.9053
400	10 351.	2968.7	3279.2	9.1003	8872.	2968.6	3279.2	9.0291	7763.	2968.6	3279.1	8.9674
420	10 659.	3000.9	3320.6	9.1609	9136.	3000.8	3320.6	9.0897	7994.	3000.8	3320.5	9.0280
440	10 967.	3033.3	3362.3	9.2202	9400.	3033.2	3362.2	9.1490	8225.	3033.2	3362.2	9.0873
460	11 275.	3066.0	3404.2	9.2781	9664.	3065.9	3404.2	9.2069	8456.	3065.9	3404.1	9.1453
480	11 583.	3098.9	3446.4	9.3349	9928.	3098.9	3446.4	9.2637	8687.	3098.8	3446.3	9.2020
500	11 891.	3132.1	3488.9	9.3906	10 192.	3132.1	3488.8	9.3194	8918.	3132.1	3488.8	9.2577
550	12 661.	3216.4	3596.2	9.5251	10 852.	3216.3	3596.2	9.4539	9495.	3216.3	3596.1	9.3922
600	13 430.	3302.3	3705.3	9.6537	11 511.	3302.3	3705.2	9.5825	10 072.	3302.3	3705.2	9.5208
650	14 200.	3390.0	3816.0	9.7770	12 171.	3390.0	3816.0	9.7059	10 649.	3390.0	3816.0	9.6442
700	14 969.	3479.5	3928.6	9.8957	12 831.	3479.5	3928.5	9.8246	11 227.	3479.4	3928.5	9.7629
750	15 739.	3570.7	4042.9	10.0103	13 490.	3570.7	4042.8	9.9391	11 804.	3570.7	4042.8	9.8775
800	16 508.	3663.7	4158.9	10.1210	14 150.	3663.7	4158.9	10.0499	12 381.	3663.7	4158.9	9.9882
850	17 277.	3758.5	4276.8	10.2284	14 809.	3758.5	4276.8	10.1572	12 958.	3758.5	4276.8	10.0956
900	18 047.	3855.0	4396.4	10.3325	15 468.	3855.0	4396.4	10.2614	13 535.	3855.0	4396.3	10.1997
950	18 816.	3953.1	4517.6	10.4337	16 128.	3953.1	4517.6	10.3626	14 112.	3953.1	4517.6	10.3009
1000	19 585.	4053.0	4640.5	10.5322	16 787.	4053.0	4640.5	10.4610	14 689.	4052.9	4640.5	10.3994
1100	21 124.	4257.4	4891.1	10.7217	18 106.	4257.4	4891.1	10.6505	15 843.	4257.4	4891.1	10.5889
1200	22 662.	4467.9	5147.7	10.9020	19 425.	4467.9	5147.7	10.8309	16 997.	4467.8	5147.7	10.7692
1300	24 201.	4683.6	5409.7	11.0740	20 744.	4683.6	5409.6	11.0029	18 151.	4683.6	5409.6	10.9412

Definitions of symbols on page 1.

10^3 v	u	h	s	10^3 v	u	h	s	10^3 v	u	h	s	t
3576.	2480.6	2641.6	7.6298	3240.	2483.9	2645.9	7.5939	2964.	2486.8	2649.8	7.5615	**Sat.**
2997.	*2400.8*	*2535.7*	*7.3034*	*2691.*	*2399.8*	*2534.3*	*7.2511*	*2440.*	*2398.7*	*2532.9*	*7.2034*	**25**
3052.	*2408.2*	*2545.5*	*7.3360*	*2740.*	*2407.1*	*2544.2*	*7.2839*	*2486.*	*2406.1*	*2542.8*	*7.2365*	**30**
3106.	*2415.6*	*2555.4*	*7.3683*	*2790.*	*2414.6*	*2554.1*	*7.3164*	*2531.*	*2413.6*	*2552.8*	*7.2691*	**35**
3161.	*2423.0*	*2565.2*	*7.4000*	*2839.*	*2422.0*	*2564.0*	*7.3483*	*2576.*	*2421.1*	*2562.8*	*7.3013*	**40**
3215.	*2430.4*	*2575.1*	*7.4313*	*2888.*	*2429.5*	*2574.0*	*7.3798*	*2621.*	*2428.6*	*2572.8*	*7.3329*	**45**
3269.	*2437.9*	*2585.0*	*7.4621*	*2937.*	*2437.0*	*2583.9*	*7.4108*	*2666.*	*2436.1*	*2582.8*	*7.3641*	**50**
3323.	*2445.3*	*2594.9*	*7.4925*	*2986.*	*2444.5*	*2593.8*	*7.4413*	*2711.*	*2443.7*	*2592.8*	*7.3948*	**55**
3377.	*2452.8*	*2604.7*	*7.5223*	*3035.*	*2452.0*	*2603.7*	*7.4713*	*2755.*	*2451.2*	*2602.8*	*7.4250*	**60**
3430.	*2460.2*	*2614.6*	*7.5516*	*3083.*	*2459.5*	*2613.6*	*7.5008*	*2799.*	*2458.7*	*2612.7*	*7.4546*	**65**
3483.	*2467.7*	*2624.4*	*7.5805*	*3131.*	*2467.0*	*2623.5*	*7.5299*	*2843.*	*2466.3*	*2622.7*	*7.4838*	**70**
3537.	*2475.1*	*2634.2*	*7.6089*	*3180.*	*2474.4*	*2633.4*	*7.5584*	*2887.*	*2473.8*	*2632.6*	*7.5125*	**75**
3590.	2482.5	2644.0	7.6369	*3228.*	*2481.9*	*2643.3*	*7.5865*	*2931.*	*2481.3*	*2642.5*	*7.5408*	**80**
3643.	2489.9	2653.8	7.6645	3275.	2489.3	2653.1	7.6142	2975.	2488.7	2652.4	7.5686	**85**
3696.	2497.3	2663.6	7.6916	3323.	2496.8	2662.9	7.6414	3018.	2496.2	2662.2	7.5959	**90**
3748.	2504.7	2673.4	7.7183	3371.	2504.2	2672.7	7.6683	3062.	2503.7	2672.1	7.6229	**95**
3801.	2512.1	2683.2	7.7447	3418.	2511.6	2682.5	7.6947	3105.	2511.1	2681.9	7.6494	**100**
3906.	2526.9	2702.7	7.7962	3513.	2526.4	2702.1	7.7465	3192.	2526.0	2701.5	7.7013	**110**
4011.	2541.6	2722.1	7.8464	3608.	2541.3	2721.6	7.7968	3278.	2540.9	2721.1	7.7518	**120**
4115.	2556.4	2741.6	7.8952	3702.	2556.0	2741.1	7.8457	3363.	2555.7	2740.7	7.8009	**130**
4219.	2571.1	2761.0	7.9429	3796.	2570.8	2760.6	7.8935	3449.	2570.5	2760.2	7.8487	**140**
4323.	2585.9	2780.5	7.9894	3889.	2585.6	2780.1	7.9401	3534.	2585.3	2779.7	7.8954	**150**
4427.	2600.7	2799.9	8.0348	3983.	2600.4	2799.6	7.9856	3619.	2600.2	2799.2	7.9410	**160**
4531.	2615.5	2819.4	8.0792	4076.	2615.2	2819.1	8.0300	3704.	2615.0	2818.8	7.9855	**170**
4634.	2630.3	2838.8	8.1227	4170.	2630.1	2838.6	8.0736	3789.	2629.9	2838.3	8.0291	**180**
4738.	2645.1	2858.4	8.1652	4263.	2645.0	2858.1	8.1162	3874.	2644.8	2857.8	8.0718	**190**
4841.	2660.0	2877.9	8.2070	4356.	2659.9	2877.7	8.1580	3959.	2659.7	2877.4	8.1136	**200**
4944.	2675.0	2897.5	8.2479	4449.	2674.8	2897.2	8.1990	4044.	2674.6	2897.0	8.1546	**210**
5048.	2689.9	2917.1	8.2881	4542.	2689.8	2916.9	8.2392	4128.	2689.6	2916.7	8.1949	**220**
5151.	2704.9	2936.7	8.3276	4635.	2704.8	2936.5	8.2786	4213.	2704.7	2936.3	8.2344	**230**
5254.	2720.0	2956.4	8.3663	4728.	2719.9	2956.2	8.3174	4297.	2719.7	2956.1	8.2732	**240**
5357.	2735.1	2976.2	8.4044	4820.	2735.0	2976.0	8.3556	4381.	2734.9	2975.8	8.3113	**250**
5460.	2750.3	2996.0	8.4419	4913.	2750.2	2995.8	8.3931	4466.	2750.0	2995.7	8.3489	**260**
5563.	2765.5	3015.8	8.4788	5006.	2765.4	3015.7	8.4300	4550.	2765.3	3015.5	8.3858	**270**
5666.	2780.7	3035.7	8.5151	5099.	2780.6	3035.6	8.4663	4634.	2780.5	3035.4	8.4221	**280**
5769.	2796.1	3055.7	8.5508	5191.	2796.0	3055.5	8.5020	4719.	2795.9	3055.4	8.4579	**290**
5872.	2811.4	3075.7	8.5860	5284.	2811.3	3075.5	8.5373	4803.	2811.3	3075.4	8.4931	**300**
6077.	2842.4	3115.9	8.6550	5469.	2842.3	3115.7	8.6062	4971.	2842.2	3115.6	8.5621	**320**
6283.	2873.5	3156.3	8.7220	5654.	2873.5	3156.2	8.6732	5140.	2873.4	3156.1	8.6291	**340**
6489.	2904.9	3196.9	8.7872	5839.	2904.9	3196.8	8.7385	5308.	2904.8	3196.8	8.6944	**360**
6694.	2936.6	3237.8	8.8509	6024.	2936.5	3237.8	8.8021	5476.	2936.5	3237.7	8.7580	**380**
6900.	2968.5	3279.0	8.9129	6209.	2968.5	3278.9	8.8642	5644.	2968.4	3278.9	8.8201	**400**
7105.	3000.7	3320.4	8.9736	6394.	3000.6	3320.4	8.9249	5813.	3000.6	3320.3	8.8808	**420**
7310.	3033.1	3362.1	9.0328	6579.	3033.1	3362.0	8.9841	5981.	3033.0	3362.0	8.9401	**440**
7516.	3065.8	3404.0	9.0908	6764.	3065.8	3404.0	9.0421	6149.	3065.7	3403.9	8.9981	**460**
7721.	3098.8	3446.2	9.1476	6949.	3098.7	3446.2	9.0989	6317.	3098.7	3446.1	9.0549	**480**
7927.	3132.0	3488.7	9.2033	7134.	3132.0	3488.7	9.1546	6485.	3131.9	3488.6	9.1106	**500**
8440.	3216.3	3596.1	9.3378	7596.	3216.2	3596.0	9.2891	6905.	3216.2	3596.0	9.2451	**550**
8953.	3302.3	3705.1	9.4664	8057.	3302.2	3705.1	9.4178	7325.	3302.2	3705.1	9.3738	**600**
9466.	3390.0	3815.9	9.5898	8519.	3389.9	3815.9	9.5412	7745.	3389.9	3815.9	9.4972	**650**
9979.	3479.4	3928.5	9.7085	8981.	3479.4	3928.5	9.6599	8164.	3479.4	3928.4	9.6159	**700**
10 492.	3570.6	4042.8	9.8231	9443.	3570.6	4042.8	9.7744	8584.	3570.6	4042.7	9.7304	**750**
11 005.	3663.7	4158.9	9.9339	9904.	3663.6	4158.9	9.8852	9004.	3663.6	4158.8	9.8412	**800**
11 518.	3758.5	4276.8	10.0412	10 366.	3758.4	4276.7	9.9926	9424.	3758.4	4276.7	9.9486	**850**
12 031.	3854.9	4396.3	10.1454	10 828.	3854.9	4396.3	10.0967	9843.	3854.9	4396.3	10.0527	**900**
12 544.	3953.1	4517.6	10.2466	11 289.	3953.1	4517.6	10.1979	10 263.	3953.1	4517.5	10.1539	**950**
13 057.	4052.9	4640.5	10.3450	11 751.	4052.9	4640.5	10.2964	10 683.	4052.9	4640.5	10.2524	**1000**
14 082.	4257.4	4891.1	10.5345	12 674.	4257.4	4891.1	10.4859	11 522.	4257.4	4891.1	10.4419	**1100**
15 108.	4467.8	5147.7	10.7149	13 597.	4467.8	5147.7	10.6662	12 361.	4467.8	5147.7	10.6222	**1200**
16 134.	4683.6	5409.6	10.8869	14 521.	4683.6	5409.6	10.8382	13 200.	4683.6	5409.6	10.7942	**1300**

Definitions of symbols on page 1.

Table 3. Vapor

p (t Sat.)	.060 (85.94)				.065 (88.01)				.070 (89.95)			
t	10^3 v	u	h	s	10^3 v	u	h	s	10^3 v	u	h	s
Sat.	2732.	2489.6	2653.5	7.5320	2535.	2492.1	2656.9	7.5048	2365.	2494.5	2660.0	7.4797
50	2440.	2435.3	2581.7	7.3212	2249.	2434.4	2580.6	7.2816	2085.	2433.5	2579.5	7.2446
55	2481.	2442.9	2591.7	7.3521	2287.	2442.0	2590.7	7.3126	2120.	2441.2	2589.6	7.2758
60	2522.	2450.4	2601.8	7.3824	2325.	2449.6	2600.8	7.3431	2156.	2448.9	2599.8	7.3065
65	2563.	2458.0	2611.8	7.4123	2363.	2457.3	2610.8	7.3731	2191.	2456.5	2609.9	7.3367
70	2603.	2465.6	2621.8	7.4416	2400.	2464.9	2620.9	7.4026	2226.	2464.2	2620.0	7.3664
75	2644.	2473.1	2631.7	7.4705	2438.	2472.4	2630.9	7.4316	2261.	2471.8	2630.1	7.3955
80	2684.	2480.6	2641.7	7.4988	2475.	2480.0	2640.9	7.4601	2296.	2479.4	2640.1	7.4241
85	2724.	2488.2	2651.6	7.5268	2512.	2487.6	2650.9	7.4882	2331.	2487.0	2650.1	7.4523
90	2764.	2495.7	2661.5	7.5542	2550.	2495.1	2660.8	7.5158	2365.	2494.6	2660.1	7.4800
95	2804.	2503.2	2671.4	7.5813	2587.	2502.6	2670.8	7.5429	2400.	2502.1	2670.1	7.5073
100	2844.	2510.6	2681.3	7.6079	2623.	2510.1	2680.7	7.5697	2434.	2509.7	2680.0	7.5341
110	2924.	2525.6	2701.0	7.6600	2697.	2525.1	2700.4	7.6219	2503.	2524.7	2699.9	7.5866
120	3003.	2540.5	2720.6	7.7106	2770.	2540.1	2720.1	7.6727	2571.	2539.7	2719.6	7.6375
130	3081.	2555.3	2740.2	7.7598	2843.	2555.0	2739.8	7.7220	2638.	2554.6	2739.3	7.6870
140	3160.	2570.2	2759.8	7.8078	2916.	2569.9	2759.4	7.7701	2706.	2569.6	2759.0	7.7351
150	3238.	2585.1	2779.4	7.8546	2988.	2584.8	2779.0	7.8170	2773.	2584.5	2778.6	7.7821
160	3317.	2599.9	2798.9	7.9002	3060.	2599.7	2798.6	7.8627	2841.	2599.4	2798.2	7.8279
170	3395.	2614.8	2818.5	7.9448	3132.	2614.5	2818.1	7.9074	2908.	2614.3	2817.8	7.8726
180	3473.	2629.7	2838.0	7.9885	3204.	2629.4	2837.7	7.9511	2975.	2629.2	2837.5	7.9164
190	3550.	2644.6	2857.6	8.0312	3276.	2644.4	2857.3	7.9939	3041.	2644.2	2857.1	7.9592
200	3628.	2659.5	2877.2	8.0731	3348.	2659.3	2876.9	8.0358	3108.	2659.1	2876.7	8.0012
210	3706.	2674.5	2896.8	8.1141	3420.	2674.3	2896.6	8.0768	3175.	2674.1	2896.4	8.0423
220	3783.	2689.5	2916.5	8.1544	3491.	2689.3	2916.3	8.1171	3241.	2689.2	2916.1	8.0826
230	3861.	2704.5	2936.2	8.1939	3563.	2704.4	2936.0	8.1567	3308.	2704.2	2935.8	8.1222
240	3938.	2719.6	2955.9	8.2328	3635.	2719.5	2955.7	8.1956	3374.	2719.3	2955.5	8.1611
250	4016.	2734.7	2975.7	8.2709	3706.	2734.6	2975.5	8.2338	3441.	2734.5	2975.3	8.1993
260	4093.	2749.9	2995.5	8.3085	3778.	2749.8	2995.3	8.2713	3507.	2749.7	2995.2	8.2369
270	4170.	2765.1	3015.4	8.3454	3849.	2765.0	3015.2	8.3083	3574.	2764.9	3015.1	8.2739
280	4248.	2780.4	3035.3	8.3818	3920.	2780.3	3035.2	8.3446	3640.	2780.2	3035.0	8.3102
290	4325.	2795.8	3055.3	8.4175	3992.	2795.7	3055.1	8.3804	3706.	2795.6	3055.0	8.3461
300	4402.	2811.2	3075.3	8.4528	4063.	2811.1	3075.2	8.4157	3772.	2811.0	3075.0	8.3813
320	4557.	2842.1	3115.5	8.5218	4206.	2842.0	3115.4	8.4847	3905.	2842.0	3115.3	8.4504
340	4711.	2873.3	3156.0	8.5889	4348.	2873.2	3155.9	8.5518	4037.	2873.2	3155.8	8.5175
360	4865.	2904.7	3196.7	8.6542	4491.	2904.7	3196.6	8.6171	4170.	2904.6	3196.5	8.5828
380	5020.	2936.4	3237.6	8.7178	4633.	2936.4	3237.5	8.6808	4302.	2936.3	3237.4	8.6465
400	5174.	2968.4	3278.8	8.7799	4775.	2968.3	3278.7	8.7429	4434.	2968.2	3278.6	8.7086
420	5328.	3000.5	3320.2	8.8406	4918.	3000.5	3320.1	8.8035	4566.	3000.4	3320.1	8.7693
440	5482.	3033.0	3361.9	8.8999	5060.	3032.9	3361.8	8.8628	4698.	3032.9	3361.8	8.8286
460	5636.	3065.7	3403.9	8.9579	5202.	3065.6	3403.8	8.9209	4831.	3065.6	3403.7	8.8866
480	5790.	3098.7	3446.1	9.0147	5345.	3098.6	3446.0	8.9777	4963.	3098.6	3446.0	8.9434
500	5944.	3131.9	3488.6	9.0704	5487.	3131.9	3488.5	9.0334	5095.	3131.8	3488.5	8.9991
520	6098.	3165.4	3531.3	9.1249	5629.	3165.4	3531.3	9.0880	5227.	3165.3	3531.2	9.0537
540	6252.	3199.2	3574.3	9.1785	5771.	3199.2	3574.3	9.1415	5359.	3199.1	3574.2	9.1073
560	6406.	3233.2	3617.6	9.2311	5913.	3233.2	3617.6	9.1941	5491.	3233.2	3617.5	9.1599
580	6560.	3267.6	3661.2	9.2828	6055.	3267.5	3661.1	9.2458	5623.	3267.5	3661.1	9.2116
600	6714.	3302.2	3705.0	9.3336	6198.	3302.1	3705.0	9.2966	5755.	3302.1	3704.9	9.2624
620	6868.	3337.0	3749.1	9.3835	6340.	3337.0	3749.1	9.3466	5887.	3337.0	3749.1	9.3123
640	7022.	3372.2	3793.5	9.4327	6482.	3372.2	3793.5	9.3957	6019.	3372.2	3793.5	9.3615
660	7176.	3407.6	3838.2	9.4811	6624.	3407.6	3838.2	9.4441	6151.	3407.6	3838.1	9.4099
680	7330.	3443.4	3883.2	9.5287	6766.	3443.3	3883.1	9.4918	6283.	3443.3	3883.1	9.4575
700	7484.	3479.4	3928.4	9.5757	6908.	3479.3	3928.4	9.5387	6415.	3479.3	3928.3	9.5045
750	7869.	3570.6	4042.7	9.6903	7263.	3570.6	4042.7	9.6533	6744.	3570.6	4042.7	9.6191
800	8254.	3663.6	4158.8	9.8010	7619.	3663.6	4158.8	9.7641	7074.	3663.6	4158.8	9.7299
850	8638.	3758.4	4276.7	9.9084	7974.	3758.4	4276.7	9.8714	7404.	3758.4	4276.7	9.8372
900	9023.	3854.9	4396.3	10.0125	8329.	3854.9	4396.3	9.9756	7734.	3854.9	4396.2	9.9414
950	9408.	3953.1	4517.5	10.1138	8684.	3953.1	4517.5	10.0768	8064.	3953.0	4517.5	10.0426
1000	9792.	4052.9	4640.4	10.2122	9039.	4052.9	4640.4	10.1753	8393.	4052.9	4640.4	10.1411
1100	10 562.	4257.4	4891.1	10.4017	9749.	4257.3	4891.0	10.3648	9053.	4257.3	4891.0	10.3306
1200	11 331.	4467.8	5147.7	10.5821	10 459.	4467.8	5147.7	10.5451	9712.	4467.8	5147.6	10.5109
1300	12 100.	4683.6	5409.6	10.7541	11 170.	4683.6	5409.6	10.7171	10 372.	4683.5	5409.6	10.6829

Definitions of symbols on page 1.

10^3 v	u	h	s	10^3 v	u	h	s	10^3 v	u	h	s	t
2217.	2496.7	2663.0	7.4564	2087.	2498.8	2665.8	7.4346	1972.1	2500.7	2668.4	7.4141	Sat.
1942.7	2432.6	2578.3	7.2100	1818.3	2431.7	2577.2	7.1775	1708.5	2430.9	2576.1	7.1467	50
1976.1	2440.4	2588.6	7.2414	1849.8	2439.5	2587.5	7.2091	1738.4	2438.7	2586.4	7.1785	55
2009.4	2448.1	2598.8	7.2723	1881.2	2447.3	2597.8	7.2401	1768.0	2446.5	2596.8	7.2097	60
2042.4	2455.8	2609.0	7.3026	1912.3	2455.0	2608.0	7.2706	1797.5	2454.3	2607.1	7.2404	65
2075.4	2463.5	2619.1	7.3324	1943.3	2462.7	2618.2	7.3006	1826.8	2462.0	2617.3	7.2705	70
2108.	2471.1	2629.2	7.3617	1974.2	2470.4	2628.4	7.3300	1856.0	2469.8	2627.5	7.3001	75
2141.	2478.8	2639.3	7.3905	2004.9	2478.1	2638.5	7.3589	1885.0	2477.5	2637.7	7.3292	80
2173.	2486.4	2649.4	7.4188	2035.5	2485.8	2648.6	7.3874	1913.9	2485.2	2647.9	7.3577	85
2206.	2494.0	2659.4	7.4466	2065.9	2493.4	2658.7	7.4153	1942.7	2492.9	2658.0	7.3858	90
2238.	2501.6	2669.4	7.4740	2096.3	2501.1	2668.8	7.4428	1971.3	2500.5	2668.1	7.4134	95
2270.	2509.2	2679.4	7.5009	2127.	2508.7	2678.8	7.4698	1999.9	2508.2	2678.1	7.4405	100
2334.	2524.2	2699.3	7.5536	2187.	2523.8	2698.7	7.5226	2056.7	2523.4	2698.2	7.4935	110
2398.	2539.3	2719.1	7.6046	2247.	2538.9	2718.6	7.5738	2113.1	2538.5	2718.1	7.5448	120
2461.	2554.3	2738.9	7.6542	2306.	2553.9	2738.4	7.6236	2169.3	2553.6	2738.0	7.5947	130
2524.	2569.3	2758.6	7.7025	2365.	2568.9	2758.2	7.6720	2225.3	2568.6	2757.8	7.6432	140
2587.	2584.2	2778.2	7.7496	2425.	2583.9	2777.9	7.7191	2281.	2583.6	2777.5	7.6904	150
2650.	2599.1	2797.9	7.7954	2484.	2598.9	2797.6	7.7651	2337.	2598.6	2797.2	7.7365	160
2713.	2614.1	2817.5	7.8403	2542.	2613.8	2817.2	7.8100	2392.	2613.6	2816.9	7.7814	170
2775.	2629.0	2837.2	7.8841	2601.	2628.8	2836.9	7.8538	2447.	2628.6	2836.6	7.8254	180
2838.	2644.0	2856.8	7.9270	2660.	2643.8	2856.6	7.8968	2503.	2643.6	2856.3	7.8684	190
2900.	2659.0	2876.5	7.9690	2718.	2658.8	2876.2	7.9388	2558.	2658.6	2876.0	7.9104	200
2962.	2674.0	2896.2	8.0101	2777.	2673.8	2895.9	7.9800	2613.	2673.6	2895.7	7.9517	210
3025.	2689.0	2915.9	8.0505	2835.	2688.9	2915.7	8.0204	2668.	2688.7	2915.4	7.9921	220
3087.	2704.1	2935.6	8.0901	2893.	2703.9	2935.4	8.0600	2722.	2703.8	2935.2	8.0318	230
3149.	2719.2	2955.4	8.1290	2951.	2719.1	2955.2	8.0990	2777.	2718.9	2955.0	8.0707	240
3211.	2734.4	2975.2	8.1673	3010.	2734.2	2975.0	8.1372	2831.	2734.1	2974.8	8.1090	250
3273.	2749.6	2995.0	8.2048	3068.	2749.5	2994.9	8.1748	2887.	2749.3	2994.7	8.1466	260
3335.	2764.8	3014.9	8.2418	3126.	2764.7	3014.8	8.2118	2942.	2764.6	3014.6	8.1837	270
3397.	2780.1	3034.9	8.2782	3184.	2780.0	3034.7	8.2482	2996.	2779.9	3034.6	8.2201	280
3459.	2795.5	3054.9	8.3140	3242.	2795.4	3054.7	8.2841	3051.	2795.3	3054.6	8.2559	290
3520.	2810.9	3074.9	8.3493	3300.	2810.8	3074.8	8.3194	3106.	2810.7	3074.7	8.2912	300
3644.	2841.9	3115.2	8.4184	3416.	2841.8	3115.1	8.3885	3215.	2841.7	3115.0	8.3603	320
3768.	2873.1	3155.7	8.4855	3532.	2873.0	3155.6	8.4556	3324.	2872.9	3155.5	8.4275	340
3891.	2904.5	3196.4	8.5508	3648.	2904.5	3196.3	8.5210	3433.	2904.4	3196.2	8.4929	360
4015.	2936.2	3237.3	8.6145	3764.	2936.2	3237.3	8.5846	3542.	2936.1	3237.2	8.5566	380
4138.	2968.2	3278.5	8.6767	3879.	2968.1	3278.5	8.6468	3651.	2968.1	3278.4	8.6187	400
4262.	3000.4	3320.0	8.7374	3995.	3000.3	3319.9	8.7075	3760.	3000.3	3319.9	8.6794	420
4385.	3032.8	3361.7	8.7967	4111.	3032.8	3361.6	8.7668	3869.	3032.7	3361.6	8.7388	440
4508.	3065.6	3403.7	8.8547	4226.	3065.5	3403.6	8.8249	3978.	3065.5	3403.5	8.7968	460
4632.	3098.5	3445.9	8.9115	4342.	3098.5	3445.8	8.8817	4086.	3098.4	3445.8	8.8537	480
4755.	3131.8	3488.4	8.9672	4458.	3131.7	3488.3	8.9374	4195.	3131.7	3488.3	8.9093	500
4878.	3165.3	3531.2	9.0218	4573.	3165.3	3531.1	8.9920	4304.	3165.2	3531.1	8.9640	520
5001.	3199.1	3574.2	9.0754	4689.	3199.0	3574.1	9.0456	4413.	3199.0	3574.1	9.0175	540
5125.	3233.1	3617.5	9.1280	4804.	3233.1	3617.4	9.0982	4521.	3233.1	3617.4	9.0702	560
5248.	3267.5	3661.1	9.1797	4920.	3267.4	3661.0	9.1499	4630.	3267.4	3661.0	9.1218	580
5371.	3302.1	3704.9	9.2305	5035.	3302.1	3704.9	9.2007	4739.	3302.0	3704.8	9.1727	600
5494.	3337.0	3749.0	9.2804	5151.	3336.9	3749.0	9.2506	4848.	3336.9	3749.0	9.2226	620
5617.	3372.1	3793.4	9.3296	5266.	3372.1	3793.4	9.2998	4956.	3372.1	3793.4	9.2718	640
5741.	3407.6	3838.1	9.3780	5382.	3407.5	3838.1	9.3482	5065.	3407.5	3838.0	9.3202	660
5864.	3443.3	3883.1	9.4257	5497.	3443.3	3883.0	9.3959	5174.	3443.2	3883.0	9.3679	680
5987.	3479.3	3928.3	9.4727	5613.	3479.3	3928.3	9.4428	5282.	3479.2	3928.3	9.4148	700
6295.	3570.5	4042.6	9.5872	5901.	3570.5	4042.6	9.5574	5554.	3570.5	4042.6	9.5294	750
6603.	3663.6	4158.7	9.6980	6190.	3663.5	4158.7	9.6682	5826.	3663.5	4158.7	9.6402	800
6910.	3758.4	4276.6	9.8054	6478.	3758.3	4276.6	9.7756	6097.	3758.3	4276.6	9.7476	850
7218.	3854.9	4396.2	9.9095	6767.	3854.8	4396.2	9.8797	6369.	3854.8	4396.2	9.8517	900
7526.	3953.0	4517.5	10.0107	7056.	3953.0	4517.5	9.9809	6641.	3953.0	4517.4	9.9530	950
7834.	4052.9	4640.4	10.1092	7344.	4052.8	4640.4	10.0794	6912.	4052.8	4640.4	10.0514	1000
8449.	4257.3	4891.0	10.2987	7921.	4257.3	4891.0	10.2689	7455.	4257.3	4891.0	10.2409	1100
9065.	4467.8	5147.6	10.4791	8498.	4467.8	5147.6	10.4493	7998.	4467.7	5147.6	10.4213	1200
9680.	4683.5	5409.6	10.6511	9075.	4683.5	5409.5	10.6213	8542.	4683.5	5409.5	10.5933	1300

Definitions of symbols on page 1.

Table 3. Vapor

p (t Sat.)	.090 (96.71)				.095 (98.20)				.10 (99.63)			
t	$10^3 v$	u	h	s	$10^3 v$	u	h	s	$10^3 v$	u	h	s
Sat.	1869.5	2502.6	2670.9	7.3949	1777.3	2504.4	2673.2	7.3767	1694.0	2506.1	2675.5	7.3594
50	*1610.9*	*2430.0*	*2574.9*	*7.1175*	*1523.6*	*2429.1*	*2573.8*	*7.0898*	*1445.0*	*2428.2*	*2572.7*	*7.0633*
55	*1639.3*	*2437.8*	*2585.4*	*7.1495*	*1550.6*	*2437.0*	*2584.3*	*7.1219*	*1470.8*	*2436.1*	*2583.2*	*7.0956*
60	*1667.4*	*2445.7*	*2595.7*	*7.1809*	*1577.4*	*2444.9*	*2594.7*	*7.1535*	*1496.4*	*2444.1*	*2593.7*	*7.1274*
65	*1695.4*	*2453.5*	*2606.1*	*7.2118*	*1604.1*	*2452.7*	*2605.1*	*7.1845*	*1521.9*	*2452.0*	*2604.2*	*7.1586*
70	*1723.2*	*2461.3*	*2616.4*	*7.2420*	*1630.6*	*2460.6*	*2615.5*	*7.2150*	*1547.1*	*2459.9*	*2614.6*	*7.1892*
75	*1750.9*	*2469.1*	*2626.7*	*7.2718*	*1656.9*	*2468.4*	*2625.8*	*7.2449*	*1572.2*	*2467.8*	*2625.0*	*7.2192*
80	*1778.4*	*2476.9*	*2636.9*	*7.3010*	*1683.0*	*2476.2*	*2636.1*	*7.2742*	*1597.2*	*2475.6*	*2635.3*	*7.2487*
85	*1805.8*	*2484.6*	*2647.1*	*7.3296*	*1709.1*	*2484.0*	*2646.4*	*7.3030*	*1622.0*	*2483.4*	*2645.6*	*7.2776*
90	*1833.1*	*2492.3*	*2657.3*	*7.3578*	*1735.0*	*2491.7*	*2656.6*	*7.3313*	*1646.8*	*2491.2*	*2655.9*	*7.3061*
95	*1860.2*	*2500.0*	*2667.4*	*7.3855*	*1760.8*	*2499.5*	*2666.7*	*7.3591*	*1671.3*	*2498.9*	*2666.1*	*7.3340*
100	1887.3	2507.7	2677.5	7.4128	1786.5	2507.2	2676.9	7.3865	1695.8	2506.7	2676.2	7.3614
110	1941.1	2522.9	2697.6	7.4659	1837.6	2522.5	2697.0	7.4398	1744.5	2522.0	2696.5	7.4149
120	1994.5	2538.1	2717.6	7.5174	1888.4	2537.7	2717.1	7.4915	1792.9	2537.3	2716.6	7.4668
130	2047.7	2553.2	2737.5	7.5674	1938.9	2552.9	2737.1	7.5416	1840.9	2552.5	2736.6	7.5170
140	2100.6	2568.3	2757.4	7.6160	1989.1	2568.0	2756.9	7.5903	1888.7	2567.7	2756.5	7.5659
150	2153.	2583.3	2777.1	7.6634	2039.	2583.0	2776.8	7.6377	1936.4	2582.8	2776.4	7.6134
160	2206.	2598.4	2796.9	7.7095	2089.	2598.1	2796.6	7.6839	1983.8	2597.8	2796.2	7.6597
170	2258.	2613.4	2816.6	7.7545	2139.	2613.1	2816.3	7.7290	2031.1	2612.9	2816.0	7.7048
180	2311.	2628.4	2836.3	7.7985	2188.	2628.2	2836.1	7.7731	2078.2	2627.9	2835.8	7.7489
190	2363.	2643.4	2856.0	7.8415	2238.	2643.2	2855.8	7.8162	2125.3	2643.0	2855.5	7.7921
200	2415.	2658.4	2875.8	7.8837	2287.	2658.2	2875.5	7.8583	2172.	2658.1	2875.3	7.8343
210	2467.	2673.5	2895.5	7.9249	2337.	2673.3	2895.3	7.8996	2219.	2673.1	2895.1	7.8756
220	2519.	2688.6	2915.2	7.9654	2386.	2688.4	2915.0	7.9401	2266.	2688.2	2914.8	7.9162
230	2571.	2703.7	2935.0	8.0051	2435.	2703.5	2934.8	7.9799	2313.	2703.4	2934.6	7.9559
240	2623.	2718.8	2954.8	8.0441	2484.	2718.7	2954.7	8.0189	2359.	2718.5	2954.5	7.9949
250	2674.	2734.0	2974.7	8.0824	2533.	2733.9	2974.5	8.0572	2406.	2733.7	2974.3	8.0333
260	2726.	2749.2	2994.6	8.1200	2582.	2749.1	2994.4	8.0949	2453.	2749.0	2994.3	8.0710
270	2778.	2764.5	3014.5	8.1571	2631.	2764.4	3014.3	8.1319	2499.	2764.3	3014.2	8.1080
280	2829.	2779.8	3034.5	8.1935	2680.	2779.7	3034.3	8.1684	2546.	2779.6	3034.2	8.1445
290	2881.	2795.2	3054.5	8.2294	2729.	2795.1	3054.4	8.2043	2592.	2795.0	3054.2	8.1804
300	2933.	2810.6	3074.6	8.2647	2778.	2810.5	3074.4	8.2396	2639.	2810.4	3074.3	8.2158
320	3036.	2841.6	3114.8	8.3338	2876.	2841.5	3114.7	8.3087	2732.	2841.5	3114.6	8.2849
340	3139.	2872.9	3155.4	8.4010	2973.	2872.8	3155.3	8.3759	2824.	2872.7	3155.2	8.3521
360	3242.	2904.3	3196.1	8.4664	3071.	2904.3	3196.0	8.4413	2917.	2904.2	3195.9	8.4175
380	3345.	2936.0	3237.1	8.5301	3169.	2936.0	3237.0	8.5051	3010.	2935.9	3236.9	8.4813
400	3448.	2968.0	3278.3	8.5923	3266.	2967.9	3278.2	8.5672	3103.	2967.9	3278.2	8.5435
420	3551.	3000.2	3319.8	8.6530	3364.	3000.2	3319.7	8.6279	3195.	3000.1	3319.6	8.6042
440	3654.	3032.7	3361.5	8.7123	3461.	3032.6	3361.4	8.6873	3288.	3032.6	3361.4	8.6636
460	3756.	3065.4	3403.5	8.7704	3558.	3065.4	3403.4	8.7454	3380.	3065.3	3403.4	8.7216
480	3859.	3098.4	3445.7	8.8272	3656.	3098.4	3445.7	8.8022	3473.	3098.3	3445.6	8.7785
500	3962.	3131.7	3488.2	8.8829	3753.	3131.6	3488.2	8.8579	3565.	3131.6	3488.1	8.8342
520	4065.	3165.2	3531.0	8.9375	3851.	3165.1	3531.0	8.9125	3658.	3165.1	3530.9	8.8888
540	4167.	3199.0	3574.0	8.9911	3948.	3198.9	3574.0	8.9661	3750.	3198.9	3574.0	8.9424
560	4270.	3233.0	3617.4	9.0437	4045.	3233.0	3617.3	9.0188	3843.	3233.0	3617.3	8.9950
580	4373.	3267.4	3660.9	9.0954	4143.	3267.4	3660.9	9.0704	3935.	3267.3	3660.9	9.0467
600	4476.	3302.0	3704.8	9.1462	4240.	3302.0	3704.8	9.1213	4028.	3301.9	3704.7	9.0976
620	4578.	3336.9	3748.9	9.1962	4337.	3336.9	3748.9	9.1712	4120.	3336.8	3748.9	9.1475
640	4681.	3372.1	3793.3	9.2454	4434.	3372.0	3793.3	9.2204	4213.	3372.0	3793.3	9.1967
660	4784.	3407.5	3838.0	9.2938	4532.	3407.5	3838.0	9.2688	4305.	3407.4	3838.0	9.2451
680	4886.	3443.2	3883.0	9.3415	4629.	3443.2	3883.0	9.3165	4397.	3443.2	3882.9	9.2928
700	4989.	3479.2	3928.2	9.3884	4726.	3479.2	3928.2	9.3635	4490.	3479.2	3928.2	9.3398
750	5245.	3570.5	4042.6	9.5030	4969.	3570.5	4042.5	9.4780	4721.	3570.4	4042.5	9.4544
800	5502.	3663.5	4158.7	9.6138	5212.	3663.5	4158.7	9.5888	4952.	3663.5	4158.6	9.5652
850	5759.	3758.3	4276.6	9.7212	5455.	3758.3	4276.6	9.6962	5183.	3758.3	4276.5	9.6725
900	6015.	3854.8	4396.2	9.8253	5698.	3854.8	4396.1	9.8004	5414.	3854.8	4396.1	9.7767
950	6272.	3953.0	4517.4	9.9266	5941.	3953.0	4517.4	9.9016	5644.	3953.0	4517.4	9.8779
1000	6528.	4052.8	4640.3	10.0251	6184.	4052.8	4640.3	10.0001	5875.	4052.8	4640.3	9.9764
1100	7041.	4257.3	4891.0	10.2145	6670.	4257.3	4891.0	10.1896	6337.	4257.3	4891.0	10.1659
1200	7554.	4467.7	5147.6	10.3949	7156.	4467.7	5147.6	10.3699	6799.	4467.7	5147.6	10.3463
1300	8067.	4683.5	5409.5	10.5669	7642.	4683.5	5409.5	10.5419	7260.	4683.5	5409.5	10.5183

Definitions of symbols on page 1.

	.11 (102.31)				.12 (104.80)				.13 (107.13)			p (t Sat.)
10^3 v	u	h	s	10^3 v	u	h	s	10^3 v	u	h	s	t
1549.5	2509.2	2679.7	7.3273	1428.4	2512.1	2683.5	7.2981	1325.4	2514.8	2687.1	7.2713	Sat.
1309.2	2426.4	2570.4	7.0136	1196.0	2424.5	2568.1	6.9678	1100.2	2422.7	2565.7	6.9251	50
1332.9	2434.4	2581.0	7.0464	1218.0	2432.7	2578.9	7.0009	1120.7	2431.0	2576.6	6.9586	55
1356.5	2442.4	2591.7	7.0785	1239.8	2440.8	2589.6	7.0334	1141.1	2439.2	2587.5	6.9915	60
1379.8	2450.5	2602.2	7.1100	1261.4	2448.9	2600.3	7.0653	1161.2	2447.4	2598.3	7.0237	65
1403.0	2458.4	2612.8	7.1410	1282.9	2457.0	2610.9	7.0965	1181.2	2455.5	2609.1	7.0553	70
1426.0	2466.4	2623.3	7.1713	1304.2	2465.0	2621.5	7.1272	1201.1	2463.7	2619.8	7.0862	75
1448.9	2474.3	2633.7	7.2011	1325.3	2473.0	2632.1	7.1572	1220.7	2471.7	2630.4	7.1166	80
1471.7	2482.2	2644.1	7.2303	1346.4	2481.0	2642.5	7.1867	1240.3	2479.8	2641.0	7.1463	85
1494.3	2490.0	2654.4	7.2589	1367.2	2488.9	2653.0	7.2156	1259.7	2487.8	2651.5	7.1754	90
1516.8	2497.9	2664.7	7.2871	1388.0	2496.8	2663.3	7.2440	1279.0	2495.7	2662.0	7.2040	95
1539.2	2505.6	2675.0	7.3147	1408.7	2504.6	2673.7	7.2718	1298.2	2503.6	2672.4	7.2321	100
1583.7	2521.1	2695.3	7.3686	1449.7	2520.2	2694.2	7.3261	1336.3	2519.3	2693.0	7.2867	110
1627.9	2536.5	2715.6	7.4207	1490.4	2535.7	2714.5	7.3785	1374.0	2534.9	2713.5	7.3395	120
1671.7	2551.8	2735.7	7.4713	1530.7	2551.1	2734.8	7.4293	1411.4	2550.3	2733.8	7.3906	130
1715.4	2567.0	2755.7	7.5203	1570.9	2566.4	2754.9	7.4786	1448.6	2565.7	2754.0	7.4401	140
1758.8	2582.2	2775.6	7.5680	1610.8	2581.6	2774.9	7.5265	1485.6	2581.0	2774.1	7.4882	150
1802.0	2597.3	2795.5	7.6145	1650.6	2596.8	2794.9	7.5731	1522.4	2596.3	2794.2	7.5349	160
1845.1	2612.4	2815.4	7.6598	1690.2	2611.9	2814.8	7.6185	1559.1	2611.5	2814.1	7.5805	170
1888.1	2627.5	2835.2	7.7040	1729.7	2627.1	2834.6	7.6629	1595.6	2626.6	2834.1	7.6250	180
1931.0	2642.6	2855.0	7.7472	1769.0	2642.2	2854.5	7.7062	1632.0	2641.8	2854.0	7.6684	190
1973.7	2657.7	2874.8	7.7895	1808.3	2657.3	2874.3	7.7486	1668.3	2657.0	2873.9	7.7109	200
2016.4	2672.8	2894.6	7.8309	1847.5	2672.5	2894.2	7.7901	1704.5	2672.1	2893.7	7.7525	210
2059.0	2687.9	2914.4	7.8715	1886.6	2687.6	2914.0	7.8308	1740.7	2687.3	2913.6	7.7932	220
2101.6	2703.1	2934.3	7.9114	1925.7	2702.8	2933.9	7.8706	1776.8	2702.5	2933.5	7.8331	230
2144.1	2718.3	2954.1	7.9504	1964.7	2718.0	2953.8	7.9098	1812.9	2717.7	2953.4	7.8723	240
2187.	2733.5	2974.0	7.9888	2004.	2733.2	2973.7	7.9483	1848.9	2733.0	2973.3	7.9108	250
2229.	2748.8	2993.9	8.0266	2043.	2748.5	2993.6	7.9860	1884.8	2748.3	2993.3	7.9486	260
2271.	2764.1	3013.9	8.0637	2081.	2763.8	3013.6	8.0231	1920.7	2763.6	3013.3	7.9857	270
2314.	2779.4	3033.9	8.1002	2120.	2779.2	3033.6	8.0596	1956.6	2779.0	3033.4	8.0223	280
2356.	2794.8	3054.0	8.1361	2159.	2794.6	3053.7	8.0956	1992.5	2794.4	3053.4	8.0583	290
2398.	2810.2	3074.1	8.1715	2198.	2810.1	3073.8	8.1310	2028.	2809.9	3073.6	8.0937	300
2483.	2841.3	3114.4	8.2406	2275.	2841.1	3114.2	8.2002	2100.	2841.0	3113.9	8.1630	320
2567.	2872.6	3155.0	8.3079	2353.	2872.4	3154.8	8.2675	2171.	2872.3	3154.5	8.2303	340
2652.	2904.1	3195.7	8.3733	2430.	2903.9	3195.5	8.3330	2243.	2903.8	3195.4	8.2958	360
2736.	2935.8	3236.7	8.4371	2508.	2935.7	3236.6	8.3968	2314.	2935.5	3236.4	8.3596	380
2820.	2967.8	3278.0	8.4993	2585.	2967.7	3277.8	8.4590	2386.	2967.5	3277.7	8.4219	400
2904.	3000.0	3319.5	8.5601	2662.	2999.9	3319.3	8.5197	2457.	2999.8	3319.2	8.4827	420
2989.	3032.5	3361.2	8.6194	2739.	3032.4	3361.1	8.5791	2528.	3032.3	3361.0	8.5421	440
3073.	3065.2	3403.2	8.6775	2816.	3065.1	3403.1	8.6372	2600.	3065.0	3403.0	8.6002	460
3157.	3098.2	3445.5	8.7344	2894.	3098.1	3445.4	8.6941	2671.	3098.1	3445.3	8.6571	480
3241.	3131.5	3488.0	8.7901	2971.	3131.4	3487.9	8.7498	2742.	3131.3	3487.8	8.7128	500
3325.	3165.0	3530.8	8.8447	3048.	3165.0	3530.7	8.8045	2813.	3164.9	3530.6	8.7675	520
3409.	3198.8	3573.9	8.8983	3125.	3198.8	3573.8	8.8581	2884.	3198.7	3573.7	8.8211	540
3493.	3232.9	3617.2	8.9510	3202.	3232.9	3617.1	8.9107	2956.	3232.8	3617.0	8.8737	560
3577.	3267.3	3660.8	9.0027	3279.	3267.2	3660.7	8.9625	3027.	3267.1	3660.6	8.9254	580
3661.	3301.9	3704.6	9.0535	3356.	3301.8	3704.6	9.0133	3098.	3301.8	3704.5	8.9763	600
3746.	3336.8	3748.8	9.1035	3433.	3336.7	3748.7	9.0633	3169.	3336.7	3748.6	9.0263	620
3830.	3371.9	3793.2	9.1527	3510.	3371.9	3793.1	9.1125	3240.	3371.8	3793.1	9.0755	640
3914.	3407.4	3837.9	9.2011	3587.	3407.4	3837.8	9.1609	3311.	3407.3	3837.8	9.1239	660
3998.	3443.1	3882.9	9.2488	3664.	3443.1	3882.8	9.2086	3382.	3443.0	3882.7	9.1716	680
4082.	3479.1	3928.1	9.2957	3741.	3479.1	3928.1	9.2555	3453.	3479.1	3928.0	9.2186	700
4292.	3570.4	4042.5	9.4103	3934.	3570.4	4042.4	9.3701	3631.	3570.3	4042.4	9.3332	750
4501.	3663.4	4158.6	9.5211	4126.	3663.4	4158.6	9.4810	3809.	3663.4	4158.5	9.4440	800
4711.	3758.2	4276.5	9.6285	4319.	3758.2	4276.5	9.5883	3986.	3758.2	4276.4	9.5514	850
4921.	3854.8	4396.1	9.7327	4511.	3854.7	4396.1	9.6925	4164.	3854.7	4396.0	9.6555	900
5131.	3952.9	4517.4	9.8339	4704.	3952.9	4517.3	9.7937	4342.	3952.9	4517.3	9.7568	950
5341.	4052.8	4640.3	9.9324	4896.	4052.7	4640.3	9.8922	4519.	4052.7	4640.2	9.8553	1000
5761.	4257.2	4890.9	10.1219	5281.	4257.2	4890.9	10.0817	4875.	4257.2	4890.9	10.0448	1100
6181.	4467.7	5147.5	10.3023	5666.	4467.7	5147.5	10.2621	5230.	4467.6	5147.5	10.2251	1200
6600.	4683.5	5409.5	10.4743	6050.	4683.4	5409.5	10.4341	5585.	4683.4	5409.4	10.3971	1300

Definitions of symbols on page 1.

Table 3. Vapor

p (t Sat.)	.14 (109.31)				.15 (111.37)				.16 (113.32)			
t	10^3 v	u	h	s	10^3 v	u	h	s	10^3 v	u	h	s
Sat.	1236.6	2517.3	2690.4	7.2464	1159.3	2519.7	2693.6	7.2233	1091.4	2521.9	2696.5	7.2017
50	*1018.0*	*2420.9*	*2563.4*	*6.8851*	*946.7*	*2419.0*	*2561.0*	*6.8473*	*884.4*	*2417.1*	*2558.6*	*6.8116*
55	*1037.3*	*2429.2*	*2574.4*	*6.9190*	*965.0*	*2427.4*	*2572.2*	*6.8817*	*901.6*	*2425.7*	*2569.9*	*6.8464*
60	*1056.4*	*2437.5*	*2585.4*	*6.9522*	*983.0*	*2435.9*	*2583.3*	*6.9153*	*918.7*	*2434.2*	*2581.2*	*6.8804*
65	*1075.3*	*2445.8*	*2596.4*	*6.9848*	*1000.8*	*2444.2*	*2594.4*	*6.9483*	*935.6*	*2442.7*	*2592.4*	*6.9137*
70	*1094.1*	*2454.1*	*2607.2*	*7.0167*	*1018.5*	*2452.6*	*2605.4*	*6.9805*	*952.4*	*2451.1*	*2603.5*	*6.9463*
75	*1112.6*	*2462.3*	*2618.0*	*7.0480*	*1036.0*	*2460.9*	*2616.3*	*7.0121*	*968.9*	*2459.5*	*2614.5*	*6.9782*
80	*1131.1*	*2470.4*	*2628.8*	*7.0786*	*1053.3*	*2469.1*	*2627.1*	*7.0430*	*985.3*	*2467.8*	*2625.4*	*7.0094*
85	*1149.4*	*2478.5*	*2639.4*	*7.1086*	*1070.5*	*2477.3*	*2637.9*	*7.0733*	*1001.6*	*2476.1*	*2636.3*	*7.0400*
90	*1167.5*	*2486.6*	*2650.0*	*7.1380*	*1087.6*	*2485.4*	*2648.6*	*7.1029*	*1017.7*	*2484.3*	*2647.1*	*7.0699*
95	*1185.6*	*2494.6*	*2660.6*	*7.1669*	*1104.6*	*2493.5*	*2659.2*	*7.1320*	*1033.7*	*2492.4*	*2657.8*	*7.0992*
100	*1203.5*	*2502.6*	*2671.1*	*7.1951*	1121.4	2501.5	2669.8	7.1605	*1049.6*	*2500.5*	*2668.4*	*7.1279*
110	1239.0	2518.4	2691.9	7.2501	*1154.8*	*2517.5*	*2690.7*	*7.2159*	*1081.0*	*2516.6*	*2689.5*	*7.1837*
120	1274.3	2534.1	2712.5	7.3032	1187.8	2533.3	2711.4	7.2693	1112.2	2532.4	2710.4	7.2374
130	1309.2	2549.6	2732.9	7.3546	1220.5	2548.9	2732.0	7.3209	1143.0	2548.2	2731.0	7.2893
140	1343.8	2565.1	2753.2	7.4043	1253.0	2564.4	2752.4	7.3709	1173.5	2563.8	2751.5	7.3395
150	1378.3	2580.4	2773.4	7.4526	1285.3	2579.8	2772.6	7.4193	1203.9	2579.3	2771.9	7.3882
160	1412.6	2595.7	2793.5	7.4995	1317.3	2595.2	2792.8	7.4665	1234.0	2594.7	2792.1	7.4354
170	1446.7	2611.0	2813.5	7.5452	1349.3	2610.5	2812.9	7.5123	1264.0	2610.0	2812.3	7.4814
180	1480.7	2626.2	2833.5	7.5898	1381.1	2625.8	2832.9	7.5570	1293.9	2625.3	2832.4	7.5263
190	1514.5	2641.4	2853.4	7.6334	1412.7	2641.0	2852.9	7.6006	1323.7	2640.6	2852.4	7.5700
200	1548.3	2656.6	2873.4	7.6759	1444.3	2656.2	2872.9	7.6433	1353.3	2655.9	2872.4	7.6127
210	1582.0	2671.8	2893.3	7.7176	1475.8	2671.5	2892.8	7.6850	1382.9	2671.1	2892.4	7.6545
220	1615.7	2687.0	2913.2	7.7584	1507.3	2686.7	2912.8	7.7259	1412.4	2686.4	2912.4	7.6955
230	1649.2	2702.2	2933.1	7.7983	1538.6	2701.9	2932.7	7.7659	1441.9	2701.7	2932.3	7.7356
240	1682.7	2717.5	2953.1	7.8376	1570.0	2717.2	2952.7	7.8052	1471.3	2716.9	2952.3	7.7749
250	1716.2	2732.7	2973.0	7.8761	1601.2	2732.5	2972.7	7.8438	1500.6	2732.2	2972.3	7.8135
260	1749.6	2748.1	2993.0	7.9139	1632.5	2747.8	2992.7	7.8817	1529.9	2747.6	2992.4	7.8514
270	1783.0	2763.4	3013.0	7.9511	1663.6	2763.2	3012.7	7.9189	1559.2	2763.0	3012.4	7.8887
280	1816.4	2778.8	3033.1	7.9877	1694.8	2778.6	3032.8	7.9555	1588.4	2778.4	3032.5	7.9254
290	1849.7	2794.2	3053.2	8.0237	1725.9	2794.0	3052.9	7.9915	1617.6	2793.8	3052.6	7.9614
300	1883.0	2809.7	3073.3	8.0592	1757.0	2809.5	3073.1	8.0270	1646.8	2809.3	3072.8	7.9969
320	1949.5	2840.8	3113.7	8.1285	1819.2	2840.6	3113.5	8.0964	1705.1	2840.5	3113.3	8.0663
340	2016.0	2872.1	3154.3	8.1958	1881.2	2872.0	3154.1	8.1638	1763.3	2871.8	3153.9	8.1337
360	2082.4	2903.6	3195.2	8.2614	1943.2	2903.5	3195.0	8.2293	1821.5	2903.4	3194.8	8.1993
380	2148.7	2935.4	3236.2	8.3252	2005.2	2935.3	3236.1	8.2932	1879.6	2935.2	3235.9	8.2632
400	2215.	2967.4	3277.5	8.3875	2067.	2967.3	3277.4	8.3555	1937.6	2967.2	3277.2	8.3255
420	2281.	2999.7	3319.1	8.4483	2129.	2999.6	3318.9	8.4163	1995.6	2999.5	3318.8	8.3864
440	2347.	3032.2	3360.8	8.5077	2191.	3032.1	3360.7	8.4757	2053.6	3032.0	3360.6	8.4458
460	2414.	3065.0	3402.9	8.5658	2253.	3064.9	3402.7	8.5339	2111.6	3064.8	3402.6	8.5040
480	2480.	3098.0	3445.2	8.6227	2314.	3097.9	3445.0	8.5908	2169.5	3097.8	3444.9	8.5609
500	2546.	3131.3	3487.7	8.6785	2376.	3131.2	3487.6	8.6466	2227.	3131.1	3487.5	8.6167
520	2612.	3164.8	3530.5	8.7332	2438.	3164.7	3530.4	8.7012	2285.	3164.7	3530.3	8.6714
540	2678.	3198.6	3573.6	8.7868	2500.	3198.6	3573.5	8.7549	2343.	3198.5	3573.4	8.7250
560	2744.	3232.7	3616.9	8.8394	2561.	3232.7	3616.8	8.8075	2401.	3232.6	3616.8	8.7777
580	2810.	3267.1	3660.5	8.8912	2623.	3267.0	3660.5	8.8593	2459.	3267.0	3660.4	8.8294
600	2876.	3301.7	3704.4	8.9420	2685.	3301.7	3704.3	8.9101	2517.	3301.6	3704.3	8.8802
620	2943.	3336.6	3748.6	8.9920	2746.	3336.6	3748.5	8.9601	2574.	3336.5	3748.4	8.9303
640	3009.	3371.8	3793.0	9.0412	2808.	3371.7	3792.9	9.0093	2632.	3371.7	3792.9	8.9795
660	3075.	3407.3	3837.7	9.0896	2870.	3407.2	3837.6	9.0577	2690.	3407.2	3837.6	9.0279
680	3141.	3443.0	3882.7	9.1373	2931.	3442.9	3882.6	9.1054	2748.	3442.9	3882.6	9.0756
700	3207.	3479.0	3927.9	9.1843	2993.	3479.0	3927.9	9.1524	2806.	3478.9	3927.8	9.1226
750	3372.	3570.3	4042.3	9.2989	3147.	3570.2	4042.3	9.2670	2950.	3570.2	4042.2	9.2372
800	3537.	3663.3	4158.5	9.4097	3301.	3663.3	4158.4	9.3779	3094.	3663.3	4158.4	9.3481
850	3702.	3758.2	4276.4	9.5171	3455.	3758.1	4276.3	9.4853	3239.	3758.1	4276.3	9.4555
900	3867.	3854.7	4396.0	9.6213	3609.	3854.6	4396.0	9.5894	3383.	3854.6	4395.9	9.5596
950	4032.	3952.9	4517.3	9.7225	3763.	3952.8	4517.2	9.6907	3528.	3952.8	4517.2	9.6609
1000	4196.	4052.7	4640.2	9.8210	3917.	4052.7	4640.2	9.7892	3672.	4052.6	4640.1	9.7594
1100	4526.	4257.2	4890.8	10.0105	4225.	4257.1	4890.8	9.9787	3960.	4257.1	4890.8	9.9489
1200	4856.	4467.6	5147.5	10.1909	4532.	4467.6	5147.5	10.1591	4249.	4467.6	5147.4	10.1293
1300	5186.	4683.4	5409.4	10.3629	4840.	4683.4	5409.4	10.3311	4538.	4683.3	5409.4	10.3013

Definitions of symbols on page 1.

	.17 (115.17)				.18 (116.93)				.19 (118.62)			p (t Sat.)
10^3 v	u	h	s	10^3 v	u	h	s	10^3 v	u	h	s	t
1031.2	2523.9	2699.2	7.1814	977.5	2525.9	2701.8	7.1623	929.3	2527.7	2704.3	7.1443	Sat.
829.3	*2415.2*	*2556.2*	*6.7777*	*780.3*	*2413.3*	*2553.8*	*6.7453*	*736.4*	*2411.4*	*2551.3*	*6.7142*	**50**
845.8	*2423.9*	*2567.7*	*6.8128*	*796.0*	*2422.1*	*2565.4*	*6.7808*	*751.5*	*2420.3*	*2563.1*	*6.7502*	**55**
862.0	*2432.5*	*2579.0*	*6.8473*	*811.6*	*2430.8*	*2576.9*	*6.8157*	*766.4*	*2429.1*	*2574.7*	*6.7855*	**60**
878.1	*2441.1*	*2590.3*	*6.8810*	*826.9*	*2439.5*	*2588.3*	*6.8497*	*781.1*	*2437.9*	*2586.3*	*6.8199*	**65**
894.0	*2449.6*	*2601.6*	*6.9139*	*842.1*	*2448.1*	*2599.7*	*6.8830*	*795.6*	*2446.6*	*2597.7*	*6.8536*	**70**
909.7	*2458.1*	*2612.7*	*6.9461*	*857.1*	*2456.6*	*2610.9*	*6.9156*	*809.9*	*2455.2*	*2609.1*	*6.8865*	**75**
925.3	*2466.5*	*2623.8*	*6.9776*	*871.9*	*2465.1*	*2622.1*	*6.9474*	*824.1*	*2463.8*	*2620.4*	*6.9186*	**80**
940.7	*2474.8*	*2634.7*	*7.0085*	*886.6*	*2473.5*	*2633.1*	*6.9785*	*838.1*	*2472.3*	*2631.5*	*6.9500*	**85**
956.0	*2483.1*	*2645.6*	*7.0386*	*901.1*	*2481.9*	*2644.1*	*7.0089*	*852.0*	*2480.7*	*2642.6*	*6.9807*	**90**
971.1	*2491.3*	*2656.4*	*7.0682*	*915.5*	*2490.2*	*2655.0*	*7.0387*	*865.8*	*2489.1*	*2653.6*	*7.0107*	**95**
986.2	*2499.5*	*2667.1*	*7.0971*	*929.8*	*2498.4*	*2665.8*	*7.0679*	*879.4*	*2497.4*	*2664.5*	*7.0401*	**100**
1016.0	*2515.6*	*2688.4*	*7.1532*	*958.1*	*2514.7*	*2687.2*	*7.1244*	*906.3*	*2513.8*	*2686.0*	*7.0970*	**110**
1045.4	2531.6	2709.3	7.2073	986.0	2530.8	2708.3	7.1788	932.9	2530.0	2707.2	7.1517	120
1074.5	2547.4	2730.1	7.2595	1013.7	2546.7	2729.2	7.2312	959.2	2546.0	2728.2	7.2044	130
1103.4	2563.1	2750.7	7.3099	1041.0	2562.4	2749.8	7.2819	985.3	2561.8	2749.0	7.2553	140
1132.0	2578.7	2771.1	7.3588	1068.2	2578.1	2770.4	7.3310	1011.1	2577.5	2769.6	7.3046	150
1160.5	2594.1	2791.4	7.4062	1095.2	2593.6	2790.7	7.3786	1036.7	2593.1	2790.0	7.3524	160
1188.8	2609.5	2811.6	7.4524	1122.0	2609.1	2811.0	7.4249	1062.1	2608.6	2810.4	7.3988	170
1217.0	2624.9	2831.8	7.4973	1148.6	2624.5	2831.2	7.4700	1087.5	2624.0	2830.6	7.4440	180
1245.1	2640.2	2851.9	7.5411	1175.2	2639.8	2851.4	7.5139	1112.7	2639.4	2850.8	7.4881	190
1273.0	2655.5	2871.9	7.5840	1201.7	2655.1	2871.4	7.5568	1137.8	2654.8	2871.0	7.5311	200
1300.9	2670.8	2891.9	7.6259	1228.1	2670.5	2891.5	7.5988	1162.8	2670.1	2891.1	7.5731	210
1328.7	2686.1	2912.0	7.6668	1254.4	2685.8	2911.5	7.6398	1187.8	2685.4	2911.1	7.6142	220
1356.5	2701.4	2932.0	7.7070	1280.6	2701.1	2931.6	7.6801	1212.7	2700.8	2931.2	7.6545	230
1384.2	2716.7	2952.0	7.7464	1306.8	2716.4	2951.6	7.7195	1237.6	2716.1	2951.3	7.6940	240
1411.9	2732.0	2972.0	7.7851	1332.9	2731.7	2971.7	7.7582	1262.3	2731.5	2971.3	7.7328	250
1439.5	2747.3	2992.1	7.8230	1359.0	2747.1	2991.7	7.7962	1287.1	2746.9	2991.4	7.7708	260
1467.0	2762.7	3012.1	7.8603	1385.1	2762.5	3011.8	7.8335	1311.8	2762.3	3011.5	7.8082	270
1494.6	2778.2	3032.2	7.8970	1411.1	2778.0	3032.0	7.8702	1336.5	2777.7	3031.7	7.8449	280
1522.1	2793.6	3052.4	7.9331	1437.1	2793.4	3052.1	7.9064	1361.1	2793.2	3051.9	7.8811	290
1549.6	2809.1	3072.6	7.9686	1463.1	2808.9	3072.3	7.9419	1385.8	2808.8	3072.1	7.9166	300
1604.5	2840.3	3113.0	8.0380	1515.0	2840.1	3112.8	8.0114	1435.0	2840.0	3112.6	7.9862	320
1659.3	2871.7	3153.7	8.1055	1566.8	2871.5	3153.5	8.0789	1484.1	2871.4	3153.3	8.0537	340
1714.0	2903.2	3194.6	8.1711	1618.6	2903.1	3194.4	8.1445	1533.1	2903.0	3194.2	8.1194	360
1768.8	2935.0	3235.7	8.2350	1670.2	2934.9	3235.6	8.2085	1582.1	2934.8	3235.4	8.1833	380
1823.4	2967.1	3277.1	8.2974	1721.9	2967.0	3276.9	8.2708	1631.0	2966.8	3276.7	8.2457	400
1878.0	2999.4	3318.6	8.3582	1773.5	2999.3	3318.5	8.3317	1680.0	2999.1	3318.3	8.3066	420
1932.6	3031.9	3360.4	8.4177	1825.1	3031.8	3360.3	8.3912	1728.8	3031.7	3360.2	8.3661	440
1987.2	3064.7	3402.5	8.4759	1876.6	3064.6	3402.4	8.4494	1777.7	3064.5	3402.3	8.4243	460
2041.7	3097.7	3444.8	8.5328	1928.1	3097.6	3444.7	8.5063	1826.5	3097.6	3444.6	8.4812	480
2096.	3131.0	3487.4	8.5886	1979.6	3130.9	3487.3	8.5621	1875.3	3130.9	3487.2	8.5370	500
2151.	3164.6	3530.2	8.6433	2031.1	3164.5	3530.1	8.6168	1924.1	3164.4	3530.0	8.5918	520
2205.	3198.4	3573.3	8.6969	2082.6	3198.4	3573.2	8.6703	1972.8	3198.3	3573.1	8.6454	540
2260.	3232.5	3616.7	8.7496	2134.0	3232.5	3616.6	8.7232	2021.6	3232.4	3616.5	8.6981	560
2314.	3266.9	3660.3	8.8013	2185.4	3266.8	3660.2	8.7749	2070.3	3266.8	3660.1	8.7499	580
2369.	3301.5	3704.2	8.8522	2237.	3301.5	3704.1	8.8258	2119.	3301.4	3704.0	8.8007	600
2423.	3336.5	3748.4	8.9022	2288.	3336.4	3748.3	8.8758	2168.	3336.3	3748.2	8.8508	620
2477.	3371.6	3792.8	8.9514	2340.	3371.6	3792.7	8.9250	2216.	3371.5	3792.7	8.9000	640
2532.	3407.1	3837.5	8.9999	2391.	3407.1	3837.4	8.9734	2265.	3407.0	3837.4	8.9484	660
2586.	3442.9	3882.5	9.0476	2442.	3442.8	3882.4	9.0211	2314.	3442.8	3882.4	8.9961	680
2641.	3478.9	3927.8	9.0946	2494.	3478.8	3927.7	9.0682	2362.	3478.8	3927.7	9.0432	700
2776.	3570.2	4042.2	9.2092	2622.	3570.1	4042.1	9.1828	2484.	3570.1	4042.1	9.1578	750
2912.	3663.2	4158.3	9.3200	2751.	3663.2	4158.3	9.2936	2606.	3663.2	4158.2	9.2686	800
3048.	3758.1	4276.3	9.4275	2879.	3758.0	4276.2	9.4010	2727.	3758.0	4276.2	9.3761	850
3184.	3854.6	4395.9	9.5316	3007.	3854.6	4395.8	9.5052	2849.	3854.5	4395.8	9.4803	900
3320.	3952.8	4517.2	9.6329	3136.	3952.7	4517.1	9.6065	2970.	3952.7	4517.1	9.5815	950
3456.	4052.6	4640.1	9.7314	3264.	4052.6	4640.1	9.7050	3092.	4052.6	4640.0	9.6800	1000
3728.	4257.1	4890.8	9.9209	3520.	4257.1	4890.7	9.8945	3335.	4257.0	4890.7	9.8695	1100
3999.	4467.5	5147.4	10.1013	3777.	4467.5	5147.4	10.0749	3578.	4467.5	5147.4	10.0499	1200
4271.	4683.3	5409.3	10.2733	4034.	4683.3	5409.3	10.2469	3821.	4683.3	5409.3	10.2219	1300

Definitions of symbols on page 1.

(25)

Table 3. Vapor

p (t Sat.)	.20 (120.23)				.22 (123.27)				.24 (126.10)			
t	10^3 v	u	h	s	10^3 v	u	h	s	10^3 v	u	h	s
Sat.	885.7	2529.5	2706.7	7.1272	810.1	2532.8	2711.0	7.0953	746.7	2535.8	2715.0	7.0663
50	696.9	2409.5	2548.9	6.6844	628.6	2405.5	2543.8	6.6279	571.6	2401.5	2538.7	6.5750
55	711.4	2418.4	2560.7	6.7208	642.1	2414.7	2556.0	6.6652	584.3	2410.9	2551.2	6.6133
60	725.8	2427.4	2572.5	6.7565	655.4	2423.9	2568.1	6.7017	596.8	2420.3	2563.5	6.6506
65	739.9	2436.2	2584.2	6.7913	668.6	2432.9	2580.0	6.7374	609.1	2429.6	2575.8	6.6871
70	753.8	2445.0	2595.8	6.8253	681.5	2441.9	2591.9	6.7721	621.1	2438.8	2587.9	6.7226
75	767.5	2453.8	2607.3	6.8586	694.2	2450.9	2603.6	6.8060	633.0	2447.9	2599.8	6.7572
80	781.1	2462.4	2618.6	6.8910	706.7	2459.7	2615.2	6.8391	644.7	2456.9	2611.7	6.7909
85	794.5	2471.0	2629.9	6.9227	719.1	2468.4	2626.7	6.8714	656.3	2465.8	2623.3	6.8238
90	807.8	2479.5	2641.1	6.9536	731.4	2477.1	2638.0	6.9029	667.7	2474.7	2634.9	6.8559
95	821.0	2488.0	2652.1	6.9839	743.5	2485.7	2649.3	6.9336	679.0	2483.4	2646.3	6.8871
100	834.0	2496.3	2663.1	7.0135	755.5	2494.2	2660.4	6.9637	690.1	2492.0	2657.7	6.9177
110	859.7	2512.8	2684.8	7.0708	779.2	2510.9	2682.4	7.0218	712.1	2509.0	2679.9	6.9766
120	885.1	2529.1	2706.2	7.1259	802.6	2527.4	2704.0	7.0776	733.7	2525.8	2701.8	7.0330
130	910.2	2545.2	2727.3	7.1789	825.6	2543.7	2725.3	7.1311	755.0	2542.2	2723.4	7.0872
140	935.0	2561.1	2748.1	7.2300	848.3	2559.8	2746.4	7.1828	776.0	2558.4	2744.7	7.1393
150	959.6	2576.9	2768.8	7.2795	870.8	2575.7	2767.3	7.2327	796.8	2574.5	2765.7	7.1896
160	984.1	2592.5	2789.3	7.3275	893.1	2591.5	2787.9	7.2810	817.4	2590.4	2786.5	7.2383
170	1008.3	2608.1	2809.8	7.3740	915.3	2607.1	2808.5	7.3278	837.8	2606.1	2807.2	7.2854
180	1032.4	2623.6	2830.1	7.4194	937.3	2622.7	2828.9	7.3734	858.1	2621.8	2827.7	7.3313
190	1056.4	2639.0	2850.3	7.4635	959.2	2638.2	2849.2	7.4178	878.3	2637.4	2848.2	7.3759
200	1080.3	2654.4	2870.5	7.5066	981.1	2653.7	2869.5	7.4611	898.3	2652.9	2868.5	7.4193
210	1104.2	2669.8	2890.6	7.5487	1002.8	2669.1	2889.7	7.5033	918.3	2668.4	2888.8	7.4618
220	1127.9	2685.1	2910.7	7.5899	1024.4	2684.5	2909.9	7.5447	938.2	2683.9	2909.0	7.5032
230	1151.6	2700.5	2930.8	7.6303	1046.0	2699.9	2930.0	7.5851	958.1	2699.3	2929.3	7.5438
240	1175.2	2715.9	2950.9	7.6698	1067.6	2715.3	2950.2	7.6248	977.8	2714.8	2949.5	7.5835
250	1198.8	2731.2	2971.0	7.7086	1089.1	2730.7	2970.3	7.6636	997.6	2730.2	2969.6	7.6225
260	1222.3	2746.6	2991.1	7.7467	1110.5	2746.2	2990.5	7.7018	1017.3	2745.7	2989.8	7.6608
270	1245.8	2762.1	3011.2	7.7841	1131.9	2761.6	3010.6	7.7393	1036.9	2761.2	3010.1	7.6983
280	1269.3	2777.5	3031.4	7.8209	1153.3	2777.1	3030.8	7.7761	1056.6	2776.7	3030.3	7.7352
290	1292.8	2793.0	3051.6	7.8570	1174.6	2792.6	3051.1	7.8123	1076.2	2792.2	3050.5	7.7715
300	1316.2	2808.6	3071.8	7.8926	1195.9	2808.2	3071.3	7.8480	1095.7	2807.8	3070.8	7.8072
320	1362.9	2839.8	3112.4	7.9622	1238.5	2839.5	3111.9	7.9176	1134.8	2839.1	3111.5	7.8769
340	1409.6	2871.2	3153.1	8.0298	1281.0	2870.9	3152.7	7.9853	1173.8	2870.6	3152.3	7.9446
360	1456.2	2902.8	3194.1	8.0955	1323.4	2902.5	3193.7	8.0510	1212.7	2902.3	3193.3	8.0104
380	1502.8	2934.7	3235.2	8.1595	1365.8	2934.4	3234.9	8.1151	1251.6	2934.2	3234.5	8.0745
400	1549.3	2966.7	3276.6	8.2218	1408.1	2966.5	3276.3	8.1775	1290.4	2966.3	3276.0	8.1370
420	1595.8	2999.0	3318.2	8.2828	1450.4	2998.8	3317.9	8.2385	1329.2	2998.6	3317.6	8.1980
440	1642.2	3031.6	3360.0	8.3423	1492.6	3031.4	3359.8	8.2980	1367.9	3031.2	3359.5	8.2576
460	1688.6	3064.4	3402.1	8.4005	1534.8	3064.2	3401.9	8.3562	1406.7	3064.0	3401.6	8.3158
480	1735.0	3097.5	3444.5	8.4575	1577.0	3097.3	3444.2	8.4132	1445.4	3097.1	3444.0	8.3729
500	1781.4	3130.8	3487.1	8.5133	1619.2	3130.6	3486.9	8.4691	1484.0	3130.5	3486.6	8.4287
520	1827.7	3164.4	3529.9	8.5680	1661.4	3164.2	3529.7	8.5238	1522.7	3164.1	3529.5	8.4835
540	1874.1	3198.2	3573.0	8.6217	1703.5	3198.1	3572.8	8.5775	1561.3	3197.9	3572.7	8.5372
560	1920.4	3232.3	3616.4	8.6744	1745.6	3232.2	3616.2	8.6302	1600.0	3232.1	3616.1	8.5899
580	1966.7	3266.7	3660.0	8.7261	1787.7	3266.6	3659.9	8.6820	1638.6	3266.5	3659.7	8.6417
600	2013.	3301.4	3704.0	8.7770	1829.8	3301.3	3703.8	8.7329	1677.2	3301.1	3703.7	8.6926
620	2059.	3336.3	3748.1	8.8270	1871.9	3336.2	3748.0	8.7829	1715.7	3336.1	3747.9	8.7426
640	2106.	3371.5	3792.6	8.8763	1914.0	3371.4	3792.5	8.8322	1754.3	3371.3	3792.3	8.7919
660	2152.	3407.0	3837.3	8.9247	1956.0	3406.9	3837.2	8.8806	1792.9	3406.8	3837.1	8.8404
680	2198.	3442.7	3882.3	8.9724	1998.1	3442.6	3882.2	8.9283	1831.4	3442.5	3882.1	8.8881
700	2244.	3478.8	3927.6	9.0194	2040.	3478.7	3927.5	8.9754	1870.0	3478.6	3927.4	8.9351
750	2360.	3570.1	4042.0	9.1341	2145.	3570.0	4041.9	9.0900	1966.3	3569.9	4041.8	9.0498
800	2475.	3663.1	4158.2	9.2449	2250.	3663.1	4158.1	9.2009	2062.7	3663.0	4158.0	9.1607
850	2591.	3758.0	4276.2	9.3524	2355.	3757.9	4276.1	9.3083	2159.0	3757.8	4276.0	9.2681
900	2706.	3854.5	4395.8	9.4566	2460.	3854.4	4395.7	9.4125	2255.	3854.4	4395.6	9.3723
950	2822.	3952.7	4517.1	9.5578	2565.	3952.6	4517.0	9.5138	2352.	3952.6	4516.9	9.4736
1000	2937.	4052.5	4640.0	9.6563	2670.	4052.5	4640.0	9.6123	2448.	4052.4	4639.9	9.5721
1100	3168.	4257.0	4890.7	9.8458	2880.	4257.0	4890.6	9.8018	2640.	4256.9	4890.6	9.7616
1200	3399.	4467.5	5147.3	10.0262	3090.	4467.4	5147.3	9.9822	2833.	4467.4	5147.2	9.9420
1300	3630.	4683.2	5409.3	10.1982	3300.	4683.2	5409.2	10.1542	3025.	4683.1	5409.2	10.1140

Definitions of symbols on page 1.

	.26 (128.73)				.28 (131.21)				.30 (133.55)			p (t Sat.)
10^3 v	u	h	s	10^3 v	u	h	s	10^3 v	u	h	s	t
692.8	2538.6	2718.7	7.0396	646.3	2541.2	2722.1	7.0149	605.8	2543.6	2725.3	6.9919	Sat.
523.2	2397.5	2533.5	6.5250	481.6	2393.3	2528.2	6.4774	445.5	2389.1	2522.7	6.4319	50
535.3	2407.1	2546.3	6.5642	493.1	2403.2	2541.3	6.5177	456.5	2399.2	2536.2	6.4732	55
547.1	2416.7	2558.9	6.6025	504.4	2413.0	2554.2	6.5569	467.3	2409.3	2549.5	6.5134	60
558.6	2426.2	2571.4	6.6398	515.4	2422.7	2567.1	6.5951	477.8	2419.2	2562.6	6.5525	65
570.0	2435.6	2583.8	6.6761	526.2	2432.4	2579.7	6.6322	488.1	2429.1	2575.5	6.5905	70
581.2	2444.9	2596.0	6.7115	536.8	2441.9	2592.2	6.6683	498.2	2438.8	2588.2	6.6273	75
592.2	2454.1	2608.1	6.7459	547.2	2451.3	2604.5	6.7034	508.1	2448.4	2600.8	6.6631	80
603.1	2463.2	2620.0	6.7793	557.4	2460.5	2616.6	6.7375	517.8	2457.8	2613.2	6.6979	85
613.8	2472.2	2631.8	6.8120	567.5	2469.7	2628.6	6.7707	527.4	2467.1	2625.4	6.7317	90
624.3	2481.1	2643.4	6.8438	577.5	2478.7	2640.4	6.8030	536.8	2476.3	2637.4	6.7646	95
634.8	2489.8	2654.9	6.8748	587.3	2487.6	2652.1	6.8345	546.1	2485.4	2649.2	6.7965	100
655.3	2507.1	2677.5	6.9346	606.6	2505.2	2675.0	6.8952	564.4	2503.2	2672.5	6.8581	110
675.5	2524.0	2699.7	6.9917	625.5	2522.3	2697.5	6.9530	582.2	2520.6	2695.2	6.9167	120
695.3	2540.7	2721.5	7.0464	644.1	2539.2	2719.5	7.0084	599.7	2537.6	2717.5	6.9726	130
714.8	2557.1	2742.9	7.0991	662.4	2555.7	2741.2	7.0615	616.9	2554.3	2739.4	7.0263	140
734.1	2573.3	2764.1	7.1498	680.4	2572.0	2762.6	7.1126	633.9	2570.8	2761.0	7.0778	150
753.3	2589.3	2785.1	7.1988	698.3	2588.2	2783.7	7.1620	650.6	2587.1	2782.3	7.1276	160
772.2	2605.1	2805.9	7.2462	716.0	2604.1	2804.6	7.2098	667.3	2603.1	2803.3	7.1757	170
791.0	2620.9	2826.6	7.2923	733.5	2620.0	2825.4	7.2561	683.7	2619.1	2824.2	7.2223	180
809.7	2636.6	2847.1	7.3371	751.0	2635.7	2846.0	7.3011	700.1	2634.9	2844.9	7.2675	190
828.3	2652.2	2867.5	7.3808	768.3	2651.4	2866.5	7.3450	716.3	2650.7	2865.6	7.3115	200
846.8	2667.7	2887.9	7.4234	785.5	2667.0	2887.0	7.3877	732.4	2666.3	2886.1	7.3545	210
865.3	2683.2	2908.2	7.4650	802.7	2682.6	2907.4	7.4295	748.5	2682.0	2906.5	7.3963	220
883.6	2698.7	2928.5	7.5057	819.8	2698.1	2927.7	7.4703	764.5	2697.6	2926.9	7.4373	230
901.9	2714.2	2948.7	7.5455	836.9	2713.7	2948.0	7.5103	780.5	2713.1	2947.3	7.4774	240
920.2	2729.7	2969.0	7.5846	853.9	2729.2	2968.3	7.5494	796.4	2728.7	2967.6	7.5166	250
938.4	2745.2	2989.2	7.6229	870.8	2744.7	2988.6	7.5878	812.2	2744.3	2987.9	7.5551	260
956.6	2760.7	3009.5	7.6605	887.7	2760.3	3008.9	7.6255	828.0	2759.8	3008.3	7.5928	270
974.7	2776.3	3029.7	7.6975	904.6	2775.9	3029.2	7.6625	843.8	2775.4	3028.6	7.6299	280
992.9	2791.9	3050.0	7.7338	921.4	2791.5	3049.5	7.6989	859.6	2791.1	3048.9	7.6664	290
1010.9	2807.5	3070.3	7.7696	938.3	2807.1	3069.8	7.7347	875.3	2806.7	3069.3	7.7022	300
1047.1	2838.8	3111.0	7.8394	971.8	2838.5	3110.6	7.8046	906.7	2838.1	3110.1	7.7722	320
1083.1	2870.3	3151.9	7.9072	1005.3	2870.0	3151.5	7.8725	938.0	2869.7	3151.1	7.8401	340
1119.1	2902.0	3192.9	7.9731	1038.8	2901.7	3192.6	7.9384	969.2	2901.4	3192.2	7.9061	360
1155.0	2933.9	3234.2	8.0372	1072.1	2933.6	3233.8	8.0026	1000.4	2933.4	3233.5	7.9704	380
1190.8	2966.0	3275.6	8.0997	1105.5	2965.8	3275.3	8.0652	1031.5	2965.6	3275.0	8.0330	400
1226.7	2998.4	3317.3	8.1607	1138.8	2998.2	3317.0	8.1262	1062.6	2998.0	3316.7	8.0941	420
1262.5	3031.0	3359.2	8.2204	1172.0	3030.8	3359.0	8.1859	1093.7	3030.6	3358.7	8.1538	440
1298.2	3063.9	3401.4	8.2786	1205.3	3063.7	3401.1	8.2442	1124.7	3063.5	3400.9	8.2121	460
1334.0	3097.0	3443.8	8.3357	1238.5	3096.8	3443.6	8.3013	1155.7	3096.6	3443.3	8.2692	480
1369.7	3130.3	3486.4	8.3916	1271.6	3130.1	3486.2	8.3572	1186.7	3130.0	3486.0	8.3251	500
1405.4	3163.9	3529.3	8.4463	1304.8	3163.8	3529.1	8.4120	1217.7	3163.6	3528.9	8.3799	520
1441.1	3197.8	3572.5	8.5001	1338.0	3197.7	3572.3	8.4657	1248.6	3197.5	3572.1	8.4337	540
1476.7	3231.9	3615.9	8.5528	1371.1	3231.8	3615.7	8.5185	1279.5	3231.7	3615.5	8.4865	560
1512.4	3266.3	3659.6	8.6046	1404.2	3266.2	3659.4	8.5703	1310.5	3266.1	3659.2	8.5383	580
1548.0	3301.0	3703.5	8.6555	1437.3	3300.9	3703.4	8.6212	1341.4	3300.8	3703.2	8.5892	600
1583.6	3336.0	3747.7	8.7056	1470.4	3335.9	3747.6	8.6713	1372.2	3335.8	3747.4	8.6393	620
1619.2	3371.2	3792.2	8.7548	1503.5	3371.1	3792.1	8.7205	1403.1	3371.0	3791.9	8.6886	640
1654.9	3406.7	3836.9	8.8033	1536.5	3406.6	3836.8	8.7690	1434.0	3406.5	3836.7	8.7371	660
1690.5	3442.5	3882.0	8.8511	1569.6	3442.4	3881.8	8.8168	1464.9	3442.3	3881.7	8.7848	680
1726.1	3478.6	3927.3	8.8981	1602.7	3478.5	3927.2	8.8638	1495.7	3478.4	3927.1	8.8319	700
1815.0	3569.9	4041.8	9.0128	1685.3	3569.8	4041.7	8.9786	1572.9	3569.8	4041.6	8.9467	750
1903.9	3663.0	4158.0	9.1237	1767.9	3662.9	4157.9	9.0895	1649.9	3662.9	4157.8	9.0576	800
1992.8	3757.8	4275.9	9.2311	1850.4	3757.7	4275.9	9.1969	1727.0	3757.7	4275.8	9.1650	850
2081.7	3854.3	4395.6	9.3353	1933.0	3854.3	4395.5	9.3011	1804.1	3854.2	4395.4	9.2692	900
2170.6	3952.5	4516.9	9.4366	2015.5	3952.5	4516.8	9.4023	1881.1	3952.4	4516.7	9.3705	950
2259.	4052.4	4639.8	9.5351	2098.	4052.3	4639.8	9.5009	1958.1	4052.3	4639.7	9.4690	1000
2437.	4256.9	4890.5	9.7246	2263.	4256.8	4890.5	9.6904	2112.1	4256.8	4890.4	9.6585	1100
2615.	4467.3	5147.2	9.9050	2428.	4467.3	5147.1	9.8708	2266.1	4467.2	5147.1	9.8389	1200
2792.	4683.1	5409.1	10.0771	2593.	4683.0	5409.1	10.0428	2420.1	4683.0	5409.0	10.0110	1300

Definitions of symbols on page 1.

(27)

Table 3. Vapor

p (t Sat.)	.32 (135.76)				.34 (137.87)				.36 (139.87)			
t	$10^3 v$	u	h	s	$10^3 v$	u	h	s	$10^3 v$	u	h	s
Sat.	570.2	2545.8	2728.3	6.9704	538.7	2547.9	2731.1	6.9502	510.6	2549.9	2733.7	6.9311
50	*413.8*	*2384.7*	*2517.1*	*6.3881*	*385.7*	*2380.3*	*2511.4*	*6.3457*	*360.6*	*2375.7*	*2505.6*	*6.3045*
55	*424.4*	*2395.1*	*2530.9*	*6.4305*	*396.0*	*2391.0*	*2525.6*	*6.3892*	*370.6*	*2386.7*	*2520.1*	*6.3492*
60	*434.8*	*2405.5*	*2544.6*	*6.4717*	*406.0*	*2401.6*	*2539.6*	*6.4315*	*380.3*	*2397.6*	*2534.5*	*6.3927*
65	*444.9*	*2415.7*	*2558.0*	*6.5117*	*415.7*	*2412.0*	*2553.4*	*6.4726*	*389.8*	*2408.3*	*2548.6*	*6.4347*
70	*454.7*	*2425.7*	*2571.2*	*6.5506*	*425.2*	*2422.3*	*2566.9*	*6.5123*	*399.0*	*2418.9*	*2562.5*	*6.4754*
75	*464.4*	*2435.7*	*2584.3*	*6.5882*	*434.5*	*2432.5*	*2580.2*	*6.5508*	*407.9*	*2429.2*	*2576.1*	*6.5148*
80	*473.8*	*2445.4*	*2597.1*	*6.6248*	*443.6*	*2442.5*	*2593.3*	*6.5881*	*416.7*	*2439.4*	*2589.4*	*6.5528*
85	*483.1*	*2455.1*	*2609.7*	*6.6602*	*452.5*	*2452.3*	*2606.1*	*6.6242*	*425.2*	*2449.5*	*2602.5*	*6.5897*
90	*492.3*	*2464.6*	*2622.1*	*6.6946*	*461.2*	*2462.0*	*2618.8*	*6.6593*	*433.6*	*2459.3*	*2615.4*	*6.6254*
95	*501.2*	*2473.9*	*2634.3*	*6.7281*	*469.8*	*2471.5*	*2631.2*	*6.6933*	*441.8*	*2469.0*	*2628.1*	*6.6600*
100	*510.1*	*2483.1*	*2646.4*	*6.7606*	*478.2*	*2480.8*	*2643.4*	*6.7263*	*449.9*	*2478.5*	*2640.5*	*6.6935*
110	*527.4*	*2501.2*	*2670.0*	*6.8230*	*494.7*	*2499.2*	*2667.4*	*6.7896*	*465.7*	*2497.1*	*2664.8*	*6.7578*
120	*544.3*	*2518.8*	*2693.0*	*6.8823*	*510.8*	*2517.0*	*2690.7*	*6.8497*	*481.0*	*2515.2*	*2688.4*	*6.8187*
130	*560.8*	*2536.0*	*2715.5*	*6.9389*	*526.5*	*2534.5*	*2713.5*	*6.9069*	*496.0*	*2532.9*	*2711.5*	*6.8765*
140	577.1	2552.9	2737.6	6.9931	542.0	2551.5	2735.8	6.9616	510.8	2550.1	2734.0	6.9318
150	593.1	2569.6	2759.4	7.0451	557.2	2568.3	2757.7	7.0141	525.2	2567.0	2756.1	6.9847
160	609.0	2585.9	2780.8	7.0952	572.2	2584.8	2779.4	7.0646	539.4	2583.7	2777.9	7.0355
170	624.6	2602.1	2802.0	7.1436	587.0	2601.1	2800.7	7.1133	553.5	2600.1	2799.4	7.0846
180	640.1	2618.2	2823.0	7.1904	601.6	2617.3	2821.8	7.1604	567.4	2616.3	2820.6	7.1320
190	655.5	2634.1	2843.9	7.2359	616.2	2633.3	2842.8	7.2061	581.2	2632.4	2841.7	7.1779
200	670.8	2649.9	2864.6	7.2801	630.6	2649.1	2863.6	7.2505	594.9	2648.4	2862.5	7.2225
210	686.0	2665.6	2885.2	7.3232	645.0	2664.9	2884.2	7.2938	608.5	2664.2	2883.3	7.2659
220	701.1	2681.3	2905.7	7.3653	659.2	2680.7	2904.8	7.3360	622.0	2680.0	2904.0	7.3083
230	716.1	2697.0	2926.1	7.4063	673.4	2696.4	2925.3	7.3772	635.5	2695.8	2924.5	7.3496
240	731.1	2712.6	2946.5	7.4465	687.6	2712.0	2945.8	7.4174	648.9	2711.5	2945.1	7.3900
250	746.1	2728.2	2966.9	7.4858	701.7	2727.7	2966.2	7.4569	662.2	2727.2	2965.6	7.4295
260	760.9	2743.8	2987.3	7.5244	715.7	2743.3	2986.7	7.4955	675.5	2742.8	2986.0	7.4682
270	775.8	2759.4	3007.7	7.5622	729.7	2759.0	3007.1	7.5334	688.8	2758.5	3006.5	7.5062
280	790.6	2775.0	3028.0	7.5994	743.7	2774.6	3027.5	7.5706	702.0	2774.2	3026.9	7.5435
290	805.4	2790.7	3048.4	7.6359	757.6	2790.3	3047.9	7.6072	715.2	2789.9	3047.3	7.5801
300	820.2	2806.3	3068.8	7.6718	771.5	2806.0	3068.3	7.6432	728.3	2805.6	3067.8	7.6161
320	849.6	2837.8	3109.7	7.7419	799.3	2837.4	3109.2	7.7133	754.6	2837.1	3108.8	7.6864
340	879.0	2869.4	3150.7	7.8099	827.0	2869.1	3150.2	7.7814	780.7	2868.8	3149.8	7.7545
360	908.3	2901.2	3191.8	7.8759	854.6	2900.9	3191.4	7.8475	806.9	2900.6	3191.1	7.8207
380	937.6	2933.1	3233.2	7.9402	882.2	2932.9	3232.8	7.9118	832.9	2932.6	3232.5	7.8851
400	966.8	2965.3	3274.7	8.0029	909.7	2965.1	3274.4	7.9745	858.9	2964.9	3274.1	7.9478
420	996.0	2997.8	3316.5	8.0640	937.2	2997.5	3316.2	8.0357	884.9	2997.3	3315.9	8.0090
440	1025.1	3030.4	3358.4	8.1237	964.6	3030.2	3358.2	8.0954	910.8	3030.0	3357.9	8.0688
460	1054.2	3063.3	3400.6	8.1821	992.0	3063.1	3400.4	8.1538	936.7	3062.9	3400.2	8.1272
480	1083.3	3096.4	3443.1	8.2392	1019.4	3096.3	3442.9	8.2110	962.6	3096.1	3442.6	8.1844
500	1112.4	3129.8	3485.8	8.2951	1046.8	3129.7	3485.6	8.2669	988.5	3129.5	3485.4	8.2404
520	1141.4	3163.5	3528.7	8.3500	1074.1	3163.3	3528.5	8.3218	1014.3	3163.2	3528.3	8.2952
540	1170.4	3197.4	3571.9	8.4037	1101.4	3197.2	3571.7	8.3756	1040.1	3197.1	3571.5	8.3490
560	1199.4	3231.6	3615.4	8.4565	1128.7	3231.4	3615.2	8.4284	1065.9	3231.3	3615.0	8.4019
580	1228.4	3266.0	3659.1	8.5084	1156.0	3265.9	3658.9	8.4802	1091.7	3265.7	3658.8	8.4537
600	1257.4	3300.7	3703.0	8.5593	1183.3	3300.6	3702.9	8.5312	1117.5	3300.4	3702.7	8.5047
620	1286.4	3335.6	3747.3	8.6094	1210.6	3335.5	3747.1	8.5813	1143.2	3335.4	3747.0	8.5548
640	1315.3	3370.9	3791.8	8.6587	1237.9	3370.8	3791.7	8.6306	1169.0	3370.7	3791.5	8.6041
660	1344.3	3406.4	3836.6	8.7072	1265.1	3406.3	3836.4	8.6791	1194.8	3406.2	3836.3	8.6526
680	1373.2	3442.2	3881.6	8.7549	1292.4	3442.1	3881.5	8.7269	1220.5	3442.0	3881.4	8.7004
700	1402.2	3478.2	3926.9	8.8020	1319.6	3478.2	3926.8	8.7739	1246.2	3478.1	3926.7	8.7475
750	1474.5	3569.6	4041.4	8.9167	1387.7	3569.5	4041.3	8.8887	1310.5	3569.4	4041.2	8.8622
800	1546.8	3662.7	4157.7	9.0277	1455.7	3662.6	4157.6	8.9996	1374.8	3662.6	4157.5	8.9732
850	1619.0	3757.6	4275.7	9.1351	1523.7	3757.5	4275.6	9.1071	1439.0	3757.5	4275.5	9.0807
900	1691.3	3854.2	4395.4	9.2394	1591.7	3854.1	4395.3	9.2113	1503.3	3854.0	4395.2	9.1849
950	1763.5	3952.4	4516.7	9.3406	1659.7	3952.3	4516.6	9.3126	1567.5	3952.3	4516.6	9.2862
1000	1835.7	4052.2	4639.7	9.4392	1727.7	4052.2	4639.6	9.4111	1631.7	4052.1	4639.5	9.3847
1100	1980.1	4256.4	4890.4	9.6287	1863.6	4256.7	4890.3	9.6007	1760.1	4256.6	4890.3	9.5743
1200	2124.5	4467.2	5147.0	9.8091	1999.5	4467.1	5147.0	9.7811	1888.4	4467.1	5146.9	9.7547
1300	2268.9	4683.0	5409.0	9.9811	2135.4	4682.9	5408.9	9.9531	2016.8	4682.9	5408.9	9.9267

Definitions of symbols on page 1.

$10^3 v$	u	h	s	$10^3 v$	u	h	s	$10^3 v$	u	h	s	t
485.3	2551.8	2736.2	6.9130	462.5	2553.6	2738.6	6.8959	441.7	2555.3	2740.8	6.8796	Sat.
338.1	2371.1	2499.5	6.2642	317.7	2366.3	2493.4	6.2248	299.1	2361.4	2487.0	6.1859	50
347.9	2382.4	2514.6	6.3103	327.3	2377.9	2508.8	6.2722	308.5	2373.3	2502.9	6.2349	55
357.3	2393.5	2529.3	6.3549	336.5	2389.4	2524.0	6.3181	317.6	2385.1	2518.5	6.2821	60
366.5	2404.5	2543.8	6.3981	345.5	2400.6	2538.8	6.3624	326.4	2396.7	2533.8	6.3276	65
375.4	2415.3	2558.0	6.4397	354.1	2411.7	2553.4	6.4051	334.8	2408.1	2548.7	6.3713	70
384.1	2425.9	2571.9	6.4800	362.6	2422.6	2567.6	6.4463	343.0	2419.2	2563.3	6.4135	75
392.5	2436.4	2585.5	6.5189	370.8	2433.2	2581.5	6.4860	351.0	2430.1	2577.5	6.4541	80
400.8	2446.6	2598.9	6.5564	378.7	2443.7	2595.2	6.5243	358.8	2440.7	2591.4	6.4932	85
408.9	2456.6	2612.0	6.5928	386.6	2453.9	2608.5	6.5614	366.4	2451.2	2605.0	6.5310	90
416.8	2466.5	2624.9	6.6280	394.2	2464.0	2621.6	6.5972	373.8	2461.4	2618.4	6.5674	95
424.5	2476.2	2637.5	6.6621	401.7	2473.8	2634.5	6.6319	381.0	2471.4	2631.4	6.6027	100
439.7	2495.1	2662.2	6.7273	416.3	2493.0	2659.5	6.6981	395.1	2490.9	2656.8	6.6699	110
454.4	2513.4	2686.1	6.7890	430.4	2511.6	2683.8	6.7605	408.7	2509.7	2681.4	6.7332	120
468.8	2531.3	2709.4	6.8475	444.2	2529.7	2707.3	6.8197	421.9	2528.0	2705.2	6.7931	130
482.8	2548.7	2732.2	6.9033	457.6	2547.3	2730.3	6.8761	434.9	2545.8	2728.5	6.8500	140
496.6	2565.8	2754.5	6.9567	470.8	2564.5	2752.8	6.9299	447.5	2563.2	2751.2	6.9043	150
510.2	2582.6	2776.4	7.0079	483.8	2581.4	2774.9	6.9815	460.0	2580.3	2773.4	6.9563	160
523.6	2599.1	2798.0	7.0573	496.6	2598.1	2796.7	7.0312	472.2	2597.0	2795.4	7.0063	170
536.8	2615.4	2819.4	7.1049	509.3	2614.5	2818.2	7.0792	484.3	2613.5	2817.0	7.0546	180
550.0	2631.6	2840.6	7.1511	521.8	2630.7	2839.5	7.1256	496.3	2629.9	2838.3	7.1012	190
563.0	2647.6	2861.5	7.1959	534.2	2646.8	2860.5	7.1706	508.2	2646.1	2859.5	7.1464	200
575.9	2663.5	2882.4	7.2395	546.5	2662.8	2881.4	7.2144	520.0	2662.1	2880.5	7.1904	210
588.7	2679.3	2903.1	7.2820	558.8	2678.7	2902.3	7.2570	531.7	2678.1	2901.4	7.2331	220
601.5	2695.2	2923.8	7.3234	570.9	2694.6	2923.0	7.2986	543.3	2694.0	2922.2	7.2748	230
614.2	2710.9	2944.3	7.3639	583.1	2710.4	2943.6	7.3392	554.9	2709.8	2942.8	7.3156	240
626.9	2726.6	2964.9	7.4036	595.1	2726.1	2964.2	7.3789	566.4	2725.6	2963.5	7.3554	250
639.5	2742.3	2985.4	7.4424	607.1	2741.9	2984.7	7.4178	577.8	2741.4	2984.1	7.3944	260
652.1	2758.1	3005.8	7.4804	619.1	2757.6	3005.2	7.4559	589.3	2757.1	3004.6	7.4326	270
664.6	2773.8	3026.3	7.5178	631.0	2773.3	3025.7	7.4933	600.7	2772.9	3025.2	7.4700	280
677.2	2789.5	3046.8	7.5545	643.0	2789.1	3046.3	7.5301	612.0	2788.7	3045.7	7.5068	290
689.6	2805.2	3067.3	7.5905	654.8	2804.8	3066.8	7.5662	623.3	2804.5	3066.3	7.5430	300
714.6	2836.8	3108.3	7.6609	678.5	2836.4	3107.8	7.6366	645.9	2836.1	3107.4	7.6135	320
739.4	2868.5	3149.4	7.7291	702.1	2868.2	3149.0	7.7049	668.5	2867.9	3148.6	7.6819	340
764.1	2900.3	3190.7	7.7953	725.7	2900.0	3190.3	7.7712	690.9	2899.8	3189.9	7.7482	360
788.8	2932.4	3232.1	7.8597	749.2	2932.1	3231.8	7.8357	713.3	2931.9	3231.5	7.8128	380
813.5	2964.6	3273.8	7.9225	772.6	2964.4	3273.4	7.8985	735.6	2964.2	3273.1	7.8756	400
838.1	2997.1	3315.6	7.9837	796.0	2996.9	3315.3	7.9598	757.9	2996.7	3315.0	7.9369	420
862.7	3029.8	3357.6	8.0435	819.4	3029.6	3357.4	8.0196	780.2	3029.4	3357.1	7.9968	440
887.3	3062.7	3399.9	8.1020	842.7	3062.6	3399.7	8.0781	802.5	3062.4	3399.4	8.0553	460
911.8	3095.9	3442.4	8.1592	866.1	3095.8	3442.2	8.1353	824.7	3095.6	3441.9	8.1125	480
936.3	3129.4	3485.1	8.2152	889.3	3129.2	3484.9	8.1913	846.9	3129.0	3484.7	8.1686	500
960.8	3163.0	3528.1	8.2701	912.6	3162.9	3527.9	8.2462	869.0	3162.7	3527.7	8.2235	520
985.3	3197.0	3571.4	8.3239	935.9	3196.8	3571.2	8.3001	891.2	3196.7	3571.0	8.2774	540
1009.7	3231.2	3614.8	8.3767	959.1	3231.0	3614.7	8.3529	913.3	3230.9	3614.5	8.3302	560
1034.1	3265.6	3658.6	8.4286	982.3	3265.5	3658.4	8.4048	935.5	3265.4	3658.3	8.3821	580
1058.6	3300.3	3702.6	8.4796	1005.5	3300.2	3702.4	8.4558	957.6	3300.1	3702.3	8.4332	600
1083.0	3335.3	3746.9	8.5297	1028.7	3335.2	3746.7	8.5059	979.7	3335.1	3746.6	8.4833	620
1107.4	3370.6	3791.4	8.5790	1051.9	3370.5	3791.3	8.5553	1001.8	3370.4	3791.1	8.5326	640
1131.8	3406.1	3836.2	8.6276	1075.1	3406.0	3836.1	8.6038	1023.9	3405.9	3835.9	8.5812	660
1156.2	3441.9	3881.3	8.6754	1098.3	3441.8	3881.1	8.6516	1045.9	3441.7	3881.0	8.6290	680
1180.6	3478.0	3926.6	8.7224	1121.5	3477.9	3926.5	8.6987	1068.0	3477.8	3926.4	8.6761	700
1241.5	3569.4	4041.1	8.8372	1179.4	3569.3	4041.0	8.8134	1123.1	3569.2	4040.9	8.7909	750
1302.4	3662.5	4157.4	8.9481	1237.2	3662.4	4157.3	8.9244	1178.3	3662.4	4157.2	8.9018	800
1363.3	3757.4	4275.5	9.0556	1295.1	3757.4	4275.4	9.0319	1233.3	3757.3	4275.3	9.0093	850
1424.1	3854.0	4395.1	9.1599	1352.9	3853.9	4395.1	9.1362	1288.4	3853.9	4395.0	9.1136	900
1485.0	3952.2	4516.5	9.2612	1410.7	3952.2	4516.4	9.2375	1343.5	3952.1	4516.4	9.2149	950
1545.8	4052.1	4639.5	9.3597	1468.5	4052.0	4639.4	9.3360	1398.5	4052.0	4639.4	9.3135	1000
1667.4	4256.6	4890.2	9.5493	1584.0	4256.5	4890.2	9.5256	1508.6	4256.5	4890.1	9.5030	1100
1789.0	4467.0	5146.9	9.7297	1699.6	4467.0	5146.8	9.7060	1618.6	4467.0	5146.8	9.6834	1200
1910.6	4682.8	5408.8	9.9017	1815.1	4682.8	5408.8	9.8780	1728.7	4682.7	5408.8	9.8555	1300

Definitions of symbols on page 1.

Table 3. Vapor

t	$10^3 v$	u	h	s	$10^3 v$	u	h	s	$10^3 v$	u	h	s
		.44 (147.10)				.46 (148.74)				.48 (150.32)		
Sat.	422.8	2556.9	2742.9	6.8641	405.5	2558.4	2744.9	6.8492	389.6	2559.8	2746.8	6.8349
100	362.2	2469.0	2628.3	6.5745	344.9	2466.5	2625.2	6.5471	329.1	2464.0	2622.0	6.5205
105	369.0	2479.0	2641.3	6.6091	351.6	2476.7	2638.4	6.5823	335.6	2474.3	2635.4	6.5563
110	375.8	2488.8	2654.1	6.6427	358.2	2486.6	2651.4	6.6164	342.0	2484.4	2648.6	6.5908
115	382.4	2498.4	2666.7	6.6752	364.6	2496.4	2664.1	6.6494	348.2	2494.3	2661.5	6.6243
120	388.9	2507.9	2679.0	6.7068	370.9	2506.0	2676.6	6.6814	354.3	2504.1	2674.2	6.6567
125	395.4	2517.2	2691.2	6.7375	377.1	2515.4	2688.9	6.7124	360.3	2513.6	2686.6	6.6882
130	401.7	2526.4	2703.1	6.7674	383.2	2524.7	2701.0	6.7427	366.3	2523.0	2698.8	6.7187
135	408.0	2535.4	2714.9	6.7965	389.3	2533.9	2712.9	6.7721	372.1	2532.3	2710.9	6.7485
140	414.1	2544.4	2726.6	6.8249	395.2	2542.9	2724.7	6.8007	377.9	2541.4	2722.8	6.7774
145	420.3	2553.2	2738.1	6.8526	401.1	2551.8	2736.3	6.8287	383.6	2550.4	2734.5	6.8056
150	426.3	2561.9	2749.5	6.8797	407.0	2560.6	2747.8	6.8560	389.2	2559.3	2746.1	6.8332
160	438.3	2579.1	2771.9	6.9321	418.5	2577.9	2770.4	6.9088	400.3	2576.7	2768.9	6.8864
170	450.1	2596.0	2794.0	6.9825	429.8	2594.9	2792.6	6.9595	411.2	2593.9	2791.3	6.9375
180	461.7	2612.6	2815.7	7.0310	441.0	2611.7	2814.5	7.0083	422.0	2610.7	2813.3	6.9866
190	473.2	2629.0	2837.2	7.0779	452.0	2628.2	2836.1	7.0555	432.6	2627.3	2835.0	7.0339
200	484.5	2645.3	2858.5	7.1233	462.9	2644.5	2857.5	7.1011	443.1	2643.7	2856.4	7.0798
210	495.8	2661.4	2879.6	7.1674	473.8	2660.7	2878.6	7.1454	453.6	2660.0	2877.7	7.1242
220	507.0	2677.4	2900.5	7.2103	484.5	2676.8	2899.7	7.1885	463.9	2676.1	2898.8	7.1675
230	518.1	2693.4	2921.4	7.2521	495.2	2692.8	2920.6	7.2304	474.1	2692.2	2919.7	7.2096
240	529.2	2709.3	2942.1	7.2930	505.8	2708.7	2941.4	7.2714	484.3	2708.1	2940.6	7.2506
250	540.2	2725.1	2962.8	7.3329	516.4	2724.6	2962.1	7.3114	494.5	2724.0	2961.4	7.2907
260	551.2	2740.9	2983.4	7.3720	526.9	2740.4	2982.8	7.3505	504.6	2739.9	2982.1	7.3300
270	562.1	2756.7	3004.0	7.4103	537.3	2756.2	3003.4	7.3889	514.6	2755.8	3002.8	7.3684
280	573.0	2772.5	3024.6	7.4478	547.8	2772.0	3024.0	7.4265	524.7	2771.6	3023.5	7.4061
290	583.9	2788.3	3045.2	7.4847	558.2	2787.9	3044.6	7.4634	534.6	2787.5	3044.1	7.4431
300	594.7	2804.1	3065.8	7.5209	568.6	2803.7	3065.2	7.4997	544.6	2803.3	3064.7	7.4794
310	605.5	2819.9	3086.3	7.5565	578.9	2819.5	3085.8	7.5353	554.5	2819.2	3085.4	7.5151
320	616.3	2835.8	3106.9	7.5915	589.3	2835.4	3106.5	7.5704	564.5	2835.1	3106.0	7.5502
330	627.1	2851.6	3127.5	7.6260	599.6	2851.3	3127.1	7.6049	574.4	2851.0	3126.7	7.5847
340	637.8	2867.5	3148.2	7.6599	609.9	2867.2	3147.8	7.6389	584.2	2866.9	3147.4	7.6187
350	648.6	2883.5	3168.9	7.6933	620.1	2883.2	3168.5	7.6724	594.1	2882.9	3168.1	7.6522
360	659.3	2899.5	3189.6	7.7263	630.4	2899.2	3189.2	7.7054	603.9	2898.9	3188.8	7.6853
370	670.0	2915.5	3210.3	7.7588	640.6	2915.3	3210.0	7.7379	613.8	2915.0	3209.6	7.7178
380	680.7	2931.6	3231.1	7.7909	650.9	2931.4	3230.8	7.7700	623.6	2931.1	3230.4	7.7500
390	691.3	2947.7	3251.9	7.8226	661.1	2947.5	3251.6	7.8017	633.4	2947.3	3251.3	7.7817
400	702.0	2963.9	3272.8	7.8538	671.3	2963.7	3272.5	7.8329	643.2	2963.5	3272.2	7.8130
420	723.3	2996.5	3314.7	7.9151	691.7	2996.2	3314.4	7.8943	662.7	2996.0	3314.1	7.8744
440	744.6	3029.2	3356.8	7.9750	712.1	3029.0	3356.6	7.9542	682.3	3028.8	3356.3	7.9343
460	765.8	3062.2	3399.2	8.0336	732.4	3062.0	3398.9	8.0128	701.8	3061.8	3398.7	7.9929
480	787.1	3095.4	3441.7	8.0908	752.7	3095.2	3441.5	8.0701	721.2	3095.1	3441.3	8.0502
500	808.2	3128.9	3484.5	8.1469	773.0	3128.7	3484.3	8.1262	740.7	3128.6	3484.1	8.1063
520	829.4	3162.6	3527.5	8.2019	793.3	3162.4	3527.3	8.1812	760.1	3162.3	3527.1	8.1613
540	850.6	3196.5	3570.6	8.2557	813.5	3196.4	3570.6	8.2351	779.5	3196.3	3570.4	8.2152
560	871.7	3230.8	3614.3	8.3086	833.7	3230.6	3614.2	8.2879	798.9	3230.5	3614.0	8.2681
580	892.8	3265.2	3658.1	8.3605	853.9	3265.1	3657.9	8.3399	818.3	3265.0	3657.8	8.3201
600	914.0	3300.0	3702.1	8.4116	874.1	3299.9	3702.0	8.3909	837.6	3299.8	3701.8	8.3711
620	935.1	3335.0	3746.4	8.4617	894.3	3334.9	3746.3	8.4411	857.0	3334.8	3746.1	8.4213
640	956.2	3370.3	3791.0	8.5110	914.5	3370.2	3790.8	8.4904	876.3	3370.1	3790.7	8.4707
660	977.2	3405.8	3835.8	8.5596	934.7	3405.7	3835.7	8.5390	895.7	3405.6	3835.6	8.5192
680	998.3	3441.6	3880.9	8.6074	954.9	3441.5	3880.8	8.5868	915.0	3441.4	3880.7	8.5671
700	1019.4	3477.7	3926.3	8.6545	975.0	3477.6	3926.1	8.6339	934.3	3477.5	3926.0	8.6142
750	1072.0	3569.1	4040.8	8.7693	1025.4	3569.1	4040.7	8.7487	982.6	3569.0	4040.6	8.7290
800	1124.7	3662.3	4157.2	8.8803	1075.7	3662.2	4157.1	8.8597	1030.9	3662.2	4157.0	8.8400
850	1177.2	3757.2	4275.2	8.9878	1126.0	3757.2	4275.1	8.9672	1079.1	3757.1	4275.1	8.9476
900	1229.8	3853.8	4394.9	9.0921	1176.3	3853.8	4394.9	9.0715	1127.3	3853.7	4394.8	9.0518
950	1282.4	3952.0	4516.3	9.1934	1226.6	3952.0	4516.2	9.1728	1175.5	3951.9	4516.2	9.1531
1000	1334.9	4051.9	4639.3	9.2919	1276.9	4051.9	4639.2	9.2714	1223.7	4051.8	4639.2	9.2517
1100	1440.0	4256.4	4890.0	9.4815	1377.4	4256.4	4890.0	9.4610	1320.0	4256.3	4889.9	9.4413
1200	1545.1	4466.9	5146.7	9.6619	1477.9	4466.9	5146.7	9.6414	1416.3	4466.8	5146.6	9.6217
1300	1650.1	4682.7	5408.7	9.8340	1578.4	4682.6	5408.7	9.8134	1512.6	4682.6	5408.6	9.7938

Definitions of symbols on page 1.

	.50 (151.86)				.52 (153.34)				.54 (154.78)			p (t Sat.)
10^3 v	u	h	s	10^3 v	u	h	s	10^3 v	u	h	s	t
374.9	2561.2	2748.7	6.8213	361.3	2562.6	2750.5	6.8081	348.7	2563.9	2752.1	6.7955	Sat.
314.6	2461.5	2618.7	6.4945	301.1	2458.9	2615.5	6.4692	288.6	2456.3	2612.1	6.4445	100
320.9	2472.0	2632.4	6.5309	307.3	2469.6	2629.3	6.5062	294.7	2467.1	2626.2	6.4821	105
327.1	2482.2	2645.8	6.5660	313.3	2480.0	2642.9	6.5419	300.5	2477.7	2640.0	6.5183	110
333.1	2492.3	2658.9	6.6000	319.2	2490.2	2656.2	6.5763	306.3	2488.1	2653.5	6.5532	115
339.1	2502.1	2671.7	6.6328	325.0	2500.2	2669.2	6.6096	311.9	2498.2	2666.7	6.5870	120
344.9	2511.8	2684.3	6.6647	330.7	2510.0	2681.9	6.6418	317.5	2508.1	2679.6	6.6196	125
350.7	2521.3	2696.7	6.6956	336.2	2519.6	2694.5	6.6731	322.9	2517.9	2692.3	6.6512	130
356.3	2530.7	2708.9	6.7256	341.7	2529.1	2706.8	6.7034	328.2	2527.5	2704.7	6.6819	135
361.9	2539.9	2720.9	6.7548	347.2	2538.4	2718.9	6.7330	333.5	2536.9	2717.0	6.7118	140
367.4	2549.0	2732.7	6.7833	352.5	2547.6	2730.9	6.7617	338.7	2546.1	2729.0	6.7408	145
372.9	2557.9	2744.4	6.8111	357.8	2556.6	2742.7	6.7898	343.8	2555.3	2740.9	6.7691	150
383.6	2575.6	2767.4	6.8648	368.2	2574.4	2765.8	6.8439	353.9	2573.2	2764.3	6.8236	160
394.2	2592.8	2789.9	6.9162	378.4	2591.7	2788.5	6.8956	363.8	2590.6	2787.1	6.8757	170
404.5	2609.7	2812.0	6.9656	388.4	2608.8	2810.8	6.9453	373.5	2607.8	2809.5	6.9257	180
414.8	2626.4	2833.8	7.0132	398.3	2625.6	2832.7	6.9932	383.1	2624.7	2831.5	6.9738	190
424.9	2642.9	2855.4	7.0592	408.1	2642.1	2854.3	7.0394	392.5	2641.3	2853.3	7.0203	200
435.0	2659.3	2876.7	7.1039	417.8	2658.5	2875.8	7.0843	401.9	2657.8	2874.8	7.0653	210
444.9	2675.5	2897.9	7.1473	427.4	2674.8	2897.0	7.1278	411.2	2674.1	2896.1	7.1090	220
454.8	2691.5	2918.9	7.1895	436.9	2690.9	2918.1	7.1702	420.4	2690.3	2917.3	7.1515	230
464.6	2707.6	2939.9	7.2307	446.4	2707.0	2939.1	7.2115	429.5	2706.4	2938.4	7.1929	240
474.4	2723.5	2960.7	7.2709	455.8	2723.0	2960.0	7.2518	438.6	2722.5	2959.3	7.2333	250
484.1	2739.4	2981.5	7.3102	465.1	2738.9	2980.8	7.2912	447.6	2738.5	2980.2	7.2728	260
493.7	2755.3	3002.2	7.3487	474.5	2754.9	3001.6	7.3298	456.6	2754.4	3001.0	7.3115	270
503.4	2771.2	3022.9	7.3865	483.7	2770.8	3022.3	7.3676	465.5	2770.3	3021.7	7.3494	280
513.0	2787.1	3043.6	7.4235	493.0	2786.7	3043.0	7.4047	474.5	2786.3	3042.5	7.3865	290
522.6	2802.9	3064.2	7.4599	502.2	2802.6	3063.7	7.4411	483.4	2802.2	3063.2	7.4230	300
532.1	2818.8	3084.9	7.4956	511.4	2818.5	3084.4	7.4769	492.2	2818.1	3083.9	7.4589	310
541.6	2834.7	3105.6	7.5308	520.6	2834.4	3105.1	7.5121	501.1	2834.0	3104.6	7.4941	320
551.2	2850.7	3126.2	7.5653	529.7	2850.3	3125.8	7.5467	509.9	2850.0	3125.4	7.5287	330
560.6	2866.6	3146.9	7.5994	538.9	2866.3	3146.5	7.5808	518.7	2866.0	3146.1	7.5629	340
570.1	2882.6	3167.7	7.6329	548.0	2882.3	3167.3	7.6144	527.5	2882.0	3166.9	7.5965	350
579.6	2898.7	3188.4	7.6660	557.1	2898.4	3188.1	7.6474	536.3	2898.1	3187.7	7.6296	360
589.0	2914.7	3209.2	7.6986	566.2	2914.5	3208.9	7.6801	545.0	2914.2	3208.5	7.6622	370
598.5	2930.8	3230.1	7.7307	575.3	2930.6	3229.7	7.7122	553.8	2930.3	3229.4	7.6944	380
607.9	2947.0	3251.0	7.7625	584.3	2946.8	3250.6	7.7440	562.5	2946.5	3250.3	7.7262	390
617.3	2963.2	3271.9	7.7938	593.4	2963.0	3271.6	7.7753	571.3	2962.8	3271.2	7.7575	400
636.1	2995.8	3313.8	7.8552	611.5	2995.6	3313.6	7.8368	588.7	2995.4	3313.3	7.8191	420
654.8	3028.6	3356.0	7.9152	629.5	3028.4	3355.8	7.8968	606.1	3028.2	3355.5	7.8791	440
673.6	3061.6	3398.4	7.9738	647.5	3061.5	3398.2	7.9555	623.4	3061.3	3397.9	7.9378	460
692.3	3094.9	3441.0	8.0312	665.5	3094.7	3440.8	8.0128	640.8	3094.6	3440.6	7.9952	480
710.9	3128.4	3483.9	8.0873	683.5	3128.2	3483.6	8.0690	658.1	3128.1	3483.4	8.0514	500
729.6	3162.1	3526.9	8.1423	701.4	3162.0	3526.7	8.1240	675.4	3161.8	3526.5	8.1064	520
748.2	3196.1	3570.2	8.1962	719.4	3196.0	3570.1	8.1780	692.6	3195.9	3569.9	8.1604	540
766.9	3230.4	3613.8	8.2491	737.3	3230.2	3613.6	8.2309	709.9	3230.1	3613.5	8.2133	560
785.5	3264.9	3657.6	8.3011	755.2	3264.8	3657.5	8.2829	727.1	3264.6	3657.3	8.2653	580
804.1	3299.6	3701.7	8.3522	773.1	3299.5	3701.5	8.3339	744.4	3299.4	3701.4	8.3164	600
822.7	3334.7	3746.0	8.4023	790.9	3334.6	3745.9	8.3841	761.6	3334.5	3745.7	8.3666	620
841.2	3370.0	3790.6	8.4517	808.8	3369.9	3790.4	8.4335	778.8	3369.8	3790.3	8.4160	640
859.8	3405.5	3835.4	8.5003	826.7	3405.4	3835.3	8.4821	796.0	3405.3	3835.2	8.4646	660
878.4	3441.4	3880.5	8.5481	844.5	3441.3	3880.4	8.5299	813.2	3441.2	3880.3	8.5124	680
896.9	3477.5	3925.9	8.5952	862.4	3477.4	3925.8	8.5770	830.4	3477.3	3925.7	8.5595	700
943.3	3568.9	4040.5	8.7101	906.9	3568.8	4040.4	8.6919	873.3	3568.8	4040.3	8.6744	750
989.6	3662.1	4156.9	8.8211	951.5	3662.0	4156.8	8.8029	916.2	3662.0	4156.7	8.7855	800
1035.9	3757.0	4275.0	8.9287	996.0	3757.0	4274.9	8.9105	959.1	3756.9	4274.8	8.8930	850
1082.2	3853.6	4394.7	9.0329	1040.5	3853.6	4394.6	9.0148	1002.0	3853.5	4394.6	8.9973	900
1128.4	3951.9	4516.1	9.1343	1085.0	3951.8	4516.0	9.1161	1044.8	3951.8	4516.0	9.0987	950
1174.7	4051.8	4639.1	9.2328	1129.5	4051.7	4639.1	9.2147	1087.6	4051.7	4639.0	9.1972	1000
1267.2	4256.3	4889.9	9.4224	1218.4	4256.3	4889.8	9.4043	1173.3	4256.2	4889.8	9.3868	1100
1359.6	4466.8	5146.6	9.6029	1307.3	4466.7	5146.5	9.5847	1258.9	4466.7	5146.5	9.5673	1200
1452.1	4682.5	5408.6	9.7749	1396.3	4682.5	5408.5	9.7568	1344.5	4682.4	5408.5	9.7393	1300

Definitions of symbols on page 1.

(31)

Table 3. Vapor

p (t Sat.)	.56 (156.17)				.58 (157.53)				.60 (158.85)			
t	$10^3 v$	u	h	s	$10^3 v$	u	h	s	$10^3 v$	u	h	s
Sat.	336.9	2565.1	2753.8	6.7832	325.9	2566.3	2755.3	6.7714	315.7	2567.4	2756.8	6.7600
100	277.0	2453.6	2608.7	6.4203	266.2	2450.9	2605.3	6.3965	256.0	2448.2	2601.8	6.3732
105	282.9	2464.7	2623.1	6.4585	272.0	2462.2	2619.9	6.4354	261.7	2459.6	2616.7	6.4127
110	288.7	2475.4	2637.1	6.4953	277.6	2473.1	2634.1	6.4728	267.3	2470.8	2631.1	6.4507
115	294.3	2486.0	2650.8	6.5307	283.1	2483.8	2648.0	6.5087	272.7	2481.6	2645.2	6.4872
120	299.8	2496.2	2664.1	6.5649	288.5	2494.2	2661.5	6.5434	277.9	2492.2	2658.9	6.5224
125	305.2	2506.3	2677.2	6.5980	293.8	2504.4	2674.8	6.5769	283.1	2502.5	2672.3	6.5563
130	310.5	2516.1	2690.0	6.6300	298.9	2514.4	2687.8	6.6093	288.1	2512.6	2685.5	6.5890
135	315.7	2525.8	2702.6	6.6610	304.0	2524.2	2700.5	6.6406	293.1	2522.5	2698.3	6.6208
140	320.8	2535.3	2715.0	6.6912	309.0	2533.8	2713.0	6.6711	297.9	2532.2	2711.0	6.6515
145	325.9	2544.7	2727.2	6.7205	313.9	2543.2	2725.3	6.7007	302.7	2541.7	2723.4	6.6814
150	330.8	2553.9	2739.2	6.7490	318.7	2552.5	2737.4	6.7295	307.5	2551.1	2735.6	6.7105
160	340.6	2571.9	2762.7	6.8040	328.3	2570.7	2761.1	6.7849	316.7	2569.5	2759.5	6.7663
170	350.2	2589.6	2785.7	6.8564	337.6	2588.5	2784.3	6.8377	325.8	2587.4	2782.8	6.8195
180	359.6	2606.8	2808.2	6.9067	346.7	2605.8	2806.9	6.8883	334.7	2604.9	2805.7	6.8705
190	368.9	2623.8	2830.4	6.9551	355.7	2622.9	2829.2	6.9370	343.4	2622.0	2828.1	6.9194
200	378.1	2640.5	2852.2	7.0018	364.6	2639.7	2851.2	6.9839	352.0	2638.9	2850.1	6.9665
210	387.1	2657.1	2873.9	7.0470	373.4	2656.3	2872.9	7.0293	360.5	2655.6	2871.9	7.0121
220	396.1	2673.4	2895.3	7.0909	382.1	2672.8	2894.4	7.0733	369.0	2672.1	2893.5	7.0562
230	405.0	2689.7	2916.5	7.1335	390.7	2689.1	2915.7	7.1160	377.3	2688.5	2914.8	7.0991
240	413.8	2705.9	2937.6	7.1750	399.2	2705.3	2936.8	7.1577	385.6	2704.7	2936.1	7.1409
250	422.6	2721.9	2958.6	7.2155	407.7	2721.4	2957.9	7.1983	393.8	2720.9	2957.2	7.1816
260	431.3	2738.0	2979.5	7.2551	416.2	2737.5	2978.8	7.2380	402.0	2737.0	2978.2	7.2214
270	440.0	2753.9	3000.3	7.2939	424.6	2753.5	2999.7	7.2768	410.2	2753.0	2999.1	7.2603
280	448.7	2769.9	3021.2	7.3318	432.9	2769.5	3020.6	7.3148	418.3	2769.0	3020.0	7.2984
290	457.3	2785.8	3041.9	7.3690	441.3	2785.4	3041.4	7.3521	426.3	2785.0	3040.8	7.3357
300	465.9	2801.8	3062.7	7.4056	449.6	2801.4	3062.2	7.3887	434.4	2801.0	3061.6	7.3724
310	474.4	2817.7	3083.4	7.4415	457.9	2817.4	3082.9	7.4246	442.4	2817.0	3082.5	7.4084
320	483.0	2833.7	3104.2	7.4767	466.1	2833.4	3103.7	7.4600	450.4	2833.0	3103.2	7.4437
330	491.5	2849.7	3124.9	7.5114	474.4	2849.4	3124.5	7.4947	458.4	2849.0	3124.0	7.4785
340	500.0	2865.7	3145.7	7.5456	482.6	2865.4	3145.3	7.5289	466.3	2865.1	3144.9	7.5127
350	508.5	2881.7	3166.5	7.5792	490.8	2881.4	3166.1	7.5625	474.2	2881.2	3165.7	7.5464
360	517.0	2897.8	3187.3	7.6124	499.0	2897.5	3186.9	7.5957	482.2	2897.3	3186.6	7.5796
370	525.4	2913.9	3208.2	7.6450	507.1	2913.7	3207.8	7.6284	490.1	2913.4	3207.4	7.6124
380	533.9	2930.1	3229.0	7.6773	515.3	2929.8	3228.7	7.6607	498.0	2929.6	3228.4	7.6446
390	542.3	2946.3	3250.0	7.7090	523.4	2946.0	3249.6	7.6925	505.8	2945.8	3249.3	7.6765
400	550.7	2962.5	3270.9	7.7404	531.6	2962.3	3270.6	7.7239	513.7	2962.1	3270.3	7.7079
420	567.5	2995.2	3313.0	7.8020	547.8	2995.0	3312.7	7.7855	529.4	2994.7	3312.4	7.7695
440	584.3	3028.0	3355.2	7.8621	564.0	3027.8	3355.0	7.8456	545.1	3027.6	3354.7	7.8297
460	601.0	3061.1	3397.7	7.9208	580.2	3060.9	3397.4	7.9043	560.8	3060.7	3397.2	7.8884
480	617.8	3094.4	3440.3	7.9782	596.4	3094.2	3440.1	7.9617	576.4	3094.0	3439.9	7.9459
500	634.5	3127.9	3483.2	8.0344	612.5	3127.8	3483.0	8.0180	592.0	3127.6	3482.8	8.0021
520	651.2	3161.7	3526.3	8.0894	628.6	3161.5	3526.1	8.0730	607.6	3161.4	3525.9	8.0572
540	667.8	3195.7	3569.7	8.1434	644.7	3195.6	3569.5	8.1270	623.1	3195.4	3569.3	8.1112
560	684.5	3230.0	3613.3	8.1964	660.8	3229.9	3613.1	8.1800	638.7	3229.7	3612.9	8.1642
580	701.1	3264.5	3657.1	8.2484	676.8	3264.4	3657.0	8.2320	654.2	3264.3	3656.8	8.2162
600	717.7	3299.3	3701.2	8.2995	692.9	3299.2	3701.1	8.2831	669.7	3299.1	3700.9	8.2674
620	734.3	3334.3	3745.6	8.3497	708.9	3334.2	3745.4	8.3334	685.3	3334.1	3745.3	8.3176
640	750.9	3369.7	3790.2	8.3991	725.0	3369.6	3790.0	8.3828	700.8	3369.5	3789.9	8.3670
660	767.5	3405.2	3835.0	8.4477	741.0	3405.1	3834.9	8.4314	716.2	3405.0	3834.8	8.4156
680	784.1	3441.1	3880.2	8.4955	757.0	3441.0	3880.1	8.4792	731.7	3440.9	3879.9	8.4635
700	800.7	3477.2	3925.6	8.5427	773.0	3477.1	3925.5	8.5264	747.2	3477.0	3925.3	8.5107
750	842.1	3568.7	4040.2	8.6576	813.0	3568.6	4040.1	8.6413	785.9	3568.5	4040.1	8.6256
800	883.5	3661.9	4156.6	8.7686	853.0	3661.8	4156.5	8.7524	824.5	3661.8	4156.5	8.7367
850	924.8	3756.9	4274.7	8.8762	892.9	3756.8	4274.7	8.8599	863.1	3756.7	4274.6	8.8442
900	966.1	3853.5	4394.5	8.9805	932.8	3853.4	4394.4	8.9643	901.7	3853.4	4394.4	8.9486
950	1007.5	3951.7	4515.9	9.0818	972.7	3951.7	4515.8	9.0656	940.3	3951.6	4515.8	9.0499
1000	1048.8	4051.6	4638.9	9.1804	1012.6	4051.6	4638.9	9.1642	978.8	4051.5	4638.8	9.1485
1100	1131.4	4256.2	4889.7	9.3700	1092.4	4256.1	4889.7	9.3538	1055.9	4256.1	4889.6	9.3381
1200	1214.0	4466.6	5146.4	9.5505	1172.1	4466.6	5146.4	9.5342	1133.0	4466.5	5146.3	9.5185
1300	1296.5	4682.4	5408.4	9.7225	1251.8	4682.3	5408.4	9.7063	1210.1	4682.3	5408.3	9.6906

Definitions of symbols on page 1.

$10^3 v$	u	h	s	$10^3 v$	u	h	s	$10^3 v$	u	h	s	t
306.0	2568.5	2758.3	6.7490	297.0	2569.6	2759.6	6.7383	288.5	2570.6	2761.0	6.7279	Sat.
												100
252.1	2457.1	2613.4	6.3905									105
257.6	2468.4	2628.1	6.4291	248.5	2466.0	2625.0	6.4078					110
262.9	2479.4	2642.4	6.4661	253.7	2477.1	2639.5	6.4454	245.0	2474.9	2636.6	6.4251	115
268.0	2490.1	2656.3	6.5018	258.8	2488.0	2653.6	6.4816	250.0	2485.9	2650.9	6.4617	120
273.1	2500.6	2669.9	6.5361	263.7	2498.6	2667.4	6.5164	254.9	2496.6	2664.9	6.4970	125
278.0	2510.8	2683.2	6.5693	268.5	2509.0	2680.8	6.5499	259.6	2507.1	2678.5	6.5310	130
282.9	2520.8	2696.2	6.6014	273.3	2519.1	2694.0	6.5824	264.2	2517.4	2691.8	6.5638	135
287.6	2530.6	2708.9	6.6324	277.9	2529.0	2706.9	6.6138	268.8	2527.4	2704.8	6.5956	140
292.3	2540.3	2721.5	6.6626	282.5	2538.8	2719.6	6.6443	273.3	2537.3	2717.6	6.6263	145
296.9	2549.7	2733.8	6.6919	287.0	2548.3	2732.0	6.6738	277.7	2546.9	2730.2	6.6562	150
305.9	2568.3	2757.9	6.7482	295.8	2567.0	2756.3	6.7306	286.3	2565.7	2754.7	6.7134	160
314.7	2586.3	2781.4	6.8018	304.4	2585.2	2780.0	6.7846	294.7	2584.0	2778.5	6.7678	170
323.4	2603.9	2804.4	6.8531	312.8	2602.9	2803.1	6.8362	302.9	2601.8	2801.8	6.8197	180
331.9	2621.1	2826.9	6.9023	321.1	2620.2	2825.7	6.8856	310.9	2619.3	2824.5	6.8694	190
340.3	2638.1	2849.1	6.9496	329.2	2637.3	2848.0	6.9332	318.9	2636.5	2846.9	6.9173	200
348.5	2654.9	2870.9	6.9954	337.3	2654.1	2870.0	6.9792	326.7	2653.4	2869.0	6.9634	210
356.7	2671.4	2892.6	7.0397	345.2	2670.7	2891.7	7.0237	334.4	2670.1	2890.8	7.0081	220
364.8	2687.8	2914.0	7.0828	353.1	2687.2	2913.2	7.0668	342.1	2686.6	2912.4	7.0514	230
372.9	2704.1	2935.3	7.1246	360.9	2703.6	2934.5	7.1089	349.7	2703.0	2933.8	7.0935	240
380.8	2720.3	2956.5	7.1655	368.7	2719.8	2955.8	7.1498	357.2	2719.3	2955.0	7.1345	250
388.8	2736.5	2977.5	7.2053	376.4	2736.0	2976.9	7.1897	364.7	2735.5	2976.2	7.1746	260
396.7	2752.6	2998.5	7.2443	384.0	2752.1	2997.9	7.2288	372.2	2751.6	2997.3	7.2137	270
404.5	2768.6	3019.4	7.2825	391.7	2768.2	3018.8	7.2670	379.6	2767.7	3018.2	7.2520	280
412.3	2784.6	3040.3	7.3199	399.2	2784.2	3039.7	7.3045	386.9	2783.8	3039.2	7.2896	290
420.1	2800.6	3061.1	7.3566	406.8	2800.3	3060.6	7.3413	394.3	2799.9	3060.1	7.3264	300
427.9	2816.7	3082.0	7.3926	414.3	2816.3	3081.5	7.3773	401.6	2815.9	3081.0	7.3625	310
435.7	2832.7	3102.8	7.4280	421.9	2832.3	3102.3	7.4128	408.9	2832.0	3101.9	7.3980	320
443.4	2848.7	3123.6	7.4628	429.3	2848.4	3123.2	7.4476	416.2	2848.1	3122.7	7.4329	330
451.1	2864.8	3144.4	7.4971	436.8	2864.5	3144.0	7.4819	423.4	2864.2	3143.6	7.4672	340
458.8	2880.9	3165.3	7.5308	444.3	2880.6	3164.9	7.5157	430.7	2880.3	3164.5	7.5010	350
466.5	2897.0	3186.2	7.5641	451.7	2896.7	3185.8	7.5490	437.9	2896.4	3185.4	7.5343	360
474.1	2913.1	3207.1	7.5968	459.1	2912.9	3206.7	7.5817	445.1	2912.6	3206.3	7.5671	370
481.8	2929.3	3228.0	7.6291	466.6	2929.1	3227.7	7.6140	452.3	2928.8	3227.3	7.5995	380
489.4	2945.5	3249.0	7.6610	474.0	2945.3	3248.6	7.6459	459.5	2945.1	3248.3	7.6314	390
497.0	2961.8	3270.0	7.6924	481.4	2961.6	3269.7	7.6774	466.6	2961.4	3269.3	7.6628	400
512.2	2994.5	3312.1	7.7541	496.1	2994.3	3311.8	7.7391	481.0	2994.1	3311.5	7.7246	420
527.4	3027.4	3354.4	7.8142	510.8	3027.2	3354.1	7.7993	495.2	3027.0	3353.9	7.7848	440
542.6	3060.5	3396.9	7.8730	525.5	3060.3	3396.7	7.8581	509.5	3060.2	3396.4	7.8437	460
557.7	3093.9	3439.6	7.9305	540.2	3093.7	3439.4	7.9156	523.7	3093.5	3439.2	7.9012	480
572.8	3127.4	3482.6	7.9868	554.8	3127.3	3482.4	7.9719	537.9	3127.1	3482.1	7.9575	500
587.9	3161.2	3525.7	8.0419	569.4	3161.1	3525.5	8.0270	552.1	3160.9	3525.3	8.0127	520
603.0	3195.3	3569.1	8.0959	584.0	3195.2	3568.9	8.0811	566.3	3195.0	3568.8	8.0667	540
618.0	3229.6	3612.8	8.1489	598.6	3229.5	3612.6	8.1341	580.4	3229.3	3612.4	8.1198	560
633.1	3264.1	3656.6	8.2010	613.2	3264.0	3656.5	8.1862	594.6	3263.9	3656.3	8.1718	580
648.1	3299.0	3700.8	8.2521	627.8	3298.8	3700.6	8.2373	608.7	3298.7	3700.5	8.2230	600
663.1	3334.0	3745.1	8.3023	642.3	3333.9	3745.0	8.2876	622.8	3333.8	3744.9	8.2732	620
678.1	3369.4	3789.8	8.3518	656.9	3369.3	3789.6	8.3370	636.9	3369.1	3789.5	8.3227	640
693.1	3404.9	3834.7	8.4004	671.4	3404.9	3834.5	8.3856	651.0	3404.8	3834.4	8.3713	660
708.1	3440.8	3879.8	8.4483	685.9	3440.7	3879.7	8.4335	665.1	3440.6	3879.6	8.4192	680
723.1	3476.9	3925.2	8.4954	700.4	3476.9	3925.1	8.4807	679.2	3476.8	3925.0	8.4664	700
760.5	3568.5	4040.0	8.6104	736.7	3568.4	4039.9	8.5956	714.3	3568.3	4039.8	8.5814	750
797.9	3661.7	4156.4	8.7215	772.9	3661.6	4156.3	8.7067	749.5	3661.6	4156.2	8.6925	800
835.2	3756.7	4274.5	8.8290	809.1	3756.6	4274.4	8.8143	784.6	3756.5	4274.4	8.8001	850
872.6	3853.3	4394.3	8.9334	845.3	3853.2	4394.2	8.9187	819.7	3853.2	4394.1	8.9044	900
909.9	3951.6	4515.7	9.0347	881.5	3951.5	4515.6	9.0200	854.7	3951.5	4515.6	9.0058	950
947.2	4051.5	4638.8	9.1333	917.6	4051.4	4638.7	9.1186	889.8	4051.4	4638.6	9.1044	1000
1021.9	4256.0	4889.6	9.3229	989.9	4256.0	4889.5	9.3083	959.9	4255.9	4889.5	9.2940	1100
1096.5	4466.5	5146.3	9.5034	1062.2	4466.4	5146.2	9.4887	1030.0	4466.4	5146.2	9.4745	1200
1171.1	4682.2	5408.3	9.6754	1134.5	4682.2	5408.2	9.6608	1100.1	4682.1	5408.2	9.6465	1300

Definitions of symbols on page 1.

(33)

Table 3. Vapor

p (t Sat.)	.68 (163.81)				.70 (164.97)				.72 (166.11)			
t	10^3 v	u	h	s	10^3 v	u	h	s	10^3 v	u	h	s
Sat.	280.5	2571.6	2762.3	6.7178	272.9	2572.5	2763.5	6.7080	265.7	2573.4	2764.7	6.6985
150	268.9	2545.5	2728.3	6.6389	260.6	2544.0	2726.5	6.6220	252.8	2542.6	2724.6	6.6055
155	273.1	2555.1	2740.8	6.6682	264.8	2553.7	2739.0	6.6516	256.9	2552.3	2737.3	6.6353
160	277.3	2564.5	2753.1	6.6967	268.9	2563.2	2751.4	6.6803	260.9	2561.9	2749.7	6.6642
165	281.4	2573.8	2765.1	6.7244	272.9	2572.6	2763.6	6.7082	264.8	2571.3	2762.0	6.6923
170	285.5	2582.9	2777.0	6.7514	276.9	2581.8	2775.6	6.7354	268.7	2580.6	2774.1	6.7198
175	289.5	2591.9	2788.8	6.7778	280.8	2590.9	2787.4	6.7620	272.6	2589.8	2786.0	6.7465
180	293.5	2600.9	2800.4	6.8037	284.7	2599.8	2799.1	6.7880	276.4	2598.8	2797.8	6.7727
185	297.5	2609.7	2812.0	6.8289	288.5	2608.7	2810.7	6.8134	280.1	2607.7	2809.4	6.7982
190	301.4	2618.4	2823.3	6.8536	292.4	2617.5	2822.2	6.8382	283.9	2616.6	2821.0	6.8232
195	305.3	2627.1	2834.6	6.8779	296.2	2626.2	2833.5	6.8626	287.6	2625.3	2832.4	6.8477
200	309.1	2635.6	2845.8	6.9017	299.9	2634.8	2844.8	6.8865	291.2	2634.0	2843.7	6.8717
210	316.7	2652.6	2868.0	6.9480	307.3	2651.9	2867.0	6.9331	298.5	2651.1	2866.0	6.9185
220	324.3	2669.4	2889.9	6.9929	314.7	2668.7	2889.0	6.9781	305.6	2668.0	2888.1	6.9636
230	331.7	2686.0	2911.5	7.0363	322.0	2685.3	2910.7	7.0217	312.7	2684.7	2909.9	7.0074
240	339.1	2702.4	2933.0	7.0786	329.2	2701.8	2932.2	7.0641	319.7	2701.2	2931.5	7.0499
250	346.5	2718.7	2954.3	7.1197	336.3	2718.2	2953.6	7.1053	326.7	2717.7	2952.9	7.0913
260	353.7	2735.0	2975.5	7.1599	343.4	2734.5	2974.9	7.1455	333.6	2734.0	2974.2	7.1316
270	361.0	2751.2	2996.6	7.1991	350.4	2750.7	2996.0	7.1848	340.5	2750.2	2995.4	7.1710
280	368.2	2767.3	3017.7	7.2375	357.4	2766.9	3017.1	7.2233	347.3	2766.4	3016.5	7.2095
290	375.3	2783.4	3038.6	7.2751	364.4	2783.0	3038.1	7.2609	354.1	2782.6	3037.5	7.2472
300	382.5	2799.5	3059.6	7.3119	371.4	2799.1	3059.1	7.2979	360.9	2798.7	3058.5	7.2842
310	389.6	2815.6	3080.5	7.3481	378.3	2815.2	3080.0	7.3341	367.6	2814.8	3079.5	7.3205
320	396.7	2831.6	3101.4	7.3836	385.2	2831.3	3100.9	7.3697	374.3	2831.0	3100.5	7.3561
330	403.8	2847.7	3122.3	7.4186	392.1	2847.4	3121.9	7.4046	381.0	2847.1	3121.4	7.3911
340	410.8	2863.8	3143.2	7.4529	398.9	2863.5	3142.8	7.4391	387.7	2863.2	3142.4	7.4255
350	417.8	2880.0	3164.1	7.4868	405.8	2879.7	3163.7	7.4729	394.3	2879.4	3163.3	7.4594
360	424.9	2896.1	3185.0	7.5201	412.6	2895.8	3184.7	7.5063	401.0	2895.6	3184.3	7.4928
370	431.9	2912.3	3206.0	7.5529	419.4	2912.1	3205.6	7.5391	407.6	2911.8	3205.3	7.5257
380	438.9	2928.5	3227.0	7.5853	426.2	2928.3	3226.6	7.5715	414.2	2928.0	3226.3	7.5581
390	445.8	2944.8	3248.0	7.6172	433.0	2944.6	3247.6	7.6035	420.8	2944.3	3247.3	7.5901
400	452.8	2961.1	3269.0	7.6487	439.7	2960.9	3268.7	7.6350	427.4	2960.6	3268.4	7.6216
420	466.7	2993.9	3311.2	7.7105	453.3	2993.7	3310.9	7.6968	440.6	2993.4	3310.6	7.6835
440	480.6	3026.8	3353.6	7.7708	466.7	3026.6	3353.3	7.7571	453.7	3026.4	3353.1	7.7438
460	494.4	3060.0	3396.2	7.8296	480.2	3059.8	3395.9	7.8160	466.8	3059.6	3395.7	7.8027
480	508.2	3093.3	3438.9	7.8872	493.6	3093.2	3438.7	7.8736	479.8	3093.0	3438.5	7.8604
500	522.0	3127.0	3481.9	7.9435	507.0	3126.8	3481.7	7.9299	492.9	3126.6	3481.5	7.9167
520	535.8	3160.8	3525.1	7.9987	520.4	3160.6	3524.9	7.9851	505.9	3160.5	3524.7	7.9719
540	549.6	3194.9	3568.6	8.0528	533.8	3194.7	3568.4	8.0392	518.9	3194.6	3568.2	8.0260
560	563.3	3229.2	3612.2	8.1058	547.1	3229.1	3612.1	8.0923	531.9	3228.9	3611.9	8.0791
580	577.0	3263.8	3656.1	8.1579	560.5	3263.7	3656.0	8.1444	544.8	3263.5	3655.8	8.1312
600	590.7	3298.6	3700.3	8.2091	573.8	3298.5	3700.2	8.1956	557.8	3298.4	3700.0	8.1824
620	604.4	3333.7	3744.7	8.2593	587.1	3333.6	3744.6	8.2458	570.8	3333.5	3744.4	8.2327
640	618.1	3369.0	3789.4	8.3088	600.4	3368.9	3789.2	8.2953	583.7	3368.8	3789.1	8.2822
660	631.8	3404.7	3834.3	8.3575	613.7	3404.6	3834.2	8.3440	596.6	3404.5	3834.0	8.3309
680	645.5	3440.5	3879.5	8.4054	627.0	3440.4	3879.3	8.3919	609.5	3440.4	3879.2	8.3788
700	659.1	3476.7	3924.9	8.4525	640.3	3476.6	3924.8	8.4391	622.4	3476.5	3924.7	8.4260
750	693.3	3568.2	4039.7	8.5675	673.4	3568.2	4039.6	8.5541	654.7	3568.1	4039.5	8.5410
800	727.4	3661.5	4156.1	8.6786	706.6	3661.4	4156.0	8.6652	686.9	3661.4	4155.9	8.6521
850	761.5	3756.5	4274.3	8.7863	739.7	3756.4	4274.2	8.7728	719.1	3756.4	4274.1	8.7598
900	795.5	3853.1	4394.1	8.8906	772.8	3853.1	4394.0	8.8772	751.3	3853.0	4393.9	8.8641
950	829.6	3951.4	4515.5	8.9920	805.9	3951.3	4515.5	8.9785	783.5	3951.3	4515.4	8.9655
1000	863.6	4051.3	4638.6	9.0906	838.9	4051.3	4638.5	9.0771	815.6	4051.2	4638.5	9.0641
1100	931.7	4255.9	4889.4	9.2802	905.0	4255.8	4889.4	9.2668	879.9	4255.8	4889.3	9.2538
1200	999.7	4466.3	5146.1	9.4607	971.1	4466.3	5146.1	9.4472	944.2	4466.2	5146.1	9.4342
1300	1067.7	4682.1	5408.2	9.6327	1037.2	4682.1	5408.1	9.6193	1008.4	4682.0	5408.1	9.6063

Definitions of symbols on page 1.

	.74 (167.23)				.76 (168.32)				.78 (169.39)			p (t Sat.)
10^3 v	u	h	s	10^3 v	u	h	s	10^3 v	u	h	s	t
258.9	2574.3	2765.9	6.6892	252.4	2575.2	2767.0	6.6802	246.3	2576.0	2768.1	6.6714	Sat.
245.4	2541.1	2722.7	6.5892	238.4	2539.6	2720.8	6.5733	231.7	2538.2	2718.9	6.5577	150
249.4	2551.0	2735.5	6.6193	242.3	2549.6	2733.7	6.6036	235.6	2548.2	2731.9	6.5883	155
253.3	2560.6	2748.1	6.6485	246.1	2559.3	2746.4	6.6330	239.3	2558.0	2744.7	6.6179	160
257.2	2570.1	2760.4	6.6768	249.9	2568.9	2758.8	6.6616	243.1	2567.7	2757.2	6.6467	165
261.0	2579.5	2772.6	6.7045	253.7	2578.3	2771.1	6.6895	246.7	2577.2	2769.6	6.6748	170
264.7	2588.7	2784.6	6.7314	257.4	2587.6	2783.2	6.7166	250.3	2586.5	2781.8	6.7021	175
268.5	2597.8	2796.5	6.7577	261.0	2596.8	2795.1	6.7430	253.9	2595.7	2793.8	6.7287	180
272.2	2606.8	2808.2	6.7834	264.6	2605.8	2806.9	6.7689	257.4	2604.8	2805.6	6.7547	185
275.8	2615.7	2819.7	6.8085	268.2	2614.7	2818.5	6.7942	260.9	2613.8	2817.3	6.7801	190
279.4	2624.4	2831.2	6.8332	271.7	2623.6	2830.1	6.8189	264.4	2622.7	2828.9	6.8050	195
283.0	2633.2	2842.6	6.8573	275.2	2632.3	2841.5	6.8432	267.8	2631.5	2840.4	6.8294	200
290.1	2650.4	2865.0	6.9042	282.1	2649.6	2864.0	6.8903	274.6	2648.8	2863.0	6.8767	210
297.1	2667.3	2887.1	6.9496	289.0	2666.6	2886.2	6.9358	281.3	2665.9	2885.3	6.9224	220
304.0	2684.1	2909.0	6.9935	295.7	2683.4	2908.2	6.9799	287.9	2682.8	2907.3	6.9666	230
310.8	2700.6	2930.7	7.0361	302.4	2700.1	2929.9	7.0226	294.4	2699.5	2929.1	7.0095	240
317.6	2717.1	2952.2	7.0776	309.0	2716.6	2951.4	7.0642	300.9	2716.0	2950.7	7.0512	250
324.4	2733.5	2973.5	7.1180	315.6	2733.0	2972.8	7.1047	307.3	2732.5	2972.2	7.0918	260
331.1	2749.8	2994.7	7.1575	322.1	2749.3	2994.1	7.1443	313.7	2748.8	2993.5	7.1314	270
337.7	2766.0	3015.9	7.1960	328.6	2765.5	3015.3	7.1829	320.0	2765.1	3014.7	7.1701	280
344.3	2782.2	3037.0	7.2338	335.1	2781.8	3036.4	7.2208	326.3	2781.3	3035.9	7.2080	290
350.9	2798.3	3058.0	7.2709	341.5	2797.9	3057.5	7.2579	332.6	2797.6	3057.0	7.2452	300
357.5	2814.5	3079.0	7.3072	347.9	2814.1	3078.5	7.2942	338.8	2813.7	3078.0	7.2816	310
364.0	2830.6	3100.0	7.3429	354.3	2830.3	3099.5	7.3300	345.1	2829.9	3099.1	7.3174	320
370.6	2846.8	3121.0	7.3779	360.7	2846.4	3120.5	7.3651	351.3	2846.1	3120.1	7.3525	330
377.1	2862.9	3141.9	7.4124	367.0	2862.6	3141.5	7.3996	357.4	2862.3	3141.1	7.3871	340
383.5	2879.1	3162.9	7.4463	373.3	2878.8	3162.5	7.4335	363.6	2878.5	3162.1	7.4211	350
390.0	2895.3	3183.9	7.4797	379.6	2895.0	3183.5	7.4670	369.8	2894.7	3183.1	7.4545	360
396.5	2911.5	3204.9	7.5126	385.9	2911.2	3204.5	7.4999	375.9	2911.0	3204.2	7.4875	370
402.9	2927.8	3225.9	7.5451	392.2	2927.5	3225.6	7.5324	382.0	2927.3	3225.2	7.5200	380
409.3	2944.1	3247.0	7.5771	398.5	2943.8	3246.6	7.5644	388.1	2943.6	3246.3	7.5520	390
415.8	2960.4	3268.1	7.6086	404.7	2960.2	3267.8	7.5960	394.2	2959.9	3267.4	7.5836	400
428.6	2993.2	3310.3	7.6705	417.2	2993.0	3310.1	7.6579	406.4	2992.8	3309.8	7.6456	420
441.3	3026.2	3352.8	7.7309	429.6	3026.0	3352.5	7.7183	418.5	3025.8	3352.3	7.7060	440
454.1	3059.4	3395.4	7.7898	442.0	3059.2	3395.2	7.7773	430.6	3059.0	3394.9	7.7650	460
466.8	3092.8	3438.3	7.8475	454.4	3092.7	3438.0	7.8349	442.7	3092.5	3437.8	7.8227	480
479.5	3126.5	3481.3	7.9039	466.8	3126.3	3481.1	7.8914	454.8	3126.1	3480.9	7.8792	500
492.2	3160.3	3524.5	7.9591	479.1	3160.2	3524.3	7.9466	466.8	3160.0	3524.1	7.9344	520
504.8	3194.5	3568.0	8.0132	491.5	3194.3	3567.8	8.0007	478.8	3194.2	3567.6	7.9886	540
517.4	3228.8	3611.7	8.0663	503.8	3228.7	3611.5	8.0539	490.8	3228.5	3611.4	8.0417	560
530.1	3263.4	3655.7	8.1184	516.1	3263.3	3655.5	8.1060	502.8	3263.2	3655.3	8.0939	580
542.7	3298.3	3699.8	8.1696	528.3	3298.1	3699.7	8.1572	514.8	3298.0	3699.5	8.1451	600
555.3	3333.4	3744.3	8.2200	540.6	3333.3	3744.1	8.2075	526.7	3333.2	3744.0	8.1954	620
567.9	3368.7	3789.0	8.2694	552.9	3368.6	3788.8	8.2570	538.7	3368.5	3788.7	8.2449	640
580.4	3404.4	3833.9	8.3181	565.1	3404.3	3833.8	8.3057	550.6	3404.2	3833.6	8.2936	660
593.0	3440.3	3879.1	8.3660	577.4	3440.2	3879.0	8.3536	562.5	3440.1	3878.9	8.3416	680
605.6	3476.4	3924.6	8.4132	589.6	3476.3	3924.4	8.4008	574.5	3476.3	3924.3	8.3888	700
637.0	3568.0	4039.4	8.5283	620.2	3567.9	4039.3	8.5159	604.3	3567.8	4039.2	8.5038	750
668.3	3661.3	4155.8	8.6394	650.7	3661.2	4155.8	8.6271	634.0	3661.1	4155.7	8.6150	800
699.7	3756.3	4274.0	8.7471	681.2	3756.2	4274.0	8.7347	663.7	3756.2	4273.9	8.7227	850
731.0	3853.0	4393.9	8.8514	711.7	3852.9	4393.8	8.8391	693.5	3852.8	4393.7	8.8270	900
762.3	3951.2	4515.3	8.9528	742.2	3951.2	4515.3	8.9405	723.2	3951.1	4515.2	8.9284	950
793.6	4051.2	4638.4	9.0514	772.7	4051.1	4638.3	9.0391	752.8	4051.1	4638.3	9.0270	1000
856.1	4255.7	4889.2	9.2411	833.6	4255.7	4889.2	9.2287	812.2	4255.6	4889.1	9.2167	1100
918.6	4466.2	5146.0	9.4215	894.5	4466.2	5146.0	9.4092	871.5	4466.1	5145.9	9.3972	1200
981.2	4682.0	5408.0	9.5936	955.3	4681.9	5408.0	9.5813	930.9	4681.9	5407.9	9.5692	1300

Definitions of symbols on page 1.

Table 3. Vapor

p (t Sat.)	.80 (170.43)				.82 (171.46)				.84 (172.47)			
t	10^3 v	u	h	s	10^3 v	u	h	s	10^3 v	u	h	s
Sat.	240.4	2576.8	2769.1	6.6628	234.9	2577.6	2770.2	6.6544	229.5	2578.3	2771.2	6.6462
150	225.4	2536.7	2717.0	6.5423	219.4	2535.1	2715.0	6.5272	213.6	2533.6	2713.0	6.5123
155	229.2	2546.8	2730.1	6.5732	223.1	2545.3	2728.3	6.5583	217.3	2543.9	2726.4	6.5437
160	232.9	2556.7	2743.0	6.6031	226.7	2555.3	2741.3	6.5885	220.9	2554.0	2739.5	6.5741
165	236.5	2566.4	2755.6	6.6321	230.3	2565.2	2754.0	6.6178	224.4	2563.9	2752.4	6.6037
170	240.1	2576.0	2768.1	6.6603	233.8	2574.8	2766.5	6.6462	227.8	2573.6	2765.0	6.6323
175	243.7	2585.4	2780.3	6.6878	237.3	2584.3	2778.9	6.6739	231.3	2583.2	2777.4	6.6602
180	247.2	2594.7	2792.4	6.7146	240.8	2593.6	2791.0	6.7008	234.6	2592.6	2789.7	6.6873
185	250.6	2603.8	2804.3	6.7408	244.1	2602.8	2803.0	6.7272	238.0	2601.8	2801.7	6.7138
190	254.0	2612.9	2816.1	6.7663	247.5	2611.9	2814.9	6.7529	241.3	2611.0	2813.6	6.7396
195	257.4	2621.8	2827.7	6.7914	250.8	2620.9	2826.6	6.7780	244.5	2620.0	2825.4	6.7649
200	260.8	2630.6	2839.3	6.8158	254.1	2629.8	2838.1	6.8026	247.7	2628.9	2837.0	6.7896
210	267.4	2648.1	2862.0	6.8634	260.6	2647.3	2861.0	6.8504	254.1	2646.5	2860.0	6.8376
220	274.0	2665.2	2884.4	6.9093	267.0	2664.5	2883.5	6.8964	260.4	2663.8	2882.5	6.8838
230	280.4	2682.1	2906.5	6.9536	273.3	2681.5	2905.6	6.9409	266.6	2680.8	2904.8	6.9285
240	286.8	2698.9	2928.3	6.9966	279.6	2698.3	2927.5	6.9841	272.7	2697.7	2926.8	6.9718
250	293.1	2715.5	2950.0	7.0384	285.8	2714.9	2949.3	7.0260	278.7	2714.4	2948.5	7.0138
260	299.4	2732.0	2971.5	7.0791	291.9	2731.4	2970.8	7.0668	284.7	2730.9	2970.1	7.0547
270	305.6	2748.3	2992.9	7.1188	298.0	2747.9	2992.2	7.1066	290.7	2747.4	2991.6	7.0946
280	311.8	2764.7	3014.1	7.1577	304.0	2764.2	3013.5	7.1454	296.6	2763.8	3012.9	7.1335
290	318.0	2780.9	3035.3	7.1956	310.1	2780.5	3034.8	7.1835	302.5	2780.1	3034.2	7.1716
300	324.1	2797.2	3056.5	7.2328	316.0	2796.8	3055.9	7.2207	308.4	2796.4	3055.4	7.2089
310	330.2	2813.4	3077.5	7.2693	322.0	2813.0	3077.0	7.2573	314.2	2812.6	3076.6	7.2455
320	336.3	2829.6	3098.6	7.3051	327.9	2829.2	3098.1	7.2931	320.0	2828.9	3097.7	7.2814
330	342.3	2845.8	3119.6	7.3403	333.9	2845.4	3119.2	7.3283	325.8	2845.1	3118.8	7.3167
340	348.4	2862.0	3140.7	7.3749	339.7	2861.7	3140.3	7.3630	331.5	2861.4	3139.8	7.3513
350	354.4	2878.2	3161.7	7.4089	345.6	2877.9	3161.3	7.3970	337.3	2877.6	3160.9	7.3854
360	360.4	2894.4	3182.7	7.4424	351.5	2894.2	3182.4	7.4306	343.0	2893.9	3182.0	7.4190
370	366.4	2910.7	3203.8	7.4754	357.3	2910.4	3203.4	7.4636	348.7	2910.2	3203.1	7.4520
380	372.4	2927.0	3224.9	7.5079	363.2	2926.7	3224.5	7.4961	354.4	2926.5	3224.2	7.4846
390	378.3	2943.3	3246.0	7.5400	369.0	2943.1	3245.7	7.5282	360.1	2942.8	3245.3	7.5167
400	384.3	2959.7	3267.1	7.5716	374.8	2959.5	3266.8	7.5598	365.8	2959.2	3266.5	7.5484
420	396.1	2992.6	3309.5	7.6336	386.4	2992.3	3309.2	7.6219	377.1	2992.1	3308.9	7.6105
440	408.0	3025.6	3352.0	7.6941	397.9	3025.4	3351.7	7.6824	388.4	3025.2	3351.4	7.6710
460	419.8	3058.9	3394.7	7.7531	409.5	3058.7	3394.4	7.7415	399.6	3058.5	3394.2	7.7301
480	431.6	3092.3	3437.6	7.8108	421.0	3092.1	3437.3	7.7992	410.9	3092.0	3437.1	7.7878
500	443.3	3126.0	3480.6	7.8673	432.4	3125.8	3480.4	7.8557	422.1	3125.7	3480.2	7.8443
520	455.1	3159.9	3523.9	7.9225	443.9	3159.7	3523.7	7.9110	433.3	3159.6	3523.5	7.8997
540	466.8	3194.0	3567.4	7.9767	455.3	3193.9	3567.3	7.9652	444.4	3193.8	3567.1	7.9539
560	478.5	3228.4	3611.2	8.0299	466.7	3228.3	3611.0	8.0183	455.6	3228.2	3610.8	8.0070
580	490.2	3263.0	3655.2	8.0820	478.2	3262.9	3655.0	8.0705	466.7	3262.8	3654.8	8.0592
600	501.8	3297.9	3699.4	8.1333	489.6	3297.8	3699.2	8.1217	477.8	3297.7	3699.1	8.1105
620	513.5	3333.0	3743.8	8.1836	500.9	3332.9	3743.7	8.1721	489.0	3332.8	3743.6	8.1609
640	525.2	3368.4	3788.6	8.2331	512.3	3368.3	3788.4	8.2216	500.1	3368.2	3788.3	8.2104
660	536.8	3404.1	3833.5	8.2818	523.7	3404.0	3833.4	8.2703	511.2	3403.9	3833.3	8.2591
680	548.4	3440.0	3878.7	8.3298	535.0	3439.9	3878.6	8.3183	522.3	3439.8	3878.5	8.3071
700	560.1	3476.2	3924.2	8.3770	546.4	3476.1	3924.1	8.3655	533.3	3476.0	3924.0	8.3543
750	589.1	3567.8	4039.1	8.4921	574.7	3567.7	4039.0	8.4806	561.0	3567.6	4038.9	8.4694
800	618.1	3661.1	4155.6	8.6033	603.0	3661.0	4155.5	8.5918	588.7	3660.9	4155.4	8.5806
850	647.1	3756.1	4273.8	8.7109	631.3	3756.0	4273.7	8.6995	616.3	3756.0	4273.7	8.6883
900	676.1	3852.8	4393.7	8.8153	659.6	3852.7	4393.6	8.8039	643.9	3852.7	4393.5	8.7927
950	705.1	3951.1	4515.1	8.9167	687.9	3951.0	4515.1	8.9053	671.5	3951.0	4515.0	8.8941
1000	734.0	4051.0	4638.2	9.0153	716.1	4051.0	4638.2	9.0039	699.0	4050.9	4638.1	8.9927
1100	791.9	4255.6	4889.1	9.2050	772.6	4255.5	4889.0	9.1936	754.2	4255.5	4889.0	9.1824
1200	849.7	4466.1	5145.9	9.3855	829.0	4466.0	5145.8	9.3740	809.3	4466.0	5145.8	9.3629
1300	907.6	4681.8	5407.9	9.5575	885.4	4681.8	5407.8	9.5461	864.4	4681.7	5407.8	9.5350

Definitions of symbols on page 1.

.86 (173.46)				.88 (174.43)				.90 (175.38)				p (t Sat.)
10^3 v	u	h	s	10^3 v	u	h	s	10^3 v	u	h	s	t
224.5	2579.1	2772.1	6.6381	219.6	2579.8	2773.0	6.6303	215.0	2580.5	2773.9	6.6226	Sat.
208.1	2532.1	2711.1	6.4976	202.9	2530.5	2709.0	6.4832	197.88	2528.9	2707.0	6.4689	150
211.7	2542.5	2724.6	6.5294	206.4	2541.0	2722.7	6.5152	201.38	2539.5	2720.8	6.5013	155
215.3	2552.6	2737.8	6.5600	209.9	2551.3	2736.0	6.5462	204.81	2549.9	2734.2	6.5325	160
218.7	2562.6	2750.7	6.5898	213.3	2561.3	2749.1	6.5761	208.17	2560.0	2747.4	6.5627	165
222.1	2572.4	2763.4	6.6187	216.7	2571.2	2761.9	6.6052	211.48	2570.0	2760.3	6.5920	170
225.5	2582.0	2776.0	6.6467	220.0	2580.9	2774.5	6.6335	214.7	2579.7	2773.0	6.6205	175
228.8	2591.5	2788.3	6.6740	223.2	2590.4	2786.9	6.6610	217.9	2589.3	2785.5	6.6481	180
232.1	2600.8	2800.4	6.7007	226.5	2599.8	2799.1	6.6878	221.1	2598.8	2797.8	6.6751	185
235.3	2610.0	2812.4	6.7267	229.6	2609.0	2811.1	6.7139	224.2	2608.1	2809.9	6.7014	190
238.5	2619.1	2824.2	6.7521	232.8	2618.2	2823.0	6.7395	227.3	2617.3	2821.8	6.7271	195
241.7	2628.1	2835.9	6.7769	235.9	2627.2	2834.8	6.7644	230.3	2626.3	2833.6	6.7522	200
247.9	2645.7	2858.9	6.8251	242.0	2644.9	2857.9	6.8128	236.4	2644.2	2856.9	6.8008	210
254.1	2663.1	2881.6	6.8715	248.0	2662.4	2880.7	6.8594	242.3	2661.7	2879.7	6.8476	220
260.1	2680.2	2903.9	6.9163	254.0	2679.5	2903.0	6.9044	248.1	2678.9	2902.2	6.8927	230
266.1	2697.1	2926.0	6.9597	259.9	2696.5	2925.2	6.9479	253.9	2695.9	2924.4	6.9364	240
272.1	2713.8	2947.8	7.0018	265.7	2713.3	2947.1	6.9902	259.6	2712.7	2946.3	6.9787	250
277.9	2730.4	2969.4	7.0428	271.4	2729.9	2968.8	7.0313	265.2	2729.4	2968.1	7.0199	260
283.8	2746.9	2991.0	7.0828	277.1	2746.4	2990.3	7.0713	270.8	2746.0	2989.7	7.0601	270
289.6	2763.3	3012.3	7.1218	282.8	2762.9	3011.8	7.1104	276.3	2762.4	3011.2	7.0992	280
295.3	2779.7	3033.6	7.1600	288.4	2779.3	3033.1	7.1487	281.9	2778.8	3032.5	7.1375	290
301.0	2796.0	3054.9	7.1974	294.0	2795.6	3054.4	7.1861	287.4	2795.2	3053.8	7.1750	300
306.7	2812.3	3076.1	7.2340	299.6	2811.9	3075.6	7.2228	292.8	2811.5	3075.1	7.2118	310
312.4	2828.5	3097.2	7.2700	305.2	2828.2	3096.7	7.2588	298.3	2827.8	3096.3	7.2478	320
318.1	2844.8	3118.3	7.3053	310.7	2844.5	3117.9	7.2941	303.7	2844.1	3117.4	7.2832	330
323.7	2861.0	3139.4	7.3400	316.2	2860.7	3139.0	7.3288	309.1	2860.4	3138.6	7.3179	340
329.3	2877.3	3160.5	7.3741	321.7	2877.0	3160.1	7.3630	314.4	2876.7	3159.7	7.3521	350
334.9	2893.6	3181.6	7.4077	327.2	2893.3	3181.2	7.3966	319.8	2893.0	3180.8	7.3858	360
340.5	2909.9	3202.7	7.4407	332.6	2909.6	3202.3	7.4297	325.1	2909.4	3202.0	7.4189	370
346.1	2926.2	3223.8	7.4733	338.1	2926.0	3223.5	7.4623	330.5	2925.7	3223.1	7.4516	380
351.6	2942.6	3245.0	7.5055	343.5	2942.3	3244.6	7.4945	335.8	2942.1	3244.3	7.4837	390
357.2	2959.0	3266.2	7.5372	349.0	2958.8	3265.8	7.5262	341.1	2958.5	3265.5	7.5155	400
368.2	2991.9	3308.6	7.5993	359.8	2991.7	3308.3	7.5884	351.7	2991.5	3308.0	7.5777	420
379.3	3025.0	3351.2	7.6598	370.6	3024.8	3350.9	7.6489	362.3	3024.6	3350.6	7.6383	440
390.3	3058.3	3393.9	7.7190	381.3	3058.1	3393.7	7.7081	372.8	3057.9	3393.4	7.6975	460
401.2	3091.8	3436.9	7.7767	392.1	3091.6	3436.6	7.7659	383.3	3091.4	3436.4	7.7553	480
412.2	3125.5	3480.0	7.8333	402.8	3125.3	3479.8	7.8224	393.8	3125.2	3479.6	7.8119	500
423.1	3159.4	3523.3	7.8886	413.5	3159.3	3523.1	7.8778	404.2	3159.1	3522.9	7.8672	520
434.0	3193.6	3566.9	7.9428	424.1	3193.5	3566.7	7.9321	414.6	3193.3	3566.5	7.9215	540
444.9	3228.0	3610.7	7.9960	434.8	3227.9	3610.5	7.9853	425.1	3227.8	3610.3	7.9747	560
455.8	3262.7	3654.7	8.0482	445.4	3262.5	3654.5	8.0375	435.5	3262.4	3654.4	8.0270	580
466.7	3297.6	3698.9	8.0995	456.0	3297.5	3698.8	8.0887	445.9	3297.3	3698.6	8.0782	600
477.6	3332.7	3743.4	8.1499	466.7	3332.6	3743.3	8.1391	456.3	3332.5	3743.1	8.1286	620
488.4	3368.1	3788.2	8.1994	477.3	3368.0	3788.0	8.1887	466.6	3367.9	3787.9	8.1782	640
499.2	3403.8	3833.1	8.2481	487.9	3403.7	3833.0	8.2374	477.0	3403.6	3832.9	8.2269	660
510.1	3439.7	3878.4	8.2961	498.5	3439.6	3878.3	8.2854	487.3	3439.5	3878.1	8.2749	680
520.9	3476.0	3923.9	8.3434	509.0	3475.9	3923.8	8.3327	497.7	3475.8	3923.7	8.3222	700
547.9	3567.6	4038.9	8.4585	535.5	3567.6	4038.8	8.4479	523.5	3567.5	4038.7	8.4374	750
574.9	3661.0	4155.4	8.5697	561.9	3660.9	4155.3	8.5591	549.3	3660.8	4155.2	8.5486	800
601.9	3755.9	4273.6	8.6774	588.2	3755.9	4273.5	8.6667	575.1	3755.8	4273.4	8.6563	850
628.9	3852.6	4393.4	8.7818	614.6	3852.5	4393.4	8.7711	600.9	3852.5	4393.3	8.7607	900
655.8	3950.9	4514.9	8.8832	640.9	3950.9	4514.9	8.8725	626.7	3950.8	4514.8	8.8621	950
682.8	4050.9	4638.0	8.9818	667.2	4050.8	4638.0	8.9712	652.4	4050.8	4637.9	8.9608	1000
736.6	4255.4	4888.9	9.1715	719.9	4255.4	4888.9	9.1609	703.9	4255.3	4888.8	9.1505	1100
790.4	4465.9	5145.7	9.3520	772.5	4465.9	5145.7	9.3413	755.3	4465.8	5145.6	9.3309	1200
844.3	4681.7	5407.7	9.5241	825.1	4681.6	5407.7	9.5134	806.7	4681.6	5407.6	9.5030	1300

Definitions of symbols on page 1.

Table 3. Vapor

p (t Sat.)	.92 (176.32)				.94 (177.24)				.96 (178.14)			
t	$10^3 v$	u	h	s	$10^3 v$	u	h	s	$10^3 v$	u	h	s
Sat.	210.5	2581.1	2774.8	6.6151	206.3	2581.8	2775.7	6.6077	202.2	2582.4	2776.5	6.6005
150	*193.09*	*2527.3*	*2705.0*	*6.4549*	*188.49*	*2525.7*	*2702.9*	*6.4410*	*184.08*	*2524.1*	*2700.8*	*6.4273*
155	*196.54*	*2538.1*	*2718.9*	*6.4875*	*191.90*	*2536.6*	*2717.0*	*6.4739*	*187.45*	*2535.1*	*2715.0*	*6.4605*
160	*199.92*	*2548.5*	*2732.4*	*6.5190*	*195.24*	*2547.1*	*2730.6*	*6.5057*	*190.74*	*2545.7*	*2728.8*	*6.4926*
165	*203.23*	*2558.7*	*2745.7*	*6.5495*	*198.50*	*2557.4*	*2744.0*	*6.5364*	*193.96*	*2556.1*	*2742.3*	*6.5236*
170	*206.49*	*2568.8*	*2758.7*	*6.5790*	*201.71*	*2567.5*	*2757.1*	*6.5662*	*197.13*	*2566.3*	*2755.5*	*6.5536*
175	*209.7*	*2578.6*	*2771.5*	*6.6077*	*204.9*	*2577.4*	*2770.0*	*6.5951*	*200.2*	*2576.3*	*2768.5*	*6.5827*
180	212.8	2588.2	2784.1	6.6355	208.0	2587.2	2782.6	6.6231	203.3	2586.1	2781.2	6.6109
185	215.9	2597.7	2796.4	6.6627	211.0	2596.7	2795.1	6.6504	206.3	2595.7	2793.7	6.6383
190	219.0	2607.1	2808.6	6.6891	214.0	2606.1	2807.3	6.6770	209.3	2605.1	2806.0	6.6651
195	222.0	2616.3	2820.6	6.7149	217.0	2615.4	2819.4	6.7030	212.2	2614.5	2818.2	6.6912
200	225.0	2625.4	2832.5	6.7401	220.0	2624.6	2831.3	6.7283	215.1	2623.7	2830.2	6.7167
210	231.0	2643.4	2855.8	6.7890	225.8	2642.6	2854.8	6.7774	220.8	2641.8	2853.8	6.7660
220	236.8	2660.9	2878.8	6.8359	231.5	2660.2	2877.8	6.8245	226.4	2659.5	2876.9	6.8133
230	242.5	2678.2	2901.3	6.8812	237.1	2677.6	2900.4	6.8700	231.9	2676.9	2899.6	6.8589
240	248.1	2695.3	2923.6	6.9250	242.6	2694.7	2922.8	6.9139	237.4	2694.1	2922.0	6.9030
250	253.7	2712.2	2945.6	6.9675	248.1	2711.6	2944.8	6.9565	242.8	2711.0	2944.1	6.9457
260	259.2	2728.9	2967.4	7.0088	253.5	2728.4	2966.7	6.9979	248.1	2727.9	2966.0	6.9872
270	264.7	2745.5	2989.0	7.0490	258.9	2745.0	2988.4	7.0382	253.4	2744.5	2987.8	7.0276
280	270.2	2762.0	3010.6	7.0883	264.3	2761.5	3010.0	7.0775	258.6	2761.1	3009.4	7.0670
290	275.6	2778.4	3032.0	7.1266	269.6	2778.0	3031.4	7.1160	263.8	2777.6	3030.8	7.1055
300	281.0	2794.8	3053.3	7.1642	274.8	2794.4	3052.8	7.1536	269.0	2794.0	3052.2	7.1432
310	286.3	2811.2	3074.6	7.2010	280.1	2810.8	3074.1	7.1904	274.1	2810.4	3073.6	7.1800
320	291.6	2827.5	3095.8	7.2370	285.3	2827.1	3095.3	7.2265	279.2	2826.8	3094.8	7.2162
330	296.9	2843.8	3117.0	7.2725	290.5	2843.5	3116.5	7.2620	284.3	2843.1	3116.1	7.2517
340	302.2	2860.1	3138.1	7.3073	295.7	2859.8	3137.7	7.2968	289.4	2859.5	3137.3	7.2866
350	307.5	2876.4	3159.3	7.3415	300.8	2876.1	3158.9	7.3311	294.5	2875.8	3158.5	7.3209
360	312.7	2892.7	3180.5	7.3752	306.0	2892.4	3180.1	7.3648	299.5	2892.2	3179.7	7.3546
370	318.0	2909.1	3201.6	7.4084	311.1	2908.8	3201.2	7.3980	304.5	2908.5	3200.9	7.3879
380	323.2	2925.4	3222.8	7.4410	316.2	2925.2	3222.4	7.4307	309.5	2924.9	3222.1	7.4206
390	328.4	2941.8	3244.0	7.4732	321.3	2941.6	3243.6	7.4629	314.5	2941.3	3243.3	7.4528
400	333.6	2958.3	3265.2	7.5050	326.4	2958.0	3264.9	7.4947	319.5	2957.8	3264.6	7.4846
420	344.0	2991.3	3307.7	7.5672	336.6	2991.0	3307.4	7.5570	329.5	2990.8	3307.1	7.5469
440	354.3	3024.4	3350.4	7.6279	346.7	3024.2	3350.1	7.6177	339.4	3024.0	3349.8	7.6077
460	364.6	3057.7	3393.2	7.6871	356.8	3057.6	3392.9	7.6769	349.3	3057.4	3392.7	7.6669
480	374.9	3091.3	3436.2	7.7449	366.8	3091.1	3435.9	7.7348	359.1	3090.9	3435.7	7.7248
500	385.1	3125.0	3479.3	7.8015	376.9	3124.9	3479.1	7.7914	369.0	3124.7	3478.9	7.7814
520	395.4	3159.0	3522.7	7.8569	386.9	3158.8	3522.5	7.8468	378.8	3158.7	3522.3	7.8369
540	405.6	3193.2	3566.3	7.9112	396.9	3193.1	3566.1	7.9011	388.6	3192.9	3566.0	7.8912
560	415.8	3227.6	3610.1	7.9644	406.9	3227.5	3610.0	7.9543	398.4	3227.4	3609.8	7.9445
580	426.0	3262.3	3654.2	8.0167	416.9	3262.2	3654.0	8.0066	408.1	3262.1	3653.9	7.9967
600	436.1	3297.2	3698.5	8.0680	426.8	3297.1	3698.3	8.0579	417.9	3297.0	3698.2	8.0481
620	446.3	3332.4	3743.0	8.1184	436.8	3332.3	3742.8	8.1083	427.6	3332.2	3742.7	8.0985
640	456.4	3367.8	3787.7	8.1679	446.7	3367.7	3787.6	8.1579	437.4	3367.6	3787.5	8.1481
660	466.6	3403.5	3832.8	8.2167	456.6	3403.4	3832.6	8.2067	447.1	3403.3	3832.5	8.1969
680	476.7	3439.4	3878.0	8.2647	466.5	3439.3	3877.9	8.2547	456.8	3439.3	3877.8	8.2449
700	486.8	3475.6	3923.5	8.3120	476.5	3475.6	3923.4	8.3019	466.5	3475.5	3923.3	8.2921
750	512.1	3567.3	4038.5	8.4271	501.2	3567.2	4038.4	8.4171	490.7	3567.2	4038.3	8.4073
800	537.4	3660.7	4155.1	8.5384	525.9	3660.6	4155.0	8.5284	515.0	3660.5	4154.9	8.5186
850	562.6	3755.7	4273.3	8.6461	550.6	3755.7	4273.3	8.6361	539.1	3755.6	4273.2	8.6263
900	587.8	3852.4	4393.2	8.7505	575.3	3852.4	4393.2	8.7405	563.3	3852.4	4393.1	8.7308
950	613.0	3950.8	4514.7	8.8519	600.0	3950.7	4514.7	8.8420	587.5	3950.7	4514.6	8.8322
1000	638.2	4050.7	4637.9	8.9506	624.6	4050.7	4637.8	8.9406	611.6	4050.6	4637.7	8.9309
1100	688.6	4255.3	4888.8	9.1403	673.9	4255.2	4888.7	9.1303	659.9	4255.2	4888.7	9.1206
1200	738.9	4465.8	5145.6	9.3208	723.2	4465.7	5145.5	9.3108	708.1	4465.7	5145.5	9.3011
1300	789.2	4681.5	5407.6	9.4929	772.4	4681.5	5407.6	9.4829	756.3	4681.4	5407.5	9.4731

Definitions of symbols on page 1.

$10^3 v$	u	h	s	$10^3 v$	u	h	s	$10^3 v$	u	h	s	t
198.23	2583.1	2777.3	6.5934	194.44	2583.6	2778.1	6.5865	185.60	2585.1	2780.0	6.5696	Sat.
												150
183.18	2533.5	2713.0	6.4473	179.07	2532.0	2711.1	6.4342					155
186.43	2544.3	2727.0	6.4797	182.29	2542.9	2725.1	6.4669	172.60	2539.2	2720.4	6.4356	160
189.61	2554.8	2740.6	6.5109	185.43	2553.4	2738.9	6.4984	175.66	2550.0	2734.5	6.4678	165
192.73	2565.0	2753.9	6.5411	188.50	2563.8	2752.3	6.5289	178.64	2560.6	2748.2	6.4989	170
195.79	2575.1	2767.0	6.5704	191.53	2573.9	2765.4	6.5584	181.56	2570.9	2761.6	6.5289	175
198.80	2584.9	2779.8	6.5988	194.50	2583.8	2778.3	6.5870	184.44	2581.0	2774.7	6.5580	180
201.77	2594.6	2792.4	6.6265	197.42	2593.6	2791.0	6.6148	187.26	2590.9	2787.5	6.5863	185
204.70	2604.2	2804.8	6.6534	200.30	2603.2	2803.5	6.6419	190.04	2600.7	2800.2	6.6137	190
207.58	2613.5	2817.0	6.6796	203.15	2612.6	2815.8	6.6682	192.79	2610.2	2812.7	6.6405	195
210.4	2622.8	2829.0	6.7052	206.0	2621.9	2827.9	6.6940	195.49	2619.7	2824.9	6.6666	200
216.1	2641.0	2852.7	6.7548	211.5	2640.2	2851.7	6.7437	200.81	2638.1	2849.0	6.7169	210
221.6	2658.8	2875.9	6.8023	216.9	2658.0	2874.9	6.7914	206.02	2656.2	2872.5	6.7651	220
227.0	2676.3	2898.7	6.8480	222.2	2675.6	2897.8	6.8374	211.14	2673.9	2895.6	6.8114	230
232.3	2693.5	2921.2	6.8922	227.5	2692.9	2920.4	6.8817	216.18	2691.3	2918.3	6.8561	240
237.6	2710.5	2943.4	6.9351	232.7	2709.9	2942.6	6.9247	221.2	2708.5	2940.7	6.8994	250
242.8	2727.3	2965.3	6.9767	237.8	2726.8	2964.6	6.9664	226.1	2725.5	2962.9	6.9414	260
248.0	2744.0	2987.1	7.0172	242.9	2743.6	2986.5	7.0069	231.0	2742.3	2984.8	6.9822	270
253.2	2760.6	3008.8	7.0567	248.0	2760.2	3008.2	7.0465	235.8	2759.1	3006.6	7.0219	280
258.3	2777.2	3030.3	7.0952	253.0	2776.7	3029.7	7.0851	240.6	2775.7	3028.3	7.0607	290
263.3	2793.6	3051.7	7.1329	257.9	2793.2	3051.2	7.1229	245.3	2792.2	3049.8	7.0986	300
268.4	2810.0	3073.1	7.1699	262.9	2809.7	3072.6	7.1599	250.1	2808.7	3071.3	7.1358	310
273.4	2826.4	3094.4	7.2061	267.8	2826.1	3093.9	7.1962	254.8	2825.2	3092.7	7.1722	320
278.4	2842.8	3115.6	7.2417	272.7	2842.5	3115.2	7.2318	259.5	2841.6	3114.1	7.2079	330
283.4	2859.2	3136.9	7.2766	277.6	2858.8	3136.4	7.2667	264.1	2858.0	3135.4	7.2429	340
288.3	2875.5	3158.1	7.3109	282.5	2875.2	3157.7	7.3011	268.8	2874.5	3156.7	7.2774	350
293.3	2891.9	3179.3	7.3447	287.3	2891.6	3178.9	7.3349	273.4	2890.9	3178.0	7.3112	360
298.2	2908.3	3200.5	7.3779	292.2	2908.0	3200.1	7.3682	278.0	2907.3	3199.2	7.3446	370
303.1	2924.7	3221.7	7.4107	297.0	2924.4	3221.4	7.4009	282.6	2923.8	3220.5	7.3774	380
308.0	2941.1	3243.0	7.4429	301.8	2940.9	3242.6	7.4332	287.2	2940.2	3241.8	7.4098	390
312.9	2957.6	3264.2	7.4748	306.6	2957.3	3263.9	7.4651	291.8	2956.7	3263.1	7.4417	400
322.7	2990.6	3306.8	7.5371	316.2	2990.4	3306.5	7.5275	300.9	2989.8	3305.8	7.5041	420
332.4	3023.8	3349.6	7.5979	325.7	3023.6	3349.3	7.5883	310.0	3023.1	3348.6	7.5650	440
342.1	3057.2	3392.4	7.6571	335.2	3057.0	3392.2	7.6476	319.1	3056.5	3391.5	7.6244	460
351.7	3090.8	3435.5	7.7151	344.7	3090.6	3435.2	7.7055	328.1	3090.1	3434.7	7.6824	480
361.4	3124.5	3478.7	7.7717	354.1	3124.4	3478.5	7.7622	337.1	3124.0	3477.9	7.7392	500
371.0	3158.5	3522.1	7.8272	363.5	3158.4	3521.9	7.8177	346.1	3158.0	3521.4	7.7947	520
380.6	3192.8	3565.8	7.8815	372.9	3192.6	3565.6	7.8720	355.1	3192.3	3565.1	7.8491	540
390.2	3227.2	3609.6	7.9348	382.3	3227.1	3609.4	7.9253	364.0	3226.8	3609.0	7.9024	560
399.8	3261.9	3653.7	7.9871	391.7	3261.8	3653.5	7.9776	373.0	3261.5	3653.1	7.9547	580
409.3	3296.9	3698.0	8.0384	401.1	3296.8	3697.9	8.0290	381.9	3296.5	3697.5	8.0061	600
418.9	3332.1	3742.6	8.0889	410.5	3332.0	3742.4	8.0794	390.8	3331.7	3742.1	8.0566	620
428.4	3367.3	3787.3	8.1384	419.8	3367.4	3787.2	8.1290	399.7	3367.2	3786.9	8.1062	640
437.9	3403.2	3832.4	8.1872	429.1	3403.1	3832.2	8.1778	408.6	3402.9	3831.9	8.1550	660
447.4	3439.2	3877.7	8.2352	438.5	3439.1	3877.5	8.2258	417.5	3438.8	3877.2	8.2031	680
457.0	3475.4	3923.2	8.2825	447.8	3475.3	3923.1	8.2731	426.4	3475.1	3922.8	8.2504	700
480.7	3567.1	4038.2	8.3977	471.1	3567.0	4038.1	8.3883	448.6	3566.8	4037.8	8.3656	750
504.4	3660.5	4154.8	8.5090	494.3	3660.4	4154.7	8.4996	470.7	3660.2	4154.5	8.4770	800
528.1	3755.5	4273.1	8.6168	517.5	3755.5	4273.0	8.6074	492.9	3755.3	4272.8	8.5847	850
551.8	3852.3	4393.0	8.7212	540.7	3852.2	4392.9	8.7118	515.0	3852.1	4392.8	8.6892	900
575.5	3950.6	4514.5	8.8226	563.9	3950.5	4514.5	8.8133	537.1	3950.4	4514.3	8.7906	950
599.1	4050.5	4637.7	8.9213	587.1	4050.5	4637.6	8.9119	559.1	4050.4	4637.5	8.8893	1000
646.4	4255.1	4888.6	9.1110	633.5	4255.1	4888.6	9.1017	603.3	4255.0	4888.4	9.0791	1100
693.6	4465.6	5145.4	9.2915	679.8	4465.6	5145.4	9.2822	647.4	4465.5	5145.2	9.2596	1200
740.9	4681.4	5407.5	9.4636	726.1	4681.3	5407.4	9.4543	691.5	4681.2	5407.3	9.4317	1300

Definitions of symbols on page 1.

Table 3. Vapor

p (t Sat.)	1.10 (184.09)				1.15 (186.07)				1.20 (187.99)			
t	10^3 v	u	h	s	10^3 v	u	h	s	10^3 v	u	h	s
Sat.	177.53	2586.4	2781.7	6.5535	170.13	2587.7	2783.3	6.5381	163.33	2588.8	2784.8	6.5233
175	*172.50*	*2567.9*	*2757.6*	*6.5004*	*164.20*	*2564.8*	*2753.6*	*6.4726*	*156.59*	*2561.6*	*2749.5*	*6.4456*
180	*175.28*	*2578.2*	*2771.0*	*6.5300*	*166.91*	*2575.2*	*2767.2*	*6.5028*	*159.23*	*2572.3*	*2763.4*	*6.4763*
185	178.02	2588.2	2784.0	6.5587	*169.57*	*2585.5*	*2780.5*	*6.5320*	*161.81*	*2582.7*	*2776.9*	*6.5060*
190	180.71	2598.1	2796.9	6.5866	172.18	2595.5	2793.5	6.5603	164.35	2592.9	2790.1	6.5348
195	183.36	2607.8	2809.5	6.6137	174.75	2605.4	2806.4	6.5878	166.84	2602.9	2803.1	6.5627
200	185.97	2617.4	2822.0	6.6401	177.28	2615.1	2819.0	6.6146	169.30	2612.8	2815.9	6.5898
205	188.56	2626.8	2834.2	6.6659	179.77	2624.6	2831.4	6.6407	171.72	2622.4	2828.5	6.6163
210	191.11	2636.1	2846.3	6.6911	182.24	2634.0	2843.6	6.6661	174.11	2631.9	2840.9	6.6420
215	193.63	2645.3	2858.3	6.7157	184.68	2643.3	2855.7	6.6910	176.46	2641.3	2853.1	6.6672
220	196.13	2654.3	2870.1	6.7398	187.09	2652.5	2867.6	6.7154	178.80	2650.6	2865.1	6.6918
225	198.60	2663.3	2881.8	6.7634	189.47	2661.5	2879.4	6.7392	181.10	2659.7	2877.1	6.7158
230	201.05	2672.2	2893.4	6.7865	191.84	2670.5	2891.1	6.7625	183.39	2668.8	2888.9	6.7394
235	203.49	2681.0	2904.9	6.8093	194.19	2679.4	2902.7	6.7855	185.66	2677.8	2900.6	6.7625
240	205.90	2689.8	2916.3	6.8316	196.51	2688.2	2914.2	6.8080	187.90	2686.6	2912.1	6.7852
245	208.30	2698.5	2927.6	6.8535	198.82	2697.0	2925.6	6.8301	190.13	2695.5	2923.6	6.8075
250	210.7	2707.1	2938.8	6.8751	201.1	2705.7	2936.9	6.8518	192.34	2704.2	2935.0	6.8294
260	215.4	2724.2	2961.1	6.9174	205.7	2722.9	2959.4	6.8943	196.73	2721.5	2957.6	6.8721
270	220.1	2741.1	2983.2	6.9584	210.2	2739.9	2981.6	6.9356	201.06	2738.7	2979.9	6.9136
280	224.7	2757.9	3005.1	6.9984	214.6	2756.8	3003.6	6.9757	205.35	2755.6	3002.1	6.9540
290	229.3	2774.6	3026.9	7.0373	219.0	2773.6	3025.4	7.0149	209.60	2772.5	3024.0	6.9933
300	233.9	2791.2	3048.5	7.0754	223.4	2790.2	3047.2	7.0531	213.8	2789.2	3045.8	7.0317
310	238.4	2807.8	3070.1	7.1127	227.8	2806.9	3068.8	7.0905	218.0	2805.9	3067.5	7.0692
320	242.9	2824.3	3091.5	7.1492	232.1	2823.4	3090.3	7.1272	222.2	2822.5	3089.1	7.1060
330	247.4	2840.8	3112.9	7.1850	236.4	2840.0	3111.8	7.1631	226.3	2839.1	3110.7	7.1420
340	251.9	2857.3	3134.3	7.2201	240.7	2856.5	3133.2	7.1983	230.4	2855.7	3132.2	7.1774
350	256.3	2873.7	3155.7	7.2547	244.9	2873.0	3154.6	7.2329	234.5	2872.2	3153.6	7.2121
360	260.7	2890.2	3177.0	7.2886	249.2	2889.4	3176.0	7.2670	238.6	2888.7	3175.1	7.2462
370	265.2	2906.6	3198.3	7.3220	253.4	2905.9	3197.4	7.3005	242.7	2905.3	3196.5	7.2797
380	269.6	2923.1	3219.6	7.3549	257.7	2922.5	3218.8	7.3334	246.7	2921.8	3217.9	7.3128
390	274.0	2939.6	3241.0	7.3873	261.9	2939.0	3240.1	7.3659	250.8	2938.4	3239.3	7.3453
400	278.3	2956.1	3262.3	7.4193	266.1	2955.5	3261.5	7.3979	254.8	2954.9	3260.7	7.3774
420	287.1	2989.3	3305.1	7.4819	274.4	2988.7	3304.3	7.4606	262.8	2988.2	3303.6	7.4401
440	295.8	3022.6	3347.9	7.5428	282.7	3022.1	3347.2	7.5216	270.8	3021.6	3346.6	7.5012
460	304.4	3056.1	3390.9	7.6023	291.0	3055.6	3390.3	7.5811	278.8	3055.1	3389.7	7.5609
480	313.1	3089.7	3434.1	7.6604	299.3	3089.3	3433.5	7.6393	286.7	3088.8	3432.9	7.6190
500	321.7	3123.6	3477.4	7.7171	307.6	3123.2	3476.9	7.6961	294.6	3122.8	3476.3	7.6759
520	330.3	3157.6	3520.9	7.7727	315.8	3157.3	3520.4	7.7517	302.5	3156.9	3519.9	7.7316
540	338.8	3191.9	3564.6	7.8272	324.0	3191.6	3564.2	7.8062	310.4	3191.2	3563.7	7.7861
560	347.4	3226.4	3608.6	7.8805	332.2	3226.1	3608.1	7.8596	318.3	3225.8	3607.7	7.8396
580	355.9	3261.2	3652.7	7.9329	340.4	3260.9	3652.3	7.9120	326.1	3260.6	3651.9	7.8920
600	364.5	3296.2	3697.1	7.9843	348.5	3295.9	3696.7	7.9634	333.9	3295.6	3696.3	7.9435
620	373.0	3331.4	3741.7	8.0348	356.7	3331.1	3741.3	8.0140	341.7	3330.9	3741.0	7.9940
640	381.5	3366.9	3786.5	8.0845	364.8	3366.6	3786.2	8.0637	349.6	3366.4	3785.9	8.0437
660	390.0	3402.6	3831.6	8.1333	373.0	3402.4	3831.3	8.1125	357.4	3402.1	3831.0	8.0926
680	398.5	3438.6	3876.9	8.1813	381.1	3438.4	3876.6	8.1606	365.2	3438.2	3876.3	8.1407
700	407.0	3474.9	3922.5	8.2287	389.2	3474.6	3922.2	8.2079	372.9	3474.4	3922.0	8.1881
750	428.1	3566.6	4037.6	8.3440	409.5	3566.4	4037.3	8.3233	392.4	3566.2	4037.1	8.3034
800	449.3	3660.1	4154.3	8.4553	429.7	3659.9	4154.1	8.4346	411.8	3659.7	4153.8	8.4148
850	470.4	3755.2	4272.6	8.5631	449.9	3755.0	4272.4	8.5425	431.1	3754.9	4272.2	8.5227
900	491.5	3851.9	4392.6	8.6676	470.1	3851.8	4392.4	8.6470	450.5	3851.6	4392.2	8.6272
950	512.6	3950.3	4514.2	8.7691	490.3	3950.1	4514.0	8.7484	469.9	3950.0	4513.8	8.7287
1000	533.7	4050.2	4637.3	8.8677	510.5	4050.1	4637.2	8.8471	489.2	4050.0	4637.0	8.8274
1100	575.8	4254.9	4888.3	9.0575	550.8	4254.7	4888.2	9.0369	527.8	4254.6	4888.0	9.0172
1200	618.0	4465.4	5145.1	9.2380	591.1	4465.2	5145.0	9.2174	566.5	4465.1	5144.9	9.1977
1300	660.1	4681.1	5407.2	9.4101	631.4	4681.0	5407.1	9.3895	605.1	4680.9	5407.0	9.3698

Definitions of symbols on page 1.

	1.25 (189.84)				**1.30** (191.64)				**1.35** (193.38)			p (t Sat.)
10^3 v	u	h	s	10^3 v	u	h	s	10^3 v	u	h	s	t
157.06	2589.9	2786.3	6.5091	151.25	2591.0	2787.6	6.4953	145.86	2591.9	2788.8	6.4821	Sat.
149.57	*2558.4*	*2745.4*	*6.4193*	*143.08*	*2555.1*	*2741.1*	*6.3935*	*137.06*	*2551.8*	*2736.8*	*6.3683*	**175**
152.15	*2569.3*	*2759.5*	*6.4506*	*145.60*	*2566.2*	*2755.5*	*6.4254*	*139.53*	*2563.1*	*2751.5*	*6.4008*	**180**
154.67	*2579.9*	*2773.2*	*6.4808*	*148.07*	*2577.0*	*2769.5*	*6.4561*	*141.94*	*2574.1*	*2765.7*	*6.4320*	**185**
157.14	2590.3	2786.7	6.5100	*150.48*	*2587.6*	*2783.2*	*6.4858*	*144.30*	*2584.8*	*2779.6*	*6.4622*	**190**
159.56	2600.4	2799.9	6.5383	152.84	2597.9	2796.6	6.5146	146.61	2595.3	2793.2	6.4914	**195**
161.95	2610.4	2812.8	6.5658	155.16	2608.0	2809.7	6.5425	148.87	2605.6	2806.5	6.5197	**200**
164.30	2620.2	2825.6	6.5926	157.45	2617.9	2822.6	6.5696	151.10	2615.6	2819.6	6.5472	**205**
166.62	2629.8	2838.1	6.6187	159.70	2627.7	2835.3	6.5960	153.29	2625.5	2832.4	6.5739	**210**
168.91	2639.3	2850.4	6.6441	161.92	2637.3	2847.8	6.6217	155.45	2635.2	2845.1	6.5999	**215**
171.17	2648.7	2862.6	6.6689	164.12	2646.7	2860.1	6.6468	157.59	2644.8	2857.5	6.6253	**220**
173.40	2657.9	2874.7	6.6932	166.29	2656.1	2872.3	6.6713	159.70	2654.2	2869.8	6.6501	**225**
175.61	2667.1	2886.6	6.7170	168.43	2665.3	2884.3	6.6953	161.78	2663.5	2881.9	6.6743	**230**
177.81	2676.1	2898.4	6.7403	170.56	2674.4	2896.2	6.7189	163.84	2672.7	2893.9	6.6980	**235**
179.98	2685.1	2910.0	6.7632	172.66	2683.5	2907.9	6.7419	165.88	2681.9	2905.8	6.7213	**240**
182.13	2693.9	2921.6	6.7856	174.75	2692.4	2919.6	6.7645	167.91	2690.9	2917.6	6.7441	**245**
184.27	2702.8	2933.1	6.8077	176.82	2701.3	2931.2	6.7867	169.91	2699.8	2929.2	6.7664	**250**
188.50	2720.2	2955.8	6.8507	180.91	2718.8	2954.0	6.8301	173.88	2717.5	2952.2	6.8100	**260**
192.69	2737.4	2978.3	6.8925	184.96	2736.2	2976.6	6.8720	177.80	2734.9	2974.9	6.8523	**270**
196.82	2754.5	3000.5	6.9330	188.95	2753.3	2999.0	6.9128	181.67	2752.2	2997.4	6.8932	**280**
200.92	2771.4	3022.6	6.9725	192.91	2770.3	3021.1	6.9525	185.50	2769.2	3019.7	6.9331	**290**
205.0	2788.2	3044.5	7.0111	196.84	2787.2	3043.1	6.9912	189.29	2786.2	3041.7	6.9720	**300**
209.0	2805.0	3066.2	7.0488	200.73	2804.0	3065.0	7.0290	193.06	2803.1	3063.7	7.0099	**310**
213.0	2821.6	3087.9	7.0856	204.60	2820.7	3086.7	7.0660	196.79	2819.8	3085.5	7.0471	**320**
217.0	2838.3	3109.5	7.1218	208.45	2837.4	3108.4	7.1022	200.51	2836.6	3107.3	7.0834	**330**
221.0	2854.9	3131.1	7.1572	212.27	2854.1	3130.0	7.1378	204.20	2853.3	3128.9	7.1190	**340**
224.9	2871.4	3152.6	7.1920	216.1	2870.7	3151.6	7.1727	207.9	2869.9	3150.6	7.1540	**350**
228.9	2888.0	3174.1	7.2262	219.9	2887.3	3173.1	7.2069	211.5	2886.6	3172.1	7.1884	**360**
232.8	2904.6	3195.5	7.2598	223.6	2903.9	3194.6	7.2406	215.2	2903.2	3193.7	7.2221	**370**
236.7	2921.1	3217.0	7.2929	227.4	2920.5	3216.1	7.2738	218.8	2919.8	3215.2	7.2554	**380**
240.6	2937.7	3238.4	7.3255	231.1	2937.1	3237.6	7.3065	222.4	2936.5	3236.7	7.2881	**390**
244.4	2954.3	3259.9	7.3576	234.9	2953.7	3259.1	7.3386	226.0	2953.1	3258.3	7.3203	**400**
252.2	2987.6	3302.8	7.4205	242.3	2987.1	3302.1	7.4016	233.2	2986.5	3301.4	7.3834	**420**
259.8	3021.1	3345.9	7.4817	249.7	3020.6	3345.2	7.4629	240.3	3020.1	3344.5	7.4447	**440**
267.5	3054.6	3389.0	7.5414	257.1	3054.2	3388.4	7.5226	247.4	3053.7	3387.8	7.5046	**460**
275.1	3088.4	3432.3	7.5996	264.4	3088.0	3431.7	7.5809	254.5	3087.5	3431.2	7.5629	**480**
282.7	3122.4	3475.8	7.6566	271.8	3121.9	3475.2	7.6379	261.6	3121.5	3474.7	7.6200	**500**
290.3	3156.5	3519.4	7.7123	279.1	3156.1	3518.9	7.6937	268.6	3155.8	3518.4	7.6758	**520**
297.9	3190.9	3563.2	7.7669	286.3	3190.5	3562.8	7.7483	275.6	3190.2	3562.3	7.7305	**540**
305.4	3225.5	3607.3	7.8203	293.6	3225.1	3606.8	7.8018	282.7	3224.8	3606.4	7.7840	**560**
313.0	3260.3	3651.5	7.8728	300.9	3260.0	3651.1	7.8543	289.6	3259.6	3650.7	7.8366	**580**
320.5	3295.3	3695.9	7.9243	308.1	3295.0	3695.6	7.9059	296.6	3294.7	3695.2	7.8881	**600**
328.0	3330.6	3740.6	7.9749	315.3	3330.3	3740.2	7.9565	303.6	3330.0	3739.9	7.9388	**620**
335.5	3366.1	3785.5	8.0246	322.5	3365.9	3785.2	8.0062	310.5	3365.6	3784.8	7.9885	**640**
343.0	3401.9	3830.7	8.0735	329.8	3401.7	3830.3	8.0551	317.5	3401.4	3830.0	8.0375	**660**
350.5	3437.9	3876.0	8.1216	337.0	3437.7	3875.7	8.1033	324.4	3437.5	3875.4	8.0856	**680**
358.0	3474.2	3921.7	8.1690	344.1	3474.0	3921.4	8.1507	331.4	3473.8	3921.1	8.1330	**700**
376.6	3566.0	4036.8	8.2844	362.1	3565.9	4036.6	8.2661	348.6	3565.7	4036.3	8.2485	**750**
395.3	3659.5	4153.6	8.3958	380.0	3659.4	4153.4	8.3776	365.9	3659.2	4153.2	8.3600	**800**
413.9	3754.7	4272.0	8.5037	397.9	3754.6	4271.9	8.4855	383.2	3754.4	4271.7	8.4679	**850**
432.5	3851.5	4392.1	8.6082	415.8	3851.3	4391.9	8.5900	400.4	3851.2	4391.7	8.5725	**900**
451.0	3949.9	4513.7	8.7097	433.7	3949.7	4513.5	8.6915	417.6	3949.6	4513.3	8.6740	**950**
469.6	4049.9	4636.9	8.8085	451.5	4049.7	4636.7	8.7903	434.8	4049.6	4636.6	8.7727	**1000**
506.7	4254.5	4887.9	8.9982	487.2	4254.4	4887.8	8.9800	469.2	4254.3	4887.6	8.9625	**1100**
543.8	4465.0	5144.8	9.1788	522.9	4464.9	5144.6	9.1606	503.5	4464.8	5144.5	9.1431	**1200**
580.9	4680.8	5406.8	9.3509	558.5	4680.6	5406.7	9.3327	537.8	4680.5	5406.6	9.3152	**1300**

Definitions of symbols on page 1.

Table 3. Vapor

p (t Sat.)	1.40 (195.07)				1.45 (196.72)				1.50 (198.32)			
t	$10^3 v$	u	h	s	$10^3 v$	u	h	s	$10^3 v$	u	h	s
Sat.	140.84	2592.8	2790.0	6.4693	136.16	2593.7	2791.1	6.4568	131.77	2594.5	2792.2	6.4448
180	*133.88*	*2559.9*	*2747.4*	*6.3766*	*128.61*	*2556.7*	*2743.2*	*6.3529*	*123.69*	*2553.4*	*2738.9*	*6.3296*
185	*136.25*	*2571.1*	*2761.9*	*6.4085*	*130.93*	*2568.1*	*2758.0*	*6.3854*	*125.97*	*2565.0*	*2754.0*	*6.3627*
190	*138.55*	*2582.0*	*2776.0*	*6.4392*	*133.20*	*2579.2*	*2772.3*	*6.4166*	*128.19*	*2576.3*	*2768.6*	*6.3944*
195	*140.81*	*2592.7*	*2789.8*	*6.4688*	*135.41*	*2590.0*	*2786.4*	*6.4467*	*130.36*	*2587.3*	*2782.9*	*6.4251*
200	143.02	2603.1	2803.3	6.4975	137.57	2600.6	2800.1	6.4758	132.48	2598.1	2796.8	6.4546
205	145.20	2613.3	2816.6	6.5254	139.70	2610.9	2813.5	6.5041	134.56	2608.6	2810.4	6.4832
210	147.34	2623.3	2829.6	6.5524	141.79	2621.1	2826.7	6.5315	136.60	2618.8	2823.7	6.5110
215	149.44	2633.1	2842.4	6.5787	143.84	2631.0	2839.6	6.5581	138.61	2628.9	2836.8	6.5379
220	151.52	2642.8	2854.9	6.6044	145.87	2640.8	2852.3	6.5840	140.59	2638.8	2849.7	6.5641
225	153.57	2652.3	2867.3	6.6294	147.87	2650.5	2864.9	6.6093	142.54	2648.5	2862.3	6.5897
230	155.60	2661.8	2879.6	6.6539	149.84	2660.0	2877.2	6.6340	144.46	2658.1	2874.8	6.6146
235	157.60	2671.0	2891.7	6.6778	151.79	2669.3	2889.4	6.6581	146.37	2667.6	2887.1	6.6390
240	159.59	2680.2	2903.7	6.7012	153.72	2678.6	2901.5	6.6818	148.25	2676.9	2899.3	6.6628
245	161.55	2689.3	2915.5	6.7242	155.63	2687.8	2913.4	6.7049	150.11	2686.2	2911.3	6.6861
250	163.50	2698.3	2927.2	6.7467	157.53	2696.8	2925.3	6.7276	151.95	2695.3	2923.3	6.7090
260	167.35	2716.1	2950.4	6.7906	161.27	2714.7	2948.6	6.7718	155.59	2713.4	2946.7	6.7535
270	171.15	2733.7	2973.3	6.8331	164.96	2732.4	2971.6	6.8145	159.18	2731.1	2969.9	6.7965
280	174.90	2751.0	2995.8	6.8743	168.60	2749.8	2994.3	6.8559	162.71	2748.6	2992.7	6.8381
290	178.61	2768.1	3018.2	6.9144	172.19	2767.1	3016.7	6.8962	166.21	2765.9	3015.3	6.8785
300	182.28	2785.2	3040.4	6.9534	175.76	2784.2	3039.0	6.9354	169.66	2783.1	3037.6	6.9179
310	185.93	2802.1	3062.4	6.9915	179.29	2801.1	3061.1	6.9736	173.09	2800.2	3059.8	6.9563
320	189.54	2818.9	3084.3	7.0287	182.79	2818.0	3083.1	7.0110	176.49	2817.1	3081.9	6.9938
330	193.14	2835.7	3106.1	7.0652	186.27	2834.9	3105.0	7.0476	179.87	2834.0	3103.8	7.0305
340	196.71	2852.5	3127.8	7.1009	189.73	2851.6	3126.8	7.0834	183.22	2850.8	3125.7	7.0664
350	200.3	2869.2	3149.5	7.1360	193.17	2868.4	3148.5	7.1186	186.56	2867.6	3147.5	7.1017
360	203.8	2885.8	3171.2	7.1704	196.60	2885.1	3170.2	7.1531	189.88	2884.4	3169.2	7.1363
370	207.3	2902.5	3192.8	7.2043	200.00	2901.8	3191.8	7.1870	193.18	2901.1	3190.9	7.1703
380	210.8	2919.2	3214.3	7.2376	203.40	2918.5	3213.4	7.2204	196.46	2917.9	3212.6	7.2037
390	214.3	2935.9	3235.9	7.2703	206.78	2935.2	3235.1	7.2532	199.74	2934.6	3234.2	7.2366
400	217.8	2952.5	3257.5	7.3026	210.1	2951.9	3256.7	7.2855	203.0	2951.3	3255.8	7.2690
420	224.7	2986.0	3300.6	7.3658	216.9	2985.4	3299.9	7.3488	209.5	2984.9	3299.1	7.3323
440	231.6	3019.5	3343.8	7.4272	223.5	3019.0	3343.1	7.4103	216.0	3018.5	3342.5	7.3940
460	238.5	3053.2	3387.1	7.4871	230.2	3052.8	3386.5	7.4703	222.4	3052.3	3385.9	7.4540
480	245.3	3087.1	3430.6	7.5456	236.8	3086.7	3430.0	7.5288	228.8	3086.2	3429.4	7.5126
500	252.1	3121.1	3474.1	7.6027	243.4	3120.7	3473.6	7.5860	235.2	3120.3	3473.1	7.5698
520	258.9	3155.4	3517.9	7.6586	249.9	3155.0	3517.4	7.6419	241.5	3154.6	3516.9	7.6258
540	265.7	3189.8	3561.8	7.7132	256.5	3189.5	3561.3	7.6966	247.8	3189.1	3560.9	7.6805
560	272.5	3224.5	3605.9	7.7668	263.0	3224.1	3605.5	7.7502	254.2	3223.8	3605.1	7.7342
580	279.2	3259.3	3650.3	7.8194	269.5	3259.0	3649.8	7.8028	260.5	3258.7	3649.4	7.7868
600	286.0	3294.4	3694.8	7.8710	276.0	3294.1	3694.4	7.8545	266.8	3293.9	3694.0	7.8385
620	292.7	3329.8	3739.5	7.9217	282.5	3329.5	3739.2	7.9052	273.0	3329.2	3738.8	7.8892
640	299.4	3365.4	3784.5	7.9715	289.0	3365.1	3784.2	7.9550	279.3	3364.8	3783.8	7.9391
660	306.1	3401.2	3829.7	8.0204	295.5	3400.9	3829.4	8.0040	285.6	3400.7	3829.1	7.9881
680	312.8	3437.2	3875.1	8.0686	301.9	3437.0	3874.8	8.0522	291.8	3436.8	3874.5	8.0363
700	319.5	3473.6	3920.8	8.1160	308.4	3473.3	3920.5	8.0996	298.1	3473.1	3920.2	8.0837
750	336.2	3565.5	4036.1	8.2315	324.5	3565.3	4035.8	8.2152	313.7	3565.1	4035.6	8.1993
800	352.8	3659.0	4153.0	8.3431	340.6	3658.9	4152.7	8.3267	329.2	3658.7	4152.5	8.3109
850	369.4	3754.2	4271.5	8.4510	356.7	3754.1	4271.3	8.4347	344.8	3753.9	4271.1	8.4189
900	386.1	3851.1	4391.5	8.5556	372.7	3850.9	4391.4	8.5393	360.3	3850.8	4391.2	8.5235
950	402.7	3949.5	4513.2	8.6571	388.7	3949.3	4513.0	8.6408	375.8	3949.2	4512.9	8.6251
1000	419.2	4049.5	4636.4	8.7559	404.8	4049.4	4636.3	8.7396	391.3	4049.2	4636.1	8.7238
1100	452.4	4254.1	4887.5	8.9457	436.8	4254.0	4887.4	8.9294	422.2	4253.9	4887.2	8.9137
1200	485.5	4464.7	5144.4	9.1262	468.8	4464.5	5144.3	9.1100	453.1	4464.4	5144.1	9.0942
1300	518.6	4680.4	5406.5	9.2984	500.8	4680.3	5406.4	9.2821	484.1	4680.2	5406.3	9.2664

Definitions of symbols on page 1.

	1.55 (199.88)				**1.60** (201.41)				**1.65** (202.89)			**p** (t Sat.)
10^3 v	u	h	s	10^3 v	u	h	s	10^3 v	u	h	s	t
127.66	2595.3	2793.1	6.4331	123.80	2596.0	2794.0	6.4218	120.16	2596.6	2794.9	6.4108	Sat.
												180
121.31	*2561.9*	*2749.9*	*6.3404*	*116.94*	*2558.7*	*2745.8*	*6.3184*					**185**
123.50	*2573.4*	*2764.8*	*6.3727*	*119.09*	*2570.5*	*2761.0*	*6.3513*	*114.94*	*2567.4*	*2757.1*	*6.3302*	**190**
125.63	*2584.6*	*2779.3*	*6.4038*	*121.18*	*2581.8*	*2775.7*	*6.3829*	*117.00*	*2579.0*	*2772.1*	*6.3624*	**195**
127.71	2595.5	2793.4	6.4338	*123.23*	*2592.9*	*2790.1*	*6.4134*	*119.02*	*2590.3*	*2786.6*	*6.3934*	**200**
129.75	2606.1	2807.2	6.4628	125.23	2603.7	2804.1	6.4428	120.98	2601.2	2800.8	6.4232	**205**
131.75	2616.6	2820.8	6.4909	127.19	2614.2	2817.7	6.4713	122.91	2611.9	2814.7	6.4521	**210**
133.71	2626.7	2834.0	6.5182	129.12	2624.6	2831.2	6.4989	124.80	2622.4	2828.3	6.4800	**215**
135.65	2636.8	2847.0	6.5447	131.01	2634.7	2844.3	6.5257	126.65	2632.6	2841.6	6.5072	**220**
137.55	2646.6	2859.8	6.5705	132.87	2644.7	2857.3	6.5518	128.48	2642.7	2854.7	6.5335	**225**
139.43	2656.3	2872.4	6.5957	134.71	2654.4	2870.0	6.5773	130.27	2652.6	2867.5	6.5592	**230**
141.29	2665.8	2884.8	6.6203	136.52	2664.1	2882.5	6.6021	132.05	2662.3	2880.2	6.5843	**235**
143.12	2675.3	2897.1	6.6443	138.32	2673.6	2894.9	6.6263	133.80	2671.9	2892.7	6.6087	**240**
144.94	2684.6	2909.2	6.6679	140.09	2683.0	2907.1	6.6500	135.53	2681.4	2905.0	6.6326	**245**
146.73	2693.8	2921.2	6.6909	141.84	2692.3	2919.2	6.6732	137.24	2690.7	2917.2	6.6560	**250**
150.28	2712.0	2944.9	6.7357	145.30	2710.6	2943.0	6.7183	140.61	2709.1	2941.2	6.7014	**260**
153.77	2729.8	2968.2	6.7789	148.69	2728.5	2966.4	6.7618	143.93	2727.2	2964.7	6.7452	**270**
157.21	2747.4	2991.1	6.8208	152.04	2746.2	2989.5	6.8039	147.19	2745.0	2987.9	6.7875	**280**
160.60	2764.8	3013.8	6.8614	155.35	2763.7	3012.3	6.8447	150.42	2762.6	3010.8	6.8285	**290**
163.97	2782.1	3036.2	6.9009	158.62	2781.1	3034.8	6.8844	153.60	2780.0	3033.5	6.8684	**300**
167.30	2799.2	3058.5	6.9395	161.86	2798.2	3057.2	6.9231	156.75	2797.3	3055.9	6.9072	**310**
170.60	2816.2	3080.6	6.9771	165.07	2815.3	3079.4	6.9609	159.88	2814.4	3078.2	6.9451	**320**
173.88	2833.2	3102.7	7.0139	168.26	2832.3	3101.5	6.9978	162.98	2831.4	3100.3	6.9822	**330**
177.13	2850.0	3124.6	7.0500	171.42	2849.2	3123.5	7.0340	166.05	2848.4	3122.4	7.0184	**340**
180.37	2866.9	3146.4	7.0855	174.56	2866.1	3145.4	7.0694	169.11	2865.3	3144.3	7.0539	**350**
183.59	2883.6	3168.2	7.1200	177.69	2882.9	3167.2	7.1042	172.15	2882.2	3166.2	7.0888	**360**
186.79	2900.4	3189.9	7.1540	180.80	2899.7	3189.0	7.1383	175.18	2899.0	3188.1	7.1230	**370**
189.98	2917.2	3211.7	7.1875	183.90	2916.5	3210.8	7.1719	178.18	2915.9	3209.9	7.1566	**380**
193.15	2934.0	3233.3	7.2205	186.98	2933.3	3232.5	7.2049	181.18	2932.7	3231.6	7.1897	**390**
196.32	2950.7	3255.0	7.2529	190.05	2950.1	3254.2	7.2374	184.17	2949.5	3253.4	7.2223	**400**
202.61	2984.3	3298.4	7.3164	196.16	2983.8	3297.6	7.3009	190.10	2983.2	3296.9	7.2859	**420**
208.87	3018.0	3341.8	7.3781	202.24	3017.5	3341.1	7.3627	196.00	3017.0	3340.4	7.3478	**440**
215.10	3051.8	3385.2	7.4382	208.28	3051.4	3384.6	7.4229	201.87	3050.9	3384.0	7.4081	**460**
221.30	3085.8	3428.8	7.4969	214.30	3085.3	3428.2	7.4816	207.71	3084.9	3427.6	7.4668	**480**
227.5	3119.9	3472.5	7.5541	220.3	3119.5	3472.0	7.5390	213.5	3119.1	3471.4	7.5242	**500**
233.6	3154.2	3516.4	7.6101	226.3	3153.9	3515.9	7.5950	219.3	3153.5	3515.4	7.5803	**520**
239.8	3188.7	3560.4	7.6650	232.2	3188.4	3559.9	7.6499	225.1	3188.0	3559.5	7.6352	**540**
245.9	3223.5	3604.6	7.7187	238.1	3223.1	3604.2	7.7036	230.9	3222.8	3603.7	7.6890	**560**
252.0	3258.4	3649.0	7.7713	244.1	3258.1	3648.6	7.7563	236.6	3257.8	3648.2	7.7418	**580**
258.1	3293.6	3693.6	7.8230	250.0	3293.3	3693.2	7.8080	242.3	3293.0	3692.9	7.7935	**600**
264.2	3329.0	3738.4	7.8738	255.9	3328.7	3738.1	7.8588	248.1	3328.4	3737.7	7.8443	**620**
270.3	3364.6	3783.5	7.9236	261.8	3364.3	3783.1	7.9087	253.8	3364.1	3782.8	7.8942	**640**
276.3	3400.4	3828.7	7.9727	267.6	3400.2	3828.4	7.9578	259.5	3400.0	3828.1	7.9433	**660**
282.4	3436.5	3874.2	8.0209	273.5	3436.3	3873.9	8.0060	265.2	3436.1	3873.6	7.9916	**680**
288.4	3472.9	3920.0	8.0684	279.4	3472.7	3919.7	8.0535	270.9	3472.5	3919.4	8.0391	**700**
303.5	3564.9	4035.3	8.1840	294.0	3564.7	4035.1	8.1692	285.1	3564.5	4034.8	8.1548	**750**
318.6	3658.5	4152.3	8.2956	308.6	3658.3	4152.1	8.2808	299.2	3658.2	4151.9	8.2664	**800**
333.6	3753.8	4270.9	8.4036	323.2	3753.6	4270.7	8.3888	313.3	3753.5	4270.5	8.3745	**850**
348.6	3850.6	4391.0	8.5082	337.7	3850.5	4390.8	8.4935	327.5	3850.3	4390.6	8.4791	**900**
363.6	3949.1	4512.7	8.6098	352.3	3948.9	4512.5	8.5951	341.6	3948.8	4512.4	8.5807	**950**
378.6	4049.1	4636.0	8.7086	366.8	4049.0	4635.8	8.6938	355.7	4048.8	4635.7	8.6795	**1000**
408.6	4253.8	4887.1	8.8984	395.8	4253.7	4887.0	8.8837	383.8	4253.5	4886.8	8.8694	**1100**
438.5	4464.3	5144.0	9.0790	424.8	4464.2	5143.9	9.0643	411.9	4464.1	5143.8	9.0500	**1200**
468.5	4680.0	5406.1	9.2512	453.8	4679.9	5406.0	9.2364	440.1	4679.8	5405.9	9.2222	**1300**

Definitions of symbols on page 1.

Table 3. Vapor

p (t Sat.)		1.70 (204.34)				1.75 (205.76)				1.80 (207.15)		
t	10^3 v	u	h	s	10^3 v	u	h	s	10^3 v	u	h	s
Sat.	116.73	2597.3	2795.7	6.4000	113.49	2597.8	2796.4	6.3896	110.42	2598.4	2797.1	6.3794
200	*115.04*	*2587.6*	*2783.2*	*6.3737*	*111.29*	*2584.9*	*2779.6*	*6.3542*	*107.75*	*2582.1*	*2776.0*	*6.3351*
205	116.98	2598.7	2797.6	6.4039	*113.20*	*2596.1*	*2794.2*	*6.3850*	*109.62*	*2593.6*	*2790.9*	*6.3663*
210	118.87	2609.5	2811.6	6.4332	115.06	2607.1	2808.5	6.4146	111.46	2604.7	2805.3	6.3964
215	120.72	2620.1	2825.4	6.4615	116.88	2617.9	2822.4	6.4433	113.25	2615.6	2819.4	6.4254
220	122.54	2630.5	2838.8	6.4890	118.67	2628.4	2836.1	6.4711	115.01	2626.2	2833.2	6.4535
225	124.33	2640.7	2852.1	6.5156	120.42	2638.7	2849.4	6.4980	116.73	2636.6	2846.7	6.4808
230	126.09	2650.7	2865.0	6.5416	122.15	2648.8	2862.5	6.5242	118.42	2646.8	2860.0	6.5073
235	127.83	2660.5	2877.8	6.5668	123.85	2658.7	2875.4	6.5498	120.09	2656.9	2873.0	6.5330
240	129.54	2670.2	2890.4	6.5915	125.53	2668.5	2888.2	6.5747	121.74	2666.7	2885.9	6.5582
245	131.24	2679.7	2902.9	6.6156	127.19	2678.1	2900.7	6.5990	123.36	2676.4	2898.5	6.5827
250	132.91	2689.2	2915.1	6.6392	128.83	2687.6	2913.1	6.6227	124.97	2686.0	2911.0	6.6066
260	136.21	2707.7	2939.3	6.6849	132.05	2706.3	2937.4	6.6688	128.12	2704.9	2935.5	6.6530
270	139.44	2725.9	2963.0	6.7290	135.21	2724.6	2961.2	6.7131	131.21	2723.3	2959.5	6.6976
280	142.63	2743.8	2986.3	6.7715	138.32	2742.6	2984.7	6.7559	134.25	2741.4	2983.0	6.7406
290	145.77	2761.5	3009.3	6.8127	141.39	2760.4	3007.8	6.7973	137.25	2759.2	3006.3	6.7822
300	148.87	2779.0	3032.1	6.8528	144.42	2777.9	3030.6	6.8375	140.21	2776.9	3029.2	6.8226
310	151.95	2796.3	3054.6	6.8918	147.41	2795.3	3053.3	6.8767	143.13	2794.3	3051.9	6.8619
320	154.99	2813.5	3077.0	6.9298	150.38	2812.5	3075.7	6.9148	146.03	2811.6	3074.5	6.9002
330	158.01	2830.6	3099.2	6.9669	153.32	2829.7	3098.0	6.9521	148.90	2828.8	3096.8	6.9376
340	161.00	2847.6	3121.3	7.0033	156.24	2846.8	3120.2	6.9886	151.74	2845.9	3119.1	6.9742
350	163.98	2864.5	3143.3	7.0389	159.14	2863.8	3142.3	7.0243	154.57	2863.0	3141.2	7.0100
360	166.94	2881.4	3165.2	7.0738	162.02	2880.7	3164.3	7.0593	157.38	2880.0	3163.3	7.0451
370	169.88	2898.3	3187.1	7.1081	164.89	2897.6	3186.2	7.0937	160.17	2896.9	3185.2	7.0796
380	172.81	2915.2	3209.0	7.1418	167.74	2914.5	3208.1	7.1274	162.95	2913.9	3207.2	7.1134
390	175.72	2932.1	3230.8	7.1750	170.58	2931.4	3229.9	7.1606	165.71	2930.8	3229.1	7.1467
400	178.62	2948.9	3252.6	7.2076	173.40	2948.3	3251.8	7.1933	168.47	2947.7	3250.9	7.1794
420	184.40	2982.7	3296.1	7.2714	179.02	2982.1	3295.4	7.2572	173.94	2981.5	3294.6	7.2434
440	190.13	3016.5	3339.7	7.3333	184.60	3016.0	3339.0	7.3192	179.37	3015.4	3338.3	7.3055
460	195.84	3050.4	3383.3	7.3937	190.15	3049.9	3382.7	7.3796	184.78	3049.5	3382.1	7.3660
480	201.51	3084.5	3427.0	7.4525	195.67	3084.0	3426.5	7.4385	190.15	3083.6	3425.9	7.4249
500	207.2	3118.7	3470.9	7.5099	201.2	3118.3	3470.3	7.4960	195.50	3117.9	3469.8	7.4825
520	212.8	3153.1	3514.9	7.5661	206.6	3152.7	3514.3	7.5522	200.83	3152.3	3513.8	7.5387
540	218.4	3187.7	3559.0	7.6210	212.1	3187.3	3558.5	7.6072	206.15	3187.0	3558.0	7.5938
560	224.0	3222.5	3603.3	7.6749	217.5	3222.1	3602.8	7.6611	21·1.44	3221.8	3602.4	7.6477
580	229.6	3257.5	3647.8	7.7276	223.0	3257.2	3647.4	7.7139	216.72	3256.9	3647.0	7.7005
600	235.2	3292.7	3692.5	7.7794	228.4	3292.4	3692.1	7.7657	222.0	3292.1	3691.7	7.7523
620	240.7	3328.1	3737.4	7.8302	233.8	3327.9	3737.0	7.8165	227.2	3327.6	3736.6	7.8032
640	246.3	3363.8	3782.5	7.8802	239.2	3363.5	3782.1	7.8665	232.5	3363.3	3781.8	7.8532
660	251.8	3399.7	3827.8	7.9293	244.6	3399.5	3827.5	7.9156	237.7	3399.2	3827.1	7.9024
680	257.3	3435.9	3873.3	7.9776	249.9	3435.6	3873.0	7.9639	243.0	3435.4	3872.7	7.9507
700	262.9	3472.2	3919.1	8.0251	255.3	3472.0	3918.8	8.0115	248.2	3471.8	3918.5	7.9983
720	268.4	3508.9	3965.1	8.0719	260.7	3508.7	3964.9	8.0583	253.4	3508.5	3964.6	8.0451
740	273.9	3545.8	4011.4	8.1180	266.0	3545.6	4011.1	8.1044	258.6	3545.4	4010.9	8.0912
760	279.4	3582.9	4057.9	8.1635	271.4	3582.7	4057.6	8.1499	263.8	3582.5	4057.4	8.1367
780	284.9	3620.3	4104.6	8.2083	276.7	3620.1	4104.4	8.1947	269.0	3620.0	4104.2	8.1816
800	290.4	3658.0	4151.6	8.2525	282.1	3657.8	4151.4	8.2390	274.2	3657.6	4151.2	8.2258
850	304.1	3753.3	4270.3	8.3606	295.4	3753.2	4270.1	8.3470	287.2	3753.0	4269.9	8.3339
900	317.8	3850.2	4390.5	8.4652	308.7	3850.1	4390.3	8.4517	300.1	3849.9	4390.1	8.4386
950	331.5	3948.7	4512.2	8.5669	322.0	3948.5	4512.1	8.5534	313.1	3948.4	4511.9	8.5403
1000	345.2	4048.7	4635.5	8.6657	335.3	4048.6	4635.4	8.6522	326.0	4048.5	4635.2	8.6391
1100	372.5	4253.4	4886.7	8.8555	361.9	4253.3	4886.6	8.8421	351.8	4253.2	4886.4	8.8290
1200	399.8	4464.0	5143.7	9.0362	388.4	4463.8	5143.5	9.0227	377.6	4463.7	5143.4	9.0096
1300	427.1	4679.7	5405.8	9.2083	414.9	4679.6	5405.7	9.1949	403.4	4679.5	5405.6	9.1818

Definitions of symbols on page 1.

	1.85 (208.51)				**1.90** (209.84)				**1.95** (211.14)			**p** (t Sat.)
10^3 v	u	h	s	10^3 v	u	h	s	10^3 v	u	h	s	t
107.51	2598.9	2797.8	6.3694	104.75	2599.4	2798.4	6.3597	102.12	2599.9	2799.0	6.3502	**Sat.**
104.38	*2579.3*	*2772.4*	*6.3162*	*101.19*	*2576.4*	*2768.7*	*6.2975*	*98.16*	*2573.5*	*2765.0*	*6.2791*	**200**
106.24	*2590.9*	*2787.5*	*6.3479*	*103.02*	*2588.3*	*2784.0*	*6.3297*	*99.97*	*2585.6*	*2780.5*	*6.3118*	**205**
108.04	2602.3	2802.1	6.3784	104.81	2599.8	2798.9	6.3607	*101.73*	*2597.2*	*2795.6*	*6.3432*	**210**
109.81	2613.3	2816.4	6.4078	106.55	2610.9	2813.4	6.3905	103.45	2608.6	2810.3	6.3734	**215**
111.54	2624.0	2830.4	6.4363	108.25	2621.8	2827.5	6.4193	105.12	2619.6	2824.6	6.4026	**220**
113.23	2634.6	2844.1	6.4638	109.92	2632.5	2841.3	6.4472	106.77	2630.4	2838.6	6.4308	**225**
114.90	2644.9	2857.5	6.4906	111.55	2642.9	2854.9	6.4742	108.38	2640.9	2852.3	6.4581	**230**
116.53	2655.0	2870.6	6.5166	113.16	2653.2	2868.2	6.5005	109.96	2651.3	2865.7	6.4847	**235**
118.15	2665.0	2883.6	6.5420	114.75	2663.2	2881.2	6.5261	111.52	2661.4	2878.9	6.5105	**240**
119.74	2674.8	2896.3	6.5667	116.31	2673.1	2894.1	6.5510	113.05	2671.4	2891.9	6.5357	**245**
121.31	2684.4	2908.9	6.5908	117.85	2682.8	2906.8	6.5754	114.57	2681.2	2904.6	6.5602	**250**
124.40	2703.4	2933.5	6.6376	120.88	2701.9	2931.6	6.6224	117.54	2700.5	2929.7	6.6076	**260**
127.43	2721.9	2957.7	6.6824	123.85	2720.6	2955.9	6.6676	120.44	2719.3	2954.1	6.6531	**270**
130.41	2740.2	2981.4	6.7257	126.76	2738.9	2979.8	6.7111	123.30	2737.7	2978.1	6.6968	**280**
133.33	2758.1	3004.8	6.7675	129.62	2756.9	3003.2	6.7531	126.10	2755.8	3001.7	6.7391	**290**
136.23	2775.8	3027.8	6.8081	132.45	2774.7	3026.4	6.7939	128.87	2773.7	3025.0	6.7800	**300**
139.08	2793.3	3050.6	6.8476	135.24	2792.3	3049.3	6.8335	131.60	2791.3	3047.9	6.8198	**310**
141.91	2810.7	3073.2	6.8860	138.01	2809.8	3072.0	6.8721	134.30	2808.8	3070.7	6.8585	**320**
144.71	2827.9	3095.7	6.9235	140.74	2827.1	3094.5	6.9097	136.98	2826.2	3093.3	6.8963	**330**
147.49	2845.1	3118.0	6.9602	143.46	2844.3	3116.9	6.9465	139.63	2843.5	3115.7	6.9332	**340**
150.25	2862.2	3140.2	6.9961	146.15	2861.4	3139.1	6.9825	142.26	2860.6	3138.0	6.9692	**350**
152.99	2879.2	3162.3	7.0313	148.83	2878.5	3161.3	7.0178	144.88	2877.7	3160.3	7.0046	**360**
155.71	2896.2	3184.3	7.0658	151.48	2895.5	3183.3	7.0524	147.47	2894.8	3182.4	7.0393	**370**
158.42	2913.2	3206.3	7.0997	154.13	2912.5	3205.4	7.0864	150.06	2911.9	3204.5	7.0733	**380**
161.12	2930.1	3228.2	7.1330	156.76	2929.5	3227.3	7.1198	152.62	2928.9	3226.5	7.1068	**390**
163.80	2947.1	3250.1	7.1658	159.38	2946.5	3249.3	7.1528	155.18	2945.9	3248.3	7.1397	**400**
169.13	2981.0	3293.9	7.2299	164.58	2980.4	3293.1	7.2168	160.26	2979.9	3292.4	7.2040	**420**
174.43	3014.9	3337.6	7.2921	169.75	3014.4	3336.9	7.2791	165.30	3013.9	3336.2	7.2664	**440**
179.69	3049.0	3381.4	7.3527	174.88	3048.5	3380.8	7.3397	170.31	3048.0	3380.1	7.3271	**460**
184.93	3083.2	3425.3	7.4117	179.99	3082.7	3424.7	7.3988	175.29	3082.3	3424.1	7.3862	**480**
190.14	3117.5	3469.2	7.4693	185.07	3117.1	3468.7	7.4565	180.25	3116.6	3468.1	7.4440	**500**
195.34	3152.0	3513.3	7.5256	190.13	3151.6	3512.8	7.5128	185.19	3151.2	3512.3	7.5003	**520**
200.51	3186.6	3557.6	7.5807	195.17	3186.3	3557.1	7.5679	190.11	3185.9	3556.6	7.5555	**540**
205.67	3221.5	3602.0	7.6346	200.20	3221.1	3601.5	7.6219	195.01	3220.8	3601.1	7.6095	**560**
210.81	3256.5	3646.5	7.6875	205.21	3256.2	3646.1	7.6748	199.89	3255.9	3645.7	7.6625	**580**
215.9	3291.8	3691.3	7.7394	210.2	3291.5	3690.9	7.7267	204.8	3291.2	3690.5	7.7144	**600**
221.1	3327.3	3736.3	7.7903	215.2	3327.0	3735.9	7.7777	209.6	3326.8	3735.5	7.7654	**620**
226.2	3363.0	3781.4	7.8403	220.2	3362.8	3781.1	7.8277	214.5	3362.5	3780.7	7.8154	**640**
231.3	3399.0	3826.8	7.8895	225.1	3398.7	3826.5	7.8769	219.3	3398.5	3826.2	7.8646	**660**
236.4	3435.2	3872.4	7.9378	230.1	3434.9	3872.1	7.9253	224.2	3434.7	3871.8	7.9130	**680**
241.4	3471.6	3918.3	7.9854	235.0	3471.4	3918.0	7.9729	229.0	3471.2	3917.7	7.9606	**700**
246.5	3508.3	3964.3	8.0322	240.0	3508.1	3964.0	8.0197	233.8	3507.8	3963.8	8.0075	**720**
251.6	3545.2	4010.6	8.0784	244.9	3545.0	4010.4	8.0659	238.6	3544.8	4010.1	8.0537	**740**
256.6	3582.4	4057.2	8.1239	249.9	3582.2	4056.9	8.1114	243.4	3582.0	4056.7	8.0992	**760**
261.7	3619.8	4103.9	8.1688	254.8	3619.6	4103.7	8.1563	248.2	3619.4	4103.5	8.1441	**780**
266.8	3657.5	4151.0	8.2130	259.7	3657.3	4150.8	8.2005	253.0	3657.1	4150.5	8.1884	**800**
279.4	3752.8	4269.7	8.3211	272.0	3752.7	4269.5	8.3087	265.0	3752.5	4269.3	8.2965	**850**
292.0	3849.8	4389.9	8.4258	284.3	3849.6	4389.8	8.4134	277.0	3849.5	4389.6	8.4013	**900**
304.6	3948.3	4511.7	8.5275	296.5	3948.1	4511.6	8.5151	288.9	3948.0	4511.4	8.5030	**950**
317.2	4048.3	4635.1	8.6263	308.8	4048.2	4634.9	8.6139	300.9	4048.1	4634.8	8.6018	**1000**
342.3	4253.1	4886.3	8.8163	333.3	4253.0	4886.2	8.8039	324.7	4252.8	4886.0	8.7918	**1100**
367.4	4463.6	5143.3	8.9969	357.7	4463.5	5143.2	8.9845	348.6	4463.4	5143.0	8.9724	**1200**
392.5	4679.3	5405.5	9.1691	382.5	4679.2	5405.3	9.1567	372.4	4679.1	5405.2	9.1446	**1300**

Definitions of symbols on page 1.

Table 3. Vapor

p (t Sat.)	2.00 (212.42)				2.05 (213.67)				2.10 (214.90)			
t	10^3 v	u	h	s	10^3 v	u	h	s	10^3 v	u	h	s
Sat.	99.63	2600.3	2799.5	6.3409	97.25	2600.7	2800.0	6.3318	94.98	2601.0	2800.5	6.3229
200	*95.27*	*2570.6*	*2761.1*	*6.2608*	*92.52*	*2567.6*	*2757.3*	*6.2427*				
205	*97.06*	*2582.8*	*2777.0*	*6.2941*	*94.29*	*2580.1*	*2773.4*	*6.2765*	*91.65*	*2577.2*	*2769.7*	*6.2591*
210	*98.80*	*2594.7*	*2792.3*	*6.3259*	*96.01*	*2592.1*	*2788.9*	*6.3089*	*93.35*	*2589.5*	*2785.5*	*6.2920*
215	100.50	2606.2	2807.2	6.3566	97.69	2603.7	2804.0	6.3399	95.01	2601.3	2800.8	6.3235
220	102.15	2617.4	2821.7	6.3861	99.32	2615.1	2818.7	6.3699	96.62	2612.8	2815.7	6.3538
225	103.77	2628.3	2835.8	6.4147	100.92	2626.1	2833.0	6.3987	98.20	2623.9	2830.2	6.3831
230	105.36	2638.9	2849.6	6.4423	102.48	2636.9	2847.0	6.4267	99.74	2634.9	2844.3	6.4113
235	106.91	2649.4	2863.2	6.4691	104.02	2647.5	2860.7	6.4538	101.25	2645.5	2858.2	6.4387
240	108.45	2659.6	2876.5	6.4952	105.52	2657.8	2874.1	6.4801	102.74	2656.0	2871.7	6.4653
245	109.95	2669.7	2889.6	6.5205	107.01	2668.0	2887.3	6.5057	104.20	2666.2	2885.0	6.4911
250	111.44	2679.6	2902.5	6.5453	108.47	2678.0	2900.3	6.5306	105.64	2676.3	2898.1	6.5162
260	114.36	2699.0	2927.7	6.5931	111.34	2697.5	2925.7	6.5788	108.45	2696.0	2923.7	6.5647
270	117.21	2717.9	2952.3	6.6388	114.13	2716.5	2950.5	6.6248	111.20	2715.2	2948.7	6.6111
280	120.01	2736.4	2976.4	6.6828	116.88	2735.2	2974.8	6.6691	113.90	2733.9	2973.1	6.6556
290	122.76	2754.6	3000.2	6.7253	119.57	2753.5	2998.6	6.7118	116.54	2752.3	2997.0	6.6985
300	125.47	2772.6	3023.5	6.7664	122.23	2771.5	3022.1	6.7531	119.15	2770.4	3020.6	6.7400
310	128.14	2790.3	3046.6	6.8063	124.85	2789.3	3045.3	6.7932	121.71	2788.3	3043.9	6.7803
320	130.79	2807.9	3069.5	6.8452	127.44	2806.9	3068.2	6.8322	124.25	2806.0	3066.9	6.8194
330	133.40	2825.3	3092.1	6.8831	130.00	2824.4	3090.9	6.8702	126.76	2823.5	3089.7	6.8576
340	136.00	2842.6	3114.6	6.9201	132.54	2841.8	3113.5	6.9073	129.25	2840.9	3112.4	6.8948
350	138.57	2859.8	3137.0	6.9563	135.06	2859.1	3135.9	6.9436	131.71	2858.3	3134.9	6.9312
360	141.13	2877.0	3159.3	6.9917	137.56	2876.3	3158.3	6.9791	134.16	2875.5	3157.2	6.9668
370	143.66	2894.1	3181.4	7.0265	140.04	2893.4	3180.5	7.0140	136.59	2892.7	3179.5	7.0017
380	146.19	2911.2	3203.6	7.0606	142.51	2910.5	3202.6	7.0482	139.00	2909.8	3201.7	7.0360
390	148.70	2928.2	3225.6	7.0941	144.96	2927.6	3224.8	7.0818	141.40	2926.9	3223.9	7.0697
400	151.20	2945.2	3247.6	7.1271	147.40	2944.6	3246.8	7.1148	143.79	2944.0	3246.0	7.1027
410	153.68	2962.3	3269.6	7.1595	149.83	2961.7	3268.8	7.1473	146.17	2961.1	3268.1	7.1353
420	156.16	2979.3	3291.6	7.1915	152.25	2978.7	3290.9	7.1793	148.54	2978.2	3290.1	7.1673
430	158.62	2996.3	3313.6	7.2229	154.66	2995.8	3312.9	7.2108	150.89	2995.3	3312.1	7.1989
440	161.08	3013.4	3335.5	7.2540	157.07	3012.9	3334.9	7.2418	153.24	3012.4	3334.2	7.2300
450	163.53	3030.5	3357.5	7.2846	159.46	3030.0	3356.9	7.2725	155.58	3029.5	3356.2	7.2607
460	165.97	3047.6	3379.5	7.3147	161.84	3047.1	3378.9	7.3027	157.91	3046.6	3378.2	7.2909
470	168.41	3064.7	3401.5	7.3445	164.22	3064.2	3400.9	7.3325	160.24	3063.8	3400.3	7.3208
480	170.84	3081.8	3423.5	7.3740	166.60	3081.4	3422.9	7.3620	162.56	3080.9	3422.3	7.3503
490	173.26	3099.0	3445.5	7.4030	168.96	3098.6	3445.0	7.3911	164.87	3098.2	3444.4	7.3794
500	175.68	3116.2	3467.6	7.4317	171.32	3115.8	3467.0	7.4198	167.18	3115.4	3466.5	7.4081
520	180.49	3150.8	3511.8	7.4882	176.03	3150.4	3511.3	7.4763	171.78	3150.0	3510.8	7.4647
540	185.29	3185.6	3556.1	7.5434	180.72	3185.2	3555.7	7.5316	176.36	3184.8	3555.2	7.5200
560	190.08	3220.5	3600.6	7.5974	185.39	3220.1	3600.2	7.5856	180.92	3219.8	3599.7	7.5741
580	194.85	3255.6	3645.3	7.6504	190.04	3255.3	3644.9	7.6387	185.47	3255.0	3644.5	7.6272
600	199.60	3290.9	3690.1	7.7024	194.69	3290.6	3689.8	7.6906	190.01	3290.3	3689.4	7.6792
620	204.35	3326.5	3735.2	7.7534	199.32	3326.2	3734.8	7.7417	194.53	3325.9	3734.4	7.7302
640	209.08	3362.2	3780.4	7.8035	203.94	3362.0	3780.1	7.7918	199.05	3361.7	3779.7	7.7804
660	213.80	3398.2	3825.9	7.8527	208.55	3398.0	3825.5	7.8410	203.55	3397.8	3825.2	7.8296
680	218.52	3434.5	3871.5	7.9011	213.16	3434.2	3871.2	7.8895	208.05	3434.0	3870.9	7.8781
700	223.2	3470.9	3917.4	7.9487	217.8	3470.7	3917.1	7.9371	212.5	3470.5	3916.8	7.9258
720	227.9	3507.6	3963.5	7.9956	222.3	3507.4	3963.2	7.9840	217.0	3507.2	3963.0	7.9727
740	232.6	3544.6	4009.8	8.0418	226.9	3544.4	4009.6	8.0303	221.5	3544.2	4009.3	8.0189
760	237.3	3581.8	4056.4	8.0874	231.5	3581.6	4056.2	8.0758	226.0	3581.4	4055.9	8.0645
780	242.0	3619.3	4103.3	8.1323	236.1	3619.1	4103.0	8.1207	230.4	3618.9	4102.8	8.1094
800	246.7	3657.0	4150.3	8.1765	240.6	3656.8	4150.1	8.1650	234.9	3656.6	4149.9	8.1537
850	258.4	3752.4	4269.1	8.2847	252.0	3752.2	4268.9	8.2732	246.0	3752.1	4268.7	8.2619
900	270.0	3849.3	4389.4	8.3895	263.4	3849.2	4389.2	8.3780	257.1	3849.1	4389.1	8.3667
950	281.7	3947.9	4511.2	8.4912	274.8	3947.7	4511.1	8.4797	268.3	3947.6	4510.9	8.4685
1000	293.3	4048.0	4634.6	8.5901	286.2	4047.8	4634.5	8.5786	279.3	4047.7	4634.3	8.5673
1100	316.6	4252.7	4885.9	8.7800	308.9	4252.6	4885.8	8.7685	301.5	4252.5	4885.6	8.7573
1200	339.8	4463.3	5142.9	8.9607	331.5	4463.1	5142.8	8.9492	323.6	4463.0	5142.7	8.9380
1300	363.1	4679.0	5405.1	9.1329	354.2	4678.9	5405.0	9.1214	345.8	4678.7	5404.9	9.1102

Definitions of symbols on page 1.

	2.15 (216.10)				2.20 (217.29)				2.25 (218.45)			p (t Sat.)
10³ v	u	h	s	10³ v	u	h	s	10³ v	u	h	s	t
92.81	2601.4	2800.9	6.3141	90.73	2601.7	2801.3	6.3056	88.75	2602.0	2801.7	6.2972	Sat.
												200
89.12	*2574.4*	*2766.0*	*6.2419*	*86.71*	*2571.5*	*2762.2*	*6.2248*					205
90.81	*2586.8*	*2782.0*	*6.2753*	*88.38*	*2584.1*	*2778.5*	*6.2587*	*86.06*	*2581.3*	*2774.9*	*6.2423*	210
92.45	*2598.8*	*2797.6*	*6.3073*	*90.01*	*2596.3*	*2794.3*	*6.2912*	*87.66*	*2593.7*	*2790.9*	*6.2753*	215
94.05	2610.4	2812.6	6.3380	91.58	2608.1	2809.5	6.3223	89.23	2605.7	2806.4	6.3068	220
95.60	2621.7	2827.3	6.3676	93.12	2619.5	2824.4	6.3523	90.75	2617.3	2821.5	6.3372	225
97.13	2632.8	2841.6	6.3962	94.63	2630.7	2838.9	6.3812	92.23	2628.6	2836.1	6.3664	230
98.62	2643.6	2855.6	6.4238	96.10	2641.6	2853.0	6.4092	93.69	2639.6	2850.4	6.3947	235
100.08	2654.1	2869.3	6.4506	97.54	2652.2	2866.8	6.4362	95.11	2650.4	2864.4	6.4220	240
101.52	2664.5	2882.7	6.4767	98.96	2662.7	2880.4	6.4626	96.51	2660.9	2878.1	6.4486	245
102.93	2674.6	2895.9	6.5021	100.35	2672.9	2893.7	6.4881	97.88	2671.3	2891.5	6.4744	250
105.71	2694.5	2921.7	6.5509	103.08	2693.0	2919.7	6.5374	100.57	2691.4	2917.7	6.5240	260
108.41	2713.8	2946.9	6.5976	105.74	2712.4	2945.0	6.5844	103.19	2711.0	2943.2	6.5713	270
111.05	2732.6	2971.4	6.6424	108.34	2731.4	2969.7	6.6294	105.74	2730.1	2968.0	6.6167	280
113.65	2751.1	2995.5	6.6855	110.89	2750.0	2993.9	6.6728	108.25	2748.8	2992.3	6.6603	290
116.21	2769.3	3019.2	6.7272	113.40	2768.2	3017.7	6.7147	110.71	2767.1	3016.3	6.7023	300
118.72	2787.3	3042.5	6.7677	115.87	2786.3	3041.2	6.7553	113.14	2785.2	3039.8	6.7431	310
121.21	2805.0	3065.7	6.8070	118.31	2804.1	3064.4	6.7947	115.54	2803.1	3063.1	6.7827	320
123.67	2822.6	3088.5	6.8452	120.72	2821.7	3087.3	6.8331	117.90	2820.9	3086.1	6.8212	330
126.11	2840.1	3111.2	6.8826	123.11	2839.3	3110.1	6.8706	120.25	2838.4	3109.0	6.8588	340
128.52	2857.5	3133.8	6.9190	125.48	2856.7	3132.7	6.9071	122.57	2855.9	3131.7	6.8955	350
130.92	2874.8	3156.2	6.9548	127.83	2874.0	3155.2	6.9429	124.87	2873.3	3154.2	6.9314	360
133.30	2892.0	3178.6	6.9898	130.16	2891.3	3177.6	6.9780	127.15	2890.5	3176.6	6.9665	370
135.66	2909.2	3200.8	7.0241	132.47	2908.5	3199.9	7.0124	129.42	2907.8	3199.0	7.0010	380
138.01	2926.3	3223.0	7.0578	134.77	2925.6	3222.1	7.0462	131.68	2925.0	3221.3	7.0349	390
140.35	2943.4	3245.2	7.0910	137.06	2942.8	3244.3	7.0794	133.92	2942.2	3243.5	7.0681	400
142.68	2960.5	3267.3	7.1236	139.34	2959.9	3266.5	7.1121	136.15	2959.3	3265.7	7.1008	410
144.99	2977.6	3289.3	7.1556	141.61	2977.0	3288.6	7.1442	138.37	2976.5	3287.8	7.1330	420
147.30	2994.7	3311.4	7.1872	143.86	2994.2	3310.7	7.1759	140.58	2993.6	3310.0	7.1647	430
149.59	3011.8	3333.5	7.2184	146.11	3011.3	3332.8	7.2071	142.79	3010.8	3332.1	7.1959	440
151.88	3029.0	3355.5	7.2491	148.35	3028.5	3354.8	7.2378	144.98	3028.0	3354.2	7.2267	450
154.16	3046.1	3377.6	7.2794	150.59	3045.6	3376.9	7.2681	147.17	3045.2	3376.3	7.2571	460
156.44	3063.3	3399.6	7.3093	152.81	3062.8	3399.0	7.2981	149.35	3062.4	3398.4	7.2871	470
158.71	3080.5	3421.7	7.3388	155.03	3080.1	3421.1	7.3276	151.52	3079.6	3420.5	7.3166	480
160.97	3097.7	3443.8	7.3680	157.25	3097.3	3443.2	7.3568	153.69	3096.9	3442.7	7.3458	490
163.23	3115.0	3465.9	7.3968	159.45	3114.6	3465.4	7.3856	155.85	3114.2	3464.8	7.3747	500
167.72	3149.7	3510.3	7.4534	163.85	3149.3	3509.8	7.4423	160.15	3148.9	3509.2	7.4314	520
172.20	3184.5	3554.7	7.5087	168.23	3184.1	3554.2	7.4976	164.44	3183.8	3553.8	7.4868	540
176.66	3219.5	3599.3	7.5629	172.60	3219.1	3598.9	7.5518	168.71	3218.8	3598.4	7.5411	560
181.11	3254.7	3644.1	7.6159	176.95	3254.4	3643.6	7.6050	172.97	3254.0	3643.2	7.5942	580
185.54	3290.1	3689.0	7.6680	181.28	3289.8	3688.6	7.6570	177.21	3289.5	3688.2	7.6463	600
189.96	3325.7	3734.1	7.7191	185.61	3325.4	3733.7	7.7081	181.44	3325.1	3733.4	7.6975	620
194.38	3361.3	3779.4	7.7692	189.92	3361.2	3779.0	7.7583	185.67	3361.0	3778.7	7.7477	640
198.78	3397.5	3824.9	7.8185	194.23	3397.3	3824.6	7.8076	189.88	3397.0	3824.2	7.7970	660
203.18	3433.8	3870.6	7.8670	198.53	3433.5	3870.3	7.8561	194.08	3433.3	3870.0	7.8455	680
207.6	3470.3	3916.5	7.9147	202.8	3470.1	3916.3	7.9039	198.28	3469.8	3916.0	7.8933	700
211.9	3507.2	3962.7	7.9616	207.1	3506.8	3962.4	7.9508	202.47	3506.6	3962.2	7.9402	720
216.3	3544.0	4009.1	8.0079	211.4	3543.8	4008.8	7.9971	206.65	3543.6	4008.6	7.9865	740
220.7	3581.2	4055.7	8.0534	215.6	3581.0	4055.5	8.0427	210.83	3580.9	4055.2	8.0321	760
225.0	3618.7	4102.6	8.0984	219.9	3618.5	4102.3	8.0876	215.00	3618.4	4102.1	8.0770	780
229.4	3656.4	4149.7	8.1427	224.2	3656.3	4149.5	8.1319	219.2	3656.1	4149.2	8.1214	800
240.3	3751.9	4268.5	8.2509	234.8	3751.8	4268.3	8.2402	229.6	3751.6	4268.1	8.2297	850
251.2	3848.9	4388.9	8.3558	245.4	3848.8	4388.7	8.3450	240.0	3848.6	4388.5	8.3345	900
262.0	3947.5	4510.8	8.4575	256.0	3947.3	4510.6	8.4468	250.3	3947.2	4510.4	8.4363	950
272.8	4047.6	4634.2	8.5564	266.6	4047.4	4634.0	8.5457	260.7	4047.3	4633.9	8.5352	1000
294.5	4252.4	4885.5	8.7464	287.8	4252.2	4885.4	8.7357	281.4	4252.1	4885.2	8.7252	1100
316.1	4462.9	5142.6	8.9271	308.9	4462.8	5142.4	8.9164	302.1	4462.7	5142.3	8.9059	1200
337.7	4678.6	5404.8	9.0593	330.1	4678.5	5404.6	9.0886	322.7	4678.4	5404.5	9.0781	1300

Definitions of symbols on page 1.

Table 3. Vapor

t	$10^3 v$	u	h	s	$10^3 v$	u	h	s	$10^3 v$	u	h	s
Sat.	86.85	2602.3	2802.0	6.2890	85.02	2602.5	2802.3	6.2809	83.27	2602.8	2802.6	6.2729
210	*83.83*	*2578.5*	*2771.3*	*6.2261*	*81.69*	*2575.7*	*2767.7*	*6.2099*	*79.63*	*2572.8*	*2763.9*	*6.1938*
215	*85.42*	*2591.1*	*2787.6*	*6.2595*	*83.27*	*2588.5*	*2784.1*	*6.2438*	*81.20*	*2585.8*	*2780.7*	*6.2283*
220	86.97	2603.2	2803.3	6.2915	*84.81*	*2600.8*	*2800.1*	*6.2763*	*82.73*	*2598.3*	*2796.8*	*6.2612*
225	88.48	2615.0	2818.5	6.3222	86.30	2612.7	2815.5	6.3074	84.21	2610.4	2812.5	6.2928
230	89.95	2626.4	2833.3	6.3518	87.75	2624.3	2830.5	6.3374	85.65	2622.1	2827.6	6.3231
235	91.38	2637.6	2847.8	6.3804	89.17	2635.5	2845.1	6.3663	87.05	2633.5	2842.4	6.3523
240	92.79	2648.5	2861.9	6.4080	90.56	2646.5	2859.4	6.3942	88.42	2644.6	2856.8	6.3805
245	94.17	2659.1	2875.7	6.4348	91.92	2657.3	2873.3	6.4212	89.77	2655.5	2870.9	6.4078
250	95.52	2669.5	2889.2	6.4609	93.26	2667.8	2887.0	6.4475	91.09	2666.1	2884.7	6.4343
260	98.17	2689.9	2915.7	6.5109	95.87	2688.3	2913.6	6.4979	93.66	2686.7	2911.5	6.4852
270	100.75	2709.6	2941.3	6.5585	98.41	2708.2	2939.4	6.5459	96.17	2706.7	2937.5	6.5335
280	103.26	2728.8	2966.3	6.6041	100.88	2727.5	2964.6	6.5918	98.60	2726.2	2962.8	6.5796
290	105.73	2747.6	2990.8	6.6479	103.31	2746.4	2989.2	6.6358	100.99	2745.2	2987.6	6.6239
300	108.15	2766.0	3014.8	6.6902	105.69	2764.9	3013.3	6.6783	103.33	2763.8	3011.8	6.6666
310	110.53	2784.2	3038.4	6.7312	108.03	2783.2	3037.1	6.7194	105.64	2782.2	3035.7	6.7079
320	112.88	2802.2	3061.8	6.7709	110.34	2801.2	3060.5	6.7593	107.91	2800.3	3059.2	6.7479
330	115.21	2820.0	3084.9	6.8096	112.62	2819.1	3083.7	6.7981	110.15	2818.1	3082.5	6.7869
340	117.51	2837.6	3107.8	6.8472	114.88	2836.7	3106.7	6.8359	112.37	2835.9	3105.6	6.8248
350	119.78	2855.1	3130.6	6.8840	117.12	2854.3	3129.5	6.8728	114.56	2853.5	3128.4	6.8618
360	122.04	2872.5	3153.2	6.9200	119.33	2871.7	3152.2	6.9089	116.74	2871.0	3151.2	6.8979
370	124.28	2889.8	3175.7	6.9553	121.53	2889.1	3174.7	6.9442	118.90	2888.4	3173.7	6.9333
380	126.51	2907.1	3198.1	6.9898	123.71	2906.4	3197.2	6.9788	121.04	2905.7	3196.2	6.9681
390	128.72	2924.3	3220.4	7.0237	125.88	2923.7	3219.5	7.0128	123.17	2923.0	3218.6	7.0021
400	130.92	2941.6	3242.7	7.0571	128.04	2940.9	3241.8	7.0462	125.28	2940.3	3241.0	7.0355
410	133.10	2958.7	3264.9	7.0898	130.18	2958.1	3264.1	7.0790	127.39	2957.6	3263.3	7.0684
420	135.28	2975.9	3287.1	7.1221	132.32	2975.4	3286.3	7.1113	129.48	2974.8	3285.5	7.1008
430	137.45	2993.1	3309.2	7.1538	134.44	2992.6	3308.5	7.1431	131.56	2992.0	3307.8	7.1326
440	139.60	3010.3	3331.4	7.1851	136.56	3009.8	3330.7	7.1744	133.64	3009.2	3330.0	7.1640
450	141.75	3027.5	3353.5	7.2159	138.67	3027.0	3352.8	7.2053	135.70	3026.5	3352.2	7.1949
460	143.90	3044.7	3375.6	7.2463	140.76	3044.2	3375.0	7.2357	137.76	3043.7	3374.4	7.2253
470	146.03	3061.9	3397.8	7.2763	142.86	3061.5	3397.2	7.2657	139.82	3061.0	3396.5	7.2554
480	148.16	3079.2	3419.9	7.3059	144.94	3078.7	3419.3	7.2954	141.86	3078.3	3418.7	7.2851
490	150.28	3096.5	3442.1	7.3351	147.02	3096.0	3441.5	7.3246	143.90	3095.6	3441.0	7.3144
500	152.40	3113.8	3464.3	7.3640	149.10	3113.4	3463.7	7.3536	145.93	3112.9	3463.2	7.3433
520	156.62	3148.5	3508.7	7.4208	153.23	3148.1	3508.2	7.4104	149.98	3147.7	3507.7	7.4002
540	160.82	3183.4	3553.3	7.4762	157.34	3183.0	3552.8	7.4659	154.02	3182.7	3552.3	7.4557
560	165.00	3218.5	3598.0	7.5305	161.44	3218.1	3597.5	7.5202	158.03	3217.8	3597.1	7.5101
580	169.17	3253.7	3642.8	7.5837	165.52	3253.4	3642.4	7.5734	162.03	3253.1	3642.0	7.5633
600	173.32	3289.2	3687.8	7.6359	169.59	3288.9	3687.4	7.6256	166.02	3288.6	3687.0	7.6155
620	177.46	3324.8	3733.0	7.6870	173.65	3324.5	3732.6	7.6768	169.99	3324.3	3732.3	7.6667
640	181.59	3360.7	3778.4	7.7372	177.70	3360.4	3778.0	7.7270	173.96	3360.2	3777.7	7.7170
660	185.72	3396.8	3823.9	7.7866	181.73	3396.5	3823.6	7.7764	177.92	3396.3	3823.3	7.7664
680	189.83	3433.1	3869.7	7.8351	185.76	3432.9	3869.4	7.8250	181.86	3432.6	3869.1	7.8150
700	193.94	3469.6	3915.7	7.8829	189.78	3469.4	3915.4	7.8727	185.80	3469.2	3915.1	7.8628
720	198.04	3506.4	3961.9	7.9299	193.80	3506.2	3961.6	7.9197	189.74	3506.0	3961.3	7.9098
740	202.13	3543.4	4008.3	7.9762	197.81	3543.2	4008.1	7.9660	193.66	3543.0	4007.8	7.9561
760	206.22	3580.7	4055.0	8.0218	201.81	3580.5	4054.7	8.0117	197.58	3580.3	4054.5	8.0018
780	210.31	3618.2	4101.9	8.0667	205.81	3618.0	4101.6	8.0566	201.50	3617.8	4101.4	8.0468
800	214.4	3655.9	4149.0	8.1111	209.8	3655.8	4148.8	8.1010	205.4	3655.6	4148.6	8.0911
850	224.6	3751.4	4267.9	8.2194	219.8	3751.3	4267.7	8.2093	215.2	3751.1	4267.5	8.1995
900	234.7	3848.5	4388.3	8.3243	229.7	3848.3	4388.2	8.3142	224.9	3848.2	4388.0	8.3044
950	244.9	3947.1	4510.3	8.4260	239.7	3946.9	4510.1	8.4160	234.7	3946.8	4510.0	8.4062
1000	255.0	4047.2	4633.7	8.5250	249.6	4047.1	4633.6	8.5149	244.4	4046.9	4633.4	8.5051
1100	275.3	4252.0	4885.1	8.7150	269.4	4251.9	4885.0	8.7050	263.8	4251.8	4884.8	8.6952
1200	295.5	4462.6	5142.2	8.8957	289.2	4462.4	5142.1	8.8857	283.2	4462.3	5142.0	8.8759
1300	315.7	4678.3	5404.4	9.0679	309.0	4678.2	5404.3	9.0579	302.6	4678.0	5404.2	9.0481

Definitions of symbols on page 1.

	2.45 (222.92)				**2.50** (223.99)				**2.55** (225.05)			p (t Sat.)
10^3 v	u	h	s	10^3 v	u	h	s	10^3 v	u	h	s	t
81.59	2603.0	2802.9	6.2651	79.98	2603.1	2803.1	6.2575	78.42	2603.3	2803.3	6.2499	Sat.
												210
79.22	*2583.1*	*2777.2*	*6.2129*	*77.31*	*2580.3*	*2773.6*	*6.1976*	*75.47*	*2577.5*	*2770.0*	*6.1823*	215
80.73	*2595.8*	*2793.5*	*6.2463*	*78.81*	*2593.2*	*2790.2*	*6.2315*	*76.96*	*2590.6*	*2786.9*	*6.2167*	220
82.20	2608.0	2809.4	6.2783	80.27	2605.6	2806.3	6.2639	*78.41*	*2603.2*	*2803.1*	*6.2496*	225
83.62	2619.9	2824.8	6.3089	81.68	2617.6	2821.8	6.2949	79.81	2615.4	2818.9	6.2811	230
85.01	2631.4	2839.7	6.3385	83.06	2629.3	2836.9	6.3248	81.17	2627.2	2834.2	6.3113	235
86.37	2642.6	2854.2	6.3670	84.40	2640.7	2851.7	6.3536	82.50	2638.7	2849.0	6.3404	240
87.70	2653.6	2868.5	6.3946	85.71	2651.7	2866.0	6.3815	83.80	2649.9	2863.6	6.3686	245
89.01	2664.3	2882.4	6.4213	87.00	2662.6	2880.1	6.4085	85.08	2660.8	2877.7	6.3958	250
91.55	2685.2	2909.5	6.4726	89.51	2683.6	2907.4	6.4601	87.56	2682.0	2905.2	6.4479	260
94.01	2705.3	2935.6	6.5212	91.95	2703.9	2933.7	6.5091	89.96	2702.4	2931.8	6.4972	270
96.42	2724.9	2961.1	6.5677	94.32	2723.5	2959.3	6.5559	92.30	2722.2	2957.6	6.5443	280
98.77	2744.0	2985.9	6.6122	96.63	2742.8	2984.3	6.6007	94.58	2741.5	2982.7	6.5893	290
101.07	2762.7	3010.3	6.6551	98.90	2761.6	3008.8	6.6438	96.81	2760.4	3007.3	6.6326	300
103.34	2781.1	3034.3	6.6966	101.13	2780.1	3032.9	6.6854	99.01	2779.0	3031.5	6.6744	310
105.57	2799.3	3057.9	6.7368	103.33	2798.3	3056.6	6.7258	101.17	2797.3	3055.3	6.7149	320
107.77	2817.2	3081.3	6.7758	105.50	2816.3	3080.1	6.7649	103.30	2815.4	3078.8	6.7543	330
109.95	2835.0	3104.4	6.8138	107.64	2834.2	3103.3	6.8031	105.41	2833.3	3102.1	6.7925	340
112.11	2852.7	3127.4	6.8509	109.76	2851.9	3126.3	6.8403	107.50	2851.1	3125.2	6.8298	350
114.25	2870.2	3150.1	6.8872	111.86	2869.5	3149.1	6.8767	109.56	2868.7	3148.1	6.8663	360
116.37	2887.7	3172.8	6.9227	113.94	2887.0	3171.8	6.9122	111.61	2886.2	3170.8	6.9020	370
118.47	2905.1	3195.3	6.9575	116.01	2904.4	3194.4	6.9471	113.64	2903.7	3193.5	6.9369	380
120.56	2922.4	3217.8	6.9916	118.06	2921.7	3216.9	6.9813	115.66	2921.1	3216.0	6.9711	390
122.64	2939.7	3240.2	7.0251	120.10	2939.1	3239.3	7.0148	117.66	2938.4	3238.5	7.0048	400
124.70	2957.0	3262.5	7.0580	122.13	2956.4	3261.7	7.0478	119.65	2955.8	3260.9	7.0378	410
126.76	2974.2	3284.8	7.0904	124.14	2973.6	3284.0	7.0803	121.63	2973.1	3283.2	7.0703	420
128.80	2991.5	3307.0	7.1223	126.15	2990.9	3306.3	7.1122	123.60	2990.4	3305.6	7.1023	430
130.84	3008.7	3329.3	7.1537	128.15	3008.2	3328.6	7.1436	125.57	3007.7	3327.9	7.1338	440
132.87	3026.0	3351.5	7.1846	130.14	3025.5	3350.8	7.1746	127.52	3025.0	3350.1	7.1648	450
134.88	3043.2	3373.7	7.2152	132.12	3042.8	3373.1	7.2052	129.47	3042.3	3372.4	7.1954	460
136.90	3060.5	3395.9	7.2453	134.10	3060.1	3395.3	7.2353	131.40	3059.6	3394.7	7.2255	470
138.90	3077.8	3418.1	7.2750	136.06	3077.4	3417.5	7.2650	133.34	3076.9	3417.0	7.2553	480
140.90	3095.2	3440.4	7.3043	138.03	3094.7	3439.8	7.2944	135.26	3094.3	3439.2	7.2847	490
142.90	3112.5	3462.6	7.3332	139.98	3112.1	3462.1	7.3234	137.18	3111.7	3461.5	7.3137	500
146.87	3147.4	3507.2	7.3901	143.88	3147.0	3506.7	7.3803	141.01	3146.6	3506.2	7.3707	520
150.82	3182.3	3551.8	7.4458	147.76	3182.0	3551.4	7.4360	144.82	3181.6	3550.9	7.4264	540
154.76	3217.5	3596.6	7.5002	151.62	3217.1	3596.2	7.4904	148.60	3216.8	3595.7	7.4809	560
158.68	3252.8	3641.6	7.5534	155.47	3252.5	3641.1	7.5438	152.38	3252.2	3640.7	7.5342	580
162.59	3288.3	3686.6	7.6057	159.30	3288.0	3686.3	7.5960	156.14	3287.7	3685.9	7.5865	600
166.49	3324.0	3731.9	7.6569	163.12	3323.7	3731.5	7.6473	159.89	3323.4	3731.2	7.6378	620
170.38	3359.9	3777.3	7.7072	166.94	3359.6	3777.0	7.6976	163.63	3359.4	3776.6	7.6882	640
174.25	3396.0	3823.0	7.7567	170.74	3395.8	3822.6	7.7471	167.36	3395.5	3822.3	7.7377	660
178.12	3432.4	3868.8	7.8052	174.53	3432.2	3868.5	7.7957	171.08	3431.9	3868.2	7.7863	680
181.98	3469.0	3914.8	7.8530	178.32	3468.7	3914.5	7.8435	174.79	3468.5	3914.2	7.8341	700
185.84	3505.8	3961.1	7.9001	182.10	3505.6	3960.8	7.8906	178.50	3505.3	3960.5	7.8812	720
189.69	3542.8	4007.5	7.9464	185.87	3542.6	4007.3	7.9369	182.20	3542.4	4007.0	7.9276	740
193.53	3580.1	4054.2	7.9921	189.64	3579.9	4054.0	7.9826	185.90	3579.7	4053.8	7.9732	760
197.37	3617.6	4101.2	8.0371	193.40	3617.4	4100.9	8.0276	189.59	3617.3	4100.7	8.0183	780
201.2	3655.5	4148.4	8.0815	197.16	3655.3	4148.2	8.0720	193.27	3655.1	4148.0	8.0627	800
210.8	3751.0	4267.4	8.1898	206.54	3750.8	4267.2	8.1804	202.47	3750.7	4267.0	8.1711	850
220.3	3848.0	4387.8	8.2947	215.90	3847.9	4387.6	8.2853	211.65	3847.8	4387.5	8.2760	900
229.9	3946.7	4509.8	8.3966	225.24	3946.5	4509.6	8.3871	220.82	3946.4	4509.5	8.3779	950
239.4	4046.8	4633.3	8.4955	234.6	4046.7	4633.1	8.4861	230.0	4046.6	4633.0	8.4768	1000
258.4	4251.6	4884.7	8.6856	253.2	4251.5	4884.6	8.6762	248.3	4251.4	4884.4	8.6670	1100
277.4	4462.2	5141.8	8.8663	271.8	4462.1	5141.7	8.8569	266.5	4462.0	5141.6	8.8477	1200
296.4	4677.9	5404.1	9.0385	290.5	4677.8	5404.0	9.0291	284.8	4677.7	5403.8	9.0199	1300

Definitions of symbols on page 1.

Table 3. Vapor

p (t Sat.)	**2.60** (226.09)				**2.65** (227.11)				**2.70** (228.12)			
t	$10^3\,v$	u	h	s	$10^3\,v$	u	h	s	$10^3\,v$	u	h	s
Sat.	76.92	2603.5	2803.5	6.2425	75.48	2603.6	2803.6	6.2352	74.09	2603.7	2803.8	6.2280
215	*73.70*	*2574.7*	*2766.3*	*6.1672*								
220	*75.18*	*2588.0*	*2783.4*	*6.2021*	*73.46*	*2585.3*	*2780.0*	*6.1876*	*71.81*	*2582.6*	*2776.5*	*6.1731*
225	*76.62*	*2600.8*	*2800.0*	*6.2354*	*74.89*	*2598.3*	*2796.7*	*6.2214*	*73.23*	*2595.8*	*2793.5*	*6.2074*
230	78.01	2613.1	2815.9	6.2673	76.27	2610.8	2812.9	6.2536	74.60	2608.4	2809.8	6.2401
235	79.36	2625.0	2831.4	6.2979	77.61	2622.9	2828.5	6.2846	75.93	2620.7	2825.7	6.2714
240	80.68	2636.6	2846.4	6.3273	78.92	2634.6	2843.7	6.3144	77.22	2632.5	2841.1	6.3015
245	81.97	2647.9	2861.1	6.3557	80.20	2646.0	2858.5	6.3431	78.49	2644.1	2856.0	6.3305
250	83.23	2659.0	2875.4	6.3832	81.44	2657.2	2873.0	6.3708	79.72	2655.3	2870.6	6.3585
260	85.68	2680.3	2903.1	6.4358	83.87	2678.7	2901.0	6.4238	82.12	2677.1	2898.8	6.4120
270	88.05	2700.9	2929.9	6.4855	86.21	2699.5	2927.9	6.4739	84.44	2698.0	2926.0	6.4624
280	90.35	2720.9	2955.8	6.5328	88.48	2719.5	2954.0	6.5215	86.68	2718.2	2952.2	6.5103
290	92.60	2740.3	2981.1	6.5781	90.70	2739.1	2979.4	6.5670	88.87	2737.8	2977.8	6.5561
300	94.81	2759.3	3005.8	6.6216	92.87	2758.2	3004.3	6.6108	91.01	2757.0	3002.8	6.6001
310	96.97	2778.0	3030.1	6.6636	95.01	2776.9	3028.7	6.6530	93.12	2775.9	3027.3	6.6425
320	99.10	2796.4	3054.0	6.7043	97.10	2795.4	3052.7	6.6938	95.18	2794.4	3051.4	6.6835
330	101.20	2814.5	3077.6	6.7437	99.17	2813.6	3076.4	6.7334	97.22	2812.7	3075.1	6.7232
340	103.27	2832.5	3101.0	6.7821	101.21	2831.6	3099.8	6.7719	99.23	2830.7	3098.6	6.7618
350	105.32	2850.3	3124.1	6.8196	103.23	2849.4	3123.0	6.8094	101.21	2848.6	3121.9	6.7995
360	107.35	2867.9	3147.0	6.8561	105.23	2867.2	3146.0	6.8461	103.18	2866.4	3145.0	6.8362
370	109.37	2885.5	3169.9	6.8918	107.21	2884.8	3168.9	6.8819	105.13	2884.0	3167.9	6.8721
380	111.36	2903.0	3192.5	6.9268	109.17	2902.3	3191.6	6.9170	107.06	2901.6	3190.7	6.9073
390	113.34	2920.4	3215.1	6.9612	111.12	2919.8	3214.2	6.9514	108.98	2919.1	3213.4	6.9417
400	115.31	2937.8	3237.6	6.9949	113.06	2937.2	3236.8	6.9851	110.88	2936.6	3236.0	6.9756
410	117.27	2955.2	3260.1	7.0280	114.98	2954.6	3259.3	7.0183	112.77	2954.0	3258.5	7.0088
420	119.22	2972.5	3282.5	7.0605	116.89	2971.9	3281.7	7.0509	114.66	2971.4	3280.9	7.0414
430	121.15	2989.8	3304.8	7.0925	118.80	2989.3	3304.1	7.0829	116.53	2988.7	3303.4	7.0735
440	123.08	3007.1	3327.2	7.1241	120.69	3006.6	3326.5	7.1145	118.39	3006.1	3325.7	7.1051
450	125.00	3024.5	3349.5	7.1551	122.58	3024.0	3348.8	7.1456	120.24	3023.5	3348.1	7.1363
460	126.91	3041.8	3371.8	7.1857	124.46	3041.3	3371.1	7.1763	122.09	3040.8	3370.5	7.1670
470	128.82	3059.1	3394.1	7.2159	126.33	3058.7	3393.4	7.2065	123.93	3058.2	3392.8	7.1973
480	130.71	3076.5	3416.4	7.2457	128.19	3076.0	3415.8	7.2364	125.76	3075.6	3415.2	7.2271
490	132.61	3093.9	3438.7	7.2752	130.05	3093.4	3438.1	7.2658	127.59	3093.0	3437.5	7.2566
500	134.49	3111.3	3461.0	7.3042	131.90	3110.9	3460.4	7.2949	129.41	3110.5	3459.9	7.2857
520	138.25	3146.2	3505.6	7.3613	135.59	3145.8	3505.1	7.3520	133.03	3145.4	3504.6	7.3429
540	141.98	3181.3	3550.4	7.4170	139.26	3180.9	3549.9	7.4078	136.64	3180.5	3549.4	7.3987
560	145.70	3216.5	3595.3	7.4715	142.91	3216.1	3594.8	7.4623	140.23	3215.8	3594.4	7.4533
580	149.41	3251.8	3640.3	7.5249	146.55	3251.5	3639.9	7.5157	143.80	3251.2	3639.5	7.5068
600	153.10	3287.4	3685.5	7.5772	150.18	3287.1	3685.1	7.5681	147.36	3286.8	3684.7	7.5591
620	156.78	3323.2	3730.8	7.6286	153.79	3322.9	3730.4	7.6195	150.91	3322.6	3730.1	7.6105
640	160.45	3359.1	3776.3	7.6789	157.39	3358.9	3776.0	7.6699	154.45	3358.6	3775.6	7.6610
660	164.11	3395.3	3822.0	7.7284	160.99	3395.1	3821.7	7.7194	157.98	3394.8	3821.3	7.7105
680	167.76	3431.7	3867.9	7.7771	164.57	3431.5	3867.6	7.7681	161.50	3431.2	3867.3	7.7592
700	171.41	3468.3	3914.0	7.8249	168.15	3468.1	3913.7	7.8159	165.01	3467.9	3913.4	7.8071
720	175.04	3505.1	3960.3	7.8720	171.72	3504.9	3960.0	7.8630	168.52	3504.7	3959.7	7.8542
740	178.68	3542.2	4006.8	7.9184	175.28	3542.0	4006.5	7.9094	172.02	3541.8	4006.3	7.9006
760	182.30	3579.5	4053.5	7.9641	178.84	3579.3	4053.3	7.9551	175.51	3579.2	4053.0	7.9463
780	185.92	3617.1	4100.5	8.0091	182.40	3616.9	4100.3	8.0002	179.00	3616.7	4100.0	7.9914
800	189.54	3655.0	4147.8	8.0536	185.94	3654.8	4147.5	8.0446	182.48	3654.6	4147.3	8.0359
850	198.56	3750.5	4266.8	8.1620	194.80	3750.4	4266.6	8.1531	191.18	3750.2	4266.4	8.1443
900	207.57	3847.6	4387.3	8.2670	203.64	3847.5	4387.1	8.2580	199.86	3847.3	4386.9	8.2493
950	216.56	3946.3	4509.3	8.3688	212.46	3946.1	4509.1	8.3599	208.52	3946.0	4509.0	8.3512
1000	225.5	4046.4	4632.8	8.4678	221.3	4046.3	4632.7	8.4589	217.2	4046.2	4632.5	8.4502
1100	243.5	4251.3	4884.3	8.6579	238.9	4251.2	4884.2	8.6490	234.4	4251.0	4884.1	8.6403
1200	261.4	4461.9	5141.5	8.8386	256.5	4461.7	5141.3	8.8298	251.7	4461.6	5141.2	8.8211
1300	279.3	4677.6	5403.7	9.0109	274.0	4677.5	5403.6	9.0020	268.9	4677.3	5403.5	8.9933

Definitions of symbols on page 1.

	2.75 (229.12)				**2.80 (230.10)**				**2.85 (231.07)**			p (t Sat.)
10^3 v	u	h	s	10^3 v	u	h	s	10^3 v	u	h	s	t
72.75	2603.8	2803.9	6.2209	71.45	2603.9	2804.0	6.2139	70.20	2604.0	2804.0	6.2070	Sat.
												215
70.21	*2579.9*	*2772.9*	*6.1587*	*68.66*	*2577.1*	*2769.3*	*6.1443*					220
71.62	*2593.2*	*2790.2*	*6.1935*	*70.07*	*2590.6*	*2786.8*	*6.1796*	*68.56*	*2588.0*	*2783.4*	*6.1659*	225
72.98	2606.1	2806.8	6.2266	*71.42*	*2603.7*	*2803.6*	*6.2132*	*69.91*	*2601.2*	*2800.5*	*6.1999*	230
74.30	2618.5	2822.8	6.2583	72.74	2616.2	2819.9	6.2454	71.22	2613.9	2816.9	6.2325	235
75.59	2630.5	2838.3	6.2888	74.01	2628.4	2835.6	6.2761	72.49	2626.2	2832.8	6.2636	240
76.84	2642.1	2853.5	6.3181	75.25	2640.2	2850.9	6.3058	73.72	2638.2	2848.3	6.2935	245
78.07	2653.5	2868.2	6.3464	76.47	2651.6	2865.7	6.3343	74.92	2649.8	2863.3	6.3224	250
80.44	2675.4	2896.6	6.4003	78.81	2673.8	2894.4	6.3887	77.25	2672.1	2892.2	6.3772	260
82.73	2696.5	2924.0	6.4511	81.08	2695.0	2922.0	6.4399	79.49	2693.5	2920.0	6.4288	270
84.95	2716.8	2950.4	6.4993	83.27	2715.5	2948.6	6.4884	81.65	2714.1	2946.8	6.4777	280
87.11	2736.6	2976.1	6.5454	85.41	2735.3	2974.5	6.5348	83.76	2734.1	2972.8	6.5243	290
89.22	2755.9	3001.2	6.5896	87.49	2754.7	2999.7	6.5792	85.82	2753.6	2998.2	6.5689	300
91.29	2774.8	3025.9	6.6321	89.53	2773.7	3024.4	6.6220	87.84	2772.7	3023.0	6.6119	310
93.33	2793.4	3050.1	6.6733	91.54	2792.4	3048.7	6.6633	89.82	2791.4	3047.4	6.6534	320
95.34	2811.7	3073.9	6.7132	93.52	2810.8	3072.7	6.7033	91.77	2809.9	3071.4	6.6935	330
97.32	2829.9	3097.5	6.7519	95.47	2829.0	3096.3	6.7422	93.69	2828.1	3095.1	6.7325	340
99.27	2847.8	3120.8	6.7897	97.40	2847.0	3119.7	6.7800	95.59	2846.2	3118.6	6.7705	350
101.21	2865.6	3143.9	6.8265	99.31	2864.8	3142.9	6.8170	97.47	2864.1	3141.9	6.8075	360
103.13	2883.3	3166.9	6.8625	101.19	2882.6	3165.9	6.8530	99.33	2881.9	3164.9	6.8437	370
105.03	2900.9	3189.7	6.8977	103.07	2900.2	3188.8	6.8883	101.17	2899.5	3187.9	6.8791	380
106.91	2918.5	3212.5	6.9323	104.92	2917.8	3211.6	6.9229	103.00	2917.1	3210.7	6.9138	390
108.79	2935.9	3235.1	6.9661	106.77	2935.3	3234.3	6.9569	104.82	2934.7	3233.4	6.9478	400
110.65	2953.4	3257.7	6.9994	108.60	2952.8	3256.9	6.9902	106.62	2952.2	3256.1	6.9811	410
112.50	2970.8	3280.2	7.0321	110.42	2970.2	3279.4	7.0230	108.41	2969.6	3278.6	7.0140	420
114.34	2988.s	3302.6	7.0643	112.23	2987.6	3301.9	7.0552	110.19	2987.1	3301.1	7.0462	430
116.17	3005.6	3325.0	7.0959	114.03	3005.0	3324.3	7.0869	111.97	3004.5	3323.6	7.0780	440
117.99	3023.0	3347.4	7.1271	115.83	3022.5	3346.8	7.1181	113.73	3021.9	3346.1	7.1092	450
119.81	3040.3	3369.8	7.1579	117.61	3039.9	3369.2	7.1489	115.49	3039.4	3368.5	7.1400	460
121.62	3057.7	3392.2	7.1882	119.39	3057.3	3391.6	7.1792	117.24	3056.8	3390.9	7.1704	470
123.42	3075.2	3414.6	7.2181	121.16	3074.7	3414.0	7.2091	118.98	3074.3	3413.4	7.2004	480
125.21	3092.6	3436.9	7.2476	122.93	3092.2	3436.3	7.2387	120.72	3091.7	3435.8	7.2299	490
127.00	3110.0	3459.3	7.2767	124.69	3109.6	3458.7	7.2678	122.45	3109.2	3458.2	7.2591	500
130.57	3145.0	3504.1	7.3339	128.19	3144.7	3503.6	7.3251	125.89	3144.3	3503.1	7.3164	520
134.11	3180.2	3549.0	7.3898	131.67	3179.8	3548.5	7.3810	129.32	3179.4	3548.0	7.3724	540
137.64	3215.4	3593.9	7.4444	135.14	3215.1	3593.5	7.4357	132.73	3214.8	3593.1	7.4271	560
141.15	3250.9	3639.1	7.4979	138.59	3250.6	3638.6	7.4892	136.12	3250.3	3638.2	7.4807	580
144.65	3286.5	3684.3	7.5503	142.03	3286.2	3683.9	7.5417	139.50	3285.9	3683.5	7.5332	600
148.13	3322.3	3729.7	7.6017	145.46	3322.1	3729.3	7.5931	142.87	3321.8	3729.0	7.5846	620
151.61	3358.3	3775.3	7.6522	148.87	3358.1	3774.9	7.6436	146.23	3357.8	3774.6	7.6351	640
155.08	3394.6	3821.0	7.7018	152.28	3394.3	3820.7	7.6932	149.58	3394.1	3820.4	7.6847	660
158.53	3431.0	3867.0	7.7505	155.68	3430.8	3866.7	7.7419	152.92	3430.5	3866.3	7.7335	680
161.99	3467.6	3913.1	7.7984	159.07	3467.4	3912.8	7.7898	156.25	3467.2	3912.5	7.7814	700
165.43	3504.5	3959.4	7.8455	162.45	3504.3	3959.2	7.8370	159.58	3504.1	3958.9	7.8286	720
168.87	3541.6	4006.0	7.8919	165.83	3541.4	4005.8	7.8834	162.90	3541.2	4005.5	7.8751	740
172.30	3579.0	4052.8	7.9377	169.20	3578.8	4052.5	7.9292	166.22	3578.6	4052.3	7.9208	760
175.73	3616.5	4099.8	7.9827	172.57	3616.4	4099.6	7.9742	169.52	3616.2	4099.3	7.9659	780
179.15	3654.4	4147.0	8.0272	175.93	3654.2	4146.8	8.0187	172.83	3654.0	4146.6	8.0104	800
187.69	3750.0	4266.2	8.1357	184.32	3749.9	4266.0	8.1272	181.08	3749.7	4265.8	8.1189	850
196.21	3847.2	4386.8	8.2407	192.69	3847.0	4386.6	8.2323	189.30	3846.9	4386.4	8.2240	900
204.72	3945.9	4508.8	8.3426	201.05	3945.7	4508.7	8.3342	197.51	3945.6	4508.5	8.3259	950
213.2	4046.0	4632.4	8.4416	209.4	4045.9	4632.2	8.4332	205.7	4045.8	4632.1	8.4249	1000
230.2	4250.9	4883.9	8.6318	226.1	4250.8	4883.8	8.6234	222.1	4250.7	4883.7	8.6151	1100
247.1	4461.5	5141.1	8.8125	242.7	4461.4	5141.0	8.8041	238.5	4461.3	5140.9	8.7959	1200
264.1	4677.2	5403.4	8.9848	259.3	4677.1	5403.3	8.9764	254.8	4677.0	5403.2	8.9681	1300

Definitions of symbols on page 1.

Table 3. Vapor

p (t Sat.)	2.90 (232.02)				2.95 (232.97)				3.00 (233.90)			
t	10^3 v	u	h	s	10^3 v	u	h	s	10^3 v	u	h	s
Sat.	68.99	2604.0	2804.1	6.2002	67.81	2604.1	2804.1	6.1935	66.68	2604.1	2804.2	6.1869
225	*67.11*	*2585.4*	*2780.0*	*6.1522*	*65.70*	*2582.7*	*2776.5*	*6.1385*	*64.34*	*2579.9*	*2773.0*	*6.1249*
230	*68.45*	*2598.8*	*2797.3*	*6.1867*	*67.04*	*2596.3*	*2794.0*	*6.1735*	*65.67*	*2593.7*	*2790.8*	*6.1604*
235	69.75	2611.7	2813.9	6.2196	68.33	2609.3	2810.9	6.2069	66.96	2607.0	2807.9	6.1942
240	71.01	2624.1	2830.0	6.2512	69.58	2621.9	2827.2	6.2388	68.20	2619.7	2824.3	6.2265
245	72.23	2636.2	2845.6	6.2814	70.80	2634.1	2843.0	6.2694	69.41	2632.1	2840.3	6.2575
250	73.43	2647.9	2860.8	6.3106	71.98	2645.9	2858.3	6.2988	70.58	2644.0	2855.8	6.2872
255	74.59	2659.3	2875.6	6.3387	73.14	2657.5	2873.2	6.3272	71.73	2655.6	2870.8	6.3158
260	75.73	2670.4	2890.0	6.3659	74.27	2668.7	2887.8	6.3546	72.85	2667.0	2885.5	6.3435
265	76.85	2681.3	2904.1	6.3922	75.37	2679.7	2902.0	6.3812	73.95	2678.0	2899.9	6.3703
270	77.95	2691.9	2918.0	6.4179	76.46	2690.4	2916.0	6.4070	75.02	2688.9	2913.9	6.3963
275	79.03	2702.4	2931.6	6.4428	77.53	2700.9	2929.7	6.4321	76.08	2699.5	2927.7	6.4216
280	80.09	2712.7	2945.0	6.4671	78.58	2711.3	2943.1	6.4566	77.12	2709.9	2941.3	6.4462
285	81.14	2722.8	2958.1	6.4908	79.62	2721.5	2956.4	6.4804	78.15	2720.2	2954.6	6.4702
290	82.18	2732.8	2971.1	6.5140	80.64	2731.5	2969.4	6.5037	79.16	2730.3	2967.7	6.4936
295	83.20	2742.7	2983.9	6.5366	81.65	2741.5	2982.3	6.5265	80.16	2740.2	2980.7	6.5165
300	84.21	2752.4	2996.6	6.5588	82.65	2751.2	2995.1	6.5488	81.14	2750.1	2993.5	6.5390
310	86.20	2771.6	3021.6	6.6020	84.62	2770.5	3020.1	6.5922	83.09	2769.4	3018.7	6.5825
320	88.15	2790.4	3046.1	6.6436	86.55	2789.4	3044.7	6.6340	84.99	2788.4	3043.4	6.6245
330	90.08	2808.9	3070.2	6.6839	88.44	2808.0	3068.9	6.6744	86.86	2807.1	3067.7	6.6651
340	91.97	2827.2	3094.0	6.7231	90.31	2826.4	3092.8	6.7137	88.71	2825.5	3091.6	6.7045
350	93.85	2845.3	3117.5	6.7611	92.16	2844.5	3116.4	6.7519	90.53	2843.7	3115.3	6.7428
360	95.70	2863.3	3140.8	6.7983	93.98	2862.5	3139.8	6.7891	92.33	2861.7	3138.7	6.7801
370	97.53	2881.1	3164.0	6.8345	95.79	2880.4	3163.0	6.8255	94.11	2879.6	3162.0	6.8165
380	99.35	2898.8	3186.9	6.8700	97.58	2898.1	3186.0	6.8610	95.87	2897.4	3185.1	6.8522
390	101.15	2916.5	3209.8	6.9047	99.35	2915.8	3208.9	6.8958	97.62	2915.1	3208.0	6.8870
400	102.93	2934.1	3232.6	6.9388	101.12	2933.4	3231.7	6.9299	99.36	2932.8	3230.9	6.9212
410	104.71	2951.6	3255.2	6.9722	102.86	2951.0	3254.4	6.9634	101.08	2950.4	3253.6	6.9548
420	106.47	2969.1	3277.8	7.0051	104.60	2968.5	3277.1	6.9964	102.79	2967.9	3276.3	6.9878
430	108.23	2986.5	3300.4	7.0374	106.33	2986.0	3299.7	7.0287	104.50	2985.4	3298.9	7.0202
440	109.97	3004.0	3322.9	7.0692	108.05	3003.5	3322.2	7.0606	106.19	3002.9	3321.5	7.0520
450	111.71	3021.4	3345.4	7.1005	109.76	3020.9	3344.7	7.0919	107.87	3020.4	3344.0	7.0834
460	113.44	3038.9	3367.9	7.1313	111.46	3038.4	3367.2	7.1228	109.55	3037.9	3366.6	7.1144
470	115.16	3056.3	3390.3	7.1618	113.16	3055.9	3389.7	7.1532	111.22	3055.4	3389.1	7.1448
480	116.88	3073.8	3412.8	7.1917	114.84	3073.4	3412.1	7.1833	112.88	3072.9	3411.5	7.1749
490	118.59	3091.3	3435.2	7.2213	116.53	3090.9	3434.6	7.2129	114.54	3090.4	3434.0	7.2046
500	120.29	3108.8	3457.6	7.2506	118.20	3108.4	3457.1	7.2421	116.19	3108.0	3456.5	7.2338
520	123.68	3143.9	3502.5	7.3079	121.54	3143.5	3502.0	7.2995	119.47	3143.1	3501.5	7.2913
540	127.05	3179.1	3547.5	7.3639	124.85	3178.7	3547.0	7.3556	122.73	3178.4	3546.6	7.3474
560	130.40	3214.4	3592.6	7.4187	128.15	3214.1	3592.2	7.4104	125.98	3213.8	3591.7	7.4022
580	133.74	3249.9	3637.8	7.4723	131.44	3249.6	3637.4	7.4640	129.21	3249.3	3637.0	7.4559
600	137.07	3285.6	3683.1	7.5248	134.71	3285.3	3682.7	7.5166	132.43	3285.0	3682.3	7.5085
620	140.38	3321.5	3728.6	7.5763	137.97	3321.2	3728.2	7.5681	135.64	3320.9	3727.9	7.5600
640	143.68	3357.6	3774.2	7.6268	141.22	3357.3	3773.9	7.6187	138.84	3357.0	3773.5	7.6106
660	146.98	3393.8	3820.0	7.6765	144.46	3393.6	3819.7	7.6683	142.03	3393.3	3819.4	7.6603
680	150.26	3430.3	3866.0	7.7252	147.69	3430.1	3865.7	7.7171	145.20	3429.8	3865.4	7.7091
700	153.54	3467.0	3912.2	7.7732	150.91	3466.8	3911.9	7.7651	148.38	3466.5	3911.7	7.7571
720	156.81	3503.9	3958.6	7.8204	154.13	3503.7	3958.4	7.8123	151.54	3503.5	3958.1	7.8043
740	160.07	3541.0	4005.2	7.8668	157.34	3540.8	4005.0	7.8588	154.70	3540.6	4004.7	7.8508
760	163.33	3578.4	4052.1	7.9126	160.54	3578.2	4051.8	7.9045	157.85	3578.0	4051.6	7.8966
780	166.58	3616.0	4099.1	7.9577	163.74	3615.8	4098.9	7.9496	161.00	3615.6	4098.6	7.9417
800	169.83	3653.9	4146.4	8.0022	166.94	3653.7	4146.2	7.9941	164.14	3653.5	4145.9	7.9862
850	177.94	3749.6	4265.6	8.1107	174.91	3749.4	4265.4	8.1027	171.98	3749.3	4265.2	8.0948
900	186.03	3846.8	4386.2	8.2158	182.86	3846.6	4386.1	8.2078	179.80	3846.5	4385.9	8.1999
950	194.10	3945.5	4508.3	8.3177	190.80	3945.3	4508.2	8.3097	187.61	3945.2	4508.0	8.3019
1000	202.2	4045.7	4631.9	8.4168	198.73	4045.5	4631.8	8.4088	195.41	4045.4	4631.6	8.4009
1100	218.3	4250.6	4883.5	8.6070	214.56	4250.5	4883.4	8.5990	210.98	4250.3	4883.3	8.5912
1200	234.3	4461.2	5140.7	8.7878	230.36	4461.0	5140.6	8.7798	226.52	4460.9	5140.5	8.7720
1300	250.4	4676.9	5403.0	8.9600	246.16	4676.8	5402.9	8.9521	242.06	4676.6	5402.8	8.9442

Definitions of symbols on page 1.

	3.05 (234.82)				3.10 (235.72)				3.15 (236.62)			p (t Sat.)
10^3 v	u	h	s	10^3 v	u	h	s	10^3 v	u	h	s	t
65.58	2604.1	2804.2	6.1804	64.52	2604.1	2804.1	6.1740	63.49	2604.1	2804.1	6.1676	Sat.
												225
64.35	*2591.2*	*2787.5*	*6.1474*	*63.06*	*2588.6*	*2784.1*	*6.1344*	*61.81*	*2586.0*	*2780.7*	*6.1214*	230
65.63	2604.6	2804.8	6.1816	*64.34*	*2602.2*	*2801.6*	*6.1691*	*63.09*	*2599.8*	*2798.5*	*6.1566*	235
66.86	2617.5	2821.5	6.2143	65.57	2615.3	2818.5	6.2022	64.31	2613.0	2815.6	6.1901	240
68.06	2630.0	2837.6	6.2456	66.76	2627.9	2834.9	6.2338	65.50	2625.8	2832.1	6.2221	245
69.23	2642.1	2853.2	6.2756	67.92	2640.1	2850.7	6.2642	66.65	2638.1	2848.1	6.2528	250
70.37	2653.8	2868.4	6.3045	69.05	2652.0	2866.0	6.2934	67.77	2650.1	2863.6	6.2823	255
71.48	2665.2	2883.2	6.3325	70.15	2663.5	2880.9	6.3215	68.86	2661.7	2878.6	6.3107	260
72.56	2676.4	2897.7	6.3595	71.23	2674.7	2895.5	6.3488	69.93	2673.1	2893.4	6.3381	265
73.63	2687.3	2911.9	6.3857	72.28	2685.7	2909.8	6.3752	70.98	2684.2	2907.7	6.3648	270
74.68	2698.0	2925.8	6.4111	73.32	2696.5	2923.8	6.4008	72.00	2695.0	2921.8	6.3906	275
75.71	2708.5	2939.4	6.4359	74.34	2707.1	2937.5	6.4257	73.01	2705.6	2935.6	6.4157	280
76.72	2718.8	2952.8	6.4601	75.34	2717.5	2951.0	6.4500	74.01	2716.1	2949.2	6.4401	285
77.72	2729.0	2966.0	6.4836	76.33	2727.7	2964.3	6.4737	74.99	2726.4	2962.6	6.4640	290
78.71	2739.0	2979.1	6.5067	77.31	2737.8	2977.4	6.4969	75.95	2736.5	2975.8	6.4873	295
79.69	2748.9	2991.9	6.5292	78.27	2747.7	2990.4	6.5196	76.91	2746.5	2988.8	6.5100	300
81.60	2768.3	3017.2	6.5730	80.17	2767.2	3015.8	6.5635	78.78	2766.1	3014.3	6.5542	310
83.48	2787.4	3042.0	6.6151	82.03	2786.4	3040.7	6.6058	80.62	2785.3	3039.3	6.5967	320
85.33	2806.1	3066.4	6.6559	83.85	2805.2	3065.1	6.6468	82.42	2804.2	3063.9	6.6378	330
87.15	2824.6	3090.4	6.6954	85.65	2823.7	3089.2	6.6864	84.20	2822.8	3088.0	6.6775	340
88.95	2842.9	3114.2	6.7338	87.42	2842.0	3113.1	6.7249	85.95	2841.2	3111.9	6.7162	350
90.73	2861.0	3137.7	6.7712	89.18	2860.2	3136.6	6.7625	87.68	2859.4	3135.6	6.7538	360
92.48	2878.9	3161.0	6.8077	90.91	2878.2	3160.0	6.7991	89.38	2877.4	3159.0	6.7905	370
94.22	2896.7	3184.1	6.8434	92.62	2896.0	3183.2	6.8348	91.08	2895.3	3182.2	6.8264	380
95.95	2914.5	3207.1	6.8784	94.32	2913.8	3206.2	6.8699	92.75	2913.1	3205.3	6.8615	390
97.66	2932.2	3230.0	6.9127	96.01	2931.5	3229.2	6.9042	94.42	2930.9	3228.3	6.8959	400
99.36	2949.8	3252.8	6.9463	97.69	2949.2	3252.0	6.9379	96.07	2948.6	3251.2	6.9296	410
101.04	2967.3	3275.5	6.9793	99.35	2966.8	3274.7	6.9709	97.71	2966.2	3274.0	6.9627	420
102.72	2984.9	3298.2	7.0117	101.00	2984.3	3297.4	7.0034	99.34	2983.8	3296.7	6.9953	430
104.39	3002.4	3320.8	7.0437	102.64	3001.9	3320.1	7.0354	100.96	3001.3	3319.4	7.0273	440
106.05	3019.9	3343.4	7.0751	104.28	3019.4	3342.7	7.0669	102.57	3018.9	3342.0	7.0588	450
107.70	3037.4	3365.9	7.1061	105.91	3036.9	3365.2	7.0979	104.17	3036.5	3364.6	7.0898	460
109.34	3054.9	3388.4	7.1366	107.53	3054.5	3387.8	7.1284	105.77	3054.0	3387.2	7.1204	470
110.98	3072.5	3410.9	7.1667	109.14	3072.0	3410.3	7.1586	107.36	3071.6	3409.7	7.1506	480
112.61	3090.0	3433.4	7.1964	110.75	3089.6	3432.9	7.1883	108.94	3089.1	3432.3	7.1803	490
114.23	3107.5	3456.0	7.2257	112.35	3107.1	3455.4	7.2176	110.52	3106.7	3454.8	7.2097	500
117.47	3142.7	3501.0	7.2832	115.53	3142.3	3500.5	7.2752	113.66	3141.9	3500.0	7.2673	520
120.68	3178.0	3546.1	7.3393	118.70	3177.6	3545.6	7.3314	116.77	3177.3	3545.1	7.3235	540
123.88	3213.4	3591.3	7.3942	121.85	3213.1	3590.8	7.3863	119.88	3212.7	3590.4	7.3785	560
127.06	3249.0	3636.5	7.4479	124.98	3248.7	3636.1	7.4400	122.96	3248.4	3635.7	7.4323	580
130.23	3284.7	3681.9	7.5005	128.10	3284.4	3681.6	7.4927	126.04	3284.1	3681.2	7.4849	600
133.39	3320.7	3727.5	7.5521	131.21	3320.4	3727.1	7.5443	129.10	3320.1	3726.8	7.5366	620
136.54	3356.8	3773.2	7.6027	134.31	3356.5	3772.9	7.5949	132.15	3356.2	3772.5	7.5872	640
139.67	3393.1	3819.1	7.6524	137.39	3392.8	3818.7	7.6446	135.19	3392.6	3818.4	7.6370	660
142.80	3429.6	3865.1	7.7012	140.47	3429.4	3864.8	7.6935	138.22	3429.1	3864.5	7.6858	680
145.92	3466.3	3911.4	7.7492	143.55	3466.1	3911.1	7.7415	141.25	3465.9	3910.8	7.7339	700
149.04	3503.3	3957.8	7.7965	146.61	3503.0	3957.5	7.7888	144.26	3502.8	3957.3	7.7812	720
152.14	3540.4	4004.5	7.8430	149.67	3540.2	4004.2	7.8353	147.28	3540.0	4004.0	7.8277	740
155.25	3577.8	4051.3	7.8888	152.72	3577.6	4051.1	7.8811	150.28	3577.4	4050.8	7.8735	760
158.34	3615.5	4098.4	7.9339	155.77	3615.3	4098.2	7.9262	153.28	3615.1	4097.9	7.9187	780
161.44	3653.3	4145.7	7.9784	158.82	3653.2	4145.5	7.9707	156.28	3653.0	4145.3	7.9632	800
169.15	3749.1	4265.0	8.0871	166.41	3748.9	4264.8	8.0794	163.76	3748.8	4264.6	8.0719	850
176.85	3846.3	4385.7	8.1922	173.98	3846.2	4385.5	8.1846	171.21	3846.0	4385.3	8.1771	900
184.53	3945.1	4507.9	8.2941	181.54	3944.9	4507.7	8.2865	178.65	3944.8	4507.5	8.2790	950
192.20	4045.3	4631.5	8.3932	189.09	4045.2	4631.3	8.3856	186.08	4045.0	4631.2	8.3781	1000
207.51	4250.2	4883.1	8.5835	204.16	4250.1	4883.0	8.5759	200.92	4250.0	4882.9	8.5684	1100
222.81	4460.8	5140.4	8.7643	219.21	4460.7	5140.3	8.7567	215.73	4460.6	5140.1	8.7492	1200
238.09	4676.5	5402.7	8.9365	234.25	4676.4	5402.6	8.9290	230.54	4676.3	5402.5	8.9215	1300

Definitions of symbols on page 1.

Table 3. Vapor

p (t Sat.)	3.20 (237.51)				3.25 (238.38)				3.30 (239.24)			
t	10^3 v	u	h	s	10^3 v	u	h	s	10^3 v	u	h	s
Sat.	62.49	2604.1	2804.1	6.1614	61.52	2604.1	2804.0	6.1552	60.57	2604.0	2803.9	6.1491
230	*60.60*	*2583.3*	*2777.2*	*6.1085*	*59.43*	*2580.6*	*2773.7*	*6.0956*				
235	*61.87*	*2597.3*	*2795.3*	*6.1442*	*60.69*	*2594.8*	*2792.0*	*6.1318*	*59.54*	*2592.3*	*2788.8*	*6.1194*
240	63.09	2610.7	2812.6	6.1781	61.91	2608.4	2809.6	6.1661	60.75	2606.1	2806.5	6.1542
245	64.27	2623.6	2829.3	6.2105	63.08	2621.5	2826.5	6.1989	61.92	2619.3	2823.6	6.1874
250	65.41	2636.1	2845.4	6.2415	64.22	2634.1	2842.8	6.2302	63.06	2632.0	2840.1	6.2190
255	66.53	2648.2	2861.1	6.2712	65.32	2646.3	2858.6	6.2603	64.15	2644.4	2856.1	6.2494
260	67.61	2659.9	2876.3	6.2999	66.40	2658.2	2874.0	6.2892	65.22	2656.4	2871.6	6.2786
265	68.67	2671.4	2891.1	6.3276	67.45	2669.7	2888.9	6.3172	66.27	2668.0	2886.7	6.3068
270	69.71	2682.6	2905.6	6.3544	68.48	2681.0	2903.5	6.3442	67.29	2679.4	2901.4	6.3340
275	70.73	2693.5	2919.8	6.3804	69.49	2692.0	2917.8	6.3704	68.29	2690.5	2915.8	6.3604
280	71.73	2704.2	2933.7	6.4057	70.48	2702.8	2931.8	6.3958	69.28	2701.3	2929.9	6.3861
285	72.71	2714.7	2947.4	6.4303	71.46	2713.4	2945.6	6.4206	70.24	2712.0	2943.8	6.4110
290	73.68	2725.1	2960.9	6.4543	72.42	2723.8	2959.1	6.4447	71.19	2722.4	2957.4	6.4352
295	74.64	2735.3	2974.1	6.4777	73.36	2734.0	2972.4	6.4683	72.13	2732.7	2970.8	6.4589
300	75.58	2745.3	2987.2	6.5006	74.30	2744.1	2985.6	6.4913	73.05	2742.9	2984.0	6.4821
310	77.44	2765.0	3012.8	6.5450	76.13	2763.9	3011.4	6.5359	74.87	2762.8	3009.9	6.5269
320	79.25	2784.3	3037.9	6.5877	77.93	2783.3	3036.6	6.5787	76.64	2782.3	3035.2	6.5699
330	81.03	2803.3	3062.6	6.6289	79.69	2802.3	3061.3	6.6201	78.38	2801.4	3060.0	6.6114
340	82.79	2821.9	3086.8	6.6688	81.42	2821.0	3085.7	6.6601	80.09	2820.1	3084.5	6.6516
350	84.51	2840.4	3110.8	6.7075	83.13	2839.5	3109.7	6.6990	81.78	2838.7	3108.6	6.6906
360	86.22	2858.6	3134.5	6.7453	84.81	2857.8	3133.4	6.7368	83.44	2857.0	3132.4	6.7285
370	87.91	2876.7	3158.0	6.7821	86.48	2875.9	3157.0	6.7737	85.09	2875.2	3156.0	6.7655
380	89.58	2894.6	3181.3	6.8180	88.12	2893.9	3180.3	6.8097	86.72	2893.2	3179.4	6.8016
390	91.23	2912.5	3204.4	6.8532	89.76	2911.8	3203.5	6.8450	88.33	2911.1	3202.6	6.8369
400	92.87	2930.2	3227.4	6.8876	91.38	2929.6	3226.6	6.8795	89.93	2929.0	3225.7	6.8715
410	94.50	2948.0	3250.4	6.9214	92.98	2947.3	3249.5	6.9134	91.51	2946.7	3248.7	6.9054
420	96.12	2965.6	3273.2	6.9546	94.58	2965.0	3272.4	6.9466	93.08	2964.4	3271.6	6.9387
430	97.73	2983.2	3295.9	6.9872	96.16	2982.7	3295.2	6.9792	94.65	2982.1	3294.5	6.9714
440	99.32	3000.8	3318.7	7.0193	97.74	3000.3	3317.9	7.0113	96.20	2999.8	3317.2	7.0035
450	100.91	3018.4	3341.3	7.0508	99.31	3017.9	3340.6	7.0429	97.75	3017.4	3339.9	7.0352
460	102.49	3036.0	3363.9	7.0819	100.86	3035.5	3363.3	7.0741	99.29	3035.0	3362.6	7.0663
470	104.07	3053.5	3386.5	7.1125	102.42	3053.1	3385.9	7.1047	100.82	3052.6	3385.3	7.0970
480	105.63	3071.1	3409.1	7.1427	103.96	3070.7	3408.5	7.1349	102.34	3070.2	3407.9	7.1273
490	107.19	3088.7	3431.7	7.1725	105.50	3088.2	3431.1	7.1647	103.86	3087.8	3430.5	7.1571
500	108.75	3106.3	3454.3	7.2019	107.03	3105.9	3453.7	7.1942	105.37	3105.4	3453.2	7.1866
520	111.84	3141.5	3499.4	7.2595	110.08	3141.2	3498.9	7.2519	108.37	3140.8	3498.4	7.2443
540	114.91	3176.9	3544.6	7.3158	113.11	3176.5	3544.1	7.3082	111.36	3176.2	3543.7	7.3007
560	117.97	3212.4	3589.9	7.3708	116.12	3212.1	3589.5	7.3632	114.33	3211.7	3589.0	7.3558
580	121.01	3248.0	3635.3	7.4246	119.12	3247.7	3634.9	7.4171	117.28	3247.4	3634.4	7.4097
600	124.04	3283.8	3680.8	7.4773	122.10	3283.5	3680.4	7.4698	120.22	3283.2	3680.0	7.4624
620	127.05	3319.8	3726.4	7.5290	125.07	3319.5	3726.0	7.5215	123.15	3319.3	3725.7	7.5142
640	130.06	3356.0	3772.2	7.5797	128.03	3355.7	3771.8	7.5722	126.07	3355.5	3771.5	7.5649
660	133.05	3392.3	3818.1	7.6294	130.98	3392.1	3817.8	7.6220	128.97	3391.8	3817.5	7.6147
680	136.04	3428.9	3864.2	7.6783	133.92	3428.7	3863.9	7.6709	131.87	3428.4	3863.6	7.6636
700	139.02	3465.7	3910.5	7.7264	136.86	3465.4	3910.2	7.7190	134.77	3465.2	3909.9	7.7117
720	141.99	3502.6	3957.0	7.7737	139.79	3502.4	3956.7	7.7663	137.65	3502.2	3956.5	7.7591
740	144.96	3539.8	4003.7	7.8202	142.71	3539.6	4003.4	7.8129	140.53	3539.4	4003.2	7.8056
760	147.92	3577.3	4050.6	7.8661	145.63	3577.1	4050.3	7.8587	143.40	3576.9	4050.1	7.8515
780	150.87	3614.9	4097.7	7.9112	148.54	3614.7	4097.5	7.9039	146.27	3614.6	4097.2	7.8967
800	153.82	3652.8	4145.1	7.9558	151.44	3652.6	4144.8	7.9485	149.13	3652.5	4144.6	7.9412
850	161.18	3748.6	4264.4	8.0645	158.69	3748.5	4264.2	8.0572	156.28	3748.3	4264.0	8.0500
900	168.53	3845.9	4385.2	8.1697	165.92	3845.8	4385.0	8.1624	163.40	3845.6	4384.8	8.1552
950	175.85	3944.6	4507.4	8.2717	173.14	3944.5	4507.2	8.2644	170.51	3944.4	4507.1	8.2572
1000	183.17	4044.9	4631.0	8.3708	180.34	4044.8	4630.9	8.3635	177.60	4044.7	4630.7	8.3564
1100	197.77	4249.9	4882.7	8.5610	194.73	4249.7	4882.6	8.5538	191.77	4249.6	4882.5	8.5467
1200	212.36	4460.5	5140.0	8.7419	209.09	4460.3	5139.9	8.7346	205.92	4460.2	5139.8	8.7275
1300	226.93	4676.2	5402.4	8.9141	223.44	4676.0	5402.2	8.9069	220.06	4675.9	5402.1	8.8998

Definitions of symbols on page 1.

10^3 v	u	h	s	10^3 v	u	h	s	10^3 v	u	h	s	t
59.66	2604.0	2803.8	6.1430	58.77	2603.9	2803.7	6.1370	57.91	2603.8	2803.6	6.1311	Sat.
												230
58.43	*2589.7*	*2785.4*	*6.1071*	*57.34*	*2587.1*	*2782.1*	*6.0947*	*56.29*	*2584.5*	*2778.7*	*6.0825*	235
59.64	*2603.7*	*2803.5*	*6.1423*	*58.55*	*2601.3*	*2800.3*	*6.1305*	*57.49*	*2598.9*	*2797.2*	*6.1187*	240
60.80	2617.1	2820.8	6.1759	59.71	2614.9	2817.9	6.1645	58.65	2612.6	2814.9	6.1531	245
61.93	2630.0	2837.4	6.2079	60.83	2627.9	2834.7	6.1969	59.76	2625.8	2832.0	6.1858	250
63.02	2642.4	2853.6	6.2386	61.92	2640.5	2851.0	6.2279	60.84	2638.5	2848.4	6.2172	255
64.08	2654.5	2869.2	6.2681	62.97	2652.7	2866.8	6.2576	61.90	2650.8	2864.4	6.2472	260
65.12	2666.3	2884.4	6.2965	64.00	2664.6	2882.2	6.2863	62.92	2662.8	2879.9	6.2761	265
66.13	2677.7	2899.3	6.3240	65.01	2676.1	2897.1	6.3140	63.92	2674.4	2895.0	6.3041	270
67.13	2688.9	2913.8	6.3506	66.00	2687.4	2911.7	6.3408	64.90	2685.8	2909.7	6.3310	275
68.10	2699.9	2928.0	6.3764	66.96	2698.4	2926.1	6.3667	65.85	2696.9	2924.1	6.3572	280
69.06	2710.6	2941.9	6.4014	67.91	2709.2	2940.1	6.3920	66.80	2707.8	2938.2	6.3826	285
70.00	2721.1	2955.6	6.4259	68.84	2719.8	2953.9	6.4166	67.72	2718.4	2952.1	6.4073	290
70.93	2731.5	2969.1	6.4497	69.76	2730.2	2967.4	6.4405	68.63	2728.9	2965.7	6.4314	295
71.84	2741.7	2982.4	6.4729	70.67	2740.5	2980.7	6.4639	69.53	2739.2	2979.1	6.4549	300
73.64	2761.7	3008.4	6.5179	72.45	2760.6	3006.9	6.5091	71.29	2759.4	3005.4	6.5004	310
75.39	2781.2	3033.8	6.5612	74.18	2780.2	3032.4	6.5525	73.01	2779.2	3031.0	6.5440	320
77.11	2800.4	3058.7	6.6028	75.88	2799.4	3057.4	6.5943	74.69	2798.5	3056.1	6.5860	330
78.81	2819.2	3083.3	6.6431	77.56	2818.3	3082.0	6.6348	76.34	2817.4	3080.8	6.6266	340
80.47	2837.8	3107.4	6.6823	79.20	2837.0	3106.3	6.6740	77.97	2836.1	3105.2	6.6659	350
82.12	2856.2	3131.3	6.7203	80.83	2855.4	3130.3	6.7122	79.58	2854.6	3129.2	6.7042	360
83.74	2874.4	3155.0	6.7574	82.43	2873.7	3154.0	6.7493	81.16	2872.9	3153.0	6.7414	370
85.35	2892.5	3178.4	6.7936	84.02	2891.8	3177.5	6.7856	82.73	2891.1	3176.5	6.7778	380
86.94	2910.5	3201.7	6.8289	85.59	2909.8	3200.8	6.8211	84.28	2909.1	3199.9	6.8133	390
88.52	2928.3	3224.9	6.8636	87.15	2927.7	3224.0	6.8558	85.82	2927.0	3223.1	6.8481	400
90.08	2946.1	3247.9	6.8976	88.70	2945.5	3247.1	6.8898	87.35	2944.9	3246.3	6.8822	410
91.64	2963.9	3270.8	6.9309	90.23	2963.3	3270.1	6.9232	88.86	2962.7	3269.3	6.9156	420
93.18	2981.6	3293.7	6.9636	91.75	2981.0	3293.0	6.9560	90.37	2980.4	3292.2	6.9485	430
94.71	2999.2	3316.5	6.9958	93.27	2998.7	3315.8	6.9883	91.86	2998.2	3315.1	6.9808	440
96.24	3016.9	3339.3	7.0275	94.77	3016.4	3338.6	7.0200	93.35	3015.8	3337.9	7.0125	450
97.75	3034.5	3362.0	7.0587	96.27	3034.0	3361.3	7.0512	94.82	3033.5	3360.6	7.0438	460
99.26	3052.1	3384.6	7.0894	97.76	3051.6	3384.0	7.0820	96.29	3051.2	3383.4	7.0746	470
100.77	3069.7	3407.3	7.1197	99.24	3069.3	3406.7	7.1123	97.76	3068.8	3406.1	7.1049	480
102.26	3087.4	3430.0	7.1496	100.71	3086.9	3429.4	7.1422	99.21	3086.5	3428.8	7.1349	490
103.75	3105.0	3452.6	7.1791	102.18	3104.6	3452.0	7.1717	100.66	3104.2	3451.5	7.1644	500
106.72	3140.4	3497.9	7.2369	105.11	3140.0	3497.3	7.2296	103.55	3139.6	3496.8	7.2223	520
109.66	3175.8	3543.2	7.2933	108.01	3175.5	3542.7	7.2860	106.41	3175.1	3542.2	7.2788	540
112.59	3211.4	3588.6	7.3484	110.90	3211.0	3588.1	7.3412	109.26	3210.7	3587.6	7.3340	560
115.50	3247.1	3634.0	7.4024	113.77	3246.8	3633.6	7.3951	112.09	3246.5	3633.2	7.3880	580
118.40	3283.0	3679.6	7.4552	116.63	3282.7	3679.2	7.4480	114.91	3282.4	3678.8	7.4409	600
121.28	3319.0	3725.3	7.5069	119.47	3318.7	3724.9	7.4997	117.72	3318.4	3724.5	7.4927	620
124.16	3355.2	3771.1	7.5577	122.31	3354.9	3770.8	7.5505	120.51	3354.7	3770.4	7.5435	640
127.03	3391.6	3817.1	7.6075	125.14	3391.3	3816.8	7.6004	123.30	3391.1	3816.5	7.5934	660
129.80	3428.2	3863.3	7.6564	127.95	3428.0	3863.0	7.6494	126.08	3427.7	3862.7	7.6424	680
132.73	3465.0	3909.6	7.7046	130.76	3464.8	3909.4	7.6975	128.85	3464.5	3909.1	7.6905	700
135.58	3502.0	3956.2	7.7519	133.57	3501.8	3955.9	7.7449	131.61	3501.6	3955.6	7.7379	720
138.41	3539.2	4002.9	7.7985	136.36	3539.0	4002.7	7.7915	134.37	3538.8	4002.4	7.7845	740
141.25	3576.7	4049.9	7.8444	139.15	3576.5	4049.6	7.8374	137.12	3576.3	4049.4	7.8304	760
144.07	3614.4	4097.0	7.8896	141.94	3614.2	4096.8	7.8826	139.87	3614.0	4096.6	7.8757	780
146.89	3652.3	4144.4	7.9341	144.72	3652.1	4144.2	7.9271	142.61	3652.0	4144.0	7.9202	800
153.93	3748.2	4263.8	8.0429	151.66	3748.0	4263.6	8.0359	149.45	3747.8	4263.4	8.0291	850
160.95	3845.5	4384.6	8.1482	158.57	3845.3	4384.5	8.1412	156.27	3845.2	4384.3	8.1343	900
167.95	3944.2	4506.9	8.2502	165.48	3944.1	4506.7	8.2432	163.07	3944.0	4506.6	8.2364	950
174.95	4044.5	4630.6	8.3493	172.37	4044.4	4630.4	8.3424	169.86	4044.3	4630.3	8.3355	1000
188.91	4249.5	4882.3	8.5396	186.12	4249.4	4882.2	8.5327	183.42	4249.3	4882.1	8.5259	1100
202.85	4460.1	5139.6	8.7205	199.86	4460.0	5139.5	8.7136	196.96	4459.9	5139.4	8.7068	1200
216.78	4675.8	5402.0	8.8928	213.59	4675.7	5401.9	8.8859	210.49	4675.6	5401.8	8.8791	1300

Definitions of symbols on page 1.

Table 3. Vapor

p (t Sat.)	3.50 (242.60)				3.55 (243.42)				3.60 (244.23)			
t	10^3 v	u	h	s	10^3 v	u	h	s	10^3 v	u	h	s
Sat.	57.07	2603.7	2803.4	6.1253	56.25	2603.6	2803.3	6.1195	55.45	2603.5	2803.1	6.1138
235	*55.26*	*2581.8*	*2775.2*	*6.0702*	*54.26*	*2579.1*	*2771.7*	*6.0579*				
240	*56.46*	*2596.4*	*2794.0*	*6.1069*	*55.46*	*2593.9*	*2790.8*	*6.0952*	*54.48*	*2591.4*	*2787.5*	*6.0835*
245	57.61	2610.3	2812.0	6.1418	56.61	2608.0	2809.0	6.1305	55.63	2605.7	2805.9	6.1192
250	58.72	2623.7	2829.2	6.1749	57.71	2621.5	2826.4	6.1640	56.73	2619.3	2823.6	6.1531
255	59.80	2636.5	2845.8	6.2065	58.79	2634.5	2843.2	6.1960	57.80	2632.5	2840.6	6.1854
260	60.85	2649.0	2861.9	6.2369	59.83	2647.1	2859.5	6.2266	58.83	2645.2	2857.0	6.2164
265	61.86	2661.1	2877.6	6.2661	60.84	2659.3	2875.3	6.2561	59.84	2657.5	2872.9	6.2461
270	62.86	2672.8	2892.8	6.2942	61.82	2671.1	2890.6	6.2844	60.82	2669.4	2888.4	6.2747
275	63.83	2684.2	2907.6	6.3214	62.79	2682.6	2905.5	6.3118	61.78	2681.1	2903.5	6.3023
280	64.78	2695.4	2922.1	6.3478	63.73	2693.9	2920.2	6.3384	62.71	2692.4	2918.2	6.3291
285	65.71	2706.4	2936.3	6.3733	64.66	2704.9	2934.5	6.3641	63.63	2703.5	2932.6	6.3550
290	66.63	2717.1	2950.3	6.3982	65.57	2715.7	2948.5	6.3891	64.53	2714.4	2946.7	6.3802
295	67.53	2727.6	2964.0	6.4224	66.46	2726.3	2962.3	6.4135	65.42	2725.0	2960.6	6.4047
300	68.42	2738.0	2977.5	6.4461	67.34	2736.8	2975.8	6.4373	66.29	2735.5	2974.2	6.4286
310	70.16	2758.3	3003.9	6.4917	69.07	2757.2	3002.4	6.4832	68.00	2756.0	3000.8	6.4747
320	71.86	2778.1	3029.6	6.5355	70.75	2777.1	3028.2	6.5271	69.67	2776.0	3026.8	6.5189
330	73.53	2797.5	3054.8	6.5777	72.40	2796.5	3053.5	6.5695	71.31	2795.5	3052.2	6.5613
340	75.17	2816.5	3079.6	6.6184	74.02	2815.6	3078.4	6.6103	72.91	2814.7	3077.2	6.6023
350	76.78	2835.3	3104.0	6.6579	75.61	2834.4	3102.9	6.6499	74.48	2833.6	3101.7	6.6421
360	78.36	2853.8	3128.1	6.6962	77.18	2853.0	3127.0	6.6884	76.03	2852.2	3126.0	6.6806
370	79.93	2872.2	3151.9	6.7336	78.73	2871.4	3150.9	6.7258	77.57	2870.7	3149.9	6.7182
380	81.48	2890.4	3175.6	6.7700	80.26	2889.7	3174.6	6.7623	79.08	2888.9	3173.6	6.7548
390	83.01	2908.4	3199.0	6.8056	81.78	2907.8	3198.1	6.7980	80.58	2907.1	3197.2	6.7905
400	84.53	2926.4	3222.3	6.8405	83.28	2925.8	3221.4	6.8330	82.06	2925.1	3220.5	6.8255
410	86.04	2944.3	3245.4	6.8746	84.77	2943.7	3244.6	6.8672	83.53	2943.1	3243.8	6.8598
420	87.54	2962.1	3268.5	6.9081	86.25	2961.5	3267.7	6.9007	84.99	2960.9	3266.9	6.8934
430	89.02	2979.9	3291.5	6.9410	87.71	2979.3	3290.7	6.9337	86.44	2978.8	3289.9	6.9264
440	90.50	2997.6	3314.4	6.9734	89.17	2997.1	3313.6	6.9660	87.88	2996.5	3312.9	6.9588
450	91.96	3015.3	3337.2	7.0052	90.62	3014.8	3336.5	6.9979	89.31	3014.3	3335.8	6.9907
460	93.42	3033.0	3360.0	7.0365	92.06	3032.5	3359.3	7.0292	90.73	3032.0	3358.7	7.0221
470	94.87	3050.7	3382.7	7.0673	93.49	3050.2	3382.1	7.0601	92.15	3049.8	3381.5	7.0530
480	96.31	3068.4	3405.5	7.0977	94.91	3067.9	3404.9	7.0905	93.55	3067.5	3404.3	7.0835
490	97.75	3086.1	3428.2	7.1277	96.33	3085.6	3427.6	7.1205	94.95	3085.2	3427.0	7.1135
500	99.18	3103.8	3450.9	7.1572	97.75	3103.3	3450.3	7.1501	96.35	3102.9	3449.8	7.1431
520	102.03	3139.2	3496.3	7.2152	100.56	3138.8	3495.8	7.2081	99.12	3138.4	3495.3	7.2012
540	104.86	3174.7	3541.7	7.2717	103.35	3174.4	3541.2	7.2647	101.88	3174.0	3540.8	7.2578
560	107.67	3210.4	3587.2	7.3270	106.12	3210.0	3586.7	7.3200	104.61	3209.7	3586.3	7.3132
580	110.46	3246.1	3632.7	7.3810	108.88	3245.8	3632.3	7.3741	107.33	3245.5	3631.9	7.3673
600	113.24	3282.1	3678.4	7.4339	111.62	3281.8	3678.0	7.4270	110.04	3281.5	3677.6	7.4202
620	116.01	3318.1	3724.2	7.4857	114.35	3317.9	3723.8	7.4789	112.74	3317.6	3723.4	7.4721
640	118.77	3354.4	3770.1	7.5366	117.07	3354.1	3769.7	7.5297	115.42	3353.9	3769.4	7.5230
660	121.52	3390.8	3816.1	7.5865	119.78	3390.6	3815.8	7.5797	118.10	3390.3	3815.5	7.5729
680	124.26	3427.5	3862.4	7.6355	122.49	3427.2	3862.1	7.6287	120.77	3427.0	3861.8	7.6220
700	126.99	3464.3	3908.8	7.6837	125.18	3464.1	3908.5	7.6769	123.42	3463.9	3908.2	7.6702
720	129.71	3501.4	3955.4	7.7311	127.87	3501.2	3955.1	7.7243	126.08	3500.9	3954.8	7.7176
740	132.43	3538.6	4002.1	7.7777	130.55	3538.4	4001.9	7.7709	128.72	3538.2	4001.6	7.7643
760	135.15	3576.1	4049.1	7.8236	133.23	3575.9	4048.9	7.8169	131.36	3575.7	4048.6	7.8102
780	137.85	3613.8	4096.3	7.8689	135.90	3613.7	4096.1	7.8621	134.00	3613.5	4095.9	7.8555
800	140.56	3651.8	4143.7	7.9134	138.57	3651.6	4143.5	7.9067	136.63	3651.4	4143.3	7.9001
850	147.30	3747.7	4263.2	8.0223	145.22	3747.5	4263.0	8.0156	143.19	3747.4	4262.8	8.0090
900	154.02	3845.0	4384.1	8.1276	151.85	3844.9	4383.9	8.1209	149.73	3844.7	4383.8	8.1143
950	160.73	3943.8	4506.4	8.2297	158.46	3943.7	4506.2	8.2230	156.25	3943.6	4506.1	8.2164
1000	167.43	4044.1	4630.1	8.3288	165.06	4044.0	4630.0	8.3222	162.77	4043.9	4629.8	8.3156
1100	180.80	4249.2	4881.9	8.5192	178.25	4249.0	4881.8	8.5125	175.77	4248.9	4881.7	8.5060
1200	194.15	4459.8	5139.3	8.7000	191.41	4459.6	5139.2	8.6934	188.75	4459.5	5139.0	8.6869
1300	207.49	4675.5	5401.7	8.8723	204.57	4675.3	5401.6	8.8657	201.72	4675.2	5401.4	8.8592

Definitions of symbols on page 1.

3.65 (245.03)				3.70 (245.82)				3.75 (246.60)				p (t Sat.)
10^3 v	u	h	s	10^3 v	u	h	s	10^3 v	u	h	s	t
54.68	2603.4	2803.0	6.1081	53.92	2603.3	2802.8	6.1025	53.19	2603.1	2802.6	6.0970	Sat.
												235
53.52	*2588.8*	*2784.2*	*6.0717*	*52.59*	*2586.2*	*2780.8*	*6.0600*	*51.69*	*2583.6*	*2777.4*	*6.0483*	240
54.67	*2603.3*	*2802.9*	*6.1079*	*53.74*	*2600.9*	*2799.7*	*6.0967*	*52.83*	*2598.5*	*2796.6*	*6.0855*	245
55.77	2617.2	2820.7	6.1423	54.84	2614.9	2817.8	6.1315	53.93	2612.7	2814.9	6.1207	250
56.84	2630.5	2837.9	6.1750	55.90	2628.4	2835.2	6.1645	54.98	2626.3	2832.5	6.1541	255
57.87	2643.3	2854.5	6.2062	56.92	2641.3	2852.0	6.1961	56.01	2639.4	2849.4	6.1860	260
58.87	2655.7	2870.6	6.2362	57.92	2653.9	2868.2	6.2264	57.00	2652.0	2865.8	6.2166	265
59.84	2667.7	2886.2	6.2651	58.89	2666.0	2883.9	6.2555	57.96	2664.3	2881.7	6.2459	270
60.79	2679.5	2901.3	6.2929	59.83	2677.8	2899.2	6.2835	58.90	2676.2	2897.1	6.2742	275
61.72	2690.9	2916.2	6.3198	60.76	2689.4	2914.2	6.3106	59.82	2687.8	2912.1	6.3015	280
62.63	2702.1	2930.7	6.3459	61.66	2700.6	2928.8	6.3369	60.72	2699.1	2926.8	6.3280	285
63.53	2713.0	2944.9	6.3713	62.55	2711.6	2943.1	6.3624	61.60	2710.2	2941.2	6.3537	290
64.41	2723.7	2958.8	6.3959	63.42	2722.4	2957.1	6.3872	62.46	2721.1	2955.3	6.3786	295
65.27	2734.3	2972.5	6.4199	64.28	2733.0	2970.9	6.4114	63.32	2731.8	2969.2	6.4029	300
66.97	2754.9	2999.3	6.4663	65.96	2753.7	2997.8	6.4580	64.98	2752.6	2996.3	6.4497	310
68.62	2774.9	3025.4	6.5106	67.60	2773.9	3024.0	6.5025	66.60	2772.8	3022.6	6.4945	320
70.24	2794.6	3050.9	6.5533	69.20	2793.6	3049.6	6.5453	68.19	2792.6	3048.3	6.5375	330
71.82	2813.8	3076.0	6.5944	70.77	2812.9	3074.7	6.5866	69.74	2812.0	3073.5	6.5789	340
73.38	2832.7	3100.6	6.6343	72.31	2831.9	3099.4	6.6266	71.27	2831.0	3098.3	6.6190	350
74.92	2851.4	3124.9	6.6730	73.83	2850.6	3123.8	6.6654	72.77	2849.8	3122.7	6.6579	360
76.43	2869.9	3148.9	6.7106	75.33	2869.2	3147.9	6.7031	74.25	2868.4	3146.8	6.6957	370
77.93	2888.2	3172.7	6.7473	76.81	2887.5	3171.7	6.7399	75.72	2886.8	3170.7	6.7326	380
79.41	2906.4	3196.3	6.7831	78.27	2905.7	3195.3	6.7758	77.17	2905.0	3194.4	6.7686	390
80.88	2924.5	3219.7	6.8182	79.72	2923.8	3218.8	6.8109	78.60	2923.2	3217.9	6.8037	400
82.33	2942.4	3243.0	6.8525	81.16	2941.8	3242.1	6.8453	80.02	2941.2	3241.3	6.8382	410
83.77	2960.4	3266.1	6.8862	82.59	2959.8	3265.3	6.8790	81.43	2959.2	3264.5	6.8720	420
85.20	2978.2	3289.2	6.9192	84.00	2977.6	3288.4	6.9121	82.83	2977.1	3287.7	6.9051	430
86.63	2996.0	3312.2	6.9517	85.40	2995.5	3311.5	6.9447	84.22	2994.9	3310.7	6.9377	440
88.04	3013.8	3335.1	6.9836	86.80	3013.3	3334.4	6.9766	85.60	3012.8	3333.7	6.9697	450
89.44	3031.5	3358.0	7.0151	88.19	3031.0	3357.3	7.0081	86.97	3030.6	3356.7	7.0012	460
90.84	3049.3	3380.8	7.0460	89.57	3048.8	3380.2	7.0391	88.33	3048.3	3379.6	7.0322	470
92.23	3067.0	3403.7	7.0765	90.94	3066.6	3403.0	7.0696	89.69	3066.1	3402.4	7.0628	480
93.61	3084.8	3426.4	7.1066	92.31	3084.3	3425.9	7.0997	91.04	3083.9	3425.3	7.0929	490
94.99	3102.5	3449.2	7.1362	93.67	3102.1	3448.7	7.1294	92.38	3101.7	3448.1	7.1226	500
97.73	3138.0	3494.7	7.1943	96.37	3137.6	3494.2	7.1876	95.05	3137.2	3493.7	7.1809	520
100.45	3173.6	3540.3	7.2510	99.06	3173.3	3539.8	7.2443	97.70	3172.9	3539.3	7.2376	540
103.15	3209.3	3585.8	7.3064	101.73	3209.0	3585.4	7.2997	100.34	3208.7	3584.9	7.2931	560
105.84	3245.2	3631.5	7.3605	104.38	3244.9	3631.1	7.3539	102.96	3244.5	3630.6	7.3473	580
108.51	3281.2	3677.2	7.4135	107.02	3280.9	3676.8	7.4069	105.56	3280.6	3676.4	7.4003	600
111.17	3317.3	3723.1	7.4654	109.64	3317.0	3722.7	7.4588	108.16	3316.7	3722.3	7.4523	620
113.82	3353.6	3769.0	7.5163	112.26	3353.3	3768.7	7.5098	110.74	3353.1	3768.4	7.5033	640
116.46	3390.1	3815.2	7.5663	114.87	3389.8	3814.8	7.5598	113.31	3389.6	3814.5	7.5533	660
119.09	3426.8	3861.5	7.6154	117.46	3426.5	3861.2	7.6089	115.88	3426.3	3860.8	7.6024	680
121.72	3463.7	3907.9	7.6636	120.05	3463.4	3907.6	7.6571	118.43	3463.2	3907.3	7.6507	700
124.33	3500.7	3954.5	7.7111	122.64	3500.5	3954.3	7.7046	120.98	3500.3	3954.0	7.6982	720
126.94	3538.0	4001.4	7.7577	125.21	3537.8	4001.1	7.7513	123.53	3537.6	4000.9	7.7449	740
129.55	3575.5	4048.4	7.8037	127.78	3575.4	4048.2	7.7972	126.06	3575.2	4047.9	7.7909	760
132.15	3613.3	4095.6	7.8490	130.35	3613.1	4095.4	7.8425	128.60	3612.9	4095.2	7.8362	780
134.74	3651.3	4143.1	7.8936	132.91	3651.1	4142.8	7.8872	131.12	3650.9	4142.6	7.8808	800
141.21	3747.2	4262.7	8.0025	139.30	3747.1	4262.5	7.9961	137.43	3746.9	4262.3	7.9898	850
147.67	3844.6	4383.6	8.1078	145.66	3844.5	4383.4	8.1014	143.71	3844.3	4383.2	8.0951	900
154.10	3943.4	4505.9	8.2100	152.01	3943.3	4505.8	8.2036	149.98	3943.2	4505.6	8.1973	950
160.53	4043.8	4629.7	8.3091	158.35	4043.6	4629.6	8.3028	156.24	4043.5	4629.4	8.2965	1000
173.36	4248.8	4881.6	8.4995	171.01	4248.7	4881.4	8.4932	168.73	4248.6	4881.3	8.4869	1100
186.17	4459.4	5138.9	8.6804	183.65	4459.3	5138.8	8.6741	181.20	4459.2	5138.7	8.6678	1200
198.96	4675.1	5401.3	8.8528	196.27	4675.0	5401.2	8.8464	193.66	4674.9	5401.1	8.8401	1300

Definitions of symbols on page 1.

Table 3. Vapor

p (t Sat.)	**3.80** (247.38)				**3.85** (248.15)				**3.90** (248.91)			
t	$10^3 v$	u	h	s	$10^3 v$	u	h	s	$10^3 v$	u	h	s
Sat.	52.47	2603.0	2802.4	6.0915	51.77	2602.8	2802.1	6.0861	51.09	2602.6	2801.9	6.0807
240	*50.80*	*2580.9*	*2774.0*	*6.0366*								
245	*51.94*	*2596.0*	*2793.4*	*6.0743*	*51.08*	*2593.6*	*2790.2*	*6.0632*	*50.23*	*2591.1*	*2787.0*	*6.0520*
250	53.04	2610.4	2812.0	6.1100	52.17	2608.1	2809.0	6.0992	51.33	2605.8	2806.0	6.0885
255	54.09	2624.2	2829.7	6.1438	53.22	2622.1	2827.0	6.1334	52.37	2619.9	2824.2	6.1231
260	55.11	2637.4	2846.8	6.1760	54.24	2635.4	2844.3	6.1660	53.39	2633.4	2841.6	6.1560
265	56.10	2650.2	2863.4	6.2068	55.22	2648.3	2860.9	6.1971	54.36	2646.5	2858.5	6.1875
270	57.05	2662.6	2879.4	6.2364	56.17	2660.8	2877.1	6.2270	55.31	2659.1	2874.8	6.2176
275	57.99	2674.6	2894.9	6.2650	57.10	2672.9	2892.8	6.2558	56.24	2671.3	2890.6	6.2466
280	58.90	2686.3	2910.1	6.2925	58.01	2684.7	2908.0	6.2835	57.14	2683.1	2906.0	6.2746
285	59.79	2697.7	2924.9	6.3191	58.90	2696.2	2922.9	6.3103	58.02	2694.7	2921.0	6.3016
290	60.67	2708.8	2939.4	6.3449	59.76	2707.4	2937.5	6.3363	58.88	2706.0	2935.7	6.3277
295	61.53	2719.8	2953.6	6.3700	60.62	2718.4	2951.8	6.3616	59.73	2717.1	2950.0	6.3531
300	62.37	2730.5	2967.5	6.3945	61.46	2729.2	2965.8	6.3861	60.56	2727.9	2964.1	6.3778
310	64.03	2751.4	2994.7	6.4415	63.09	2750.2	2993.2	6.4334	62.19	2749.1	2991.6	6.4254
320	65.63	2771.7	3021.2	6.4865	64.69	2770.7	3019.7	6.4786	63.77	2769.6	3018.3	6.4707
330	67.20	2791.6	3047.0	6.5296	66.24	2790.6	3045.6	6.5219	65.31	2789.6	3044.3	6.5143
340	68.74	2811.0	3072.3	6.5712	67.77	2810.1	3071.0	6.5636	66.82	2809.2	3069.8	6.5561
350	70.25	2830.2	3097.1	6.6115	69.26	2829.3	3096.0	6.6040	68.30	2828.4	3094.8	6.5966
360	71.74	2849.0	3121.6	6.6505	70.74	2848.2	3120.5	6.6431	69.76	2847.4	3119.4	6.6358
370	73.21	2867.6	3145.8	6.6884	72.19	2866.9	3144.8	6.6811	71.20	2866.1	3143.8	6.6740
380	74.66	2886.1	3169.8	6.7253	73.62	2885.3	3168.8	6.7182	72.62	2884.6	3167.8	6.7111
390	76.09	2904.4	3193.5	6.7614	75.04	2903.7	3192.6	6.7543	74.02	2903.0	3191.7	6.7473
400	77.51	2922.5	3217.1	6.7967	76.44	2921.9	3216.2	6.7896	75.41	2921.2	3215.3	6.7827
410	78.91	2940.6	3240.5	6.8312	77.83	2940.0	3239.6	6.8242	76.78	2939.4	3238.8	6.8173
420	80.31	2958.6	3263.8	6.8650	79.21	2958.0	3263.0	6.8581	78.14	2957.4	3262.2	6.8513
430	81.69	2976.5	3286.9	6.8982	80.58	2976.0	3286.2	6.8914	79.49	2975.4	3285.4	6.8846
440	83.06	2994.4	3310.0	6.9308	81.93	2993.9	3309.3	6.9240	80.84	2993.3	3308.6	6.9173
450	84.42	3012.2	3333.0	6.9629	83.28	3011.7	3332.4	6.9561	82.17	3011.2	3331.7	6.9494
460	85.78	3030.1	3356.0	6.9944	84.62	3029.6	3355.3	6.9877	83.49	3029.1	3354.7	6.9811
470	87.13	3047.9	3378.9	7.0255	85.95	3047.4	3378.3	7.0188	84.81	3046.9	3377.7	7.0122
480	88.47	3065.6	3401.8	7.0561	87.28	3065.2	3401.2	7.0494	86.12	3064.7	3400.6	7.0429
490	89.80	3083.4	3424.7	7.0862	88.60	3083.0	3424.1	7.0796	87.42	3082.6	3423.5	7.0731
500	91.13	3101.2	3447.5	7.1160	89.91	3100.8	3447.0	7.1094	88.72	3100.4	3446.4	7.1029
520	93.77	3136.8	3493.2	7.1742	92.52	3136.5	3492.6	7.1677	91.30	3136.1	3492.1	7.1613
540	96.39	3172.5	3538.8	7.2311	95.10	3172.2	3538.3	7.2246	93.85	3171.8	3537.8	7.2182
560	98.99	3208.3	3584.5	7.2866	97.67	3208.0	3584.0	7.2801	96.39	3207.6	3583.6	7.2737
580	101.58	3244.2	3630.2	7.3408	100.23	3243.9	3629.8	7.3344	98.92	3243.6	3629.4	7.3281
600	104.15	3280.3	3676.0	7.3939	102.77	3280.0	3675.6	7.3875	101.43	3279.7	3675.2	7.3812
620	106.71	3316.5	3722.0	7.4459	105.30	3316.2	3721.6	7.4395	103.93	3315.9	3721.2	7.4333
640	109.26	3352.8	3768.0	7.4969	107.82	3352.5	3767.7	7.4906	106.42	3352.3	3767.3	7.4843
660	111.80	3389.3	3814.2	7.5469	110.33	3389.1	3813.9	7.5406	108.90	3388.8	3813.5	7.5344
680	114.33	3426.1	3860.5	7.5961	112.83	3425.8	3860.2	7.5898	111.37	3425.6	3859.9	7.5836
700	116.86	3463.0	3907.0	7.6443	115.32	3462.8	3906.8	7.6381	113.83	3462.5	3906.5	7.6319
720	119.38	3500.1	3953.7	7.6918	117.81	3499.9	3953.5	7.6856	116.28	3499.7	3953.2	7.6794
740	121.89	3537.4	4000.6	7.7386	120.29	3537.2	4000.3	7.7323	118.73	3537.0	4000.1	7.7262
760	124.39	3575.0	4047.7	7.7846	122.76	3574.8	4047.4	7.7783	121.17	3574.6	4047.2	7.7722
780	126.89	3612.7	4094.9	7.8299	125.23	3612.6	4094.7	7.8237	123.61	3612.4	4094.5	7.8175
800	129.39	3650.7	4142.4	7.8745	127.69	3650.6	4142.2	7.8683	126.05	3650.4	4142.0	7.8622
850	135.61	3746.7	4262.1	7.9835	133.84	3746.6	4261.9	7.9773	132.11	3746.4	4261.7	7.9712
900	141.81	3844.2	4383.1	8.0889	139.96	3844.0	4382.9	8.0827	138.16	3843.9	4382.7	8.0767
950	148.00	3943.0	4505.4	8.1910	146.07	3942.9	4505.3	8.1849	144.19	3942.8	4505.1	8.1788
1000	154.18	4043.4	4629.3	8.2903	152.17	4043.3	4629.1	8.2841	150.21	4043.1	4629.0	8.2781
1100	166.50	4248.4	4881.2	8.4807	164.34	4248.3	4881.0	8.4746	162.23	4248.2	4880.9	8.4685
1200	178.81	4459.1	5138.6	8.6616	176.49	4459.0	5138.4	8.6555	174.23	4458.8	5138.3	8.6495
1300	191.11	4674.8	5401.0	8.8339	188.63	4674.6	5400.9	8.8278	186.21	4674.5	5400.8	8.8218

Definitions of symbols on page 1.

	3.95 (249.66)				4.0 (250.40)				4.1 (251.87)			p (t Sat.)
10^3 v	u	h	s	10^3 v	u	h	s	10^3 v	u	h	s	t
50.43	2602.5	2801.7	6.0754	49.78	2602.3	2801.4	6.0701	48.53	2601.9	2800.9	6.0597	Sat.
												240
49.41	*2588.5*	*2783.7*	*6.0409*	*48.60*	*2585.9*	*2780.3*	*6.0297*	*47.04*	*2580.7*	*2773.6*	*6.0074*	245
50.50	*2603.5*	*2802.9*	*6.0779*	*49.69*	*2601.1*	*2799.9*	*6.0672*	*48.13*	*2596.3*	*2793.6*	*6.0459*	250
51.55	2617.7	2821.3	6.1129	50.74	2615.5	2818.5	6.1026	49.17	2611.1	2812.7	6.0822	255
52.55	2631.4	2839.0	6.1461	51.74	2629.4	2836.3	6.1362	50.17	2625.2	2830.9	6.1165	260
53.53	2644.6	2856.0	6.1779	52.71	2642.7	2853.5	6.1683	51.14	2638.8	2848.5	6.1493	265
54.47	2657.3	2872.4	6.2083	53.65	2655.5	2870.1	6.1990	52.07	2651.9	2865.4	6.1805	270
55.39	2669.6	2888.4	6.2375	54.57	2667.9	2886.2	6.2285	52.98	2664.5	2881.7	6.2105	275
56.29	2681.6	2903.9	6.2657	55.46	2680.0	2901.8	6.2568	53.86	2676.7	2897.6	6.2393	280
57.16	2693.2	2919.0	6.2929	56.33	2691.7	2917.0	6.2842	54.72	2688.7	2913.0	6.2671	285
58.02	2704.6	2933.8	6.3192	57.18	2703.2	2931.9	6.3108	55.56	2700.3	2928.1	6.2940	290
58.86	2715.7	2948.2	6.3448	58.02	2714.4	2946.4	6.3365	56.39	2711.6	2942.8	6.3200	295
59.69	2726.6	2962.4	6.3696	58.84	2725.3	2960.7	6.3615	57.20	2722.7	2957.2	6.3453	300
61.30	2747.9	2990.0	6.4174	60.44	2746.7	2988.5	6.4095	58.77	2744.3	2985.3	6.3939	310
62.87	2768.5	3016.8	6.4630	61.99	2767.4	3015.4	6.4553	60.30	2765.2	3012.5	6.4401	320
64.40	2788.6	3043.0	6.5067	63.51	2787.6	3041.6	6.4991	61.79	2785.6	3038.9	6.4843	330
65.89	2808.3	3068.5	6.5487	64.99	2807.3	3067.3	6.5413	63.25	2805.4	3064.8	6.5268	340
67.36	2827.5	3093.6	6.5893	66.45	2826.7	3092.5	6.5821	64.68	2824.9	3090.1	6.5678	350
68.81	2846.6	3118.3	6.6286	67.88	2845.7	3117.2	6.6215	66.08	2844.1	3115.0	6.6075	360
70.23	2865.3	3142.7	6.6669	69.29	2864.6	3141.7	6.6598	67.47	2863.0	3139.6	6.6460	370
71.63	2883.9	3166.8	6.7041	70.68	2883.2	3165.9	6.6971	68.83	2881.7	3163.9	6.6834	380
73.02	2902.3	3190.7	6.7404	72.05	2901.6	3189.8	6.7335	70.18	2900.2	3188.0	6.7200	390
74.39	2920.6	3214.4	6.7758	73.41	2919.9	3213.6	6.7690	71.51	2918.6	3211.8	6.7557	400
75.75	2938.7	3238.0	6.8105	74.75	2938.1	3237.1	6.8038	72.83	2936.9	3235.5	6.7906	410
77.10	2956.8	3261.4	6.8445	76.09	2956.2	3260.6	6.8379	74.13	2955.0	3259.0	6.8247	420
78.44	2974.8	3284.7	6.8779	77.41	2974.3	3283.9	6.8713	75.43	2973.1	3282.4	6.8582	430
79.76	2992.8	3307.8	6.9107	78.72	2992.2	3307.1	6.9041	76.71	2991.2	3305.7	6.8911	440
81.08	3010.7	3331.0	6.9428	80.02	3010.2	3330.3	6.9363	77.99	3009.1	3328.9	6.9235	450
82.39	3028.6	3354.0	6.9745	81.32	3028.1	3353.4	6.9680	79.25	3027.1	3352.0	6.9552	460
83.69	3046.4	3377.0	7.0057	82.61	3046.0	3376.4	6.9992	80.51	3045.0	3375.1	6.9865	470
84.99	3064.3	3400.0	7.0364	83.89	3063.8	3399.4	7.0299	81.77	3062.9	3398.1	7.0173	480
86.28	3082.1	3422.9	7.0666	85.16	3081.7	3422.3	7.0602	83.01	3080.8	3421.1	7.0476	490
87.56	3100.0	3445.8	7.0964	86.43	3099.5	3445.3	7.0901	84.25	3098.7	3444.1	7.0776	500
90.11	3135.7	3491.6	7.1549	88.95	3135.3	3491.1	7.1486	86.71	3134.5	3490.0	7.1362	520
92.64	3171.4	3537.3	7.2118	91.45	3171.1	3536.9	7.2056	89.16	3170.3	3535.9	7.1933	540
95.15	3207.3	3583.1	7.2675	93.93	3206.9	3582.7	7.2612	91.58	3206.3	3581.7	7.2490	560
97.64	3243.3	3628.9	7.3218	96.39	3242.9	3628.5	7.3156	93.99	3242.3	3627.7	7.3035	580
100.12	3279.4	3674.8	7.3750	98.85	3279.1	3674.4	7.3688	96.39	3278.5	3673.7	7.3568	600
102.59	3315.6	3720.8	7.4271	101.29	3315.3	3720.5	7.4210	98.77	3314.8	3719.7	7.4089	620
105.05	3352.0	3767.0	7.4781	103.72	3351.8	3766.6	7.4720	101.15	3351.2	3765.9	7.4601	640
107.50	3388.6	3813.2	7.5282	106.13	3388.3	3812.9	7.5222	103.51	3387.8	3812.2	7.5102	660
109.94	3425.4	3859.6	7.5774	108.55	3425.1	3859.3	7.5714	105.86	3424.7	3858.7	7.5595	680
112.37	3462.3	3906.2	7.6258	110.95	3462.1	3905.9	7.6198	108.21	3461.6	3905.3	7.6079	700
114.79	3499.5	3952.9	7.6733	113.34	3499.3	3952.6	7.6673	110.55	3498.8	3952.1	7.6555	720
117.21	3536.8	3999.8	7.7201	115.73	3536.6	3999.6	7.7141	112.88	3536.2	3999.1	7.7023	740
119.63	3574.4	4046.9	7.7661	118.12	3574.2	4046.7	7.7601	115.21	3573.8	4046.2	7.7484	760
122.03	3612.2	4094.2	7.8115	120.50	3612.0	4094.0	7.8055	117.53	3611.7	4093.5	7.7938	780
124.44	3650.2	4141.7	7.8562	122.87	3650.0	4141.5	7.8502	119.85	3649.7	4141.1	7.8385	800
130.43	3746.3	4261.5	7.9652	128.79	3746.1	4261.3	7.9593	125.63	3745.8	4260.9	7.9476	850
136.40	3843.7	4382.5	8.0707	134.69	3843.6	4382.3	8.0647	131.39	3843.3	4382.0	8.0531	900
142.36	3942.6	4505.0	8.1728	140.57	3942.5	4504.8	8.1669	137.13	3942.2	4504.5	8.1553	950
148.30	4043.0	4628.8	8.2721	146.45	4042.9	4628.7	8.2662	142.86	4042.6	4628.4	8.2546	1000
160.17	4248.1	4880.8	8.4626	158.17	4248.0	4880.6	8.4567	154.30	4247.7	4880.4	8.4451	1100
172.02	4458.7	5138.2	8.6435	169.87	4458.6	5138.1	8.6376	165.72	4458.4	5137.8	8.6261	1200
183.86	4674.4	5400.6	8.8159	181.56	4674.3	5400.5	8.8100	177.13	4674.1	5400.3	8.7984	1300

Definitions of symbols on page 1.

Table 3. Vapor

p (t Sat.)	4.2 (253.31)				4.3 (254.73)				4.4 (256.12)			
t	$10^3 v$	u	h	s	$10^3 v$	u	h	s	$10^3 v$	u	h	s
Sat.	47.33	2601.5	2800.3	6.0495	46.19	2601.0	2799.7	6.0394	45.10	2600.6	2799.0	6.0296
250	*46.64*	*2591.3*	*2787.2*	*6.0246*	*45.21*	*2586 3*	*2780.6*	*6.0033*	43.83	2581.1	2773.9	5.9819
255	47.68	2606.5	2806.8	6.0618	46.25	2601.9	2800.7	6.0414	*44.87*	*2597.1*	*2794.5*	*6.0211*
260	48.68	2621.0	2825.4	6.0970	47.24	2616.7	2819.8	6.0774	45.87	2612.3	2814.1	6.0580
265	49.64	2634.8	2843.3	6.1304	48.20	2630.8	2838.1	6.1115	46.82	2626.7	2832.8	6.0928
270	50.56	2648.2	2860.5	6.1622	49.12	2644.4	2855.6	6.1440	47.74	2640.6	2850.7	6.1259
275	51.46	2661.0	2877.2	6.1927	50.01	2657.5	2872.6	6.1750	48.63	2653.9	2867.9	6.1575
280	52.34	2673.5	2893.3	6.2220	50.88	2670.2	2889.0	6.2048	49.49	2666.8	2884.6	6.1878
285	53.19	2685.6	2909.0	6.2502	51.72	2682.5	2904.9	6.2334	50.32	2679.3	2900.7	6.2169
290	54.02	2697.4	2924.2	6.2774	52.55	2694.4	2920.4	6.2611	51.14	2691.4	2916.4	6.2449
295	54.83	2708.9	2939.2	6.3038	53.35	2706.1	2935.5	6.2878	51.93	2703.2	2931.7	6.2719
300	55.63	2720.1	2953.8	6.3294	54.14	2717.4	2950.2	6.3137	52.71	2714.7	2946.7	6.2981
310	57.19	2741.9	2982.1	6.3784	55.67	2739.5	2978.9	6.3633	54.23	2737.1	2975.7	6.3483
320	58.69	2763.0	3009.6	6.4251	57.16	2760.8	3006.6	6.4103	55.69	2758.6	3003.6	6.3958
330	60.16	2783.5	3036.2	6.4697	58.60	2781.5	3033.5	6.4553	57.12	2779.4	3030.7	6.4411
340	61.59	2803.5	3062.2	6.5125	60.01	2801.5	3059.7	6.4984	58.50	2799.7	3057.2	6.4846
350	63.00	2823.2	3087.8	6.5537	61.39	2821.4	3085.4	6.5400	59.86	2819.6	3083.0	6.5264
360	64.38	2842.4	3112.8	6.5937	62.75	2840.8	3110.6	6.5801	61.19	2839.1	3108.4	6.5668
370	65.73	2861.5	3137.5	6.6324	64.08	2859.9	3135.4	6.6190	62.50	2858.3	3133.3	6.6059
380	67.07	2880.2	3161.9	6.6700	65.39	2878.8	3160.0	6.6569	63.79	2877.3	3158.0	6.6439
390	68.39	2898.8	3186.1	6.7067	66.69	2897.5	3184.2	6.6937	65.07	2896.1	3182.3	6.6810
400	69.70	2917.3	3210.0	6.7425	67.97	2916.0	3208.3	6.7297	66.32	2914.7	3206.5	6.7171
410	70.99	2935.6	3233.8	6.7776	69.24	2934.4	3232.1	6.7648	67.57	2933.1	3230.4	6.7524
420	72.27	2953.8	3257.4	6.8119	70.49	2952.7	3255.8	6.7992	68.80	2951.5	3254.2	6.7869
430	73.54	2972.0	3280.8	6.8455	71.74	2970.8	3279.3	6.8330	70.02	2969.7	3277.8	6.8207
440	74.80	2990.1	3304.2	6.8785	72.97	2989.0	3302.7	6.8661	71.23	2987.9	3301.3	6.8539
450	76.04	3008.1	3327.5	6.9109	74.19	3007.0	3326.1	6.8986	72.43	3006.0	3324.7	6.8865
460	77.29	3026.1	3350.7	6.9427	75.41	3025.1	3349.3	6.9305	73.62	3024.1	3348.0	6.9185
470	78.52	3044.0	3373.8	6.9741	76.62	3043.1	3372.5	6.9619	74.80	3042.1	3371.2	6.9500
480	79.74	3062.0	3396.9	7.0050	77.82	3061.1	3395.7	6.9929	75.98	3060.1	3394.4	6.9810
490	80.96	3079.9	3420.0	7.0354	79.01	3079.0	3418.8	7.0233	77.15	3078.1	3417.6	7.0115
500	82.18	3097.8	3443.0	7.0653	80.20	3097.0	3441.8	7.0533	78.31	3096.1	3440.7	7.0416
510	83.38	3115.8	3466.0	7.0949	81.38	3114.9	3464.9	7.0830	79.47	3114.1	3463.8	7.0713
520	84.59	3133.7	3489.0	7.1240	82.56	3132.9	3487.9	7.1122	80.62	3132.1	3486.8	7.1006
530	85.78	3151.6	3511.9	7.1528	83.73	3150.9	3510.9	7.1410	81.77	3150.1	3509.9	7.1294
540	86.98	3169.6	3534.9	7.1812	84.90	3168.9	3533.9	7.1695	82.91	3168.1	3532.9	7.1579
550	88.16	3187.6	3557.9	7.2093	86.06	3186.9	3556.9	7.1976	84.05	3186.1	3556.0	7.1861
560	89.35	3205.6	3580.8	7.2371	87.22	3204.9	3579.9	7.2254	85.18	3204.2	3579.0	7.2139
570	90.53	3223.6	3603.8	7.2645	88.37	3222.9	3602.9	7.2528	86.32	3222.3	3602.1	7.2414
580	91.70	3241.7	3626.8	7.2916	89.52	3241.0	3626.0	7.2800	87.44	3240.4	3625.1	7.2686
590	92.88	3259.7	3649.8	7.3184	90.67	3259.1	3649.0	7.3068	88.57	3258.5	3648.2	7.2955
600	94.05	3277.9	3672.9	7.3449	91.82	3277.3	3672.1	7.3334	89.69	3276.7	3671.3	7.3221
620	96.38	3314.2	3719.0	7.3972	94.10	3313.6	3718.2	7.3857	91.92	3313.1	3717.5	7.3744
640	98.70	3350.7	3765.2	7.4484	96.36	3350.2	3764.5	7.4369	94.14	3349.6	3763.8	7.4257
660	101.01	3387.3	3811.6	7.4986	98.62	3386.8	3810.9	7.4872	96.35	3386.3	3810.3	7.4760
680	103.31	3424.2	3858.1	7.5479	100.87	3423.7	3857.5	7.5365	98.55	3423.2	3856.8	7.5254
700	105.60	3461.2	3904.7	7.5963	103.11	3460.8	3904.1	7.5850	100.74	3460.3	3903.6	7.5739
720	107.89	3498.4	3951.5	7.6439	105.35	3498.0	3951.0	7.6327	102.93	3497.6	3950.4	7.6216
740	110.17	3535.8	3998.5	7.6908	107.58	3535.4	3998.0	7.6795	105.11	3535.0	3997.5	7.6685
760	112.44	3573.4	4045.7	7.7369	109.80	3573.1	4045.2	7.7257	107.28	3572.7	4044.7	7.7147
780	114.71	3611.3	4093.1	7.7823	112.02	3610.9	4092.6	7.7711	109.45	3610.6	4092.1	7.7601
800	116.98	3649.3	4140.6	7.8270	114.23	3649.0	4140.2	7.8159	111.62	3648.7	4139.8	7.8049
850	122.62	3745.5	4260.5	7.9362	119.75	3745.2	4260.1	7.9251	117.01	3744.9	4259.7	7.9142
900	128.24	3843.0	4381.6	8.0417	125.25	3842.7	4381.3	8.0306	122.38	3842.4	4380.9	8.0198
950	133.85	3942.0	4504.2	8.1440	130.73	3941.7	4503.8	8.1329	127.74	3941.4	4503.5	8.1221
1000	139.45	4042.4	4628.1	8.2433	136.20	4042.1	4627.8	8.2322	133.09	4041.9	4627.5	8.2214
1100	150.62	4247.5	4880.1	8.4338	147.11	4247.3	4879.8	8.4228	143.76	4247.0	4879.6	8.4120
1200	161.77	4458.1	5137.6	8.6148	158.01	4457.9	5137.4	8.6038	154.42	4457.7	5137.1	8.5930
1300	172.91	4673.8	5400.1	8.7872	168.89	4673.6	5399.8	8.7762	165.06	4673.4	5399.6	8.7654

Definitions of symbols on page 1.

	4.5 (257.49)				4.6 (258.83)				4.7 (260.15)			
10^3 v	u	h	s	10^3 v	u	h	s	10^3 v	u	h	s	t
44.06	2600.1	2798.3	6.0198	43.06	2599.5	2797.6	6.0103	42.10	2599.0	2796.8	6.0008	Sat.
												250
43.56	*2592.2*	*2788.2*	*6.0008*	*42.29*	*2587.2*	*2781.7*	*5.9804*	*41.07*	*2582.1*	*2775.1*	*5.9599*	255
44.55	2607.8	2808.3	6.0385	43.28	2603.2	2802.3	6.0191	*42.07*	*2598.5*	*2796.2*	*5.9997*	260
45.50	2622.6	2827.3	6.0742	44.23	2618.3	2821.8	6.0556	43.02	2614.0	2816.2	6.0370	265
46.41	2636.7	2845.6	6.1079	45.15	2632.8	2840.5	6.0900	43.93	2628.8	2835.2	6.0722	270
47.30	2650.3	2863.2	6.1401	46.02	2646.6	2858.3	6.1228	44.80	2642.9	2853.5	6.1056	275
48.15	2663.4	2880.1	6.1709	46.87	2660.0	2875.6	6.1541	45.64	2656.5	2871.0	6.1375	280
48.98	2676.1	2896.5	6.2004	47.69	2672.8	2892.2	6.1841	46.46	2669.6	2887.9	6.1680	285
49.79	2688.4	2912.4	6.2288	48.50	2685.3	2908.4	6.2130	47.26	2682.2	2904.3	6.1972	290
50.58	2700.4	2928.0	6.2563	49.28	2697.5	2924.1	6.2408	48.03	2694.5	2920.3	6.2254	295
51.35	2712.0	2943.1	6.2828	50.04	2709.3	2939.5	6.2676	48.79	2706.5	2935.8	6.2526	300
52.84	2734.6	2972.4	6.3335	51.52	2732.1	2969.1	6.3189	50.25	2729.6	2965.8	6.3044	310
54.29	2756.3	3000.6	6.3815	52.95	2754.0	2997.6	6.3673	51.66	2751.7	2994.5	6.3534	320
55.69	2777.3	3028.0	6.4272	54.33	2775.2	3025.2	6.4134	53.03	2773.1	3022.4	6.3999	330
57.06	2797.8	3054.6	6.4710	55.68	2795.9	3052.0	6.4576	54.36	2793.9	3049.4	6.4443	340
58.40	2817.8	3080.6	6.5131	57.00	2816.0	3078.2	6.4999	55.65	2814.2	3075.8	6.4870	350
59.71	2837.4	3106.1	6.5537	58.29	2835.8	3103.9	6.5408	56.92	2834.1	3101.6	6.5281	360
60.99	2856.8	3131.2	6.5930	59.55	2855.2	3129.1	6.5804	58.17	2853.6	3127.0	6.5679	370
62.26	2875.8	3156.0	6.6312	60.80	2874.3	3154.0	6.6188	59.40	2872.8	3152.0	6.6065	380
63.51	2894.7	3180.5	6.6684	62.03	2893.3	3178.6	6.6561	60.60	2891.8	3176.7	6.6440	390
64.75	2913.3	3204.7	6.7047	63.24	2912.0	3202.9	6.6925	61.80	2910.7	3201.1	6.6806	400
65.97	2931.8	3228.7	6.7401	64.44	2930.6	3227.0	6.7281	62.98	2929.3	3225.3	6.7162	410
67.18	2950.3	3252.5	6.7747	65.63	2949.0	3250.9	6.7628	64.14	2947.8	3249.3	6.7511	420
68.37	2968.6	3276.2	6.8087	66.80	2967.4	3274.7	6.7969	65.30	2966.3	3273.1	6.7853	430
69.56	2986.8	3299.8	6.8420	67.97	2985.7	3298.3	6.8303	66.44	2984.6	3296.9	6.8188	440
70.74	3005.0	3323.3	6.8746	69.12	3003.9	3321.9	6.8630	67.57	3002.9	3320.4	6.8516	450
71.91	3023.1	3346.6	6.9067	70.27	3022.1	3345.3	6.8952	68.70	3021.1	3343.9	6.8839	460
73.07	3041.2	3370.0	6.9383	71.41	3040.2	3368.7	6.9269	69.82	3039.2	3367.4	6.9156	470
74.22	3059.2	3393.2	6.9694	72.54	3058.3	3392.0	6.9580	70.93	3057.4	3390.7	6.9468	480
75.37	3077.2	3416.4	7.0000	73.66	3076.4	3415.2	6.9887	72.03	3075.5	3414.0	6.9775	490
76.51	3095.3	3439.6	7.0301	74.78	3094.4	3438.4	7.0189	73.13	3093.5	3437.3	7.0078	500
77.64	3113.3	3462.7	7.0599	75.89	3112.5	3461.6	7.0486	74.22	3111.6	3460.5	7.0376	510
78.77	3131.3	3485.8	7.0892	77.00	3130.5	3484.7	7.0780	75.31	3129.7	3483.7	7.0671	520
79.90	3149.3	3508.9	7.1181	78.10	3148.6	3507.8	7.1070	76.39	3147.8	3506.8	7.0961	530
81.02	3167.4	3531.9	7.1467	79.20	3166.6	3531.0	7.1356	77.46	3165.9	3530.0	7.1247	540
82.13	3185.4	3555.0	7.1749	80.29	3184.7	3554.1	7.1638	78.54	3184.0	3553.1	7.1530	550
83.24	3203.5	3578.1	7.2027	81.38	3202.8	3577.2	7.1917	79.60	3202.1	3576.3	7.1810	560
84.35	3221.6	3601.2	7.2303	82.47	3220.9	3600.3	7.2193	80.67	3220.3	3599.4	7.2086	570
85.45	3239.7	3624.3	7.2575	83.55	3239.1	3623.4	7.2466	81.73	3238.4	3622.5	7.2359	580
86.55	3257.9	3647.4	7.2844	84.63	3257.2	3646.5	7.2735	82.78	3256.6	3645.7	7.2629	590
87.65	3276.0	3670.5	7.3110	85.70	3275.4	3669.7	7.3002	83.84	3274.8	3668.9	7.2895	600
89.83	3312.5	3716.8	7.3634	87.84	3311.9	3716.0	7.3526	85.93	3311.4	3715.3	7.3421	620
92.01	3349.1	3763.1	7.4148	89.97	3348.6	3762.4	7.4040	88.02	3348.0	3761.7	7.3935	640
94.17	3385.8	3809.6	7.4651	92.09	3385.3	3809.0	7.4544	90.10	3384.8	3808.3	7.4440	660
96.33	3422.8	3856.2	7.5145	94.20	3422.3	3855.6	7.5039	92.17	3421.8	3855.0	7.4935	680
98.47	3459.9	3903.0	7.5631	96.30	3459.4	3902.4	7.5525	94.23	3459.0	3901.8	7.5421	700
100.61	3497.1	3949.9	7.6108	98.40	3496.7	3949.4	7.6002	96.28	3496.3	3948.8	7.5899	720
102.75	3534.6	3997.0	7.6577	100.49	3534.2	3996.5	7.6472	98.32	3533.8	3995.9	7.6369	740
104.87	3572.3	4044.2	7.7039	102.57	3571.9	4043.7	7.6934	100.37	3571.5	4043.3	7.6831	760
107.00	3610.2	4091.7	7.7494	104.65	3609.8	4091.2	7.7389	102.40	3609.5	4090.7	7.7287	780
109.11	3648.3	4139.3	7.7942	106.72	3648.0	4138.9	7.7838	104.43	3647.6	4138.4	7.7735	800
114.39	3744.6	4259.3	7.9035	111.89	3744.2	4258.9	7.8931	109.49	3743.9	4258.5	7.8829	850
119.65	3842.2	4380.6	8.0091	117.03	3841.9	4380.2	7.9988	114.53	3841.6	4379.9	7.9886	900
124.89	3941.2	4503.2	8.1115	122.17	3940.9	4502.9	8.1011	119.56	3940.6	4502.5	8.0910	950
130.13	4041.6	4627.2	8.2108	127.29	4041.4	4626.9	8.2005	124.57	4041.1	4626.6	8.1904	1000
140.56	4246.8	4879.3	8.4015	137.50	4246.6	4879.1	8.3912	134.57	4246.3	4878.8	8.3811	1100
150.98	4457.5	5136.9	8.5825	147.70	4457.2	5136.6	8.5722	144.55	4457.0	5136.4	8.5621	1200
161.39	4673.1	5399.4	8.7549	157.88	4672.9	5399.2	8.7446	154.52	4672.7	5398.9	8.7345	1300

Definitions of symbols on page 1.

Table 3. Vapor

p (t Sat.)		4.8 (261.45)				4.9 (262.73)				5.0 (263.99)		
t	$10^3 v$	u	h	s	$10^3 v$	u	h	s	$10^3 v$	u	h	s
Sat.	41.18	2598.4	2796.0	5.9916	40.29	2597.8	2795.2	5.9824	39.44	2597.1	2794.3	5.9734
255	*39.89*	*2576.8*	*2768.3*	*5.9394*								
260	*40.90*	*2593.7*	*2790.0*	*5.9802*	*39.77*	*2588.8*	*2783.6*	*5.9607*	*38.67*	*2583.7*	*2777.1*	*5.9411*
265	41.85	2609.6	2810.4	6.0184	40.72	2605.1	2804.6	5.9999	39.63	2600.4	2798.6	5.9813
270	42.75	2624.7	2829.9	6.0544	41.62	2620.5	2824.5	6.0367	40.54	2616.2	2818.9	6.0189
275	43.62	2639.1	2848.5	6.0885	42.49	2635.2	2843.4	6.0714	41.41	2631.3	2838.3	6.0544
280	44.46	2652.9	2866.3	6.1209	43.33	2649.3	2861.6	6.1044	42.24	2645.6	2856.8	6.0880
285	45.28	2666.2	2883.6	6.1519	44.14	2662.8	2879.1	6.1359	43.04	2659.4	2874.6	6.1201
290	46.07	2679.1	2900.2	6.1816	44.92	2675.9	2896.0	6.1661	43.82	2672.7	2891.8	6.1507
295	46.83	2691.6	2916.4	6.2102	45.69	2688.6	2912.4	6.1951	44.58	2685.5	2908.4	6.1801
300	47.58	2703.7	2932.1	6.2377	46.43	2700.8	2928.3	6.2230	45.32	2698.0	2924.5	6.2084
310	49.03	2727.0	2962.4	6.2902	47.86	2724.5	2959.0	6.2760	46.74	2721.9	2955.6	6.2621
320	50.43	2749.4	2991.5	6.3396	49.24	2747.1	2988.4	6.3260	48.10	2744.7	2985.3	6.3125
330	51.78	2771.0	3019.5	6.3865	50.57	2768.9	3016.7	6.3733	49.42	2766.7	3013.8	6.3603
340	53.09	2792.0	3046.8	6.4313	51.87	2790.0	3044.1	6.4185	50.70	2788.0	3041.5	6.4058
350	54.37	2812.4	3073.3	6.4743	53.13	2810.5	3070.9	6.4617	51.94	2808.7	3068.4	6.4493
360	55.62	2832.4	3099.3	6.5157	54.36	2830.7	3097.0	6.5034	53.16	2828.9	3094.7	6.4912
370	56.84	2852.0	3124.8	6.5557	55.57	2850.4	3122.7	6.5436	54.35	2848.8	3120.5	6.5317
380	58.05	2871.3	3150.0	6.5944	56.76	2869.8	3148.0	6.5825	55.52	2868.3	3145.9	6.5708
390	59.24	2890.4	3174.8	6.6321	57.93	2889.0	3172.9	6.6204	56.68	2887.6	3171.0	6.6089
400	60.41	2909.3	3199.3	6.6688	59.09	2908.0	3197.5	6.6572	57.81	2906.6	3195.7	6.6459
410	61.57	2928.0	3223.6	6.7046	60.23	2926.8	3221.9	6.6932	58.93	2925.5	3220.2	6.6820
420	62.72	2946.6	3247.7	6.7396	61.35	2945.4	3246.0	6.7283	60.04	2944.2	3244.4	6.7172
430	63.85	2965.1	3271.6	6.7739	62.47	2963.9	3270.0	6.7627	61.14	2962.8	3268.5	6.7517
440	64.98	2983.5	3295.4	6.8075	63.57	2982.4	3293.9	6.7964	62.23	2981.3	3292.4	6.7855
450	66.09	3001.8	3319.0	6.8404	64.67	3000.7	3317.6	6.8294	63.30	2999.7	3316.2	6.8186
460	67.20	3020.0	3342.6	6.8728	65.75	3019.0	3341.2	6.8619	64.37	3018.0	3339.9	6.8511
470	68.29	3038.3	3366.1	6.9046	66.83	3037.3	3364.8	6.8937	65.43	3036.3	3363.5	6.8831
480	69.38	3056.4	3389.5	6.9359	67.90	3055.5	3388.2	6.9251	66.48	3054.6	3387.0	6.9145
490	70.47	3074.6	3412.8	6.9666	68.97	3073.7	3411.6	6.9559	67.53	3072.8	3410.4	6.9454
500	71.55	3092.7	3436.1	6.9970	70.03	3091.8	3435.0	6.9863	68.57	3091.0	3433.8	6.9759
510	72.62	3110.8	3459.4	7.0269	71.08	3110.0	3458.3	7.0163	69.60	3109.1	3457.1	7.0059
520	73.68	3128.9	3482.6	7.0563	72.12	3128.1	3481.5	7.0458	70.63	3127.3	3480.5	7.0355
530	74.74	3147.0	3505.8	7.0854	73.17	3146.3	3504.8	7.0749	71.65	3145.5	3503.7	7.0646
540	75.80	3165.2	3529.0	7.1141	74.20	3164.4	3528.0	7.1037	72.67	3163.7	3527.0	7.0934
550	76.85	3183.3	3552.2	7.1424	75.23	3182.6	3551.2	7.1320	73.68	3181.8	3550.3	7.1218
560	77.90	3201.4	3575.3	7.1704	76.26	3200.7	3574.4	7.1601	74.69	3200.0	3573.5	7.1499
570	78.94	3219.6	3598.5	7.1981	77.29	3218.9	3597.6	7.1878	75.70	3218.3	3596.7	7.1776
580	79.98	3237.8	3621.7	7.2254	78.31	3237.1	3620.8	7.2151	76.70	3236.5	3620.0	7.2050
590	81.02	3256.0	3644.9	7.2524	79.32	3255.4	3644.0	7.2422	77.70	3254.7	3643.2	7.2321
600	82.05	3274.2	3668.1	7.2791	80.34	3273.6	3667.3	7.2689	78.69	3273.0	3666.5	7.2589
620	84.11	3310.8	3714.5	7.3317	82.35	3310.2	3713.8	7.3216	80.67	3309.7	3713.0	7.3116
640	86.15	3347.5	3761.0	7.3832	84.36	3347.0	3760.3	7.3731	82.64	3346.4	3759.6	7.3632
660	88.19	3384.3	3807.6	7.4337	86.36	3383.8	3807.0	7.4237	84.60	3383.3	3806.3	7.4138
680	90.22	3421.3	3854.4	7.4833	88.35	3420.9	3853.8	7.4733	86.55	3420.4	3853.1	7.4634
700	92.23	3458.5	3901.2	7.5319	90.33	3458.1	3900.7	7.5219	88.49	3457.6	3900.1	7.5122
720	94.25	3495.9	3948.3	7.5797	92.30	3495.4	3947.7	7.5698	90.43	3495.0	3947.2	7.5601
740	96.25	3533.4	3995.4	7.6268	94.26	3533.0	3994.9	7.6169	92.36	3532.6	3994.4	7.6071
760	98.25	3571.2	4042.8	7.6730	96.22	3570.8	4042.3	7.6632	94.28	3570.4	4041.8	7.6535
780	100.25	3609.1	4090.3	7.7186	98.18	3608.7	4089.8	7.7087	96.20	3608.4	4089.3	7.6991
800	102.24	3647.3	4138.0	7.7635	100.13	3646.9	4137.5	7.7536	98.11	3646.6	4137.1	7.7440
850	107.19	3743.6	4258.1	7.8729	104.99	3743.3	4257.7	7.8631	102.87	3743.0	4257.4	7.8535
900	112.13	3841.3	4379.5	7.9786	109.83	3841.0	4379.2	7.9689	107.62	3840.7	4378.8	7.9593
950	117.05	3940.4	4502.2	8.0810	114.65	3940.1	4501.9	8.0713	112.35	3939.8	4501.6	8.0618
1000	121.96	4040.9	4626.3	8.1805	119.47	4040.6	4626.0	8.1707	117.07	4040.4	4625.7	8.1612
1100	131.76	4246.1	4878.5	8.3712	129.07	4245.8	4878.3	8.3615	126.48	4245.6	4878.0	8.3520
1200	141.54	4456.8	5136.2	8.5522	138.65	4456.5	5135.9	8.5426	135.87	4456.3	5135.7	8.5331
1300	151.31	4672.4	5398.7	8.7247	148.22	4672.2	5398.5	8.7150	145.26	4672.0	5398.2	8.7055

Definitions of symbols on page 1.

	5.1 (265.23)				5.2 (266.45)				5.3 (267.66)			p (t Sat.)
10^3 v	u	h	s	10^3 v	u	h	s	10^3 v	u	h	s	t
38.62	2596.5	2793.5	5.9645	37.83	2595.8	2792.6	5.9557	37.07	2595.1	2791.6	5.9470	Sat.
												255
37.62	*2578.5*	*2770.4*	*5.9214*									260
38.58	*2595.7*	*2792.5*	*5.9627*	*37.56*	*2590.9*	*2786.2*	*5.9440*	*36.58*	*2586.0*	*2779.8*	*5.9252*	265
39.49	2611.9	2813.3	6.0012	38.48	2607.5	2807.6	5.9834	37.50	2603.0	2801.7	5.9656	270
40.36	2627.2	2833.1	6.0374	39.34	2623.2	2827.7	6.0204	38.37	2619.0	2822.3	6.0035	275
41.19	2641.9	2851.9	6.0717	40.17	2638.1	2847.0	6.0554	39.20	2634.2	2842.0	6.0391	280
41.99	2655.9	2870.1	6.1043	40.97	2652.4	2865.4	6.0886	39.99	2648.8	2860.7	6.0729	285
42.76	2669.4	2887.5	6.1354	41.74	2666.1	2883.2	6.1202	40.76	2662.7	2878.7	6.1050	290
43.52	2682.4	2904.4	6.1652	42.49	2679.3	2900.3	6.1504	41.50	2676.2	2896.1	6.1358	295
44.25	2695.1	2920.7	6.1939	43.22	2692.1	2916.8	6.1795	42.23	2689.1	2912.9	6.1652	300
45.66	2719.3	2952.1	6.2482	44.62	2716.6	2948.6	6.2345	43.61	2714.0	2945.1	6.2209	310
47.01	2742.4	2982.1	6.2992	45.96	2740.0	2978.9	6.2860	44.94	2737.6	2975.7	6.2729	320
48.31	2764.5	3010.9	6.3474	47.24	2762.4	3008.0	6.3346	46.22	2760.2	3005.1	6.3220	330
49.57	2786.0	3038.8	6.3932	48.49	2784.0	3036.1	6.3809	47.45	2782.0	3033.4	6.3686	340
50.80	2806.8	3065.9	6.4371	49.71	2805.0	3063.5	6.4251	48.65	2803.1	3060.9	6.4131	350
52.00	2827.2	3092.4	6.4793	50.89	2825.5	3090.1	6.4675	49.82	2823.7	3087.8	6.4558	360
53.18	2847.2	3118.4	6.5200	52.05	2845.5	3116.2	6.5084	50.96	2843.9	3114.0	6.4970	370
54.33	2866.8	3143.9	6.5593	53.19	2865.3	3141.8	6.5480	52.09	2863.7	3139.8	6.5368	380
55.47	2886.1	3169.0	6.5975	54.31	2884.7	3167.1	6.5863	53.19	2883.3	3165.2	6.5753	390
56.59	2905.3	3193.9	6.6347	55.41	2903.9	3192.0	6.6236	54.28	2902.5	3190.2	6.6128	400
57.69	2924.2	3218.4	6.6709	56.50	2922.9	3216.7	6.6600	55.35	2921.6	3215.0	6.6493	410
58.78	2943.0	3242.8	6.7063	57.57	2941.7	3241.1	6.6955	56.41	2940.5	3239.5	6.6849	420
59.86	2961.6	3266.9	6.7409	58.63	2960.5	3265.4	6.7302	57.45	2959.3	3263.8	6.7197	430
60.93	2980.2	3290.9	6.7747	59.69	2979.0	3289.4	6.7642	58.49	2977.9	3287.9	6.7538	440
61.99	2998.6	3314.8	6.8080	60.73	2997.6	3313.3	6.7975	59.51	2996.5	3311.9	6.7872	450
63.04	3017.0	3338.5	6.8406	61.76	3016.0	3337.1	6.8302	60.53	3015.0	3335.8	6.8200	460
64.08	3035.3	3362.2	6.8726	62.79	3034.4	3360.8	6.8623	61.54	3033.4	3359.5	6.8522	470
65.12	3053.6	3385.7	6.9041	63.80	3052.7	3384.5	6.8939	62.54	3051.7	3383.2	6.8838	480
66.14	3071.9	3409.2	6.9351	64.81	3071.0	3408.0	6.9249	63.53	3070.1	3406.8	6.9149	490
67.17	3090.1	3432.6	6.9656	65.82	3089.2	3431.5	6.9555	64.52	3088.4	3430.3	6.9456	500
68.18	3108.3	3456.0	6.9957	66.82	3107.5	3454.9	6.9856	65.50	3106.6	3453.8	6.9758	510
69.19	3126.5	3479.4	7.0253	67.81	3125.7	3478.3	7.0153	66.48	3124.9	3477.2	7.0055	520
70.20	3144.7	3502.7	7.0545	68.80	3143.9	3501.7	7.0446	67.45	3143.2	3500.6	7.0348	530
71.20	3162.9	3526.0	7.0834	69.78	3162.2	3525.0	7.0735	68.42	3161.4	3524.0	7.0637	540
72.19	3181.1	3549.3	7.1118	70.76	3180.4	3548.3	7.1020	69.38	3179.7	3547.4	7.0923	550
73.18	3199.3	3572.6	7.1399	71.73	3198.6	3571.7	7.1301	70.34	3197.9	3570.7	7.1205	560
74.17	3217.6	3595.8	7.1677	72.70	3216.9	3595.0	7.1579	71.29	3216.2	3594.1	7.1483	570
75.15	3235.8	3619.1	7.1951	73.67	3235.2	3618.3	7.1854	72.24	3234.5	3617.4	7.1758	580
76.13	3254.1	3642.4	7.2222	74.63	3253.5	3641.6	7.2125	73.18	3252.8	3640.7	7.2030	590
77.11	3272.4	3665.7	7.2491	75.59	3271.8	3664.9	7.2394	74.13	3271.2	3664.1	7.2299	600
79.05	3309.1	3712.3	7.3018	77.50	3308.5	3711.5	7.2922	76.00	3307.9	3710.8	7.2828	620
80.99	3345.9	3758.9	7.3535	79.40	3345.3	3758.2	7.3439	77.87	3344.8	3757.5	7.3345	640
82.91	3382.8	3805.7	7.4041	81.29	3382.3	3805.0	7.3946	79.72	3381.8	3804.3	7.3853	660
84.82	3419.9	3852.5	7.4538	83.17	3419.4	3851.9	7.4443	81.57	3419.0	3851.3	7.4350	680
86.73	3457.2	3899.5	7.5026	85.04	3456.7	3898.9	7.4931	83.41	3456.3	3898.3	7.4839	700
88.63	3494.6	3946.6	7.5505	86.90	3494.2	3946.1	7.5411	85.24	3493.7	3945.5	7.5319	720
90.52	3532.2	3993.9	7.5976	88.76	3531.8	3993.3	7.5882	87.06	3531.4	3992.8	7.5790	740
92.41	3570.0	4041.3	7.6440	90.61	3569.6	4040.8	7.6346	88.88	3569.2	4040.3	7.6255	760
94.29	3608.0	4088.9	7.6896	92.46	3607.6	4088.4	7.6803	90.69	3607.3	4088.0	7.6711	780
96.17	3646.2	4136.7	7.7345	94.30	3645.9	4136.2	7.7252	92.50	3645.5	4135.8	7.7161	800
100.84	3742.7	4257.0	7.8441	98.89	3742.4	4256.6	7.8348	97.01	3742.0	4256.2	7.8258	850
105.50	3840.4	4378.5	7.9499	103.46	3840.1	4378.1	7.9407	101.49	3839.9	4377.8	7.9317	900
110.14	3939.6	4501.3	8.0524	108.01	3939.3	4500.9	8.0432	105.96	3939.0	4500.6	8.0342	950
114.76	4040.1	4625.4	8.1519	112.55	4039.9	4625.1	8.1427	110.42	4039.6	4624.8	8.1337	1000
124.00	4245.4	4877.8	8.3427	121.61	4245.1	4877.5	8.3335	119.31	4244.9	4877.2	8.3246	1100
133.21	4456.1	5135.4	8.5238	130.65	4455.8	5135.2	8.5147	128.18	4455.6	5135.0	8.5057	1200
142.41	4671.7	5398.0	8.6962	139.67	4671.5	5397.8	8.6871	137.04	4671.3	5397.6	8.6782	1300

Definitions of symbols on page 1.

Table 3. Vapor

p (t Sat.)	5.4 (268.84)				5.5 (270.02)				5.6 (271.17)			
t	$10^3 v$	u	h	s	$10^3 v$	u	h	s	$10^3 v$	u	h	s
Sat.	36.34	2594.4	2790.7	5.9385	35.63	2593.7	2789.7	5.9300	34.95	2592.9	2788.6	5.9217
265	*35.63*	*2580.9*	*2773.3*	*5.9063*	*34.71*	*2575.7*	*2766.6*	*5.8873*				
270	36.55	2598.3	2795.7	5.9478	*35.63*	*2593.6*	*2789.6*	*5.9299*	*34.74*	*2588.8*	*2783.3*	*5.9119*
275	37.42	2614.8	2816.8	5.9865	36.50	2610.4	2811.2	5.9695	35.62	2606.0	2805.5	5.9525
280	38.25	2630.3	2836.9	6.0229	37.34	2626.3	2831.7	6.0066	36.45	2622.2	2826.4	5.9904
285	39.05	2645.1	2856.0	6.0573	38.13	2641.4	2851.1	6.0417	37.24	2637.6	2846.2	6.0261
290	39.81	2659.3	2874.3	6.0900	38.89	2655.8	2869.8	6.0749	38.01	2652.3	2865.2	6.0600
295	40.55	2673.0	2891.9	6.1211	39.63	2669.7	2887.7	6.1066	38.74	2666.4	2883.4	6.0921
300	41.27	2686.1	2909.0	6.1510	40.34	2683.1	2905.0	6.1369	39.45	2680.0	2900.9	6.1229
310	42.65	2711.3	2941.6	6.2074	41.71	2708.5	2938.0	6.1940	40.81	2705.8	2934.3	6.1807
320	43.96	2735.1	2972.5	6.2600	43.02	2732.7	2969.3	6.2472	42.11	2730.2	2966.0	6.2345
330	45.23	2757.9	3002.1	6.3096	44.27	2755.7	2999.2	6.2972	43.35	2753.4	2996.2	6.2850
340	46.45	2779.9	3030.7	6.3565	45.48	2777.9	3028.0	6.3446	44.54	2775.8	3025.2	6.3328
350	47.63	2801.2	3058.4	6.4014	46.65	2799.3	3055.9	6.3897	45.70	2797.4	3053.3	6.3782
360	48.79	2822.0	3085.4	6.4444	47.79	2820.2	3083.1	6.4330	46.83	2818.4	3080.7	6.4218
370	49.92	2842.3	3111.8	6.4857	48.91	2840.6	3109.6	6.4746	47.93	2839.0	3107.4	6.4636
380	51.02	2862.2	3137.7	6.5257	50.00	2860.6	3135.7	6.5148	49.01	2859.1	3133.6	6.5040
390	52.11	2881.8	3163.2	6.5644	51.08	2880.4	3161.3	6.5537	50.07	2878.9	3159.3	6.5431
400	53.18	2901.2	3188.4	6.6021	52.13	2899.8	3186.5	6.5915	51.12	2898.4	3184.7	6.5811
410	54.24	2920.3	3213.2	6.6387	53.17	2919.0	3211.5	6.6283	52.14	2917.7	3209.7	6.6180
420	55.28	2939.3	3237.8	6.6745	54.20	2938.0	3236.2	6.6642	53.16	2936.8	3234.5	6.6540
430	56.31	2958.1	3262.2	6.7094	55.22	2956.9	3260.6	6.6992	54.16	2955.8	3259.0	6.6892
440	57.33	2976.8	3286.4	6.7436	56.22	2975.7	3284.9	6.7335	55.15	2974.6	3283.4	6.7236
450	58.34	2995.4	3310.5	6.7771	57.22	2994.3	3309.0	6.7671	56.13	2993.3	3307.6	6.7573
460	59.35	3013.9	3334.4	6.8099	58.20	3012.9	3333.0	6.8001	57.10	3011.9	3331.6	6.7903
470	60.34	3032.4	3358.2	6.8422	59.18	3031.4	3356.9	6.8324	58.06	3030.4	3355.6	6.8227
480	61.32	3050.8	3381.9	6.8739	60.15	3049.9	3380.7	6.8642	59.02	3048.9	3379.4	6.8546
490	62.30	3069.2	3405.6	6.9051	61.11	3068.3	3404.4	6.8954	59.97	3067.3	3403.2	6.8859
500	63.27	3087.5	3429.2	6.9358	62.07	3086.6	3428.0	6.9262	60.91	3085.7	3426.8	6.9167
510	64.24	3105.8	3452.7	6.9660	63.02	3105.0	3451.6	6.9565	61.84	3104.1	3450.5	6.9471
520	65.20	3124.1	3476.2	6.9958	63.96	3123.3	3475.1	6.9863	62.77	3122.5	3474.0	6.9770
530	66.15	3142.4	3499.6	7.0252	64.90	3141.6	3498.6	7.0158	63.70	3140.8	3497.5	7.0065
540	67.10	3160.7	3523.0	7.0542	65.84	3159.9	3522.0	7.0448	64.62	3159.2	3521.0	7.0355
550	68.05	3179.0	3546.4	7.0828	66.77	3178.2	3545.5	7.0734	65.53	3177.5	3544.5	7.0642
560	68.99	3197.3	3569.8	7.1110	67.70	3196.6	3568.9	7.1017	66.45	3195.8	3567.9	7.0925
570	69.93	3215.6	3593.2	7.1389	68.62	3214.9	3592.3	7.1296	67.35	3214.2	3591.4	7.1205
580	70.86	3233.9	3616.5	7.1664	69.54	3233.2	3615.7	7.1572	68.26	3232.6	3614.8	7.1481
590	71.79	3252.2	3639.9	7.1937	70.45	3251.6	3639.1	7.1844	69.16	3250.9	3638.2	7.1754
600	72.72	3270.6	3663.3	7.2206	71.36	3270.0	3662.4	7.2114	70.05	3269.3	3661.6	7.2024
620	74.56	3307.4	3710.0	7.2735	73.17	3306.8	3709.3	7.2644	71.84	3306.2	3708.5	7.2554
640	76.40	3344.3	3756.8	7.3253	74.98	3343.7	3756.1	7.3163	73.61	3343.2	3755.4	7.3074
660	78.22	3381.3	3803.7	7.3761	76.77	3380.8	3803.0	7.3671	75.37	3380.3	3802.4	7.3582
680	80.03	3418.5	3850.7	7.4259	78.55	3418.0	3850.0	7.4169	77.12	3417.5	3849.4	7.4081
700	81.84	3455.8	3897.7	7.4748	80.33	3455.4	3897.2	7.4659	78.87	3454.9	3896.6	7.4571
720	83.64	3493.3	3945.0	7.5228	82.09	3492.9	3944.4	7.5139	80.61	3492.5	3943.8	7.5052
740	85.43	3531.0	3992.3	7.5700	83.85	3530.6	3991.8	7.5612	82.34	3530.2	3991.3	7.5524
760	87.22	3568.8	4039.8	7.6165	85.61	3568.5	4039.3	7.6076	84.06	3568.1	4038.8	7.5989
780	89.00	3606.9	4087.5	7.6622	87.36	3606.5	4087.0	7.6533	85.78	3606.2	4086.6	7.6447
800	90.77	3645.2	4135.3	7.7072	89.11	3644.8	4134.9	7.6984	87.50	3644.5	4134.4	7.6897
850	95.20	3741.7	4255.8	7.8169	93.45	3741.4	4255.4	7.8081	91.77	3741.1	4255.0	7.7995
900	99.60	3839.6	4377.4	7.9228	97.78	3839.3	4377.1	7.9141	96.02	3839.0	4376.7	7.9055
950	103.99	3938.8	4500.3	8.0254	102.09	3938.5	4500.0	8.0167	100.26	3938.2	4499.7	8.0082
1000	108.37	4039.3	4624.5	8.1249	106.39	4039.1	4624.2	8.1163	104.48	4038.8	4623.9	8.1077
1100	117.09	4244.7	4877.0	8.3158	114.96	4244.4	4876.7	8.3072	112.90	4244.2	4876.5	8.2987
1200	125.80	4455.4	5134.7	8.4969	123.51	4455.2	5134.5	8.4883	121.31	4454.9	5134.2	8.4799
1300	134.50	4671.0	5397.3	8.6694	132.06	4670.8	5397.1	8.6608	129.70	4670.6	5396.9	8.6523

Definitions of symbols on page 1.

	5.7 (272.31)				5.8 (273.43)				5.9 (274.54)			p (t Sat.)
$10^3 v$	u	h	s	$10^3 v$	u	h	s	$10^3 v$	u	h	s	t
34.29	2592.1	2787.6	5.9134	33.65	2591.3	2786.5	5.9052	33.04	2590.5	2785.5	5.8972	Sat.
												265
33.88	*2583.8*	*2777.0*	*5.8938*	*33.04*	*2578.8*	*2770.4*	*5.8756*	*32.23*	*2573.6*	*2763.7*	*5.8573*	270
34.76	2601.5	2799.6	5.9354	33.93	2596.9	2793.7	5.9183	33.12	2592.2	2787.6	5.9010	275
35.59	2618.1	2821.0	5.9742	34.76	2613.9	2815.5	5.9579	33.96	2609.6	2809.9	5.9416	280
36.39	2633.8	2841.2	6.0106	35.56	2629.9	2836.1	5.9951	34.75	2625.9	2831.0	5.9795	285
37.15	2648.8	2860.5	6.0450	36.32	2645.1	2855.8	6.0301	35.51	2641.5	2851.0	6.0152	290
37.88	2663.1	2879.0	6.0777	37.05	2659.7	2874.6	6.0633	36.24	2656.3	2870.1	6.0490	295
38.59	2676.9	2896.8	6.1089	37.75	2673.7	2892.7	6.0950	36.95	2670.5	2888.5	6.0812	300
39.94	2703.0	2930.7	6.1675	39.10	2700.2	2927.0	6.1544	38.29	2697.4	2923.2	6.1413	310
41.23	2727.7	2962.7	6.2219	40.38	2725.2	2959.3	6.2094	39.55	2722.6	2956.0	6.1970	320
42.46	2751.2	2993.2	6.2729	41.60	2748.9	2990.1	6.2609	40.76	2746.6	2987.1	6.2490	330
43.64	2773.7	3022.5	6.3211	42.77	2771.6	3019.7	6.3095	41.93	2769.5	3016.9	6.2980	340
44.79	2795.5	3050.8	6.3669	43.91	2793.5	3048.2	6.3556	43.05	2791.6	3045.6	6.3445	350
45.91	2816.6	3078.3	6.4107	45.01	2814.9	3075.9	6.3997	44.15	2813.0	3073.5	6.3889	360
47.00	2837.3	3105.2	6.4528	46.09	2835.6	3102.9	6.4421	45.21	2834.0	3100.7	6.4315	370
48.06	2857.5	3131.5	6.4934	47.14	2856.0	3129.4	6.4829	46.25	2854.4	3127.3	6.4725	380
49.11	2877.4	3157.3	6.5327	48.18	2876.0	3155.4	6.5224	47.27	2874.5	3153.4	6.5122	390
50.14	2897.0	3182.8	6.5708	49.19	2895.6	3181.0	6.5607	48.28	2894.3	3179.1	6.5507	400
51.15	2916.4	3208.0	6.6079	50.19	2915.1	3206.2	6.5979	49.26	2913.8	3204.4	6.5880	410
52.15	2935.6	3232.8	6.6440	51.18	2934.3	3231.2	6.6342	50.24	2933.1	3229.5	6.6244	420
53.14	2954.6	3257.5	6.6793	52.15	2953.4	3255.9	6.6696	51.20	2952.2	3254.3	6.6599	430
54.11	2973.4	3281.9	6.7138	53.11	2972.3	3280.4	6.7042	52.15	2971.2	3278.9	6.6947	440
55.08	2992.2	3306.2	6.7476	54.07	2991.1	3304.7	6.7380	53.09	2990.0	3303.3	6.7286	450
56.04	3010.9	3330.3	6.7807	55.01	3009.8	3328.9	6.7713	54.02	3008.8	3327.5	6.7619	460
56.99	3029.4	3354.3	6.8132	55.95	3028.4	3352.9	6.8038	54.94	3027.5	3351.6	6.7946	470
57.93	3048.0	3378.1	6.8451	56.87	3047.0	3376.9	6.8358	55.86	3046.1	3375.6	6.8267	480
58.86	3066.4	3402.0	6.8765	57.79	3065.5	3400.7	6.8673	56.76	3064.6	3399.5	6.8582	490
59.79	3084.9	3425.7	6.9074	58.71	3084.0	3424.5	6.8983	57.66	3083.1	3423.3	6.8892	500
60.71	3103.3	3449.3	6.9378	59.62	3102.4	3448.2	6.9287	58.56	3101.6	3447.1	6.9197	510
61.63	3121.7	3472.9	6.9678	60.52	3120.9	3471.9	6.9587	59.45	3120.0	3470.8	6.9498	520
62.54	3140.0	3496.5	6.9973	61.41	3139.3	3495.5	6.9883	60.33	3138.5	3494.4	6.9794	530
63.44	3158.4	3520.0	7.0264	62.31	3157.7	3519.0	7.0175	61.21	3156.9	3518.0	7.0086	540
64.34	3176.8	3543.5	7.0552	63.19	3176.1	3542.6	7.0462	62.08	3175.3	3541.6	7.0375	550
65.24	3195.1	3567.0	7.0835	64.08	3194.4	3566.1	7.0746	62.95	3193.7	3565.2	7.0659	560
66.13	3213.5	3590.5	7.1115	64.96	3212.8	3589.6	7.1027	63.82	3212.2	3588.7	7.0940	570
67.02	3231.9	3613.9	7.1392	65.83	3231.3	3613.1	7.1304	64.68	3230.6	3612.2	7.1217	580
67.91	3250.3	3637.4	7.1665	66.70	3249.7	3636.5	7.1577	65.54	3249.0	3635.7	7.1491	590
68.79	3268.7	3660.8	7.1935	67.57	3268.1	3660.0	7.1848	66.39	3267.5	3659.2	7.1762	600
70.54	3305.6	3707.7	7.2466	69.30	3305.1	3707.0	7.2379	68.09	3304.5	3706.2	7.2294	620
72.29	3342.6	3754.7	7.2986	71.01	3342.1	3754.0	7.2900	69.78	3341.6	3753.3	7.2815	640
74.02	3379.8	3801.7	7.3495	72.72	3379.3	3801.0	7.3409	71.46	3378.8	3800.4	7.3325	660
75.74	3417.0	3848.8	7.3994	74.41	3416.6	3848.2	7.3909	73.13	3416.1	3847.5	7.3825	680
77.46	3454.5	3896.0	7.4485	76.10	3454.0	3895.4	7.4400	74.79	3453.5	3894.8	7.4316	700
79.17	3492.0	3943.3	7.4966	77.78	3491.6	3942.7	7.4881	76.44	3491.2	3942.2	7.4798	720
80.87	3529.8	3990.7	7.5439	79.46	3529.4	3990.2	7.5354	78.09	3529.0	3989.7	7.5272	740
82.57	3567.7	4038.3	7.5904	81.13	3567.3	4037.8	7.5820	79.73	3566.9	4037.3	7.5737	760
84.26	3605.8	4086.1	7.6362	82.79	3605.4	4085.6	7.6278	81.37	3605.1	4085.1	7.6196	780
85.95	3644.1	4134.0	7.6812	84.45	3643.8	4133.6	7.6729	83.00	3643.4	4133.1	7.6647	800
90.15	3740.8	4254.6	7.7911	88.58	3740.5	4254.2	7.7828	87.06	3740.2	4253.8	7.7746	850
94.32	3838.7	4376.4	7.8971	92.69	3838.4	4376.0	7.8889	91.10	3838.1	4375.6	7.8807	900
98.49	3938.0	4499.3	7.9998	96.78	3937.7	4499.0	7.9915	95.13	3937.4	4498.7	7.9834	950
102.64	4038.6	4623.6	8.0994	100.86	4038.3	4623.3	8.0911	99.15	4038.1	4623.0	8.0831	1000
110.92	4244.0	4876.2	8.2903	109.00	4243.7	4875.9	8.2821	107.15	4243.5	4875.7	8.2741	1100
119.18	4454.7	5134.0	8.4715	117.12	4454.5	5133.8	8.4633	115.13	4454.2	5133.5	8.4553	1200
127.42	4670.3	5396.7	8.6440	125.23	4670.1	5396.4	8.6358	123.11	4669.9	5396.2	8.6278	1300

Definitions of symbols on page 1.

Table 3. Vapor

p (t Sat.)	6.0 (275.64)				6.1 (276.72)				6.2 (277.78)			
t	$10^3 v$	u	h	s	$10^3 v$	u	h	s	$10^3 v$	u	h	s
Sat.	32.44	2589.7	2784.3	5.8892	31.86	2588.9	2783.2	5.8813	31.30	2588.0	2782.1	5.8734
275	*32.33*	*2587.4*	*2781.4*	*5.8837*	*31.56*	*2582.4*	*2775.0*	*5.8663*	*30.82*	*2577.4*	*2768.5*	*5.8487*
280	33.17	2605.2	2804.2	5.9252	32.41	2600.7	2798.4	5.9088	31.67	2596.1	2792.5	5.8923
285	33.97	2621.9	2825.7	5.9640	33.21	2617.8	2820.4	5.9484	32.48	2613.6	2815.0	5.9328
290	34.73	2637.7	2846.1	6.0003	33.97	2633.9	2841.2	5.9855	33.24	2630.1	2836.2	5.9706
295	35.46	2652.8	2865.6	6.0347	34.70	2649.3	2861.0	6.0204	33.97	2645.7	2856.3	6.0062
300	36.16	2667.2	2884.2	6.0674	35.40	2664.0	2879.9	6.0536	34.67	2660.6	2875.5	6.0399
310	37.50	2694.5	2919.5	6.1284	36.73	2691.6	2915.7	6.1155	35.99	2688.7	2911.8	6.1026
320	38.76	2720.0	2952.6	6.1846	37.98	2717.4	2949.1	6.1724	37.24	2714.8	2945.7	6.1602
330	39.96	2744.2	2984.0	6.2372	39.18	2741.9	2980.9	6.2254	38.42	2739.5	2977.7	6.2138
340	41.11	2767.4	3014.0	6.2866	40.32	2765.2	3011.2	6.2753	39.56	2763.1	3008.3	6.2641
350	42.23	2789.6	3043.0	6.3335	41.43	2787.7	3040.4	6.3225	40.65	2785.7	3037.7	6.3117
360	43.31	2811.2	3071.1	6.3782	42.50	2809.4	3068.7	6.3675	41.71	2807.6	3066.2	6.3570
370	44.36	2832.3	3098.4	6.4210	43.54	2830.6	3096.2	6.4107	42.75	2828.9	3093.9	6.4004
380	45.39	2852.8	3125.2	6.4623	44.56	2851.2	3123.1	6.4522	43.75	2849.6	3120.9	6.4421
390	46.40	2873.0	3151.4	6.5021	45.56	2871.5	3149.4	6.4922	44.74	2870.0	3147.4	6.4824
400	47.39	2892.9	3177.2	6.5408	46.54	2891.5	3175.3	6.5310	45.71	2890.0	3173.4	6.5213
410	48.37	2912.4	3202.7	6.5783	47.50	2911.1	3200.9	6.5687	46.66	2909.8	3199.1	6.5592
420	49.33	2931.8	3227.8	6.6148	48.45	2930.6	3226.1	6.6053	47.60	2929.3	3224.4	6.5959
430	50.28	2951.0	3252.7	6.6504	49.39	2949.8	3251.1	6.6411	48.52	2948.6	3249.5	6.6318
440	51.22	2970.0	3277.3	6.6853	50.31	2968.9	3275.8	6.6760	49.44	2967.8	3274.3	6.6669
450	52.14	2988.9	3301.8	6.7193	51.23	2987.9	3300.3	6.7102	50.34	2986.8	3298.9	6.7011
460	53.06	3007.7	3326.1	6.7527	52.13	3006.7	3324.7	6.7437	51.23	3005.7	3323.3	6.7347
470	53.97	3026.5	3350.3	6.7855	53.03	3025.5	3348.9	6.7765	52.12	3024.5	3347.6	6.7676
480	54.87	3045.1	3374.3	6.8176	53.92	3044.2	3373.1	6.8087	53.00	3043.2	3371.8	6.7999
490	55.77	3063.7	3398.3	6.8492	54.80	3062.8	3397.1	6.8404	53.87	3061.9	3395.8	6.8317
500	56.65	3082.2	3422.2	6.8803	55.68	3081.4	3421.0	6.8715	54.73	3080.5	3419.8	6.8629
510	57.53	3100.7	3446.0	6.9109	56.55	3099.9	3444.8	6.9022	55.59	3099.1	3443.7	6.8936
520	58.41	3119.2	3469.7	6.9410	57.41	3118.4	3468.6	6.9323	56.44	3117.6	3467.5	6.9238
530	59.28	3137.7	3493.4	6.9707	58.27	3136.9	3492.3	6.9621	57.29	3136.1	3491.3	6.9536
540	60.15	3156.1	3517.0	6.9999	59.12	3155.4	3516.0	6.9914	58.13	3154.6	3515.0	6.9829
550	61.01	3174.6	3540.6	7.0288	59.97	3173.9	3539.7	7.0203	58.96	3173.1	3538.7	7.0119
560	61.86	3193.0	3564.2	7.0573	60.81	3192.3	3563.3	7.0488	59.80	3191.6	3562.4	7.0404
570	62.72	3211.5	3587.8	7.0854	61.65	3210.8	3586.9	7.0770	60.62	3210.1	3586.0	7.0686
580	63.57	3229.9	3611.3	7.1132	62.49	3229.3	3610.5	7.1048	61.45	3228.6	3609.6	7.0965
590	64.41	3248.4	3634.9	7.1406	63.32	3247.8	3634.0	7.1322	62.27	3247.1	3633.2	7.1240
600	65.25	3266.9	3658.4	7.1677	64.15	3266.3	3657.6	7.1594	63.08	3265.7	3656.8	7.1511
620	66.93	3303.9	3705.5	7.2210	65.80	3303.3	3704.7	7.2127	64.71	3302.8	3704.0	7.2046
640	68.59	3341.0	3752.6	7.2731	67.44	3340.5	3751.9	7.2649	66.32	3339.9	3751.1	7.2568
660	70.24	3378.2	3799.7	7.3242	69.07	3377.7	3799.0	7.3160	67.93	3377.2	3798.4	7.3080
680	71.89	3415.6	3846.9	7.3743	70.68	3415.1	3846.3	7.3661	69.52	3414.6	3845.7	7.3581
700	73.52	3453.1	3894.2	7.4234	72.29	3452.6	3893.6	7.4153	71.11	3452.2	3893.0	7.4073
720	75.15	3490.7	3941.6	7.4716	73.90	3490.3	3941.1	7.4636	72.69	3489.9	3940.5	7.4556
740	76.77	3528.5	3989.2	7.5190	75.49	3528.1	3988.7	7.5110	74.26	3527.7	3988.1	7.5031
760	78.39	3566.5	4036.9	7.5656	77.08	3566.1	4036.4	7.5576	75.82	3565.8	4035.9	7.5497
780	80.00	3604.7	4084.7	7.6114	78.67	3604.3	4084.2	7.6035	77.38	3604.0	4083.7	7.5956
800	81.60	3643.1	4132.7	7.6566	80.25	3642.7	4132.2	7.6486	78.94	3642.4	4131.8	7.6408
850	85.60	3739.8	4253.4	7.7666	84.18	3739.5	4253.0	7.7587	82.81	3739.2	4252.6	7.7509
900	89.58	3837.8	4375.3	7.8727	88.10	3837.6	4374.9	7.8648	86.66	3837.3	4374.6	7.8571
950	93.54	3937.2	4498.4	7.9755	91.99	3936.9	4498.1	7.9676	90.50	3936.6	4497.7	7.9599
1000	97.49	4037.8	4622.7	8.0751	95.88	4037.6	4622.5	8.0673	94.33	4037.3	4622.2	8.0596
1100	105.36	4243.3	4875.4	8.2661	103.63	4243.0	4875.2	8.2583	101.95	4242.8	4874.9	8.2507
1200	113.21	4454.0	5133.3	8.4474	111.36	4453.8	5133.1	8.4396	109.56	4453.6	5132.8	8.4319
1300	121.06	4669.6	5396.0	8.6199	119.07	4669.4	5395.8	8.6121	117.15	4669.2	5395.5	8.6045

Definitions of symbols on page 1.

	6.3 (278.84)				6.4 (279.88)				6.5 (280.91)			p (t Sat.)
10³ v	u	h	s	10³ v	u	h	s	10³ v	u	h	s	t
30.76	2587.1	2780.9	5.8657	30.23	2586.2	2779.7	5.8580	29.72	2585.3	2778.5	5.8504	Sat.
30.09	*2572.2*	*2761.8*	*5.8310*	*29.38*	*2566.9*	*2755.0*	*5.8131*					275
30.95	2591.5	2786.4	5.8757	30.25	2586.7	2780.3	5.8590	*29.57*	*2581.8*	*2774.0*	*5.8422*	280
31.76	2609.4	2809.4	5.9171	31.06	2605.0	2803.8	5.9014	30.38	2600.6	2798.1	5.8856	285
32.52	2626.2	2831.1	5.9557	31.83	2622.2	2825.9	5.9408	31.15	2618.1	2820.6	5.9258	290
33.25	2642.1	2851.6	5.9919	32.56	2638.4	2846.8	5.9777	31.88	2634.6	2841.9	5.9634	295
33.95	2657.2	2871.1	6.0262	33.26	2653.8	2866.6	6.0125	32.58	2650.3	2862.1	5.9989	300
35.27	2685.7	2907.9	6.0898	34.57	2682.7	2904.0	6.0771	33.90	2679.6	2900.0	6.0643	310
36.51	2712.2	2942.2	6.1481	35.81	2709.5	2938.7	6.1361	35.13	2706.8	2935.1	6.1241	320
37.69	2737.1	2974.6	6.2023	36.98	2734.7	2971.4	6.1908	36.29	2732.3	2968.2	6.1794	330
38.82	2760.9	3005.4	6.2530	38.10	2758.7	3002.5	6.2420	37.40	2756.5	2999.6	6.2311	340
39.90	2783.7	3035.1	6.3010	39.18	2781.7	3032.4	6.2904	38.47	2779.7	3029.7	6.2798	350
40.95	2805.7	3063.8	6.3466	40.22	2803.9	3061.3	6.3363	39.50	2802.0	3058.8	6.3261	360
41.98	2827.1	3091.6	6.3903	41.23	2825.4	3089.3	6.3802	40.51	2823.7	3087.0	6.3703	370
42.97	2848.0	3118.8	6.4322	42.22	2846.4	3116.6	6.4224	41.48	2844.8	3114.5	6.4127	380
43.95	2868.5	3145.4	6.4727	43.18	2867.0	3143.4	6.4630	42.44	2865.5	3141.4	6.4535	390
44.91	2888.6	3171.5	6.5118	44.13	2887.2	3169.6	6.5024	43.38	2885.8	3167.7	6.4930	400
45.85	2908.5	3197.3	6.5498	45.06	2907.1	3195.5	6.5405	44.30	2905.8	3193.7	6.5313	410
46.78	2928.0	3222.7	6.5867	45.98	2926.8	3221.0	6.5775	45.20	2925.5	3219.3	6.5685	420
47.69	2947.4	3247.8	6.6227	46.88	2946.2	3246.2	6.6137	46.10	2945.0	3244.6	6.6048	430
48.59	2966.6	3272.7	6.6578	47.77	2965.5	3271.2	6.6489	46.98	2964.3	3269.7	6.6401	440
49.48	2985.7	3297.4	6.6922	48.65	2984.6	3296.0	6.6834	47.85	2983.5	3294.5	6.6747	450
50.37	3004.6	3321.9	6.7259	49.52	3003.6	3320.5	6.7171	48.71	3002.5	3319.1	6.7085	460
51.24	3023.5	3346.3	6.7589	50.39	3022.5	3344.9	6.7502	49.56	3021.5	3343.6	6.7417	470
52.10	3042.2	3370.5	6.7912	51.24	3041.3	3369.2	6.7827	50.40	3040.3	3367.9	6.7742	480
52.96	3060.9	3394.6	6.8230	52.09	3060.0	3393.4	6.8146	51.24	3059.1	3392.2	6.8062	490
53.82	3079.6	3418.6	6.8543	52.93	3078.7	3417.5	6.8459	52.07	3077.8	3416.3	6.8376	500
54.66	3098.2	3442.6	6.8851	53.76	3097.4	3441.4	6.8767	52.89	3096.5	3440.3	6.8684	510
55.50	3116.8	3466.4	6.9154	54.59	3116.0	3465.3	6.9070	53.71	3115.1	3464.3	6.8988	520
56.33	3135.3	3490.2	6.9452	55.41	3134.5	3489.2	6.9369	54.52	3133.8	3488.2	6.9288	530
57.16	3153.9	3514.0	6.9746	56.23	3153.1	3513.0	6.9664	55.33	3152.3	3512.0	6.9583	540
57.99	3172.4	3537.7	7.0036	57.05	3171.7	3536.8	6.9954	56.13	3170.9	3535.8	6.9874	550
58.81	3190.9	3561.4	7.0322	57.85	3190.2	3560.5	7.0241	56.93	3189.5	3559.5	7.0161	560
59.63	3209.4	3585.1	7.0604	58.66	3208.8	3584.2	7.0523	57.72	3208.1	3583.3	7.0444	570
60.44	3228.0	3608.7	7.0883	59.46	3227.3	3607.9	7.0803	58.51	3226.6	3607.0	7.0723	580
61.25	3246.5	3632.4	7.1158	60.26	3245.9	3631.5	7.1078	59.30	3245.2	3630.7	7.0999	590
62.05	3265.0	3656.0	7.1430	61.05	3264.4	3655.2	7.1351	60.08	3263.8	3654.3	7.1272	600
63.65	3302.2	3703.2	7.1965	62.63	3301.6	3702.4	7.1886	61.64	3301.0	3701.7	7.1808	620
65.24	3339.4	3750.4	7.2488	64.20	3338.9	3749.7	7.2410	63.18	3338.3	3749.0	7.2332	640
66.82	3376.7	3797.7	7.3000	65.76	3376.2	3797.0	7.2922	64.72	3375.7	3796.4	7.2845	660
68.39	3414.2	3845.0	7.3502	67.30	3413.7	3844.4	7.3425	66.25	3413.2	3843.8	7.3348	680
69.96	3451.7	3892.5	7.3995	68.84	3451.3	3891.9	7.3917	67.76	3450.8	3891.3	7.3841	700
71.51	3489.5	3940.0	7.4478	70.37	3489.0	3939.4	7.4401	69.27	3488.6	3938.9	7.4325	720
73.06	3527.3	3987.6	7.4953	71.90	3526.9	3987.1	7.4876	70.78	3526.5	3986.6	7.4801	740
74.60	3565.4	4035.4	7.5420	73.42	3565.0	4034.9	7.5343	72.27	3564.6	4034.4	7.5268	760
76.14	3603.6	4083.3	7.5879	74.93	3603.2	4082.8	7.5803	73.77	3602.9	4082.3	7.5728	780
77.67	3642.0	4131.3	7.6331	76.44	3641.7	4130.9	7.6255	75.25	3641.3	4130.5	7.6180	800
81.49	3738.9	4252.3	7.7432	80.20	3738.6	4251.9	7.7357	78.96	3738.3	4251.5	7.7282	850
85.28	3837.0	4374.2	7.8495	83.94	3836.7	4373.9	7.8420	82.64	3836.4	4373.5	7.8346	900
89.06	3936.4	4497.4	7.9523	87.66	3936.1	4497.1	7.9448	86.30	3935.8	4496.8	7.9374	950
92.82	4037.1	4621.9	8.0520	91.37	4036.8	4621.6	8.0445	89.95	4036.6	4621.3	8.0372	1000
100.33	4242.6	4874.6	8.2431	98.76	4242.3	4874.4	8.2357	97.24	4242.1	4874.1	8.2284	1100
107.82	4453.3	5132.6	8.4244	106.13	4453.1	5132.3	8.4170	104.50	4452.9	5132.1	8.4097	1200
115.29	4669.0	5395.3	8.5969	113.49	4668.7	5395.1	8.5895	111.75	4668.5	5394.8	8.5822	1300

Definitions of symbols on page 1.

Table 3. Vapor

p (t Sat.)	6.6 (281.93)				6.7 (282.93)				6.8 (283.93)			
t	$10^3 v$	u	h	s	$10^3 v$	u	h	s	$10^3 v$	u	h	s
Sat.	29.22	2584.4	2777.2	5.8428	28.74	2583.4	2776.0	5.8353	28.27	2582.5	2774.7	5.8279
280	*28.90*	*2576.8*	*2767.5*	*5.8253*	*28.25*	*2571.7*	*2760.9*	*5.8081*	*27.61*	*2566.4*	*2754.1*	*5.7908*
285	29.72	2596.0	2792.2	5.8697	29.08	2591.4	2786.2	5.8537	28.44	2586.7	2780.1	5.8376
290	30.49	2614.0	2815.2	5.9108	29.85	2609.8	2809.8	5.8957	29.23	2605.5	2804.2	5.8806
295	31.23	2630.8	2836.9	5.9491	30.59	2627.0	2831.9	5.9349	29.96	2623.0	2826.8	5.9205
300	31.92	2646.8	2857.5	5.9852	31.29	2643.2	2852.8	5.9715	30.66	2639.6	2848.1	5.9579
310	33.24	2676.6	2895.9	6.0517	32.60	2673.4	2891.9	6.0390	31.97	2670.3	2887.7	6.0264
320	34.46	2704.1	2931.5	6.1122	33.82	2701.3	2927.9	6.1003	33.19	2698.5	2924.2	6.0885
330	35.62	2729.8	2964.9	6.1681	34.97	2727.4	2961.7	6.1568	34.34	2724.9	2958.4	6.1456
340	36.73	2754.3	2996.7	6.2202	36.07	2752.0	2993.7	6.2095	35.43	2749.8	2990.7	6.1988
350	37.79	2777.6	3027.0	6.2694	37.12	2775.6	3024.3	6.2590	36.48	2773.5	3021.6	6.2487
360	38.81	2800.1	3056.3	6.3160	38.14	2798.3	3053.8	6.3059	37.48	2796.4	3051.3	6.2960
370	39.80	2822.0	3084.7	6.3604	39.12	2820.2	3082.3	6.3507	38.46	2818.5	3080.0	6.3410
380	40.77	2843.2	3112.3	6.4031	40.08	2841.6	3110.1	6.3935	39.41	2839.9	3107.9	6.3841
390	41.72	2864.0	3139.3	6.4441	41.02	2862.4	3137.3	6.4348	40.34	2860.9	3135.2	6.4256
400	42.65	2884.4	3165.8	6.4838	41.94	2882.9	3163.9	6.4747	41.25	2881.5	3162.0	6.4656
410	43.56	2904.4	3191.9	6.5222	42.84	2903.1	3190.1	6.5133	42.14	2901.7	3188.3	6.5044
420	44.45	2924.2	3217.6	6.5596	43.72	2922.9	3215.9	6.5507	43.02	2921.6	3214.2	6.5420
430	45.33	2943.8	3243.0	6.5959	44.60	2942.6	3241.4	6.5872	43.88	2941.3	3239.7	6.5786
440	46.20	2963.2	3268.1	6.6314	45.46	2962.0	3266.6	6.6228	44.73	2960.8	3265.0	6.6143
450	47.06	2982.4	3293.0	6.6661	46.31	2981.3	3291.5	6.6576	45.57	2980.2	3290.1	6.6492
460	47.91	3001.5	3317.7	6.7000	47.15	3000.4	3316.3	6.6916	46.40	2999.4	3314.9	6.6833
470	48.76	3020.5	3342.3	6.7333	47.98	3019.5	3340.9	6.7249	47.22	3018.4	3339.6	6.7167
480	49.59	3039.4	3366.6	6.7659	48.80	3038.4	3365.4	6.7576	48.04	3037.4	3364.1	6.7495
490	50.42	3058.2	3390.9	6.7979	49.62	3057.3	3389.7	6.7897	48.84	3056.3	3388.4	6.7816
500	51.23	3076.9	3415.1	6.8294	50.43	3076.0	3413.9	6.8212	49.64	3075.2	3412.7	6.8132
510	52.05	3095.6	3439.2	6.8603	51.23	3094.8	3438.0	6.8522	50.43	3093.9	3436.9	6.8443
520	52.85	3114.3	3463.2	6.8907	52.03	3113.5	3462.1	6.8828	51.22	3112.7	3461.0	6.8749
530	53.66	3133.0	3487.1	6.9207	52.82	3132.2	3486.0	6.9128	52.00	3131.4	3485.0	6.9050
540	54.45	3151.6	3511.0	6.9503	53.60	3150.8	3510.0	6.9424	52.78	3150.1	3508.9	6.9346
550	55.24	3170.2	3534.8	6.9794	54.38	3169.5	3533.8	6.9716	53.55	3168.7	3532.9	6.9638
560	56.03	3188.8	3558.6	7.0081	55.16	3188.1	3557.7	7.0003	54.32	3187.4	3556.7	6.9926
570	56.82	3207.4	3582.4	7.0365	55.93	3206.7	3581.5	7.0287	55.08	3206.0	3580.6	7.0211
580	57.60	3226.0	3606.1	7.0645	56.70	3225.3	3605.2	7.0568	55.84	3224.6	3604.4	7.0491
590	58.37	3244.6	3629.8	7.0921	57.47	3243.9	3629.0	7.0844	56.59	3243.3	3628.1	7.0769
600	59.14	3263.2	3653.5	7.1194	58.23	3262.6	3652.7	7.1118	57.35	3261.9	3651.9	7.1042
620	60.68	3300.4	3700.9	7.1731	59.75	3299.9	3700.2	7.1655	58.84	3299.3	3699.4	7.1580
640	62.20	3337.8	3748.3	7.2256	61.25	3337.2	3747.6	7.2180	60.32	3336.7	3746.9	7.2106
660	63.72	3375.2	3795.7	7.2769	62.74	3374.7	3795.0	7.2694	61.80	3374.2	3794.4	7.2620
680	65.22	3412.7	3843.2	7.3272	64.23	3412.2	3842.5	7.3198	63.26	3411.7	3841.9	7.3124
700	66.72	3450.4	3890.7	7.3766	65.70	3449.9	3890.1	7.3692	64.71	3449.5	3889.5	7.3619
720	68.20	3488.2	3938.3	7.4250	67.17	3487.7	3937.8	7.4177	66.16	3487.3	3937.2	7.4104
740	69.69	3526.1	3986.0	7.4726	68.63	3525.7	3985.5	7.4653	67.60	3525.3	3985.0	7.4580
760	71.16	3564.2	4033.9	7.5194	70.08	3563.8	4033.4	7.5121	69.04	3563.4	4032.9	7.5048
780	72.63	3602.5	4081.9	7.5654	71.53	3602.1	4081.4	7.5581	70.47	3601.8	4080.9	7.5509
800	74.10	3641.0	4130.0	7.6106	72.98	3640.6	4129.6	7.6034	71.89	3640.2	4129.1	7.5962
850	77.75	3737.9	4251.1	7.7209	76.58	3737.6	4250.7	7.7137	75.44	3737.3	4250.3	7.7066
900	81.37	3836.1	4373.2	7.8273	80.15	3835.8	4372.8	7.8201	78.96	3835.5	4372.5	7.8130
950	84.98	3935.6	4496.5	7.9302	83.71	3935.3	4496.1	7.9230	82.47	3935.0	4495.8	7.9160
1000	88.58	4036.3	4621.0	8.0299	87.26	4036.1	4620.7	8.0228	85.97	4035.8	4620.4	8.0158
1100	95.76	4241.9	4873.9	8.2211	94.33	4241.6	4873.6	8.2140	92.93	4241.4	4873.4	8.2070
1200	102.91	4452.6	5131.9	8.4025	101.37	4452.4	5131.6	8.3954	99.88	4452.2	5131.4	8.3884
1300	110.05	4668.3	5394.6	8.5750	108.41	4668.0	5394.4	8.5679	106.82	4667.8	5394.2	8.5610

Definitions of symbols on page 1.

10^3 v	u	h	s	10^3 v	u	h	s	10^3 v	u	h	s	t
27.81	2581.5	2773.4	5.8206	27.37	2580.5	2772.1	5.8133	26.94	2579.5	2770.8	5.8061	Sat.
26.98	*2561.0*	*2747.2*	*5.7734*									**280**
27.83	2581.9	2773.9	5.8214	*27.23*	*2576.9*	*2767.5*	*5.8050*	*26.64*	*2571.8*	*2761.0*	*5.7885*	**285**
28.62	2601.1	2798.5	5.8654	28.02	2596.6	2792.8	5.8501	27.44	2592.1	2786.9	5.8347	**290**
29.36	2619.0	2821.6	5.9062	28.76	2615.0	2816.3	5.8917	28.18	2610.8	2810.9	5.8773	**295**
30.06	2635.9	2843.3	5.9442	29.47	2632.2	2838.4	5.9305	28.89	2628.3	2833.5	5.9168	**300**
31.37	2667.1	2883.5	6.0138	30.78	2663.9	2879.3	6.0013	30.20	2660.6	2875.0	5.9887	**310**
32.58	2695.7	2920.5	6.0767	31.99	2692.9	2916.8	6.0650	31.41	2690.0	2913.0	6.0533	**320**
33.73	2722.4	2955.1	6.1345	33.13	2719.8	2951.7	6.1234	32.55	2717.3	2948.4	6.1123	**330**
34.81	2747.5	2987.7	6.1881	34.21	2745.2	2984.7	6.1776	33.62	2742.9	2981.6	6.1670	**340**
35.85	2771.5	3018.8	6.2385	35.24	2769.4	3016.0	6.2283	34.65	2767.3	3013.3	6.2182	**350**
36.85	2794.5	3048.7	6.2861	36.23	2792.5	3046.2	6.2763	35.63	2790.6	3043.6	6.2666	**360**
37.82	2816.7	3077.6	6.3314	37.19	2814.9	3075.3	6.3219	36.59	2813.1	3072.9	6.3125	**370**
38.76	2838.3	3105.7	6.3748	38.13	2836.6	3103.5	6.3655	37.51	2835.0	3101.3	6.3563	**380**
39.68	2859.4	3133.2	6.4165	39.04	2857.8	3131.1	6.4074	38.42	2856.3	3129.0	6.3984	**390**
40.58	2880.0	3160.0	6.4567	39.93	2878.6	3158.1	6.4478	39.30	2877.1	3156.2	6.4390	**400**
41.46	2900.3	3186.4	6.4956	40.80	2899.0	3184.6	6.4869	40.16	2897.6	3182.8	6.4783	**410**
42.33	2920.4	3212.4	6.5334	41.66	2919.1	3210.7	6.5248	41.01	2917.8	3209.0	6.5163	**420**
43.18	2940.1	3238.1	6.5701	42.51	2938.9	3236.4	6.5617	41.85	2937.7	3234.8	6.5533	**430**
44.02	2959.7	3263.5	6.6059	43.34	2958.5	3261.9	6.5976	42.67	2957.3	3260.3	6.5894	**440**
44.86	2979.1	3288.6	6.6409	44.16	2978.0	3287.1	6.6327	43.49	2976.8	3285.6	6.6246	**450**
45.68	2998.3	3313.5	6.6751	44.97	2997.2	3312.1	6.6670	44.29	2996.2	3310.6	6.6590	**460**
46.49	3017.4	3338.2	6.7086	45.78	3016.4	3336.8	6.7006	45.08	3015.4	3335.5	6.6926	**470**
47.29	3036.5	3362.8	6.7414	46.57	3035.5	3361.5	6.7335	45.87	3034.5	3360.2	6.7256	**480**
48.09	3055.4	3387.2	6.7737	47.36	3054.5	3386.0	6.7658	46.65	3053.5	3384.7	6.7580	**490**
48.88	3074.3	3411.5	6.8053	48.14	3073.4	3410.3	6.7975	47.42	3072.5	3409.1	6.7898	**500**
49.66	3093.1	3435.7	6.8365	48.91	3092.2	3434.6	6.8287	48.18	3091.4	3433.5	6.8210	**510**
50.44	3111.8	3459.9	6.8671	49.68	3111.0	3458.8	6.8594	48.94	3110.2	3457.7	6.8518	**520**
51.21	3130.6	3483.9	6.8972	50.44	3129.8	3482.9	6.8896	49.69	3129.0	3481.8	6.8820	**530**
51.98	3149.3	3507.9	6.9269	51.20	3148.5	3506.9	6.9193	50.44	3147.8	3505.9	6.9118	**540**
52.74	3168.0	3531.9	6.9562	51.95	3167.2	3530.9	6.9486	51.19	3166.5	3529.9	6.9412	**550**
53.50	3186.7	3555.8	6.9850	52.70	3185.9	3554.8	6.9775	51.92	3185.2	3553.9	6.9701	**560**
54.25	3205.3	3579.6	7.0135	53.44	3204.6	3578.7	7.0061	52.66	3203.9	3577.8	6.9987	**570**
55.00	3224.0	3603.5	7.0416	54.18	3223.3	3602.6	7.0342	53.39	3222.7	3601.7	7.0269	**580**
55.74	3242.6	3627.3	7.0694	54.92	3242.0	3626.4	7.0620	54.12	3241.4	3625.6	7.0547	**590**
56.49	3261.3	3651.1	7.0968	55.65	3260.7	3650.3	7.0894	54.84	3260.1	3649.4	7.0821	**600**
57.96	3298.7	3698.6	7.1506	57.11	3298.1	3697.9	7.1433	56.28	3297.5	3697.1	7.1361	**620**
59.43	3336.1	3746.2	7.2032	58.55	3335.6	3745.4	7.1960	57.70	3335.0	3744.7	7.1889	**640**
60.88	3373.6	3793.7	7.2547	59.99	3373.1	3793.0	7.2476	59.12	3372.6	3792.4	7.2405	**660**
62.32	3411.3	3841.3	7.3052	61.41	3410.8	3840.7	7.2981	60.53	3410.3	3840.0	7.2910	**680**
63.76	3449.0	3888.9	7.3547	62.83	3448.5	3888.3	7.3476	61.92	3448.1	3887.7	7.3405	**700**
65.19	3486.9	3936.6	7.4032	64.24	3486.4	3936.1	7.3961	63.31	3486.0	3935.5	7.3892	**720**
66.61	3524.9	3984.5	7.4509	65.64	3524.5	3983.9	7.4438	64.70	3524.1	3983.4	7.4369	**740**
68.02	3563.0	4032.4	7.4977	67.04	3562.7	4031.9	7.4907	66.08	3562.3	4031.4	7.4838	**760**
69.43	3601.4	4080.5	7.5438	68.43	3601.0	4080.0	7.5368	67.45	3600.6	4079.5	7.5299	**780**
70.84	3639.9	4128.7	7.5892	69.81	3639.5	4128.2	7.5822	68.82	3639.2	4127.8	7.5753	**800**
74.33	3737.0	4249.9	7.6996	73.26	3736.7	4249.5	7.6926	72.22	3736.4	4249.1	7.6858	**850**
77.81	3835.3	4372.1	7.8060	76.69	3835.0	4371.8	7.7991	75.60	3834.7	4371.4	7.7923	**900**
81.27	3934.8	4495.5	7.9090	80.10	3934.5	4495.2	7.9021	78.96	3934.2	4494.9	7.8954	**950**
84.71	4035.6	4620.1	8.0088	83.50	4035.3	4619.8	8.0020	82.32	4035.1	4619.5	7.9953	**1000**
91.58	4241.2	4873.1	8.2001	90.27	4240.9	4872.8	8.1933	89.00	4240.7	4872.6	8.1866	**1100**
98.43	4452.0	5131.1	8.3815	97.03	4451.7	5130.9	8.3747	95.66	4451.5	5130.7	8.3680	**1200**
105.27	4667.6	5393.9	8.5541	103.77	4667.3	5393.7	8.5473	102.31	4667.1	5393.5	8.5406	**1300**

Definitions of symbols on page 1.

Table 3. Vapor

p (t Sat.)	7.2 (287.79)				7.3 (288.74)				7.4 (289.67)			
t	$10^3 v$	u	h	s	$10^3 v$	u	h	s	$10^3 v$	u	h	s
Sat.	26.52	2578.5	2769.4	5.7989	26.11	2577.4	2768.0	5.7917	25.71	2576.4	2766.7	5.7847
285	*26.06*	*2566.7*	*2754.3*	*5.7718*	*25.49*	*2561.3*	*2747.4*	*5.7549*	*24.94*	*2555.9*	*2740.4*	*5.7378*
290	26.87	2587.4	2780.9	5.8192	26.31	2582.7	2774.7	5.8036	25.76	2577.8	2768.4	5.7878
295	27.62	2606.6	2805.4	5.8627	27.07	2602.3	2799.9	5.8481	26.53	2597.9	2794.2	5.8334
300	28.33	2624.5	2828.4	5.9030	27.78	2620.5	2823.3	5.8892	27.24	2616.5	2818.1	5.8753
310	29.64	2657.3	2870.7	5.9761	29.10	2653.9	2866.3	5.9636	28.56	2650.5	2861.9	5.9510
320	30.85	2687.1	2909.2	6.0416	30.30	2684.2	2905.4	6.0300	29.77	2681.2	2901.5	6.0184
330	31.98	2714.7	2945.0	6.1014	31.43	2712.1	2941.5	6.0904	30.89	2709.5	2938.1	6.0795
340	33.05	2740.6	2978.5	6.1566	32.49	2738.2	2975.4	6.1462	31.95	2735.9	2972.3	6.1359
350	34.07	2765.2	3010.5	6.2082	33.51	2763.0	3007.6	6.1983	32.96	2760.9	3004.8	6.1884
360	35.05	2788.7	3041.0	6.2569	34.48	2786.7	3038.4	6.2473	33.93	2784.8	3035.8	6.2378
370	36.00	2811.3	3070.5	6.3031	35.42	2809.5	3068.1	6.2938	34.86	2807.7	3065.7	6.2846
380	36.91	2833.3	3099.1	6.3472	36.33	2831.6	3096.9	6.3382	35.77	2830.0	3094.6	6.3292
390	37.81	2854.7	3127.0	6.3896	37.22	2853.2	3124.9	6.3807	36.65	2851.6	3122.8	6.3720
400	38.68	2875.7	3154.2	6.4303	38.09	2874.2	3152.2	6.4217	37.50	2872.7	3150.3	6.4132
410	39.54	2896.2	3180.9	6.4697	38.94	2894.9	3179.1	6.4613	38.35	2893.5	3177.2	6.4529
420	40.38	2916.5	3207.2	6.5080	39.77	2915.2	3205.5	6.4997	39.17	2913.9	3203.7	6.4914
430	41.21	2936.4	3233.1	6.5451	40.59	2935.2	3231.5	6.5369	39.98	2934.0	3229.8	6.5288
440	42.03	2956.2	3258.8	6.5812	41.39	2955.0	3257.2	6.5732	40.78	2953.8	3255.6	6.5652
450	42.83	2975.7	3284.1	6.6165	42.19	2974.6	3282.6	6.6086	41.57	2973.5	3281.1	6.6007
460	43.62	2995.1	3309.2	6.6510	42.98	2994.1	3307.8	6.6432	42.35	2993.0	3306.4	6.6354
470	44.41	3014.4	3334.1	6.6848	43.75	3013.4	3332.8	6.6770	43.12	3012.3	3331.4	6.6693
480	45.19	3033.5	3358.9	6.7179	44.52	3032.6	3357.6	6.7102	43.88	3031.6	3356.3	6.7026
490	45.95	3052.6	3383.5	6.7503	45.28	3051.7	3382.2	6.7427	44.63	3050.7	3381.0	6.7352
500	46.72	3071.6	3407.9	6.7822	46.04	3070.7	3406.7	6.7746	45.37	3069.8	3405.5	6.7671
510	47.47	3090.5	3432.3	6.8135	46.78	3089.6	3431.2	6.8060	46.11	3088.8	3430.0	6.7986
520	48.22	3109.4	3456.6	6.8443	47.53	3108.5	3455.5	6.8368	46.85	3107.7	3454.4	6.8295
530	48.97	3128.2	3480.8	6.8746	48.26	3127.4	3479.7	6.8672	47.57	3126.6	3478.6	6.8599
540	49.71	3147.0	3504.9	6.9044	48.99	3146.2	3503.9	6.8971	48.30	3145.4	3502.8	6.8898
550	50.44	3165.8	3528.9	6.9338	49.72	3165.0	3528.0	6.9265	49.01	3164.3	3527.0	6.9193
560	51.17	3184.5	3552.9	6.9628	50.44	3183.8	3552.0	6.9556	49.73	3183.1	3551.0	6.9484
570	51.90	3203.2	3576.9	6.9914	51.16	3202.6	3576.0	6.9842	50.43	3201.9	3575.1	6.9771
580	52.62	3222.0	3600.8	7.0196	51.87	3221.3	3600.0	7.0125	51.14	3220.6	3599.1	7.0054
590	53.34	3240.7	3624.7	7.0475	52.58	3240.1	3623.9	7.0403	51.84	3239.4	3623.0	7.0333
600	54.05	3259.4	3648.6	7.0750	53.28	3258.8	3647.8	7.0679	52.54	3258.2	3647.0	7.0609
620	55.47	3296.9	3696.3	7.1290	54.69	3296.4	3695.6	7.1220	53.92	3295.8	3694.8	7.1150
640	56.88	3334.5	3744.0	7.1818	56.08	3333.9	3743.3	7.1748	55.30	3333.4	3742.6	7.1679
660	58.28	3372.1	3791.7	7.2334	57.46	3371.6	3791.0	7.2265	56.66	3371.1	3790.3	7.2197
680	59.67	3409.8	3839.4	7.2840	58.83	3409.3	3838.8	7.2772	58.01	3408.8	3838.1	7.2704
700	61.05	3447.6	3887.2	7.3336	60.19	3447.2	3886.6	7.3268	59.36	3446.7	3886.0	7.3200
720	62.42	3485.6	3935.0	7.3823	61.55	3485.1	3934.4	7.3755	60.70	3484.7	3933.9	7.3687
740	63.78	3523.6	3982.9	7.4300	62.89	3523.2	3982.4	7.4233	62.03	3522.8	3981.8	7.4166
760	65.14	3561.9	4030.9	7.4770	64.24	3561.5	4030.4	7.4702	63.35	3561.1	4029.9	7.4636
780	66.50	3600.3	4079.1	7.5231	65.57	3599.9	4078.6	7.5164	64.67	3599.5	4078.1	7.5098
800	67.85	3638.8	4127.3	7.5685	66.91	3638.5	4126.9	7.5618	65.99	3638.1	4126.5	7.5552
850	71.20	3736.1	4248.7	7.6791	70.22	3735.7	4248.3	7.6724	69.26	3735.4	4247.9	7.6659
900	74.54	3834.4	4371.1	7.7856	73.51	3834.1	4370.7	7.7790	72.51	3833.8	4370.4	7.7725
950	77.86	3934.0	4494.5	7.8887	76.79	3933.7	4494.2	7.8821	75.74	3933.4	4493.9	7.8756
1000	81.17	4034.8	4619.2	7.9886	80.05	4034.6	4618.9	7.9821	78.96	4034.3	4618.6	7.9756
1100	87.76	4240.5	4872.3	8.1800	86.55	4240.2	4872.1	8.1734	85.38	4240.0	4871.8	8.1670
1200	94.33	4451.3	5130.4	8.3614	93.03	4451.0	5130.2	8.3549	91.78	4450.8	5130.0	8.3484
1300	100.89	4666.9	5393.3	8.5340	99.51	4666.6	5393.0	8.5275	98.16	4666.4	5392.8	8.5211

Definitions of symbols on page 1.

7.5 (290.59)				7.6 (291.50)				7.7 (292.41)				p (t Sat.)
10^3 v	u	h	s	10^3 v	u	h	s	10^3 v	u	h	s	t
25.32	2575.3	2765.2	5.7776	24.94	2574.3	2763.8	5.7707	24.57	2573.2	2762.4	5.7637	Sat.
												285
_25.23_	_2572.8_	_2762.0_	_5.7719_	_24.70_	_2567.7_	_2755.5_	_5.7558_	_24.19_	_2562.5_	_2748.7_	_5.7395_	**290**
26.00	2593.4	2788.4	5.8186	25.48	2588.8	2782.5	5.8036	24.97	2584.2	2776.5	5.7886	**295**
26.72	2612.5	2812.9	5.8614	26.21	2608.3	2807.5	5.8474	25.70	2604.1	2802.0	5.8334	**300**
28.04	2647.1	2857.4	5.9384	27.53	2643.6	2852.8	5.9258	27.03	2640.0	2848.2	5.9132	**310**
29.25	2678.2	2897.6	6.0067	28.74	2675.2	2893.6	5.9952	28.24	2672.1	2889.6	5.9836	**320**
30.37	2706.8	2934.6	6.0686	29.86	2704.1	2931.0	6.0578	29.36	2701.4	2927.5	6.0470	**330**
31.42	2733.5	2969.2	6.1256	30.91	2731.1	2966.0	6.1153	30.41	2728.7	2962.8	6.1051	**340**
32.43	2758.7	3001.9	6.1785	31.91	2756.6	2999.0	6.1688	31.40	2754.4	2996.1	6.1590	**350**
33.39	2782.8	3033.2	6.2283	32.86	2780.8	3030.6	6.2189	32.35	2778.8	3027.9	6.2096	**360**
34.32	2805.9	3063.3	6.2755	33.78	2804.1	3060.8	6.2664	33.27	2802.2	3058.4	6.2574	**370**
35.21	2828.3	3092.4	6.3204	34.68	2826.6	3090.1	6.3115	34.15	2824.9	3087.8	6.3028	**380**
36.09	2850.0	3120.7	6.3634	35.54	2848.4	3118.6	6.3548	35.01	2846.9	3116.4	6.3462	**390**
36.94	2871.3	3148.3	6.4047	36.39	2869.8	3146.3	6.3963	35.85	2868.3	3144.3	6.3880	**400**
37.77	2892.1	3175.4	6.4446	37.21	2890.7	3173.5	6.4364	36.67	2889.3	3171.6	6.4283	**410**
38.59	2912.5	3202.0	6.4833	38.02	2911.2	3200.2	6.4752	37.47	2909.9	3198.4	6.4672	**420**
39.39	2932.7	3228.2	6.5208	38.82	2931.5	3226.5	6.5129	38.26	2930.2	3224.8	6.5050	**430**
40.19	2952.6	3254.0	6.5573	39.60	2951.5	3252.4	6.5495	39.04	2950.3	3250.9	6.5418	**440**
40.96	2972.4	3279.6	6.5929	40.38	2971.2	3278.1	6.5852	39.80	2970.1	3276.6	6.5776	**450**
41.73	2991.9	3304.9	6.6277	41.14	2990.8	3303.5	6.6201	40.56	2989.8	3302.0	6.6126	**460**
42.49	3011.3	3330.0	6.6617	41.89	3010.3	3328.7	6.6542	41.30	3009.3	3327.3	6.6468	**470**
43.25	3030.6	3355.0	6.6950	42.63	3029.6	3353.6	6.6876	42.04	3028.6	3352.3	6.6802	**480**
43.99	3049.8	3379.7	6.7277	43.37	3048.8	3378.5	6.7203	42.77	3047.9	3377.2	6.7130	**490**
44.73	3068.9	3404.3	6.7598	44.10	3068.0	3403.1	6.7525	43.49	3067.1	3401.9	6.7452	**500**
45.46	3087.9	3428.9	6.7913	44.82	3087.0	3427.7	6.7840	44.20	3086.2	3426.5	6.7769	**510**
46.18	3106.9	3453.3	6.8223	45.54	3106.0	3452.1	6.8150	44.91	3105.2	3451.0	6.8079	**520**
46.90	3125.8	3477.6	6.8527	46.25	3125.0	3476.5	6.8456	45.62	3124.2	3475.4	6.8385	**530**
47.62	3144.7	3501.8	6.8827	46.96	3143.9	3500.8	6.8756	46.32	3143.1	3499.8	6.8686	**540**
48.33	3163.5	3526.0	6.9122	47.66	3162.8	3525.0	6.9052	47.01	3162.0	3524.0	6.8983	**550**
49.03	3182.4	3550.1	6.9414	48.36	3181.6	3549.1	6.9344	47.70	3180.9	3548.2	6.9275	**560**
49.73	3201.2	3574.2	6.9701	49.05	3200.5	3573.2	6.9631	48.38	3199.8	3572.3	6.9563	**570**
50.43	3220.0	3598.2	6.9984	49.74	3219.3	3597.3	6.9915	49.06	3218.6	3596.4	6.9847	**580**
51.12	3238.8	3622.2	7.0264	50.42	3238.1	3621.3	7.0195	49.74	3237.5	3620.5	7.0127	**590**
51.81	3257.6	3646.2	7.0540	51.10	3256.9	3645.3	7.0471	50.41	3256.3	3644.5	7.0404	**600**
53.18	3295.2	3694.0	7.1082	52.46	3294.6	3693.3	7.1014	51.75	3294.0	3692.5	7.0947	**620**
54.54	3332.8	3741.9	7.1611	53.80	3332.3	3741.1	7.1544	53.08	3331.7	3740.4	7.1478	**640**
55.88	3370.5	3789.7	7.2129	55.13	3370.0	3789.0	7.2063	54.39	3369.5	3788.3	7.1997	**660**
57.22	3408.3	3837.5	7.2637	56.45	3407.9	3836.9	7.2570	55.70	3407.4	3836.2	7.2505	**680**
58.55	3446.2	3885.4	7.3134	57.76	3445.8	3884.8	7.3068	56.99	3445.3	3884.2	7.3003	**700**
59.87	3484.3	3933.3	7.3621	59.07	3483.8	3932.7	7.3556	58.28	3483.4	3932.2	7.3491	**720**
61.19	3522.4	3981.5	7.4100	60.37	3522.0	3980.8	7.4035	59.57	3521.6	3980.3	7.3970	**740**
62.50	3560.7	4029.4	7.4570	61.66	3560.3	4028.9	7.4505	60.84	3559.9	4028.4	7.4441	**760**
63.80	3599.2	4077.7	7.5032	62.95	3598.8	4077.2	7.4968	62.12	3598.4	4076.7	7.4904	**780**
65.10	3637.8	4126.0	7.5487	64.23	3637.4	4125.6	7.5423	63.38	3637.1	4125.1	7.5359	**800**
68.32	3735.1	4247.5	7.6594	67.42	3734.8	4247.1	7.6530	66.53	3734.5	4246.8	7.6467	**850**
71.53	3833.5	4370.0	7.7661	70.58	3833.2	4369.7	7.7597	69.66	3833.0	4369.3	7.7534	**900**
74.72	3933.2	4493.6	7.8692	73.73	3932.9	4493.3	7.8629	72.77	3932.6	4493.0	7.8566	**950**
77.90	4034.1	4618.3	7.9692	76.87	4033.8	4618.0	7.9629	75.87	4033.6	4617.8	7.9566	**1000**
84.24	4239.8	4871.5	8.1606	83.13	4239.5	4871.3	8.1543	82.04	4239.3	4871.0	8.1481	**1100**
90.55	4450.6	5129.7	8.3421	89.36	4450.4	5129.5	8.3358	88.20	4450.1	5129.3	8.3296	**1200**
96.85	4666.2	5392.6	8.5147	95.58	4666.0	5392.4	8.5085	94.34	4665.7	5392.1	8.5023	**1300**

Definitions of symbols on page 1.

Table 3. Vapor

p (t Sat.)		7.8 (293.30)				7.9 (294.19)				8.0 (295.06)		
t	10^3 v	u	h	s	10^3 v	u	h	s	10^3 v	u	h	s
Sat.	24.21	2572.1	2760.9	5.7568	23.86	2570.9	2759.5	5.7500	23.52	2569.8	2758.0	5.7432
295	24.48	2579.4	2770.3	5.7734	23.99	2574.5	2764.0	5.7581	*23.51*	*2569.5*	*2757.6*	*5.7426*
300	25.21	2599.8	2796.5	5.8192	24.73	2595.4	2790.8	5.8049	24.26	2590.9	2785.0	5.7906
305	25.90	2618.7	2820.7	5.8614	25.42	2614.7	2815.6	5.8480	24.95	2610.7	2810.3	5.8345
310	26.55	2636.4	2843.5	5.9006	26.07	2632.8	2838.7	5.8879	25.61	2629.0	2833.9	5.8752
315	27.17	2653.1	2865.0	5.9373	26.69	2649.7	2860.6	5.9253	26.23	2646.3	2856.1	5.9132
320	27.76	2669.0	2885.5	5.9720	27.28	2665.8	2881.4	5.9604	26.82	2662.7	2877.2	5.9489
325	28.32	2684.1	2905.1	6.0049	27.85	2681.2	2901.2	5.9937	27.39	2678.2	2897.3	5.9826
330	28.87	2698.7	2923.9	6.0362	28.40	2695.9	2920.3	6.0254	27.93	2693.1	2916.6	6.0147
335	29.40	2712.7	2942.0	6.0662	28.93	2710.1	2938.6	6.0557	28.46	2707.5	2935.2	6.0453
340	29.92	2726.2	2959.6	6.0949	29.44	2723.8	2956.3	6.0848	28.97	2721.3	2953.1	6.0747
345	30.42	2739.4	2976.6	6.1226	29.94	2737.0	2973.6	6.1127	29.47	2734.7	2970.5	6.1029
350	30.91	2752.2	2993.2	6.1493	30.42	2749.9	2990.3	6.1397	29.95	2747.7	2987.3	6.1301
360	31.85	2776.8	3025.2	6.2003	31.36	2774.8	3022.5	6.1911	30.89	2772.7	3019.8	6.1819
370	32.76	2800.4	3055.9	6.2484	32.27	2798.5	3053.4	6.2395	31.79	2796.7	3051.0	6.2306
380	33.64	2823.2	3085.5	6.2941	33.14	2821.4	3083.3	6.2855	32.66	2819.7	3081.0	6.2769
390	34.49	2845.3	3114.3	6.3378	33.99	2843.7	3112.2	6.3294	33.50	2842.0	3110.0	6.3211
400	35.33	2866.8	3142.3	6.3797	34.81	2865.3	3140.3	6.3716	34.32	2863.8	3138.3	6.3634
410	36.14	2887.9	3169.8	6.4202	35.62	2886.5	3167.9	6.4122	35.12	2885.1	3166.0	6.4042
420	36.93	2908.6	3196.7	6.4593	36.41	2907.3	3194.9	6.4514	35.90	2905.9	3193.1	6.4437
430	37.72	2929.0	3223.2	6.4972	37.19	2927.7	3221.5	6.4895	36.67	2926.4	3219.8	6.4819
440	38.49	2949.1	3249.3	6.5341	37.95	2947.9	3247.7	6.5265	37.42	2946.7	3246.1	6.5190
450	39.24	2969.0	3275.1	6.5700	38.70	2967.8	3273.6	6.5626	38.17	2966.7	3272.0	6.5551
460	39.99	2988.7	3300.6	6.6051	39.44	2987.6	3299.2	6.5977	38.90	2986.5	3297.7	6.5904
470	40.73	3008.2	3325.9	6.6394	40.17	3007.2	3324.5	6.6321	39.62	3006.2	3323.1	6.6249
480	41.46	3027.6	3351.0	6.6729	40.89	3026.7	3349.7	6.6657	40.34	3025.7	3348.4	6.6586
490	42.18	3046.9	3375.9	6.7058	41.60	3046.0	3374.7	6.6987	41.05	3045.1	3373.4	6.6916
500	42.89	3066.2	3400.7	6.7381	42.31	3065.3	3399.5	6.7310	41.75	3064.3	3398.3	6.7240
510	43.60	3085.3	3425.4	6.7698	43.01	3084.4	3424.2	6.7628	42.44	3083.5	3423.1	6.7558
520	44.30	3104.4	3449.9	6.8009	43.71	3103.5	3448.8	6.7940	43.13	3102.7	3447.7	6.7871
530	45.00	3123.4	3474.4	6.8315	44.40	3122.6	3473.3	6.8246	43.81	3121.8	3472.2	6.8178
540	45.69	3142.3	3498.7	6.8617	45.08	3141.6	3497.7	6.8548	44.48	3140.8	3496.7	6.8481
550	46.38	3161.3	3523.0	6.8914	45.76	3160.5	3522.0	6.8846	45.16	3159.8	3521.0	6.8778
560	47.06	3180.2	3547.2	6.9206	46.43	3179.5	3546.3	6.9139	45.82	3178.7	3545.3	6.9072
570	47.73	3199.1	3571.4	6.9495	47.10	3198.4	3570.5	6.9428	46.49	3197.7	3569.6	6.9361
580	48.41	3218.0	3595.5	6.9779	47.77	3217.3	3594.7	6.9712	47.14	3216.6	3593.8	6.9646
590	49.08	3236.8	3619.6	7.0060	48.43	3236.2	3618.8	6.9993	47.80	3235.5	3617.9	6.9928
600	49.74	3255.7	3643.7	7.0337	49.09	3255.1	3642.9	7.0271	48.45	3254.4	3642.0	7.0206
610	50.41	3274.6	3667.7	7.0611	49.74	3273.9	3666.9	7.0545	49.10	3273.3	3666.1	7.0480
620	51.07	3293.4	3691.7	7.0881	50.40	3292.8	3691.0	7.0816	49.74	3292.2	3690.2	7.0751
630	51.72	3312.3	3715.7	7.1148	51.04	3311.7	3715.0	7.1083	50.38	3311.2	3714.2	7.1019
640	52.38	3331.2	3739.7	7.1412	51.69	3330.6	3739.0	7.1347	51.02	3330.1	3738.3	7.1283
650	53.03	3350.1	3763.7	7.1673	52.33	3349.5	3763.0	7.1609	51.66	3349.0	3762.3	7.1545
660	53.67	3369.0	3787.7	7.1932	52.98	3368.5	3787.0	7.1867	52.29	3368.0	3786.3	7.1804
670	54.32	3387.9	3811.6	7.2187	53.61	3387.4	3811.0	7.2123	52.93	3386.9	3810.3	7.2060
680	54.97	3406.9	3835.6	7.2440	54.25	3406.4	3835.0	7.2376	53.56	3405.9	3834.3	7.2313
690	55.61	3425.9	3859.6	7.2691	54.89	3425.4	3859.0	7.2627	54.18	3424.9	3858.4	7.2564
700	56.25	3444.9	3883.6	7.2938	55.52	3444.4	3883.0	7.2875	54.81	3443.9	3882.4	7.2812
720	57.52	3483.0	3931.6	7.3427	56.78	3482.5	3931.1	7.3364	56.05	3482.1	3930.5	7.3301
740	58.79	3521.2	3979.7	7.3907	58.03	3520.8	3979.2	7.3844	57.29	3520.4	3978.7	7.3782
760	60.05	3559.5	4027.9	7.4378	59.28	3559.2	4027.4	7.4315	58.52	3558.8	4026.9	7.4253
780	61.31	3598.1	4076.2	7.4841	60.52	3597.7	4075.8	7.4778	59.75	3597.3	4075.3	7.4717
800	62.56	3636.7	4124.7	7.5296	61.75	3636.4	4124.2	7.5234	60.97	3636.0	4123.8	7.5173
850	65.67	3734.2	4246.4	7.6405	64.83	3733.8	4246.0	7.6343	64.01	3733.5	4245.6	7.6282
900	68.76	3832.7	4369.0	7.7472	67.88	3832.4	4368.6	7.7411	67.02	3832.1	4368.3	7.7351
950	71.83	3932.4	4492.6	7.8505	70.91	3932.1	4492.3	7.8444	70.02	3931.8	4492.0	7.8384
1000	74.89	4033.3	4617.5	7.9505	73.94	4033.1	4617.2	7.9444	73.01	4032.8	4616.9	7.9384
1100	80.99	4239.1	4870.8	8.1420	79.96	4238.8	4870.5	8.1360	78.96	4238.6	4870.3	8.1300
1200	87.07	4449.9	5129.0	8.3235	85.96	4449.7	5128.8	8.3175	84.89	4449.5	5128.5	8.3115
1300	93.13	4665.5	5391.9	8.4962	91.95	4665.3	5391.7	8.4902	90.80	4665.0	5391.5	8.4842

Definitions of symbols on page 1.

10^3 v	u	h	s	10^3 v	u	h	s	10^3 v	u	h	s	t
23.18	2568.7	2756.4	5.7364	22.86	2567.5	2754.9	5.7297	22.54	2566.3	2753.4	5.7230	Sat.
23.04	*2564.4*	*2751.0*	*5.7269*	*22.57*	*2559.2*	*2744.3*	*5.7110*	*22.12*	*2553.8*	*2737.3*	*5.6949*	**295**
23.79	2586.4	2779.1	5.7761	23.34	2581.7	2773.1	5.7615	22.89	2576.9	2766.9	5.7467	**300**
24.50	2606.5	2804.9	5.8210	24.05	2602.3	2799.5	5.8074	23.60	2598.1	2794.0	5.7937	**305**
25.15	2625.3	2829.0	5.8624	24.71	2621.4	2824.0	5.8496	24.27	2617.5	2819.0	5.8368	**310**
25.78	2642.8	2851.6	5.9010	25.33	2639.3	2847.0	5.8889	24.90	2635.7	2842.4	5.8767	**315**
26.37	2659.4	2873.0	5.9373	25.92	2656.2	2868.7	5.9257	25.49	2652.9	2864.4	5.9141	**320**
26.93	2675.2	2893.4	5.9715	26.49	2672.2	2889.4	5.9604	26.06	2669.1	2885.4	5.9492	**325**
27.48	2690.3	2912.9	6.0040	27.04	2687.5	2909.2	5.9933	26.60	2684.6	2905.4	5.9826	**330**
28.01	2704.8	2931.7	6.0349	27.56	2702.1	2928.1	6.0246	27.13	2699.4	2924.6	6.0143	**335**
28.52	2718.8	2949.8	6.0646	28.07	2716.3	2946.4	6.0546	27.64	2713.7	2943.1	6.0445	**340**
29.01	2732.3	2967.3	6.0931	28.56	2729.9	2964.2	6.0833	28.13	2727.5	2961.0	6.0736	**345**
29.49	2745.4	2984.3	6.1205	29.05	2743.2	2981.3	6.1110	28.61	2740.9	2978.3	6.1015	**350**
30.43	2770.7	3017.1	6.1727	29.97	2768.6	3014.4	6.1636	29.53	2766.5	3011.6	6.1546	**360**
31.32	2794.8	3048.4	6.2218	30.86	2792.9	3045.9	6.2131	30.41	2791.0	3043.4	6.2044	**370**
32.18	2817.9	3078.6	6.2684	31.72	2816.2	3076.3	6.2599	31.27	2814.4	3073.9	6.2515	**380**
33.02	2840.4	3107.8	6.3128	32.55	2838.8	3105.7	6.3046	32.09	2837.1	3103.5	6.2964	**390**
33.83	2862.3	3136.3	6.3553	33.35	2860.7	3134.2	6.3473	32.89	2859.2	3132.2	6.3394	**400**
34.62	2883.6	3164.1	6.3963	34.14	2882.2	3162.2	6.3885	33.67	2880.8	3160.2	6.3807	**410**
35.40	2904.6	3191.3	6.4359	34.91	2903.2	3189.5	6.4282	34.44	2901.9	3187.7	6.4206	**420**
36.16	2925.2	3218.1	6.4742	35.67	2923.9	3216.4	6.4667	35.19	2922.6	3214.7	6.4592	**430**
36.91	2945.5	3244.4	6.5115	36.41	2944.3	3242.8	6.5041	35.92	2943.1	3241.2	6.4967	**440**
37.65	2965.5	3270.5	6.5477	37.14	2964.4	3269.0	6.5404	36.65	2963.2	3267.4	6.5332	**450**
38.37	2985.4	3296.2	6.5831	37.86	2984.3	3294.8	6.5759	37.36	2983.2	3293.3	6.5688	**460**
39.09	3005.1	3321.7	6.6176	38.57	3004.1	3320.3	6.6105	38.06	3003.0	3319.0	6.6035	**470**
39.80	3024.6	3347.0	6.6514	39.27	3023.7	3345.7	6.6444	38.76	3022.7	3344.4	6.6375	**480**
40.50	3044.1	3372.1	6.6846	39.97	3043.1	3370.9	6.6776	39.45	3042.2	3369.6	6.6708	**490**
41.19	3063.4	3397.1	6.7170	40.65	3062.5	3395.9	6.7102	40.13	3061.6	3394.6	6.7034	**500**
41.88	3082.6	3421.9	6.7489	41.33	3081.8	3420.7	6.7421	40.80	3080.9	3419.5	6.7354	**510**
42.56	3101.8	3446.6	6.7802	42.01	3101.0	3445.4	6.7735	41.47	3100.1	3444.3	6.7668	**520**
43.23	3120.9	3471.1	6.8110	42.68	3120.1	3470.1	6.8043	42.13	3119.3	3469.0	6.7977	**530**
43.90	3140.0	3495.6	6.8413	43.34	3139.2	3494.6	6.8347	42.79	3138.4	3493.6	6.8281	**540**
44.57	3159.0	3520.0	6.8712	44.00	3158.3	3519.0	6.8646	43.44	3157.5	3518.0	6.8580	**550**
45.23	3178.0	3544.4	6.9005	44.65	3177.3	3543.4	6.8940	44.08	3176.6	3542.4	6.8875	**560**
45.88	3197.0	3568.6	6.9295	45.30	3196.3	3567.7	6.9230	44.73	3195.6	3566.8	6.9166	**570**
46.54	3215.9	3592.9	6.9581	45.94	3215.3	3592.0	6.9516	45.36	3214.6	3591.1	6.9452	**580**
47.18	3234.9	3617.1	6.9863	46.58	3234.3	3616.2	6.9798	46.00	3233.6	3615.3	6.9735	**590**
47.83	3253.8	3641.2	7.0141	47.22	3253.2	3640.4	7.0077	46.63	3252.5	3639.6	7.0014	**600**
48.47	3272.7	3665.3	7.0416	47.85	3272.1	3664.5	7.0352	47.26	3271.5	3663.7	7.0289	**610**
49.11	3291.7	3689.4	7.0687	48.49	3291.1	3688.7	7.0624	47.88	3290.5	3687.9	7.0561	**620**
49.74	3310.6	3713.5	7.0955	49.11	3310.0	3712.8	7.0892	48.50	3309.5	3712.0	7.0830	**630**
50.37	3329.6	3737.6	7.1220	49.74	3329.0	3736.9	7.1157	49.12	3328.4	3736.1	7.1095	**640**
51.00	3348.5	3761.6	7.1482	50.36	3348.0	3760.9	7.1420	49.73	3347.4	3760.2	7.1358	**650**
51.63	3367.5	3785.7	7.1741	50.98	3367.0	3785.0	7.1679	50.35	3366.4	3784.3	7.1617	**660**
52.25	3386.5	3809.7	7.1997	51.60	3386.0	3809.1	7.1935	50.96	3385.5	3808.4	7.1874	**670**
52.88	3405.5	3833.8	7.2251	52.21	3405.0	3833.1	7.2189	51.57	3404.5	3832.5	7.2128	**680**
53.50	3424.5	3857.8	7.2502	52.83	3424.0	3857.2	7.2440	52.17	3423.6	3856.6	7.2380	**690**
54.11	3443.6	3881.9	7.2751	53.44	3443.1	3881.3	7.2689	52.78	3442.6	3880.7	7.2628	**700**
55.35	3481.7	3930.0	7.3240	54.66	3481.3	3929.5	7.3179	53.98	3480.9	3928.9	7.3119	**720**
56.57	3520.0	3978.2	7.3721	55.87	3519.6	3977.7	7.3660	55.18	3519.2	3977.2	7.3600	**740**
57.79	3558.5	4026.5	7.4193	57.07	3558.1	4026.0	7.4133	56.37	3557.7	4025.5	7.4073	**760**
59.00	3597.0	4074.9	7.4657	58.27	3596.7	4074.4	7.4597	57.55	3596.3	4074.0	7.4537	**780**
60.21	3635.7	4123.4	7.5113	59.46	3635.4	4123.0	7.5053	58.73	3635.0	4122.5	7.4994	**800**
63.21	3733.2	4245.2	7.6222	62.43	3732.9	4244.8	7.6163	61.67	3732.6	4244.4	7.6104	**850**
66.19	3831.8	4367.9	7.7291	65.37	3831.5	4367.6	7.7232	64.58	3831.2	4367.2	7.7174	**900**
69.15	3931.6	4491.7	7.8324	68.30	3931.3	4491.4	7.8265	67.47	3931.0	4491.0	7.8207	**950**
72.10	4032.6	4616.6	7.9325	71.22	4032.3	4616.3	7.9266	70.35	4032.1	4616.0	7.9208	**1000**
77.98	4238.4	4870.6	8.1241	77.03	4238.1	4869.7	8.1183	76.09	4237.9	4869.5	8.1125	**1100**
83.84	4449.2	5128.3	8.3056	82.81	4449.0	5128.1	8.2998	81.81	4448.8	5127.8	8.2941	**1200**
89.68	4664.8	5391.2	8.4783	88.59	4664.6	5391.0	8.4725	87.52	4664.3	5390.8	8.4668	**1300**

Definitions of symbols on page 1.

Table 3. Vapor

p (t Sat.)	8.4 (298.49)				8.5 (299.33)				8.6 (300.16)			
t	10^3 v	u	h	s	10^3 v	u	h	s	10^3 v	u	h	s
Sat.	22.22	2565.2	2751.8	5.7164	21.92	2564.0	2750.3	5.7098	21.62	2562.8	2748.7	5.7032
300	22.45	2572.1	2760.7	5.7319	22.02	2567.1	2754.3	5.7168	_21.59_	_2562.0_	_2747.7_	_5.7015_
305	23.17	2593.7	2788.3	5.7799	22.75	2589.3	2782.6	5.7660	22.33	2584.7	2776.7	5.7520
310	23.84	2613.6	2813.9	5.8239	23.42	2609.6	2808.6	5.8109	23.01	2605.5	2803.3	5.7978
315	24.47	2632.1	2837.7	5.8645	24.05	2628.4	2832.9	5.8523	23.64	2624.7	2828.0	5.8400
320	25.07	2649.5	2860.1	5.9025	24.65	2646.1	2855.7	5.8908	24.24	2642.7	2851.2	5.8792
325	25.63	2666.0	2881.3	5.9382	25.22	2662.9	2877.2	5.9270	24.81	2659.7	2873.1	5.9159
330	26.18	2681.7	2901.6	5.9719	25.76	2678.8	2897.8	5.9612	25.36	2675.8	2893.9	5.9506
335	26.70	2696.7	2921.0	6.0040	26.29	2694.0	2917.4	5.9937	25.88	2691.2	2913.8	5.9834
340	27.21	2711.2	2939.7	6.0346	26.79	2708.6	2936.3	6.0246	26.39	2706.0	2932.9	6.0147
345	27.70	2725.1	2957.8	6.0639	27.28	2722.7	2954.6	6.0543	26.88	2720.2	2951.3	6.0446
350	28.18	2738.6	2975.3	6.0921	27.76	2736.3	2972.2	6.0827	27.35	2733.9	2969.1	6.0734
360	29.10	2764.5	3008.9	6.1456	28.67	2762.3	3006.1	6.1366	28.26	2760.2	3003.3	6.1277
370	29.98	2789.1	3040.9	6.1958	29.55	2787.1	3038.3	6.1872	29.13	2785.2	3035.8	6.1786
380	30.82	2812.7	3071.6	6.2432	30.39	2810.9	3069.3	6.2349	29.97	2809.2	3066.9	6.2267
390	31.64	2835.5	3101.3	6.2884	31.21	2833.9	3099.1	6.2803	30.78	2832.2	3096.9	6.2723
400	32.44	2857.7	3130.2	6.3316	32.00	2856.2	3128.2	6.3237	31.56	2854.6	3126.1	6.3160
410	33.21	2879.4	3158.4	6.3731	32.77	2877.9	3156.4	6.3654	32.33	2876.5	3154.5	6.3579
420	33.97	2900.5	3185.9	6.4131	33.52	2899.2	3184.1	6.4057	33.08	2897.8	3182.3	6.3982
430	34.72	2921.4	3213.0	6.4519	34.26	2920.1	3211.3	6.4446	33.81	2918.8	3209.6	6.4373
440	35.45	2941.9	3239.6	6.4895	34.98	2940.7	3238.0	6.4823	34.53	2939.4	3236.4	6.4752
450	36.17	2962.1	3265.9	6.5261	35.69	2961.0	3264.4	6.5190	35.23	2959.8	3262.8	6.5120
460	36.87	2982.2	3291.9	6.5618	36.40	2981.1	3290.4	6.5548	35.93	2980.0	3289.0	6.5479
470	37.57	3002.0	3317.6	6.5966	37.09	3001.0	3316.2	6.5897	36.61	2999.9	3314.8	6.5829
480	38.26	3021.7	3343.1	6.6307	37.77	3020.7	3341.7	6.6238	37.29	3019.7	3340.4	6.6171
490	38.94	3041.2	3368.3	6.6640	38.44	3040.3	3367.1	6.6573	37.96	3039.3	3365.8	6.6506
500	39.61	3060.7	3393.4	6.6967	39.11	3059.8	3392.2	6.6900	38.62	3058.8	3391.0	6.6834
510	40.28	3080.0	3418.4	6.7288	39.77	3079.2	3417.2	6.7221	39.28	3078.3	3416.1	6.7156
520	40.94	3099.3	3443.2	6.7602	40.43	3098.5	3442.1	6.7537	39.92	3097.6	3441.0	6.7472
530	41.60	3118.5	3467.9	6.7912	41.08	3117.7	3466.8	6.7847	40.57	3116.9	3465.8	6.7783
540	42.25	3137.7	3492.5	6.8217	41.72	3136.9	3491.5	6.8152	41.20	3136.1	3490.5	6.8088
550	42.89	3156.8	3517.1	6.8516	42.36	3156.0	3516.1	6.8452	41.84	3155.3	3515.1	6.8389
560	43.53	3175.8	3541.5	6.8811	42.99	3175.1	3540.5	6.8748	42.46	3174.4	3539.6	6.8685
570	44.17	3194.9	3565.9	6.9102	43.62	3194.2	3565.0	6.9039	43.09	3193.5	3564.0	6.8977
580	44.80	3213.9	3590.2	6.9389	44.25	3213.2	3589.3	6.9327	43.71	3212.5	3588.4	6.9265
590	45.43	3232.9	3614.5	6.9672	44.87	3232.3	3613.6	6.9610	44.32	3231.6	3612.8	6.9548
600	46.05	3251.9	3638.7	6.9951	45.48	3251.3	3637.9	6.9889	44.93	3250.6	3637.1	6.9828
610	46.67	3270.9	3662.9	7.0227	46.10	3270.3	3662.1	7.0165	45.54	3269.7	3661.3	7.0104
620	47.29	3289.9	3687.1	7.0499	46.71	3289.3	3686.3	7.0438	46.15	3288.7	3685.5	7.0377
630	47.90	3308.9	3711.2	7.0768	47.32	3308.3	3710.5	7.0707	46.75	3307.7	3709.8	7.0647
640	48.51	3327.9	3735.4	7.1034	47.92	3327.3	3734.7	7.0973	47.35	3326.7	3733.9	7.0913
650	49.12	3346.9	3759.5	7.1296	48.53	3346.3	3758.8	7.1236	47.94	3345.8	3758.1	7.1176
660	49.73	3365.9	3783.6	7.1556	49.13	3365.4	3782.9	7.1496	48.54	3364.8	3782.3	7.1436
670	50.33	3384.9	3807.7	7.1813	49.72	3384.4	3807.1	7.1753	49.13	3383.9	3806.4	7.1694
680	50.94	3403.9	3831.8	7.2067	50.32	3403.5	3831.2	7.2008	49.72	3403.0	3830.5	7.1948
690	51.54	3423.0	3855.9	7.2319	50.91	3422.5	3855.3	7.2259	50.31	3422.1	3854.7	7.2200
700	52.13	3442.1	3880.0	7.2568	51.51	3441.6	3879.4	7.2509	50.89	3441.2	3878.8	7.2450
720	53.32	3480.4	3928.3	7.3059	52.68	3479.9	3927.7	7.3000	52.06	3479.5	3927.2	7.2941
740	54.51	3518.7	3976.6	7.3540	53.85	3518.3	3976.1	7.3482	53.21	3517.9	3975.5	7.3423
760	55.69	3557.2	4025.0	7.4013	55.02	3556.8	4024.5	7.3955	54.37	3556.4	4024.0	7.3897
780	56.86	3595.8	4073.4	7.4478	56.18	3595.4	4072.9	7.4420	55.51	3595.1	4072.5	7.4362
800	58.02	3634.6	4122.0	7.4935	57.33	3634.2	4121.5	7.4877	56.65	3633.9	4121.1	7.4819
850	60.92	3732.3	4244.0	7.6046	60.20	3731.9	4243.6	7.5988	59.49	3731.6	4243.2	7.5931
900	63.80	3830.9	4366.9	7.7116	63.04	3830.7	4366.5	7.7059	62.30	3830.4	4366.2	7.7002
950	66.66	3930.8	4490.7	7.8150	65.87	3930.5	4490.4	7.8093	65.10	3930.2	4490.1	7.8037
1000	69.51	4031.8	4615.7	7.9151	68.69	4031.6	4615.4	7.9095	67.88	4031.3	4615.1	7.9039
1100	75.19	4237.7	4869.2	8.1068	74.30	4237.4	4869.0	8.1012	73.43	4237.2	4868.7	8.0956
1200	80.84	4448.5	5127.6	8.2884	79.89	4448.3	5127.4	8.2828	78.96	4448.1	5127.1	8.2772
1300	86.48	4664.1	5390.6	8.4611	85.47	4663.9	5390.3	8.4555	84.47	4663.7	5390.1	8.4499

Definitions of symbols on page 1.

	8.7 (300.98)				8.8 (301.80)				8.9 (302.60)			p (t Sat.)
10^3 v	u	h	s	10^3 v	u	h	s	10^3 v	u	h	s	t
21.33	2561.5	2747.1	5.6967	21.04	2560.3	2745.4	5.6902	20.76	2559.0	2743.8	5.6837	Sat.
_21.17_	_2556.8_	_2741.0_	_5.6861_	_20.76_	_2551.4_	_2734.1_	_5.6704_	_20.35_	_2545.9_	_2727.0_	_5.6544_	300
21.92	2580.1	2770.8	5.7378	21.51	2575.3	2764.7	5.7235	21.12	2570.5	2758.4	5.7090	305
22.60	2601.3	2797.9	5.7846	22.20	2597.1	2792.5	5.7714	21.81	2592.7	2786.9	5.7580	310
23.24	2620.9	2823.1	5.8276	22.85	2617.1	2818.1	5.8152	22.46	2613.2	2813.0	5.8027	315
23.84	2639.2	2846.7	5.8675	23.45	2635.7	2842.1	5.8558	23.07	2632.1	2837.4	5.8440	320
24.41	2656.5	2868.9	5.9048	24.02	2653.2	2864.6	5.8936	23.64	2649.9	2860.3	5.8824	325
24.96	2672.8	2890.0	5.9399	24.57	2669.8	2886.0	5.9292	24.19	2666.7	2882.0	5.9185	330
25.48	2688.4	2910.1	5.9731	25.09	2685.6	2906.4	5.9629	24.71	2682.7	2902.6	5.9526	335
25.99	2703.3	2929.4	6.0048	25.60	2700.7	2925.9	5.9949	25.22	2698.0	2g22.4	5.9850	340
26.48	2717.7	2948.0	6.0350	26.08	2715.2	2944.7	6.0254	25.70	2712.7	2941.4	6.0159	345
26.95	2731.6	2966.0	6.0640	26.56	2729.2	2962.9	6.0547	26.17	2726.8	2959.8	6.0454	350
27.86	2758.1	3000.5	6.1188	27.46	2756.0	2997.6	6.1100	27.08	2753.8	2994.8	6.1012	360
28.72	2783.3	3033.2	6.1701	28.33	2781.3	3030.6	6.1616	27.94	2779.3	3028.0	6.1532	370
29.56	2807.4	3064.5	6.2185	29.15	2805.6	3062.1	6.2103	28.76	2803.8	3059.7	6.2022	380
30.36	2830.6	3094.7	6.2644	29.95	2828.9	3092.5	6.2565	29.56	2827.2	3090.3	6.2486	390
31.14	2853.1	3124.0	6.3082	30.73	2851.5	3122.0	6.3006	30.33	2850.0	3119.9	6.2929	400
31.90	2875.0	3152.6	6.3503	31.48	2873.6	3150.6	6.3428	31.08	2872.1	3148.7	6.3354	410
32.64	2896.5	3180.5	6.3909	32.22	2895.1	3178.7	6.3836	31.81	2893.7	3176.8	6.3763	420
33.37	2917.5	3207.9	6.4301	32.94	2916.2	3206.1	6.4229	32.52	2914.9	3204.4	6.4158	430
34.08	2938.2	3234.8	6.4681	33.65	2937.0	3233.1	6.4610	33.23	2935.8	3231.5	6.4541	440
34.78	2958.7	3261.3	6.5050	34.35	2957.5	3259.7	6.4981	33.92	2956.3	3258.2	6.4912	450
35.47	2978.9	3287.5	6.5410	35.03	2977.8	3286.0	6.5342	34.59	2976.6	3284.5	6.5274	460
36.15	2998.9	3313.4	6.5761	35.70	2997.8	3312.0	6.5694	35.26	2996.7	3310.6	6.5627	470
36.82	3018.7	3339.1	6.6104	36.37	3017.7	3337.7	6.6038	35.92	3016.7	3336.4	6.5972	480
37.49	3038.4	3364.5	6.6440	37.03	3037.4	3363.2	6.6374	36.57	3036.4	3361.9	6.6309	490
38.14	3057.9	3389.8	6.6769	37.68	3057.0	3388.5	6.6704	37.22	3056.1	3387.3	6.6639	500
38.79	3077.4	3414.9	6.7091	38.32	3076.5	3413.7	6.7027	37.85	3075.6	3412.5	6.6963	510
39.43	3096.8	3439.8	6.7408	38.95	3095.9	3438.7	6.7344	38.49	3095.1	3437.6	6.7281	520
40.07	3116.1	3464.7	6.7719	39.58	3115.2	3463.6	6.7656	39.11	3114.4	3462.5	6.7594	530
40.70	3135.3	3489.4	6.8025	40.21	3134.5	3488.4	6.7963	39.73	3133.7	3487.3	6.7901	540
41.33	3154.5	3514.1	6.8326	40.83	3153.7	3513.1	6.8264	40.34	3153.0	3512.0	6.8203	550
41.95	3173.7	3538.6	6.8623	41.45	3172.9	3537.6	6.8561	40.95	3172.2	3536.7	6.8500	560
42.57	3192.8	3563.1	6.8915	42.06	3192.1	3562.2	6.8854	41.56	3191.4	3561.2	6.8793	570
43.18	3211.9	3587.5	6.9203	42.66	3211.2	3586.6	6.9142	42.16	3210.5	3585.7	6.9082	580
43.79	3230.9	3611.9	6.9487	43.27	3230.3	3611.0	6.9427	42.76	3229.6	3610.2	6.9367	590
44.39	3250.0	3636.2	6.9767	43.87	3249.4	3635.4	6.9707	43.35	3248.7	3634.6	6.9648	600
45.00	3269.1	3660.5	7.0044	44.46	3268.4	3659.7	6.9984	43.94	3267.8	3658.9	6.9925	610
45.59	3288.1	3684.8	7.0317	45.06	3287.5	3684.0	7.0258	44.53	3286.9	3683.2	7.0199	620
46.19	3307.1	3709.0	7.0587	45.65	3306.6	3708.3	7.0528	45.11	3306.0	3707.5	7.0469	630
46.78	3326.2	3733.2	7.0853	46.23	3325.6	3732.5	7.0795	45.69	3325.1	3731.8	7.0736	640
47.37	3345.2	3757.4	7.1117	46.82	3344.7	3756.7	7.1058	46.27	3344.2	3756.0	7.1000	650
47.96	3364.3	3781.6	7.1377	47.40	3363.8	3780.9	7.1319	46.85	3363.3	3780.2	7.1261	660
48.55	3383.4	3805.7	7.1635	47.98	3382.9	3805.1	7.1577	47.42	3382.4	3804.4	7.1519	670
49.13	3402.5	3829.9	7.1890	48.56	3402.0	3829.3	7.1832	47.99	3401.5	3828.6	7.1775	680
49.71	3421.6	3854.1	7.2142	49.13	3421.1	3853.5	7.2084	48.56	3420.6	3852.8	7.2027	690
50.29	3440.7	3878.2	7.2392	49.70	3440.3	3877.6	7.2334	49.13	3439.8	3877.0	7.2277	700
51.44	3479.0	3926.6	7.2884	50.84	3478.6	3926.0	7.2826	50.26	3478.2	3925.5	7.2770	720
52.59	3517.5	3975.0	7.3366	51.98	3517.1	3974.5	7.3309	51.38	3516.6	3973.9	7.3253	740
53.73	3556.0	4023.5	7.3840	53.11	3555.6	4023.0	7.3783	52.50	3555.2	4022.5	7.3727	760
54.86	3594.7	4072.0	7.4305	54.23	3594.3	4071.5	7.4249	53.61	3594.0	4071.1	7.4193	780
55.99	3633.5	4120.7	7.4763	55.34	3633.2	4120.2	7.4707	54.71	3632.8	4119.8	7.4651	800
58.80	3731.3	4242.8	7.5875	58.12	3731.0	4242.4	7.5820	57.46	3730.7	4242.0	7.5765	850
61.58	3830.1	4365.8	7.6947	60.87	3829.8	4365.5	7.6891	60.18	3829.5	4365.1	7.6837	900
64.35	3930.0	4489.8	7.7981	63.61	3929.7	4489.5	7.7926	62.89	3929.4	4489.1	7.7872	950
67.10	4031.1	4614.8	7.8983	66.33	4030.8	4614.5	7.8929	65.58	4030.6	4614.3	7.8875	1000
72.58	4237.0	4868.5	8.0901	71.76	4236.8	4868.2	8.0847	70.95	4236.5	4868.0	8.0793	1100
78.05	4447.9	5126.9	8.2717	77.16	4447.6	5126.7	8.2663	76.29	4447.4	5126.4	8.2610	1200
83.50	4663.4	5389.9	8.4445	82.55	4663.2	5389.7	8.4390	81.63	4663.0	5389.4	8.4337	1300

Definitions of symbols on page 1.

Table 3. Vapor

p (t Sat.)	9.0 (303.40)				9.1 (304.20)				9.2 (304.99)			
t	10^3 v	u	h	s	10^3 v	u	h	s	10^3 v	u	h	s
Sat.	20.48	2557.8	2742.1	5.6772	20.22	2556.5	2740.5	5.6708	19.952	2555.2	2738.8	5.6644
300	*19.948*	*2540.2*	*2719.7*	*5.6382*								
305	20.723	2565.5	2752.1	5.6944	20.336	2560.5	2745.5	5.6796	19.955	2555.3	2738.9	5.6646
310	21.428	2588.3	2781.2	5.7446	21.049	2583.8	2775.4	5.7310	20.676	2579.2	2769.5	5.7173
315	22.080	2609.2	2807.9	5.7902	21.706	2605.1	2802.6	5.7775	21.339	2601.0	2797.3	5.7648
320	22.690	2628.5	2832.7	5.8322	22.320	2624.8	2827.9	5.8203	21.956	2621.1	2823.0	5.8084
325	23.27	2646.6	2856.0	5.8712	22.90	2643.2	2851.5	5.8600	22.54	2639.7	2847.1	5.8487
330	23.81	2663.6	2877.9	5.9078	23.45	2660.5	2873.8	5.8971	23.09	2657.3	2869.7	5.8864
335	24.34	2679.8	2898.8	5.9423	23.97	2676.9	2895.0	5.9321	23.61	2673.9	2891.2	5.9218
340	24.84	2695.3	2918.8	5.9751	24.47	2692.5	2915.2	5.9652	24.11	2689.8	2911.6	5.9553
345	25.33	2710.1	2938.1	6.0063	24.96	2707.5	2934.7	5.9968	24.60	2704.9	2931.3	5.9872
350	25.80	2724.4	2956.6	6.0361	25.43	2722.0	2953.4	6.0269	25.07	2719.5	2950.2	6.0177
360	26.70	2751.6	2991.9	6.0924	26.33	2749.4	2989.0	6.0836	25.96	2747.2	2986.1	6.0749
370	27.55	2777.4	3025.4	6.1448	27.18	2775.4	3022.7	6.1365	26.81	2773.4	3020.1	6.1281
380	28.37	2801.9	3057.3	6.1941	28.00	2800.1	3054.9	6.1861	27.63	2798.3	3052.4	6.1781
390	29.17	2825.6	3088.0	6.2408	28.78	2823.9	3085.8	6.2331	28.41	2822.2	3083.5	6.2254
400	29.93	2848.4	3117.8	6.2854	29.55	2846.8	3115.7	6.2778	29.17	2845.3	3113.6	6.2704
410	30.68	2870.6	3146.7	6.3280	30.29	2869.2	3144.8	6.3207	29.90	2867.7	3142.8	6.3134
420	31.40	2892.4	3175.0	6.3691	31.01	2891.0	3173.1	6.3619	30.62	2889.6	3171.3	6.3548
430	32.11	2913.6	3202.7	6.4087	31.71	2912.3	3200.9	6.4017	31.32	2911.0	3199.2	6.3948
440	32.81	2934.6	3229.9	6.4471	32.40	2933.3	3228.2	6.4403	32.01	2932.1	3226.6	6.4334
450	33.50	2955.2	3256.6	6.4844	33.08	2954.0	3255.1	6.4777	32.68	2952.8	3253.5	6.4710
460	34.17	2975.5	3283.1	6.5207	33.75	2974.4	3281.6	6.5141	33.34	2973.3	3280.1	6.5075
470	34.83	2995.7	3309.2	6.5561	34.41	2994.6	3307.8	6.5495	34.00	2993.6	3306.3	6.5430
480	35.49	3015.7	3335.0	6.5906	35.06	3014.6	3333.7	6.5842	34.64	3013.6	3332.3	6.5778
490	36.13	3035.5	3360.7	6.6245	35.70	3034.5	3359.4	6.6181	35.28	3033.5	3358.1	6.6117
500	36.77	3055.2	3386.1	6.6576	36.33	3054.2	3384.9	6.6512	35.90	3053.3	3383.6	6.6450
510	37.40	3074.7	3411.3	6.6900	36.96	3073.8	3410.2	6.6838	36.53	3072.9	3409.0	6.6776
520	38.03	3094.2	3436.5	6.7219	37.58	3093.4	3435.3	6.7157	37.14	3092.5	3434.2	6.7095
530	38.65	3113.6	3461.4	6.7532	38.19	3112.8	3460.3	6.7470	37.75	3112.0	3459.3	6.7409
540	39.26	3132.9	3486.3	6.7839	38.80	3132.2	3485.2	6.7778	38.35	3131.4	3484.2	6.7718
550	39.87	3152.2	3511.0	6.8142	39.40	3151.5	3510.0	6.8082	38.95	3150.7	3509.0	6.8022
560	40.47	3171.5	3535.7	6.8440	40.00	3170.7	3534.7	6.8380	39.54	3170.0	3533.8	6.8321
570	41.07	3190.7	3560.3	6.8733	40.60	3189.9	3559.4	6.8674	40.13	3189.2	3558.4	6.8615
580	41.67	3209.8	3584.8	6.9022	41.19	3209.1	3583.9	6.8963	40.72	3208.5	3583.0	6.8905
590	42.26	3229.0	3609.3	6.9308	41.77	3228.3	3608.4	6.9249	41.30	3227.6	3607.6	6.9191
600	42.85	3248.1	3633.7	6.9589	42.36	3247.5	3632.9	6.9531	41.87	3246.8	3632.1	6.9473
610	43.43	3267.2	3658.1	6.9867	42.93	3266.6	3657.3	6.9808	42.45	3266.0	3656.5	6.9751
620	44.01	3286.3	3682.4	7.0141	43.51	3285.7	3681.7	7.0083	43.02	3285.1	3680.9	7.0026
630	44.59	3305.4	3706.8	7.0411	44.08	3304.8	3706.0	7.0354	43.59	3304.3	3705.2	7.0297
640	45.17	3324.5	3731.0	7.0679	44.65	3324.0	3730.3	7.0621	44.15	3323.4	3729.6	7.0565
650	45.74	3343.6	3755.3	7.0943	45.22	3343.1	3754.6	7.0886	44.71	3342.5	3753.9	7.0830
660	46.31	3362.7	3779.5	7.1204	45.79	3362.2	3778.9	7.1147	45.27	3361.7	3778.2	7.1091
670	46.88	3381.9	3803.8	7.1462	46.35	3381.4	3803.1	7.1406	45.83	3380.8	3802.5	7.1350
680	47.44	3401.0	3828.0	7.1718	46.91	3400.5	3827.4	7.1662	46.38	3400.0	3826.7	7.1606
690	48.01	3420.2	3852.2	7.1971	47.47	3419.7	3851.6	7.1915	46.93	3419.2	3851.0	7.1859
700	48.57	3439.3	3876.5	7.2221	48.02	3438.9	3875.9	7.2165	47.49	3438.4	3875.3	7.2110
720	49.69	3477.7	3924.9	7.2714	49.13	3477.3	3924.4	7.2659	48.58	3476.9	3923.8	7.2604
740	50.80	3516.2	3973.4	7.3197	50.23	3515.8	3972.9	7.3142	49.67	3515.4	3972.4	7.3088
760	51.90	3554.8	4022.0	7.3672	51.32	3554.5	4021.5	7.3617	50.75	3554.1	4021.0	7.3563
780	53.00	3593.6	4070.6	7.4138	52.41	3593.2	4070.1	7.4084	51.83	3592.8	4069.6	7.4030
800	54.09	3632.5	4119.3	7.4596	53.49	3632.1	4118.9	7.4542	52.90	3631.8	4118.4	7.4488
850	56.81	3730.3	4241.6	7.5710	56.18	3730.0	4241.2	7.5657	55.56	3729.7	4240.8	7.5603
900	59.50	3829.2	4364.8	7.6783	58.84	3828.9	4364.4	7.6729	58.20	3828.6	4364.1	7.6676
950	62.18	3929.2	4488.8	7.7818	61.49	3928.9	4488.5	7.7765	60.82	3928.6	4488.2	7.7713
1000	64.85	4030.3	4614.0	7.8821	64.13	4030.1	4613.7	7.8768	63.43	4029.8	4613.4	7.8716
1100	70.16	4236.3	4867.7	8.0740	69.38	4236.1	4867.4	8.0687	68.63	4235.8	4867.2	8.0635
1200	75.44	4447.2	5126.2	8.2556	74.61	4447.0	5126.0	8.2504	73.80	4446.7	5125.7	8.2452
1300	80.72	4662.7	5389.2	8.4284	79.83	4662.5	5389.0	8.4231	78.97	4662.3	5388.8	8.4180

Definitions of symbols on page 1.

	9.3 (305.77)				**9.4** (306.54)				**9.5** (307.31)			
10^3 v	u	h	s	10^3 v	u	h	s	10^3 v	u	h	s	t
19.695	2553.9	2737.1	5.6580	19.442	2552.6	2735.4	5.6517	19.194	2551.3	2733.6	5.6454	Sat.
												300
19.578	*2549.9*	*2732.0*	*5.6493*	*19.205*	*2544.5*	*2725.0*	*5.6338*	*18.837*	*2538.8*	*2717.8*	*5.6180*	305
20.308	2574.6	2763.4	5.7034	19.946	2569.8	2757.3	5.6894	19.589	2564.9	2751.0	5.6752	310
20.977	2596.8	2791.9	5.7520	20.622	2592.5	2786.4	5.7391	20.271	2588.2	2780.8	5.7261	315
21.599	2617.3	2818.1	5.7964	21.247	2613.4	2813.1	5.7844	20.902	2609.5	2808.0	5.7722	320
22.18	2636.3	2842.5	5.8374	21.83	2632.7	2838.0	5.8261	21.49	2629.1	2833.3	5.8147	325
22.73	2654.1	2865.5	5.8756	22.39	2650.8	2861.3	5.8649	22.05	2647.5	2857.0	5.8541	330
23.26	2670.9	2887.3	5.9115	22.91	2667.9	2883.3	5.9013	22.57	2664.9	2879.3	5.8910	335
23.76	2687.0	2908.0	5.9455	23.42	2684.2	2904.3	5.9356	23.08	2681.3	2900.6	5.9257	340
24.25	2702.3	2927.8	5.9777	23.90	2699.7	2924.3	5.9682	23.56	2697.0	2920.8	5.9587	345
24.71	2717.1	2946.9	6.0085	24.37	2714.6	2943.6	5.9993	24.03	2712.1	2940.3	5.9901	350
25.61	2745.0	2983.2	6.0662	25.26	2742.8	2980.2	6.0575	24.92	2740.5	2977.3	6.0489	360
26.45	2771.4	3017.4	6.1198	26.10	2769.3	3014.7	6.1116	25.76	2767.3	3012.0	6.1034	370
27.26	2796.4	3050.0	6.1701	26.91	2794.6	3047.5	6.1622	26.56	2792.7	3045.1	6.1543	380
28.04	2820.5	3081.3	6.2177	27.68	2818.8	3079.0	6.2100	27.33	2817.0	3076.7	6.2025	390
28.80	2843.7	3111.5	6.2629	28.43	2842.1	3109.4	6.2555	28.08	2840.5	3107.3	6.2482	400
29.53	2866.2	3140.8	6.3062	29.16	2864.7	3138.9	6.2990	28.80	2863.2	3136.9	6.2918	410
30.24	2888.2	3169.4	6.3478	29.87	2886.8	3167.6	6.3407	29.51	2885.4	3165.7	6.3338	420
30.94	2909.7	3197.4	6.3879	30.56	2908.4	3195.7	6.3810	30.19	2907.1	3193.9	6.3742	430
31.62	2930.9	3224.9	6.4267	31.24	2929.6	3223.3	6.4199	30.87	2928.4	3221.6	6.4132	440
32.29	2951.7	3251.9	6.4643	31.90	2950.5	3250.4	6.4577	31.53	2949.3	3248.8	6.4511	450
32.95	2972.2	3278.6	6.5009	32.56	2971.1	3277.1	6.4944	32.17	2970.0	3275.6	6.4879	460
33.59	2992.5	3304.9	6.5366	33.20	2991.4	3303.5	6.5302	32.81	2990.4	3302.1	6.5238	470
34.23	3012.6	3331.0	6.5714	33.83	3011.6	3329.6	6.5651	33.44	3010.6	3328.3	6.5588	480
34.86	3032.6	3356.8	6.6054	34.46	3031.6	3355.5	6.5992	34.06	3030.6	3354.2	6.5930	490
35.49	3052.4	3382.4	6.6388	35.08	3051.4	3381.1	6.6326	34.67	3050.5	3379.9	6.6265	500
36.10	3072.1	3407.8	6.6714	35.69	3071.2	3406.6	6.6653	35.28	3070.3	3405.4	6.6593	510
36.71	3091.6	3433.0	6.7035	36.29	3090.8	3431.9	6.6974	35.88	3089.9	3430.8	6.6914	520
37.31	3111.1	3458.2	6.7349	36.89	3110.3	3457.1	6.7289	36.47	3109.5	3456.0	6.7230	530
37.91	3130.6	3483.1	6.7658	37.48	3129.8	3482.1	6.7599	37.06	3129.0	3481.0	6.7540	540
38.50	3149.9	3508.0	6.7962	38.07	3149.2	3507.0	6.7904	37.64	3148.4	3506.0	6.7845	550
39.09	3169.3	3532.8	6.8262	38.65	3168.5	3531.8	6.8203	38.22	3167.8	3530.9	6.8146	560
39.68	3188.5	3557.5	6.8556	39.23	3187.8	3556.6	6.8499	38.79	3187.1	3555.6	6.8441	570
40.25	3207.8	3582.1	6.8847	39.80	3207.1	3581.2	6.8789	39.36	3206.4	3580.3	6.8732	580
40.83	3227.0	3606.7	6.9133	40.37	3226.3	3605.8	6.9076	39.93	3225.7	3605.0	6.9019	590
41.40	3246.2	3631.2	6.9415	40.94	3245.5	3630.4	6.9359	40.49	3244.9	3629.5	6.9303	600
41.97	3265.4	3655.7	6.9694	41.50	3264.7	3654.9	6.9638	41.05	3264.1	3654.1	6.9582	610
42.54	3284.5	3680.1	6.9969	42.06	3283.9	3679.3	6.9913	41.60	3283.3	3678.5	6.9857	620
43.10	3303.7	3704.5	7.0241	42.62	3303.1	3703.7	7.0185	42.15	3302.5	3703.0	7.0130	630
43.66	3322.8	3728.9	7.0509	43.17	3322.3	3728.1	7.0453	42.70	3321.7	3727.4	7.0398	640
44.21	3342.0	3753.2	7.0774	43.73	3341.5	3752.5	7.0719	43.25	3340.9	3751.8	7.0664	650
44.77	3361.2	3777.5	7.1036	44.27	3360.6	3776.8	7.0981	43.79	3360.1	3776.1	7.0926	660
45.32	3380.3	3801.8	7.1295	44.82	3379.8	3801.1	7.1240	44.33	3379.3	3800.5	7.1186	670
45.87	3399.5	3826.1	7.1551	45.36	3399.0	3825.5	7.1497	44.87	3398.5	3824.8	7.1443	680
46.42	3418.7	3850.4	7.1805	45.91	3418.2	3849.8	7.1750	45.41	3417.8	3849.1	7.1696	690
46.96	3437.9	3874.7	7.2055	46.45	3437.5	3874.1	7.2001	45.94	3437.0	3873.5	7.1948	700
48.04	3476.4	3923.2	7.2549	47.52	3476.0	3922.7	7.2496	47.01	3475.5	3922.1	7.2442	720
49.12	3515.0	3971.8	7.3034	48.59	3514.6	3971.3	7.2980	48.06	3514.2	3970.8	7.2927	740
50.19	3553.7	4020.5	7.3509	49.65	3553.3	4020.0	7.3456	49.11	3552.9	4019.5	7.3403	760
51.26	3592.5	4069.2	7.3976	50.70	3592.1	4068.7	7.3923	50.16	3591.7	4068.2	7.3871	780
52.32	3631.4	4118.0	7.4435	51.75	3631.0	4117.5	7.4383	51.20	3630.7	4117.1	7.4330	800
54.95	3729.4	4240.5	7.5551	54.36	3729.1	4240.1	7.5499	53.78	3728.8	4239.7	7.5447	850
57.56	3828.3	4363.7	7.6624	56.95	3828.1	4363.3	7.6572	56.34	3827.8	4363.0	7.6521	900
60.16	3928.4	4487.9	7.7661	59.51	3928.1	4487.5	7.7609	58.88	3927.8	4487.2	7.7558	950
62.74	4029.6	4613.1	7.8664	62.07	4029.3	4612.8	7.8613	61.41	4029.1	4612.5	7.8562	1000
67.89	4235.6	4866.9	8.0583	67.16	4235.4	4866.7	8.0532	66.45	4235.1	4866.4	8.0482	1100
73.01	4446.5	5125.5	8.2401	72.23	4446.3	5125.3	8.2350	71.47	4446.1	5125.0	8.2299	1200
78.12	4662.1	5388.6	8.4128	77.29	4661.8	5388.3	8.4077	76.48	4661.6	5388.1	8.4027	1300

Definitions of symbols on page 1.

Table 3.   Vapor

p (t Sat.)	9.6 (308.07)				9.7 (308.83)				9.8 (309.58)			
t	10^3 v	u	h	s	10^3 v	u	h	s	10^3 v	u	h	s
Sat.	18.951	2549.9	2731.9	5.6391	18.713	2548.6	2730.1	5.6328	18.480	2547.2	2728.3	5.6265
305	*18.472*	*2533.0*	*2710.4*	*5.6020*	*18.110*	*2527.1*	*2702.7*	*5.5856*				
310	19.236	2559.9	2744.5	5.6608	18.887	2554.7	2737.9	5.6462	18.543	2549.5	2731.2	5.6314
315	19.926	2583.8	2775.0	5.7129	19.586	2579.2	2769.2	5.6997	19.251	2574.6	2763.3	5.6862
320	20.562	2605.5	2802.9	5.7600	20.227	2601.4	2797.6	5.7478	19.897	2597.3	2792.3	5.7354
325	21.15	2625.5	2828.6	5.8032	20.82	2621.8	2823.8	5.7917	20.50	2618.1	2819.0	5.7802
330	21.71	2644.2	2852.6	5.8432	21.38	2640.8	2848.2	5.8324	21.06	2637.4	2843.8	5.8215
335	22.24	2661.8	2875.3	5.8807	21.91	2658.7	2871.2	5.8703	21.59	2655.5	2867.1	5.8600
340	22.74	2678.4	2896.8	5.9159	22.42	2675.5	2893.0	5.9060	22.10	2672.6	2889.2	5.8961
345	23.23	2694.3	2917.3	5.9492	22.90	2691.6	2913.8	5.9397	22.58	2688.9	2910.2	5.9302
350	23.69	2709.5	2937.0	5.9809	23.37	2707.0	2933.6	5.9718	23.05	2704.4	2930.3	5.9626
360	24.58	2738.3	2974.3	6.0403	24.25	2736.0	2971.3	6.0317	23.93	2733.7	2968.2	6.0231
370	25.42	2765.3	3009.3	6.0952	25.09	2763.2	3006.6	6.0870	24.77	2761.1	3003.8	6.0789
380	26.22	2790.8	3042.6	6.1465	25.89	2789.0	3040.1	6.1387	25.56	2787.1	3037.6	6.1309
390	26.99	2815.3	3074.4	6.1949	26.65	2813.6	3072.1	6.1874	26.32	2811.8	3069.8	6.1799
400	27.73	2838.9	3105.1	6.2408	27.39	2837.3	3103.0	6.2336	27.06	2835.7	3100.8	6.2263
410	28.45	2861.8	3134.9	6.2847	28.11	2860.3	3132.9	6.2777	27.77	2858.7	3130.9	6.2706
420	29.15	2884.0	3163.8	6.3268	28.80	2882.6	3162.0	6.3199	28.46	2881.2	3160.1	6.3131
430	29.83	2905.8	3192.2	6.3674	29.48	2904.4	3190.4	6.3607	29.13	2903.1	3188.6	6.3540
440	30.50	2927.1	3219.9	6.4066	30.14	2925.9	3218.3	6.4000	29.79	2924.6	3216.6	6.3935
450	31.16	2948.1	3247.2	6.4446	30.79	2946.9	3245.6	6.4381	30.44	2945.7	3244.1	6.4317
460	31.80	2968.8	3274.1	6.4815	31.43	2967.7	3272.6	6.4752	31.07	2966.6	3271.1	6.4689
470	32.43	2989.3	3300.7	6.5175	32.06	2988.2	3299.2	6.5112	31.70	2987.2	3297.8	6.5050
480	33.06	3009.6	3326.9	6.5526	32.68	3008.5	3325.5	6.5464	32.31	3007.5	3324.2	6.5403
490	33.67	3029.6	3352.9	6.5869	33.29	3028.7	3351.6	6.5808	32.92	3027.7	3350.3	6.5747
500	34.28	3049.6	3378.7	6.6204	33.90	3048.6	3377.4	6.6144	33.52	3047.7	3376.2	6.6084
510	34.88	3069.4	3404.2	6.6533	34.49	3068.5	3403.0	6.6473	34.11	3067.6	3401.8	6.6414
520	35.48	3089.1	3429.6	6.6855	35.08	3088.2	3428.5	6.6796	34.69	3087.3	3427.3	6.6738
530	36.06	3108.7	3454.9	6.7171	35.66	3107.8	3453.8	6.7113	35.27	3107.0	3452.7	6.7055
540	36.65	3128.2	3480.0	6.7482	36.24	3127.4	3478.9	6.7424	35.85	3126.6	3477.9	6.7367
550	37.22	3147.6	3505.0	6.7788	36.82	3146.9	3504.0	6.7730	36.41	3146.1	3503.0	6.7674
560	37.80	3167.0	3529.9	6.8088	37.38	3166.3	3528.9	6.8032	36.98	3165.6	3527.9	6.7975
570	38.37	3186.4	3554.7	6.8384	37.95	3185.7	3553.8	6.8328	37.54	3185.0	3552.8	6.8272
580	38.93	3205.7	3579.4	6.8676	38.51	3205.0	3578.5	6.8620	38.09	3204.3	3577.6	6.8565
590	39.49	3225.0	3604.1	6.8963	39.06	3224.3	3603.2	6.8908	38.64	3223.7	3602.4	6.8853
600	40.05	3244.3	3628.7	6.9247	39.61	3243.6	3627.9	6.9192	39.19	3243.0	3627.0	6.9137
610	40.60	3263.5	3653.3	6.9526	40.16	3262.9	3652.4	6.9471	39.73	3262.3	3651.6	6.9417
620	41.15	3282.7	3677.8	6.9802	40.71	3282.1	3677.0	6.9748	40.27	3281.5	3676.2	6.9694
630	41.70	3301.9	3702.2	7.0075	41.25	3301.4	3701.5	7.0020	40.81	3300.8	3700.7	6.9967
640	42.24	3321.2	3726.7	7.0344	41.79	3320.6	3725.9	7.0290	41.34	3320.0	3725.2	7.0236
650	42.78	3340.4	3751.1	7.0610	42.32	3339.8	3750.4	7.0556	41.88	3339.3	3749.7	7.0503
660	43.32	3359.6	3775.5	7.0872	42.86	3359.1	3774.8	7.0819	42.40	3358.5	3774.1	7.0766
670	43.86	3378.8	3799.8	7.1132	43.39	3378.3	3799.2	7.1079	42.93	3377.8	3798.5	7.1026
680	44.39	3398.0	3824.2	7.1389	43.92	3397.5	3823.5	7.1336	43.45	3397.0	3822.9	7.1283
690	44.92	3417.3	3848.5	7.1643	44.44	3416.8	3847.9	7.1590	43.98	3416.3	3847.3	7.1538
700	45.45	3436.5	3872.9	7.1894	44.97	3436.1	3872.3	7.1842	44.50	3435.6	3871.7	7.1790
720	46.50	3475.1	3921.5	7.2390	46.01	3474.7	3921.0	7.2337	45.53	3474.2	3920.4	7.2286
740	47.55	3513.7	3970.2	7.2875	47.05	3513.3	3969.7	7.2823	46.56	3512.9	3969.2	7.2772
760	48.59	3552.5	4019.0	7.3351	48.08	3552.1	4018.5	7.3300	47.58	3551.7	4018.0	7.3249
780	49.63	3591.3	4067.8	7.3819	49.10	3591.0	4067.3	7.3768	48.59	3590.6	4066.8	7.3717
800	50.66	3630.3	4116.6	7.4279	50.12	3630.0	4116.2	7.4228	49.60	3629.6	4115.7	7.4177
850	53.21	3728.4	4239.3	7.5396	52.66	3728.1	4238.9	7.5345	52.11	3727.8	4238.5	7.5295
900	55.75	3827.5	4362.6	7.6470	55.17	3827.2	4362.3	7.6420	54.60	3826.9	4361.9	7.6370
950	58.26	3927.6	4486.9	7.7508	57.66	3927.3	4486.6	7.7458	57.06	3927.0	4486.3	7.7408
1000	60.77	4028.8	4612.2	7.8512	60.14	4028.6	4611.9	7.8462	59.52	4028.3	4611.6	7.8413
1100	65.76	4234.9	4866.2	8.0432	65.08	4234.7	4865.9	8.0382	64.41	4234.4	4865.7	8.0333
1200	70.72	4445.8	5124.8	8.2249	69.99	4445.6	5124.5	8.2200	69.28	4445.4	5124.3	8.2151
1300	75.68	4661.4	5387.9	8.3977	74.90	4661.1	5387.7	8.3928	74.14	4660.9	5387.4	8.3879

Definitions of symbols on page 1.

10^3 v	u	h	s	10^3 v	u	h	s	10^3 v	u	h	s	t
18.251	2545.8	2726.5	5.6203	18.026	2544.4	2724.7	5.6141	17.588	2541.6	2721.0	5.6017	Sat.
												305
18.202	*2544.1*	*2724.3*	*5.6164*	*17.864*	*2538.5*	*2717.1*	*5.6011*	*17.197*	*2526.9*	*2702.3*	*5.5697*	310
18.920	2569.9	2757.2	5.6727	18.593	2565.1	2751.0	5.6589	17.950	2555.1	2738.2	5.6309	315
19.573	2593.1	2786.9	5.7229	19.253	2588.8	2781.3	5.7103	18.626	2580.0	2770.0	5.6848	320
20.18	2614.3	2814.0	5.7685	19.861	2610.4	2809.1	5.7568	19.244	2602.5	2798.8	5.7332	325
20.74	2633.9	2839.3	5.8106	20.429	2630.4	2834.7	5.7996	19.819	2623.2	2825.4	5.7774	330
21.27	2652.3	2862.9	5.8496	20.964	2649.1	2858.7	5.8392	20.358	2642.5	2850.1	5.8183	335
21.78	2669.6	2885.3	5.8862	21.472	2666.6	2881.4	5.8763	20.869	2660.6	2873.4	5.8564	340
22.27	2686.1	2906.5	5.9207	21.957	2683.3	2902.9	5.9112	21.355	2677.6	2895.5	5.8922	345
22.73	2701.8	2926.9	5.9535	22.42	2699.2	2923.4	5.9443	21.82	2693.9	2916.5	5.9261	350
23.62	2731.4	2965.2	6.0145	23.31	2729.1	2962.1	6.0060	22.70	2724.4	2955.9	5.9889	360
24.45	2759.0	3001.1	6.0707	24.14	2756.9	2998.3	6.0626	23.53	2752.7	2992.7	6.0465	370
25.24	2785.2	3035.0	6.1231	24.93	2783.2	3032.5	6.1154	24.31	2779.4	3027.4	6.1000	380
26.00	2810.1	3067.5	6.1724	25.68	2808.3	3065.1	6.1650	25.06	2804.8	3060.4	6.1503	390
26.73	2834.0	3098.7	6.2191	26.41	2832.4	3096.5	6.2120	25.78	2829.1	3092.1	6.1977	400
27.44	2857.2	3128.9	6.2636	27.11	2855.7	3126.8	6.2567	26.48	2852.7	3122.8	6.2429	410
28.12	2879.8	3158.2	6.3063	27.80	2878.4	3156.3	6.2995	27.16	2875.5	3152.5	6.2861	420
28.79	2901.8	3186.9	6.3473	28.46	2900.4	3185.1	6.3407	27.82	2897.8	3181.5	6.3276	430
29.45	2923.4	3214.9	6.3870	29.11	2922.1	3213.2	6.3805	28.46	2919.6	3209.9	6.3677	440
30.09	2944.6	3242.5	6.4253	29.75	2943.4	3240.9	6.4190	29.09	2941.0	3237.7	6.4064	450
30.72	2965.4	3269.6	6.4626	30.38	2964.3	3268.1	6.4564	29.71	2962.0	3265.1	6.4440	460
31.34	2986.1	3296.4	6.4989	30.99	2985.0	3294.9	6.4927	30.32	2982.8	3292.0	6.4806	470
31.95	3006.5	3322.8	6.5342	31.60	3005.4	3321.4	6.5282	30.91	3003.4	3318.7	6.5162	480
32.55	3026.7	3349.0	6.5687	32.20	3025.7	3347.7	6.5628	31.50	3023.7	3345.1	6.5510	490
33.15	3046.8	3374.9	6.6025	32.79	3045.8	3373.7	6.5966	32.08	3043.9	3371.2	6.5850	500
33.74	3066.7	3400.7	6.6356	33.37	3065.8	3399.5	6.6297	32.66	3064.0	3397.1	6.6182	510
34.32	3086.5	3426.2	6.6680	33.94	3085.6	3425.1	6.6622	33.22	3083.9	3422.8	6.6508	520
34.89	3106.2	3451.6	6.6998	34.51	3105.3	3450.5	6.6941	33.78	3103.7	3448.3	6.6828	530
35.46	3125.8	3476.8	6.7310	35.08	3125.0	3475.8	6.7254	34.34	3123.4	3473.7	6.7142	540
36.02	3145.3	3502.0	6.7617	35.64	3144.6	3500.9	6.7561	34.89	3143.0	3498.9	6.7451	550
36.58	3164.8	3527.0	6.7919	36.19	3164.1	3526.0	6.7864	35.44	3162.6	3524.0	6.7754	560
37.13	3184.3	3551.9	6.8217	36.74	3183.5	3550.9	6.8162	35.98	3182.1	3549.1	6.8053	570
37.68	3203.6	3576.7	6.8509	37.29	3203.0	3575.8	6.8455	36.51	3201.6	3574.0	6.8347	580
38.23	3223.0	3601.5	6.8798	37.83	3222.3	3600.6	6.8744	37.05	3221.0	3598.9	6.8637	590
38.77	3242.3	3626.2	6.9083	38.37	3241.7	3625.3	6.9029	37.57	3240.4	3623.7	6.8922	600
39.31	3261.6	3650.8	6.9363	38.90	3261.0	3650.0	6.9310	38.10	3259.8	3648.4	6.9204	610
39.85	3280.9	3675.4	6.9640	39.43	3280.3	3674.6	6.9587	38.62	3279.1	3673.1	6.9482	620
40.38	3300.2	3700.0	6.9913	39.96	3299.6	3699.2	6.9860	39.14	3298.5	3697.7	6.9756	630
40.91	3319.5	3724.5	7.0183	40.48	3318.9	3723.7	7.0131	39.66	3317.8	3722.3	7.0027	640
41.44	3338.7	3749.0	7.0450	41.01	3338.2	3748.2	7.0398	40.17	3337.1	3746.8	7.0294	650
41.96	3358.0	3773.4	7.0713	41.52	3357.5	3772.7	7.0661	40.68	3356.4	3771.4	7.0559	660
42.48	3377.3	3797.8	7.0974	42.04	3376.8	3797.2	7.0922	41.19	3375.7	3795.9	7.0820	670
43.00	3396.5	3822.3	7.1231	42.56	3396.1	3821.6	7.1180	41.69	3395.1	3820.3	7.1078	680
43.52	3415.8	3846.7	7.1486	43.07	3415.4	3846.0	7.1435	42.20	3414.4	3844.8	7.1333	690
44.03	3435.1	3871.1	7.1738	43.58	3434.7	3870.5	7.1687	42.70	3433.7	3869.3	7.1586	700
45.06	3473.8	3919.8	7.2234	44.60	3473.3	3919.3	7.2183	43.70	3472.4	3918.2	7.2083	720
46.08	3512.5	3968.6	7.2721	45.60	3512.1	3968.1	7.2670	44.69	3511.2	3967.0	7.2571	740
47.09	3551.3	4017.5	7.3198	46.61	3550.9	4017.0	7.3148	45.67	3550.1	4016.0	7.3049	760
48.09	3590.2	4066.3	7.3666	47.60	3589.8	4065.9	7.3617	46.65	3589.1	4064.9	7.3518	780
49.09	3629.3	4115.3	7.4127	48.59	3628.9	4114.8	7.4077	47.62	3628.2	4113.9	7.3979	800
51.58	3727.5	4238.1	7.5245	51.05	3727.2	4237.7	7.5196	50.04	3726.5	4236.9	7.5099	850
54.04	3826.6	4361.6	7.6321	53.49	3826.3	4361.2	7.6272	52.43	3825.8	4360.5	7.6176	900
56.48	3926.8	4486.0	7.7359	55.91	3926.5	4485.6	7.7311	54.81	3926.0	4485.0	7.7215	950
58.92	4028.1	4611.3	7.8364	58.32	4027.8	4611.0	7.8315	57.17	4027.3	4610.5	7.8220	1000
63.76	4234.2	4865.4	8.0285	63.12	4234.0	4865.1	8.0237	61.87	4233.5	4864.6	8.0142	1100
68.58	4445.2	5124.1	8.2103	67.89	4444.9	5123.8	8.2055	66.56	4444.5	5123.4	8.1961	1200
73.39	4660.7	5387.2	8.3831	72.65	4660.5	5387.0	8.3783	71.23	4660.0	5386.5	8.3689	1300

Definitions of symbols on page 1.

Table 3. Vapor

p (t Sat.)	10.4 (313.96)				10.6 (315.38)				10.8 (316.77)			
t	10^3 v	u	h	s	10^3 v	u	h	s	10^3 v	u	h	s
Sat.	17.167	2538.7	2717.3	5.5894	16.760	2535.8	2713.5	5.5771	16.367	2532.8	2709.6	5.5649
315	17.321	2544.6	2724.8	5.6021	*16.703*	*2533.6*	*2710.6*	*5.5723*	*16.094*	*2521.9*	*2695.7*	*5.5413*
320	18.015	2570.9	2758.2	5.6587	17.418	2561.3	2745.9	5.6321	16.833	2551.3	2733.1	5.6047
325	18.645	2594.4	2788.3	5.7092	18.061	2585.9	2777.4	5.6849	17.492	2577.1	2766.1	5.6600
330	19.227	2615.9	2815.8	5.7551	18.653	2608.2	2806.0	5.7325	18.094	2600.4	2795.8	5.7095
335	19.771	2635.7	2841.4	5.7973	19.203	2628.8	2832.4	5.7760	18.651	2621.7	2823.1	5.7546
340	20.285	2654.3	2865.3	5.8364	19.721	2647.9	2857.0	5.8164	19.173	2641.4	2848.5	5.7962
345	20.773	2671.8	2887.9	5.8732	20.211	2665.9	2880.2	5.8540	19.667	2659.9	2872.3	5.8349
350	21.24	2688.5	2909.4	5.9078	20.68	2683.0	2902.2	5.8895	20.14	2677.4	2894.8	5.8712
360	22.12	2719.6	2949.6	5.9719	21.56	2714.7	2943.2	5.9549	21.02	2709.8	2936.8	5.9379
370	22.94	2748.4	2987.0	6.0305	22.38	2744.0	2981.3	6.0145	21.84	2739.6	2975.5	5.9986
380	23.72	2775.5	3022.2	6.0848	23.16	2771.5	3017.0	6.0696	22.61	2767.5	3011.7	6.0545
390	24.47	2801.2	3055.7	6.1356	23.90	2797.6	3050.9	6.1211	23.34	2793.9	3046.0	6.1067
400	25.18	2825.8	3087.7	6.1836	24.61	2822.5	3083.3	6.1696	24.05	2819.1	3078.8	6.1558
410	25.87	2849.6	3118.7	6.2292	25.29	2846.5	3114.6	6.2157	24.73	2843.4	3110.4	6.2023
420	26.54	2872.6	3148.7	6.2728	25.95	2869.7	3144.8	6.2597	25.38	2866.8	3140.9	6.2467
430	27.20	2895.1	3177.9	6.3147	26.60	2892.3	3174.3	6.3019	26.02	2889.6	3170.6	6.2892
440	27.83	2917.0	3206.5	6.3550	27.23	2914.4	3203.0	6.3425	26.64	2911.9	3199.6	6.3302
450	28.45	2938.6	3234.5	6.3940	27.84	2936.1	3231.2	6.3818	27.25	2933.7	3228.0	6.3697
460	29.06	2959.7	3262.0	6.4319	28.44	2957.4	3259.0	6.4199	27.85	2955.1	3255.9	6.4080
470	29.66	2980.7	3289.1	6.4686	29.04	2978.5	3286.2	6.4568	28.43	2976.3	3283.3	6.4452
480	30.25	3001.3	3315.9	6.5044	29.62	2999.2	3313.2	6.4928	29.00	2997.1	3310.4	6.4813
490	30.83	3021.8	3342.4	6.5394	30.19	3019.8	3339.8	6.5279	29.57	3017.8	3337.1	6.5166
500	31.41	3042.0	3368.6	6.5735	30.75	3040.1	3366.1	6.5622	30.13	3038.2	3363.6	6.5511
510	31.97	3062.1	3394.6	6.6069	31.31	3060.3	3392.2	6.5958	30.68	3058.5	3389.8	6.5847
520	32.53	3082.1	3420.4	6.6397	31.86	3080.4	3418.1	6.6286	31.22	3078.6	3415.8	6.6177
530	33.08	3102.0	3446.1	6.6718	32.41	3100.3	3443.8	6.6608	31.76	3098.6	3441.6	6.6501
540	33.63	3121.8	3471.5	6.7033	32.95	3120.2	3469.4	6.6925	32.29	3118.5	3467.3	6.6818
550	34.17	3141.5	3496.9	6.7342	33.48	3139.9	3494.8	6.7235	32.82	3138.4	3492.8	6.7130
560	34.71	3161.1	3522.1	6.7647	34.01	3159.6	3520.1	6.7541	33.34	3158.1	3518.1	6.7436
570	35.24	3180.7	3547.2	6.7946	34.53	3179.2	3545.3	6.7841	33.85	3177.8	3543.4	6.7737
580	35.77	3200.2	3572.2	6.8241	35.05	3198.8	3570.4	6.8137	34.36	3197.4	3568.5	6.8034
590	36.29	3219.7	3597.1	6.8532	35.57	3218.3	3595.4	6.8428	34.87	3217.0	3593.6	6.8326
600	36.81	3239.1	3622.0	6.8818	36.08	3237.8	3620.3	6.8715	35.38	3236.5	3618.6	6.8614
610	37.33	3258.5	3646.8	6.9100	36.59	3257.3	3645.1	6.8998	35.88	3256.0	3643.5	6.8897
620	37.84	3277.9	3671.5	6.9379	37.10	3276.7	3669.9	6.9277	36.37	3275.5	3668.3	6.9177
630	38.35	3297.3	3696.2	6.9653	37.60	3296.1	3694.6	6.9553	36.87	3294.9	3693.1	6.9453
640	38.86	3316.6	3720.8	6.9925	38.10	3315.5	3719.3	6.9824	37.36	3314.4	3717.9	6.9726
650	39.37	3336.0	3745.4	7.0193	38.59	3334.9	3744.0	7.0093	37.85	3333.8	3742.6	6.9995
660	39.87	3355.4	3770.0	7.0458	39.09	3354.3	3768.6	7.0358	38.33	3353.2	3767.2	7.0261
670	40.37	3374.7	3794.5	7.0719	39.58	3373.7	3793.2	7.0620	38.82	3372.6	3791.9	7.0523
680	40.86	3394.1	3819.1	7.0978	40.07	3393.1	3817.8	7.0879	39.30	3392.1	3816.5	7.0783
690	41.36	3413.4	3843.6	7.1234	40.55	3412.5	3842.3	7.1136	39.78	3411.5	3841.1	7.1039
700	41.85	3432.8	3868.1	7.1487	41.04	3431.9	3866.9	7.1389	40.25	3430.9	3865.6	7.1293
720	42.83	3471.6	3917.0	7.1985	42.00	3470.7	3915.9	7.1888	41.20	3469.8	3914.8	7.1793
740	43.81	3510.4	3966.0	7.2473	42.96	3509.6	3964.9	7.2377	42.14	3508.7	3963.9	7.2282
760	44.77	3549.3	4015.0	7.2951	43.91	3548.5	4014.0	7.2856	43.07	3547.7	4013.0	7.2762
780	45.73	3588.3	4064.0	7.3421	44.85	3587.6	4063.0	7.3326	44.00	3586.8	4062.1	7.3233
800	46.69	3627.5	4113.0	7.3883	45.79	3626.8	4112.2	7.3788	44.93	3626.1	4111.3	7.3696
850	49.06	3725.9	4236.1	7.5004	48.12	3725.3	4235.3	7.4910	47.22	3724.6	4234.6	7.4819
900	51.41	3825.2	4359.8	7.6081	50.43	3824.6	4359.1	7.5989	49.48	3824.0	4358.4	7.5898
950	53.74	3925.5	4484.4	7.7121	52.72	3924.9	4483.7	7.7029	51.73	3924.4	4483.1	7.6938
1000	56.06	4026.8	4609.9	7.8127	55.00	4026.3	4609.3	7.8035	53.97	4025.8	4608.7	7.7945
1100	60.68	4233.1	4864.1	8.0049	59.53	4232.6	4863.6	7.9958	58.42	4232.1	4863.1	7.9868
1200	65.28	4444.0	5122.9	8.1868	64.04	4443.6	5122.4	8.1777	62.86	4443.1	5122.0	8.1688
1300	69.86	4659.5	5386.1	8.3596	68.54	4659.1	5385.7	8.3505	67.28	4658.6	5385.2	8.3416

Definitions of symbols on page 1.

	11.0 (318.15)				11.2 (319.50)				11.4 (320.84)			p (t Sat.)
10^3 v	u	h	s	10^3 v	u	h	s	10^3 v	u	h	s	t
15.987	2529.8	2705.6	5.5527	15.620	2526.7	2701.6	5.5406	15.264	2523.5	2697.5	5.5285	Sat.
15.490	*2509.4*	*2679.8*	*5.5089*									315
16.259	2540.8	2719.6	5.5764	15.694	2529.7	2705.5	5.5472	*15.135*	*2518.0*	*2690.6*	*5.5168*	320
16.937	2568.0	2754.3	5.6346	16.393	2558.5	2742.1	5.6086	15.859	2548.5	2729.3	5.5818	325
17.550	2592.2	2785.3	5.6862	17.020	2583.8	2774.5	5.6625	16.502	2575.1	2763.2	5.6383	330
18.115	2614.3	2813.6	5.7329	17.594	2606.8	2803.8	5.7110	17.086	2599.0	2793.8	5.6888	335
18.643	2634.7	2839.8	5.7758	18.127	2627.8	2830.9	5.7553	17.626	2620.8	2821.7	5.7345	340
19.139	2653.7	2864.3	5.8156	18.628	2647.4	2856.0	5.7962	18.131	2641.0	2847.7	5.7767	345
19.611	2671.6	2887.3	5.8527	19.101	2665.8	2879.7	5.8343	18.607	2659.8	2871.9	5.8158	350
20.492	2704.7	2930.2	5.9209	19.985	2699.6	2923.5	5.9040	19.493	2694.5	2916.7	5.8870	360
21.311	2735.1	2969.6	5.9827	20.802	2730.6	2963.6	5.9669	20.310	2726.0	2957.6	5.9511	370
22.080	2763.5	3006.4	6.0395	21.569	2759.4	3001.0	6.0245	21.075	2755.3	2995.5	6.0097	380
22.810	2790.2	3041.1	6.0923	22.296	2786.5	3036.2	6.0781	21.798	2782.7	3031.2	6.0639	390
23.51	2815.7	3074.3	6.1420	22.99	2812.3	3069.8	6.1283	22.49	2808.8	3065.2	6.1148	400
24.18	2840.2	3106.2	6.1890	23.66	2837.0	3102.0	6.1758	23.15	2833.8	3097.7	6.1628	410
24.83	2863.8	3137.0	6.2338	24.30	2860.9	3133.1	6.2210	23.79	2857.9	3129.1	6.2084	420
25.46	2886.8	3166.9	6.2766	24.93	2884.1	3163.3	6.2642	24.41	2881.3	3159.6	6.2520	430
26.08	2909.3	3196.1	6.3179	25.54	2906.6	3192.7	6.3058	25.01	2904.0	3189.2	6.2938	440
26.68	2931.2	3224.7	6.3577	26.13	2928.8	3221.4	6.3458	25.60	2926.3	3218.2	6.3341	450
27.27	2952.8	3252.8	6.3962	26.71	2950.5	3249.7	6.3846	26.18	2948.1	3246.6	6.3731	460
27.85	2974.0	3280.4	6.4336	27.29	2971.8	3277.4	6.4222	26.74	2969.6	3274.5	6.4110	470
28.41	2995.0	3307.6	6.4700	27.85	2992.9	3304.8	6.4588	27.30	2990.8	3302.0	6.4477	480
28.97	3015.7	3334.4	6.5054	28.40	3013.7	3331.8	6.4944	27.84	3011.7	3329.1	6.4835	490
29.52	3036.3	3361.0	6.5400	28.94	3034.3	3358.5	6.5292	28.38	3032.4	3355.9	6.5184	500
30.07	3056.6	3387.4	6.5738	29.48	3054.8	3384.9	6.5631	28.91	3052.9	3382.5	6.5526	510
30.60	3076.8	3413.5	6.6070	30.00	3075.1	3411.1	6.5964	29.43	3073.3	3408.8	6.5859	520
31.13	3096.9	3439.4	6.6394	30.53	3095.2	3437.1	6.6290	29.94	3093.5	3434.9	6.6187	530
31.66	3116.9	3465.1	6.6713	31.04	3115.3	3463.0	6.6609	30.45	3113.6	3460.8	6.6507	540
32.17	3136.8	3490.7	6.7026	31.55	3135.2	3488.6	6.6923	30.96	3133.7	3486.6	6.6822	550
32.69	3156.6	3516.2	6.7333	32.06	3155.1	3514.2	6.7231	31.46	3153.6	3512.2	6.7131	560
33.20	3176.3	3541.5	6.7635	32.56	3174.9	3539.6	6.7535	31.95	3173.4	3537.7	6.7435	570
33.70	3196.0	3566.7	6.7933	33.06	3194.6	3564.9	6.7833	32.44	3193.2	3563.0	6.7735	580
34.20	3215.6	3591.8	6.8225	33.55	3214.3	3590.1	6.8126	32.93	3212.9	3588.3	6.8029	590
34.70	3235.2	3616.9	6.8514	34.04	3233.9	3615.2	6.8416	33.41	3232.6	3613.5	6.8319	600
35.19	3254.8	3641.9	6.8798	34.53	3253.5	3640.2	6.8701	33.89	3252.3	3638.6	6.8605	610
35.68	3274.3	3666.8	6.9079	35.01	3273.1	3665.2	6.8982	34.36	3271.9	3663.6	6.8887	620
36.17	3293.8	3691.6	6.9356	35.49	3292.6	3690.1	6.9259	34.84	3291.4	3688.6	6.9165	630
36.65	3313.3	3716.4	6.9629	35.97	3312.1	3715.0	6.9533	35.30	3311.0	3713.5	6.9439	640
37.13	3332.8	3741.2	6.9898	36.44	3331.6	3739.8	6.9803	35.77	3330.5	3738.3	6.9710	650
37.61	3352.2	3765.9	7.0165	36.91	3351.1	3764.5	7.0070	36.23	3350.1	3763.1	6.9977	660
38.08	3371.7	3790.6	7.0428	37.38	3370.6	3789.3	7.0334	36.70	3369.6	3787.9	7.0241	670
38.56	3391.1	3815.3	7.0688	37.84	3390.1	3814.0	7.0594	37.15	3389.1	3812.7	7.0502	680
39.03	3410.6	3839.9	7.0945	38.31	3409.6	3838.6	7.0852	37.61	3408.6	3837.4	7.0760	690
39.50	3430.0	3864.5	7.1199	38.77	3429.1	3863.3	7.1106	38.06	3428.2	3862.1	7.1015	700
40.43	3469.0	3913.7	7.1700	39.69	3468.1	3912.6	7.1608	38.97	3467.2	3911.4	7.1517	720
41.35	3508.0	3962.9	7.2190	40.60	3507.1	3961.8	7.2098	39.86	3506.3	3960.7	7.2009	740
42.27	3547.0	4012.0	7.2670	41.50	3546.2	4011.0	7.2580	40.75	3545.4	4010.0	7.2490	760
43.19	3586.2	4061.2	7.3142	42.40	3585.4	4060.3	7.3052	41.64	3584.7	4059.3	7.2963	780
44.09	3625.4	4110.4	7.3605	43.29	3624.7	4109.5	7.3515	42.51	3624.0	4108.6	7.3427	800
46.34	3724.0	4233.8	7.4728	45.50	3723.4	4233.0	7.4640	44.69	3722.7	4232.2	7.4552	850
48.57	3823.4	4357.7	7.5808	47.69	3822.9	4357.0	7.5720	46.85	3822.3	4356.3	7.5633	900
50.78	3923.9	4482.5	7.6849	49.87	3923.3	4481.8	7.6762	48.98	3922.8	4481.2	7.6676	950
52.98	4025.3	4608.1	7.7856	52.03	4024.8	4607.6	7.7769	51.11	4024.3	4607.0	7.7684	1000
57.36	4231.7	4862.6	7.9780	56.33	4231.2	4862.1	7.9694	55.34	4230.8	4861.6	7.9609	1100
61.71	4442.7	5121.5	8.1600	60.61	4442.2	5121.0	8.1514	59.54	4441.8	5120.6	8.1429	1200
66.05	4658.2	5384.8	8.3329	64.88	4657.7	5384.3	8.3243	63.74	4657.3	5383.9	8.3158	1300

Definitions of symbols on page 1.

Table 3. Vapor

p (t Sat.)	11.6 (322.16)				11.8 (323.46)				12.0 (324.75)			
t	10^3 v	u	h	s	10^3 v	u	h	s	10^3 v	u	h	s
Sat.	14.920	2520.3	2693.4	5.5165	14.587	2517.0	2689.2	5.5045	14.263	2513.7	2684.9	5.4924
320	*14.580*	*2505.6*	*2674.7*	*5.4850*								
325	15.333	2538.1	2715.9	5.5543	14.815	2527.1	2701.9	5.5258	14.301	2515.4	2687.0	5.4961
330	15.994	2566.1	2751.6	5.6137	15.496	2556.6	2739.5	5.5884	15.007	2546.8	2726.9	5.5624
335	16.590	2591.0	2783.4	5.6662	16.105	2582.7	2772.7	5.6432	15.631	2574.0	2761.6	5.6198
340	17.138	2613.6	2812.4	5.7136	16.662	2606.1	2802.7	5.6924	16.198	2598.4	2792.8	5.6708
345	17.648	2634.4	2839.1	5.7570	17.178	2627.6	2830.3	5.7372	16.720	2620.6	2821.3	5.7171
350	18.128	2653.8	2864.0	5.7972	17.662	2647.5	2855.9	5.7785	17.209	2641.2	2847.7	5.7596
360	19.017	2689.2	2909.8	5.8701	18.555	2683.9	2902.8	5.8531	18.106	2678.4	2895.7	5.8361
370	19.834	2721.4	2951.5	5.9354	19.372	2716.7	2945.3	5.9196	18.925	2711.9	2939.0	5.9039
380	20.597	2751.1	2990.0	5.9949	20.134	2746.9	2984.4	5.9801	19.685	2742.6	2978.8	5.9654
390	21.317	2779.0	3026.2	6.0499	20.851	2775.1	3021.2	6.0359	20.400	2771.2	3016.0	6.0220
400	22.00	2805.4	3060.6	6.1014	21.53	2801.9	3056.0	6.0880	21.08	2798.3	3051.3	6.0747
410	22.66	2830.6	3093.5	6.1498	22.19	2827.4	3089.2	6.1370	21.73	2824.1	3084.8	6.1242
420	23.30	2854.9	3125.2	6.1959	22.82	2851.9	3121.2	6.1835	22.35	2848.9	3117.1	6.1711
430	23.91	2878.5	3155.9	6.2398	23.43	2875.7	3152.1	6.2278	22.96	2872.8	3148.3	6.2158
440	24.51	2901.4	3185.7	6.2820	24.02	2898.8	3182.2	6.2703	23.55	2896.1	3178.7	6.2586
450	25.09	2923.8	3214.9	6.3226	24.60	2921.3	3211.6	6.3112	24.12	2918.8	3208.2	6.2998
460	25.66	2945.8	3243.5	6.3619	25.16	2943.4	3240.3	6.3507	24.68	2941.1	3237.2	6.3396
470	26.22	2967.4	3271.5	6.3999	25.71	2965.2	3268.6	6.3889	25.22	2962.9	3265.6	6.3780
480	26.77	2988.7	3299.2	6.4369	26.25	2986.6	3296.3	6.4261	25.76	2984.4	3293.5	6.4154
490	27.30	3009.7	3326.4	6.4728	26.79	3007.7	3323.7	6.4622	26.28	3005.6	3321.0	6.4517
500	27.83	3030.5	3353.4	6.5079	27.31	3028.6	3350.8	6.4975	26.80	3026.6	3348.2	6.4871
510	28.36	3051.1	3380.1	6.5422	27.83	3049.3	3377.6	6.5319	27.31	3047.4	3375.1	6.5217
520	28.87	3071.6	3406.5	6.5757	28.33	3069.8	3404.1	6.5655	27.81	3068.0	3401.8	6.5555
530	29.38	3091.9	3432.7	6.6085	28.84	3090.1	3430.4	6.5985	28.31	3088.4	3428.2	6.5885
540	29.88	3112.0	3458.7	6.6407	29.33	3110.4	3456.5	6.6308	28.80	3108.7	3454.4	6.6209
550	30.38	3132.1	3484.5	6.6723	29.82	3130.5	3482.4	6.6624	29.29	3128.9	3480.4	6.6527
560	30.87	3152.1	3510.2	6.7033	30.31	3150.6	3508.2	6.6936	29.77	3149.0	3506.2	6.6840
570	31.36	3172.0	3535.8	6.7338	30.79	3170.5	3533.8	6.7242	30.24	3169.0	3531.9	6.7146
580	31.84	3191.8	3561.2	6.7638	31.27	3190.4	3559.4	6.7542	30.71	3189.0	3557.5	6.7448
590	32.32	3211.6	3586.5	6.7933	31.74	3210.2	3584.8	6.7838	31.18	3208.9	3583.0	6.7745
600	32.80	3231.3	3611.8	6.8224	32.21	3230.0	3610.1	6.8130	31.64	3228.7	3608.3	6.8037
610	33.27	3251.0	3636.9	6.8510	32.67	3249.7	3635.3	6.8417	32.10	3248.4	3633.6	6.8325
620	33.74	3270.6	3662.0	6.8793	33.14	3269.4	3660.4	6.8700	32.55	3268.2	3658.8	6.8608
630	34.20	3290.2	3687.0	6.9071	33.59	3289.1	3685.5	6.8979	33.01	3287.9	3683.9	6.8888
640	34.67	3309.8	3712.0	6.9346	34.05	3308.7	3710.5	6.9254	33.45	3307.5	3709.0	6.9164
650	35.13	3329.4	3736.9	6.9617	34.50	3328.3	3735.4	6.9526	33.90	3327.2	3734.0	6.9436
660	35.58	3349.0	3761.7	6.9885	34.95	3347.9	3760.3	6.9794	34.34	3346.8	3759.0	6.9705
670	36.04	3368.5	3786.5	7.0149	35.40	3367.5	3785.2	7.0059	34.79	3366.4	3783.9	6.9971
680	36.49	3388.1	3811.3	7.0411	35.85	3387.0	3810.0	7.0321	35.22	3386.0	3808.7	7.0233
690	36.94	3407.6	3836.1	7.0669	36.29	3406.6	3834.8	7.0580	35.66	3405.6	3833.6	7.0492
700	37.39	3427.1	3860.8	7.0925	36.73	3426.2	3859.6	7.0836	36.10	3425.2	3858.4	7.0749
720	38.27	3466.2	3910.2	7.1427	37.60	3465.3	3909.1	7.1339	36.96	3464.5	3907.9	7.1253
740	39.16	3505.4	3959.6	7.1919	38.47	3504.5	3958.5	7.1832	37.81	3503.7	3957.4	7.1746
760	40.03	3544.6	4008.9	7.2402	39.34	3543.8	4007.9	7.2315	38.66	3543.0	4006.9	7.2230
780	40.90	3583.8	4058.3	7.2875	40.19	3583.1	4057.3	7.2789	39.51	3582.4	4056.4	7.2704
800	41.77	3623.2	4107.7	7.3339	41.04	3622.5	4106.8	7.3254	40.34	3621.8	4105.9	7.3170
850	43.91	3722.1	4231.4	7.4466	43.15	3721.4	4230.6	7.4382	42.42	3720.8	4229.8	7.4299
900	46.03	3821.7	4355.6	7.5548	45.24	3821.1	4354.9	7.5465	44.47	3820.6	4354.2	7.5382
950	48.13	3922.3	4480.6	7.6591	47.31	3921.7	4480.0	7.6508	46.51	3921.2	4479.3	7.6426
1000	50.22	4023.8	4606.4	7.7599	49.36	4023.3	4605.8	7.7517	48.53	4022.8	4605.3	7.7435
1100	54.38	4230.3	4861.1	7.9525	53.45	4229.8	4860.6	7.9443	52.56	4229.4	4860.1	7.9362
1200	58.51	4441.3	5120.1	8.1346	57.52	4440.9	5119.7	8.1264	56.56	4440.5	5119.2	8.1183
1300	62.64	4656.8	5383.4	8.3075	61.58	4656.4	5383.0	8.2993	60.55	4655.9	5382.6	8.2913

Definitions of symbols on page 1.

	12.5 (327.89)				**13.0 (330.93)**				**13.5 (333.88)**			**p** (t Sat.)
10^3 v	u	h	s	10^3 v	u	h	s	10^3 v	u	h	s	t
13.495	2505.1	2673.8	5.4624	12.780	2496.1	2662.2	5.4323	12.112	2486.7	2650.2	5.4021	Sat.
												320
13.021	*2482.7*	*2645.4*	*5.4151*									325
13.810	2519.9	2692.5	5.4935	*12.631*	*2488.7*	*2652.9*	*5.4169*					330
14.484	2551.0	2732.0	5.5587	13.376	2525.2	2699.1	5.4931	12.287	2495.7	2661.5	5.4208	335
15.081	2578.1	2766.6	5.6154	14.016	2555.9	2738.1	5.5570	12.989	2531.2	2706.6	5.4946	340
15.624	2602.4	2797.7	5.6659	14.587	2582.8	2772.4	5.6127	13.598	2561.4	2745.0	5.5570	345
16.126	2624.6	2826.2	5.7118	15.108	2606.9	2803.3	5.6626	14.143	2588.0	2778.9	5.6117	350
17.039	2664.3	2877.3	5.7932	16.041	2649.5	2858.1	5.7497	15.103	2633.9	2837.8	5.7055	360
17.863	2699.6	2922.8	5.8645	16.873	2686.8	2906.1	5.8250	15.947	2673.4	2888.7	5.7852	370
18.622	2731.6	2964.4	5.9287	17.634	2720.3	2949.6	5.8921	16.712	2708.6	2934.2	5.8555	380
19.332	2761.4	3003.0	5.9873	18.341	2751.2	2989.6	5.9530	17.418	2740.8	2975.9	5.9188	390
20.00	2789.3	3039.3	6.0417	19.007	2780.1	3027.2	6.0091	18.081	2770.6	3014.7	5.9768	400
20.64	2815.8	3073.9	6.0926	19.640	2807.3	3062.7	6.0615	18.707	2798.7	3051.2	6.0307	410
21.26	2841.2	3106.9	6.1407	20.246	2833.4	3096.6	6.1107	19.306	2825.4	3086.0	6.0813	420
21.85	2865.7	3138.8	6.1864	20.830	2858.4	3129.1	6.1574	19.880	2850.9	3119.3	6.1290	430
22.43	2889.4	3169.7	6.2300	21.394	2882.6	3160.7	6.2020	20.435	2875.6	3151.5	6.1744	440
22.99	2912.5	3199.8	6.2719	21.94	2906.1	3191.3	6.2446	20.97	2899.6	3182.7	6.2179	450
23.53	2935.1	3229.2	6.3123	22.47	2929.0	3221.2	6.2857	21.50	2922.9	3213.1	6.2596	460
24.06	2957.2	3258.0	6.3513	23.00	2951.5	3250.4	6.3253	22.00	2945.7	3242.8	6.2998	470
24.59	2979.0	3286.4	6.3892	23.51	2973.6	3279.1	6.3637	22.50	2968.1	3271.9	6.3387	480
25.10	3000.5	3314.2	6.4260	24.01	2995.3	3307.4	6.4009	22.99	2990.1	3300.5	6.3764	490
25.60	3021.7	3341.8	6.4618	24.50	3016.8	3335.2	6.4371	23.47	3011.8	3328.6	6.4131	500
26.10	3042.7	3368.9	6.4967	24.98	3038.0	3362.7	6.4725	23.94	3033.2	3356.4	6.4488	510
26.59	3063.5	3395.8	6.5309	25.45	3059.0	3389.9	6.5069	24.40	3054.4	3383.9	6.4837	520
27.07	3084.1	3422.5	6.5642	25.92	3079.8	3416.8	6.5407	24.86	3075.4	3411.0	6.5177	530
27.54	3104.6	3448.9	6.5969	26.38	3100.4	3443.5	6.5736	25.31	3096.3	3438.0	6.5510	540
28.01	3125.0	3475.2	6.6290	26.84	3121.0	3469.9	6.6060	25.75	3116.9	3464.6	6.5836	550
28.48	3145.2	3501.2	6.6605	27.29	3141.4	3496.2	6.6377	26.19	3137.5	3491.1	6.6156	560
28.94	3165.4	3527.1	6.6914	27.74	3161.7	3522.3	6.6688	26.63	3157.9	3517.4	6.6470	570
29.40	3185.4	3552.9	6.7218	28.18	3181.9	3548.2	6.6994	27.06	3178.3	3543.5	6.6778	580
29.85	3205.4	3578.5	6.7516	28.62	3202.0	3574.0	6.7295	27.48	3198.5	3569.5	6.7081	590
30.29	3225.4	3604.0	6.7810	29.05	3222.1	3599.7	6.7591	27.90	3218.7	3595.4	6.7379	600
30.74	3245.3	3629.5	6.8100	29.48	3242.1	3625.3	6.7883	28.32	3238.8	3621.2	6.7672	610
31.18	3265.1	3654.8	6.8385	29.91	3262.0	3650.8	6.8170	28.73	3258.9	3646.8	6.7960	620
31.62	3284.9	3680.1	6.8667	30.33	3281.9	3676.2	6.8452	29.14	3278.9	3672.3	6.8245	630
32.05	3304.7	3705.3	6.8944	30.75	3301.8	3701.6	6.8731	29.55	3298.9	3697.8	6.8525	640
32.48	3324.4	3730.4	6.9218	31.17	3321.6	3726.8	6.9007	29.95	3318.8	3723.2	6.8802	650
32.91	3344.1	3755.5	6.9488	31.58	3341.4	3752.0	6.9278	30.36	3338.7	3748.5	6.9075	660
33.33	3363.8	3780.5	6.9755	32.00	3361.2	3777.1	6.9546	30.75	3358.6	3773.8	6.9344	670
33.76	3383.5	3805.5	7.0018	32.40	3381.0	3802.2	6.9811	31.15	3378.4	3799.0	6.9610	680
34.18	3403.2	3830.4	7.0279	32.81	3400.7	3827.3	7.0072	31.55	3398.3	3824.1	6.9872	690
34.60	3422.9	3855.3	7.0536	33.22	3420.5	3852.3	7.0331	31.94	3418.1	3849.3	7.0132	700
35.43	3462.2	3905.1	7.1042	34.02	3460.0	3902.2	7.0838	32.72	3457.7	3899.4	7.0642	720
36.26	3501.6	3954.8	7.1537	34.82	3499.5	3952.1	7.1335	33.49	3497.3	3949.4	7.1140	740
37.07	3541.0	4004.4	7.2022	35.61	3539.0	4001.9	7.1822	34.25	3537.0	3999.3	7.1628	760
37.89	3580.4	4054.0	7.2498	36.39	3578.5	4051.6	7.2299	35.01	3576.6	4049.2	7.2107	780
38.69	3620.0	4103.6	7.2965	37.17	3618.2	4101.4	7.2767	35.76	3616.3	4099.1	7.2576	800
40.69	3719.2	4227.9	7.4096	39.10	3717.6	4225.9	7.3901	37.62	3716.0	4223.9	7.3713	850
42.67	3819.1	4352.5	7.5182	41.00	3817.7	4350.7	7.4988	39.46	3816.2	4349.0	7.4802	900
44.63	3919.9	4477.7	7.6227	42.89	3918.6	4476.2	7.6035	41.29	3917.2	4474.6	7.5851	950
46.58	4021.6	4603.8	7.7237	44.77	4020.4	4602.4	7.7047	43.10	4019.1	4600.9	7.6863	1000
50.45	4228.2	4858.8	7.9165	48.50	4227.1	4857.6	7.8976	46.69	4226.0	4856.3	7.8794	1100
54.30	4439.3	5118.0	8.0987	52.20	4438.2	5116.9	8.0799	50.27	4437.1	5115.7	8.0617	1200
58.13	4654.8	5381.4	8.2717	55.90	4653.6	5380.3	8.2529	53.83	4652.5	5379.2	8.2348	1300

Definitions of symbols on page 1.

Table 3. Vapor

p (t Sat.)	14.0 (336.75)				14.5 (339.53)				15.0 (342.24)			
t	10^3 v	u	h	s	10^3 v	u	h	s	10^3 v	u	h	s
Sat.	11.485	2476.8	2637.6	5.3717	10.895	2466.4	2624.4	5.3410	10.337	2455.5	2610.5	5.3098
335	*11.185*	*2460.5*	*2617.1*	*5.3381*								
340	11.983	2503.3	2671.1	5.4266	10.973	2470.8	2629.9	5.3500	*9.917*	*2430.6*	*2579.4*	*5.2590*
345	12.644	2537.9	2714.9	5.4978	11.712	2511.6	2681.5	5.4337	10.784	2481.4	2643.2	5.3628
350	13.221	2567.5	2752.6	5.5585	12.334	2545.2	2724.0	5.5023	11.470	2520.4	2692.4	5.4421
360	14.218	2617.4	2816.5	5.6602	13.377	2599.9	2793.9	5.6136	12.574	2581.2	2769.8	5.5653
370	15.078	2659.5	2870.6	5.7450	14.258	2644.9	2851.7	5.7042	13.482	2629.6	2831.8	5.6625
380	15.849	2696.5	2918.4	5.8188	15.039	2683.9	2902.0	5.7818	14.275	2670.9	2885.0	5.7446
390	16.557	2730.0	2961.8	5.8847	15.749	2718.9	2947.3	5.8506	14.990	2707.5	2932.3	5.8165
400	17.216	2760.9	3001.9	5.9448	16.408	2751.0	2988.9	5.9129	15.649	2740.7	2975.5	5.8811
410	17.838	2789.8	3039.6	6.0003	17.026	2780.8	3027.7	5.9701	16.265	2771.6	3015.5	5.9402
420	18.430	2817.2	3075.3	6.0521	17.612	2808.9	3064.3	6.0234	16.847	2800.5	3053.2	5.9949
430	18.997	2843.4	3109.4	6.1010	18.172	2835.7	3099.2	6.0734	17.401	2827.9	3089.0	6.0461
440	19.543	2868.6	3142.2	6.1474	18.711	2861.5	3132.8	6.1207	17.933	2854.2	3123.2	6.0945
450	20.07	2893.0	3174.0	6.1916	19.231	2886.3	3165.1	6.1658	18.445	2879.5	3156.2	6.1404
460	20.58	2916.7	3204.9	6.2340	19.735	2910.4	3196.6	6.2090	18.941	2904.0	3188.2	6.1843
470	21.08	2939.8	3235.0	6.2749	20.225	2933.9	3227.2	6.2505	19.423	2927.9	3219.3	6.2265
480	21.57	2962.5	3264.5	6.3143	20.704	2956.9	3257.1	6.2905	19.893	2951.2	3249.6	6.2671
490	22.05	2984.8	3293.5	6.3526	21.171	2979.5	3286.5	6.3292	20.351	2974.1	3279.4	6.3063
500	22.52	3006.8	3322.0	6.3897	21.63	3001.7	3315.3	6.3667	20.80	2996.6	3308.6	6.3443
510	22.98	3028.4	3350.1	6.4258	22.08	3023.6	3343.7	6.4033	21.24	3018.7	3337.3	6.3812
520	23.43	3049.8	3377.8	6.4610	22.52	3045.2	3371.7	6.4388	21.67	3040.5	3365.6	6.4172
530	23.87	3071.0	3405.3	6.4953	22.95	3066.6	3399.4	6.4735	22.10	3062.1	3393.6	6.4522
540	24.31	3092.0	3432.4	6.5289	23.38	3087.8	3426.9	6.5074	22.51	3083.5	3421.3	6.4864
550	24.74	3112.9	3459.3	6.5618	23.80	3108.8	3454.0	6.5406	22.93	3104.7	3448.6	6.5199
560	25.17	3133.6	3486.0	6.5941	24.22	3129.7	3480.9	6.5731	23.33	3125.8	3475.8	6.5527
570	25.59	3154.2	3512.5	6.6257	24.63	3150.4	3507.6	6.6050	23.73	3146.7	3502.7	6.5848
580	26.01	3174.7	3538.8	6.6567	25.04	3171.1	3534.1	6.6362	24.13	3167.4	3529.4	6.6163
590	26.42	3195.1	3565.0	6.6872	25.44	3191.6	3560.5	6.6669	24.52	3188.1	3555.9	6.6472
600	26.83	3215.4	3591.1	6.7172	25.84	3212.0	3586.7	6.6971	24.91	3208.6	3582.3	6.6776
610	27.24	3235.6	3617.0	6.7467	26.23	3232.4	3612.8	6.7268	25.29	3229.1	3608.5	6.7075
620	27.64	3255.8	3642.8	6.7758	26.62	3252.7	3638.7	6.7560	25.67	3249.5	3634.6	6.7368
630	28.04	3275.9	3668.5	6.8044	27.01	3272.9	3664.5	6.7848	26.05	3269.8	3660.6	6.7658
640	28.43	3296.0	3694.1	6.8326	27.40	3293.0	3690.3	6.8131	26.43	3290.1	3686.5	6.7943
650	28.83	3316.0	3719.6	6.8603	27.78	3313.2	3715.9	6.8411	26.80	3310.3	3712.3	6.8224
660	29.22	3336.0	3745.0	6.8878	28.15	3333.3	3741.5	6.8686	27.16	3330.5	3738.0	6.8500
670	29.60	3355.9	3770.4	6.9148	28.53	3353.3	3767.0	6.8958	27.53	3350.7	3763.6	6.8774
680	29.99	3375.9	3795.7	6.9415	28.90	3373.3	3792.4	6.9226	27.89	3370.8	3789.2	6.9043
690	30.37	3395.8	3821.0	6.9679	29.27	3393.3	3817.8	6.9491	28.25	3390.8	3814.6	6.9309
700	30.75	3415.7	3846.2	6.9939	29.64	3413.3	3843.1	6.9753	28.61	3410.9	3840.1	6.9572
720	31.50	3455.5	3896.5	7.0451	30.37	3453.2	3893.6	7.0267	29.32	3450.9	3890.8	7.0087
740	32.25	3495.2	3946.7	7.0952	31.10	3493.1	3944.0	7.0769	30.02	3490.9	3941.3	7.0591
760	32.99	3535.0	3996.8	7.1441	31.82	3532.9	3994.3	7.1260	30.72	3530.9	3991.7	7.1084
780	33.72	3574.7	4046.9	7.1921	32.53	3572.8	4044.5	7.1741	31.41	3570.9	4042.1	7.1567
800	34.45	3614.5	4096.9	7.2392	33.23	3612.7	4094.6	7.2213	32.10	3610.9	4092.4	7.2040
850	36.25	3714.4	4222.0	7.3531	34.98	3712.8	4220.0	7.3355	33.79	3711.2	4218.0	7.3184
900	38.03	3814.8	4347.2	7.4622	36.70	3813.3	4345.5	7.4448	35.46	3811.9	4343.8	7.4279
950	39.79	3915.9	4473.0	7.5672	38.41	3914.6	4471.5	7.5499	37.11	3913.3	4469.9	7.5332
1000	41.54	4017.9	4599.5	7.6685	40.10	4016.7	4598.1	7.6514	38.75	4015.4	4596.6	7.6348
1100	45.02	4224.8	4855.1	7.8618	43.46	4223.7	4853.8	7.8447	42.00	4222.6	4852.6	7.8283
1200	48.47	4436.0	5114.6	8.0442	46.80	4434.9	5113.4	8.0272	45.23	4433.8	5112.3	8.0108
1300	51.91	4651.4	5378.1	8.2173	50.12	4650.3	5377.0	8.2004	48.45	4649.1	5376.0	8.1840

Definitions of symbols on page 1.

	15.5 (344.87)				16.0 (347.44)				16.5 (349.93)			p (t Sat.)
10^3 v	u	h	s	10^3 v	u	h	s	10^3 v	u	h	s	t
9.809	2443.9	2596.0	5.2780	9.306	2431.7	2580.6	5.2455	8.825	2418.8	2564.4	5.2121	Sat.
												335
												340
9.832	2445.3	2597.7	5.2809	*8.793*	*2398.6*	*2539.3*	*5.1788*					345
10.615	2492.4	2656.9	5.3762	9.749	2459.7	2615.7	5.3020	8.838	2419.6	2565.5	5.2138	350
11.801	2561.0	2743.9	5.5147	11.053	2539.0	2715.8	5.4614	10.320	2514.7	2685.0	5.4043	360
12.744	2613.4	2810.9	5.6199	12.038	2596.3	2788.9	5.5760	11.361	2578.0	2765.5	5.5305	370
13.552	2657.2	2867.3	5.7069	12.867	2643.0	2848.9	5.6685	12.214	2628.1	2829.7	5.6295	380
14.274	2695.6	2916.9	5.7822	13.598	2683.4	2900.9	5.7477	12.956	2670.7	2884.5	5.7128	390
14.935	2730.2	2961.7	5.8493	14.262	2719.4	2947.6	5.8175	13.625	2708.3	2933.1	5.7856	400
15.549	2762.1	3003.1	5.9103	14.876	2752.4	2990.4	5.8806	14.240	2742.5	2977.4	5.8510	410
16.128	2791.9	3041.8	5.9666	15.452	2783.1	3030.3	5.9386	14.814	2774.1	3018.5	5.9106	420
16.677	2820.0	3078.5	6.0191	15.997	2811.9	3067.9	5.9924	15.356	2803.7	3057.1	5.9659	430
17.203	2846.8	3113.5	6.0686	16.518	2839.4	3103.7	6.0429	15.873	2831.8	3093.7	6.0176	440
17.709	2872.6	3147.1	6.1154	17.018	2865.7	3138.0	6.0907	16.367	2858.6	3128.7	6.0663	450
18.198	2897.6	3179.7	6.1601	17.500	2891.1	3171.1	6.1362	16.843	2884.5	3162.4	6.1126	460
18.672	2921.8	3211.3	6.2029	17.967	2915.7	3203.2	6.1797	17.304	2909.5	3195.0	6.1569	470
19.134	2945.5	3242.1	6.2441	18.421	2939.7	3234.4	6.2215	17.751	2933.9	3226.8	6.1993	480
19.584	2968.7	3272.2	6.2838	18.864	2963.2	3265.0	6.2618	18.186	2957.6	3257.7	6.2401	490
20.02	2991.4	3301.8	6.3223	19.296	2986.2	3294.9	6.3007	18.611	2980.9	3288.0	6.2796	500
20.45	3013.8	3330.8	6.3597	19.719	3008.8	3324.3	6.3385	19.027	3003.8	3317.8	6.3178	510
20.88	3035.9	3359.5	6.3960	20.133	3031.1	3353.3	6.3752	19.434	3026.4	3347.0	6.3549	520
21.29	3057.7	3387.7	6.4314	20.541	3053.1	3381.8	6.4110	19.833	3048.6	3375.8	6.3910	530
21.70	3079.2	3415.6	6.4659	20.941	3074.9	3410.0	6.4459	20.225	3070.6	3404.3	6.4262	540
22.11	3100.6	3443.2	6.4997	21.34	3096.5	3437.8	6.4799	20.61	3092.3	3432.4	6.4606	550
22.50	3121.8	3470.6	6.5327	21.72	3117.8	3465.4	6.5132	20.99	3113.8	3460.2	6.4941	560
22.89	3142.9	3497.7	6.5651	22.11	3139.0	3492.7	6.5458	21.37	3135.2	3487.7	6.5270	570
23.28	3163.8	3524.6	6.5968	22.48	3160.1	3519.8	6.5778	21.74	3156.4	3515.0	6.5592	580
23.66	3184.6	3551.3	6.6279	22.86	3181.0	3546.8	6.6091	22.10	3177.5	3542.1	6.5908	590
24.04	3205.2	3577.9	6.6585	23.23	3201.8	3573.5	6.6399	22.46	3198.4	3569.1	6.6218	600
24.42	3225.8	3604.3	6.6886	23.59	3222.6	3600.1	6.6702	22.82	3219.3	3595.8	6.6522	610
24.79	3246.4	3630.6	6.7182	23.95	3243.2	3626.5	6.6999	23.17	3240.0	3622.4	6.6821	620
25.15	3266.8	3656.7	6.7473	24.31	3263.7	3652.8	6.7292	23.52	3260.7	3648.8	6.7116	630
25.52	3287.2	3682.7	6.7759	24.67	3284.2	3678.9	6.7580	23.87	3281.2	3675.1	6.7405	640
25.88	3307.5	3708.6	6.8041	25.02	3304.6	3705.0	6.7864	24.21	3301.8	3701.3	6.7691	650
26.24	3327.8	3734.5	6.8320	25.37	3325.0	3730.9	6.8143	24.55	3322.2	3727.4	6.7972	660
26.59	3348.0	3760.2	6.8594	25.72	3345.3	3756.8	6.8419	24.89	3342.6	3753.4	6.8249	670
26.95	3368.2	3785.9	6.8865	26.06	3365.6	3782.6	6.8691	25.23	3363.0	3779.3	6.8522	680
27.30	3388.3	3811.5	6.9132	26.40	3385.8	3808.3	6.8959	25.56	3383.3	3805.1	6.8791	690
27.65	3408.5	3837.0	6.9395	26.74	3406.0	3833.9	6.9224	25.89	3403.6	3830.8	6.9057	700
28.34	3448.7	3887.9	6.9913	27.41	3446.4	3885.0	6.9744	26.55	3444.1	3882.1	6.9579	720
29.02	3488.8	3938.6	7.0419	28.08	3486.7	3935.9	7.0251	27.19	3484.5	3933.2	7.0088	740
29.70	3528.9	3989.2	7.0913	28.74	3526.9	3986.7	7.0747	27.83	3524.8	3984.1	7.0586	760
30.37	3569.0	4039.7	7.1397	29.39	3567.1	4037.3	7.1233	28.47	3565.2	4034.9	7.1073	780
31.03	3609.1	4090.1	7.1872	30.03	3607.3	4087.8	7.1708	29.10	3605.5	4085.6	7.1550	800
32.67	3709.6	4216.0	7.3019	31.63	3708.0	4214.1	7.2858	30.65	3706.4	4212.1	7.2702	850
34.29	3810.5	4342.0	7.4116	33.20	3809.0	4340.3	7.3957	32.18	3807.6	4338.5	7.3803	900
35.90	3911.9	4468.3	7.5170	34.76	3910.6	4466.8	7.5013	33.69	3909.3	4465.2	7.4860	950
37.49	4014.2	4595.2	7.6187	36.30	4013.0	4593.8	7.6031	35.19	4011.7	4592.4	7.5879	1000
40.64	4221.4	4851.3	7.8123	39.36	4220.3	4850.1	7.7969	38.16	4219.2	4848.8	7.7819	1100
43.77	4432.7	5111.1	7.9950	42.40	4431.6	5110.0	7.9796	41.11	4430.5	5108.9	7.9646	1200
46.89	4648.0	5374.9	8.1682	45.43	4646.9	5373.8	8.1528	44.06	4645.8	5372.7	8.1379	1300

Definitions of symbols on page 1.

Table 3. Vapor

p (t Sat.)	17.0 (352.37)				17.5 (354.75)				18.0 (357.06)			
t	10^3 v	u	h	s	10^3 v	u	h	s	10^3 v	u	h	s
Sat.	8.364	2405.0	2547.2	5.1777	7.920	2390.2	2528.8	5.1419	7.489	2374.3	2509.1	5.1044
350	*7.787*	*2364.2*	*2496.6*	*5.0966*								
355	8.856	2438.7	2589.2	5.2448	7.976	2394.3	2533.8	5.1499	*6.882*	*2327.3*	*2451.2*	*5.0123*
360	9.593	2487.6	2650.7	5.3422	8.857	2456.3	2611.3	5.2728	8.090	2418.9	2564.5	5.1922
365	10.190	2526.0	2699.3	5.4187	9.518	2501.1	2667.6	5.3614	8.847	2473.0	2632.2	5.2987
370	10.706	2558.5	2740.5	5.4830	10.070	2537.4	2713.6	5.4331	9.447	2514.3	2684.4	5.3801
375	11.168	2586.9	2776.8	5.5392	10.554	2568.5	2753.2	5.4944	9.959	2548.7	2727.9	5.4476
380	11.589	2612.5	2809.5	5.5895	10.990	2596.0	2788.3	5.5485	10.413	2578.5	2766.0	5.5060
385	11.980	2635.9	2839.5	5.6353	11.391	2620.9	2820.2	5.5972	10.827	2605.2	2800.1	5.5581
390	12.346	2657.5	2867.4	5.6775	11.764	2643.8	2849.7	5.6417	11.208	2629.5	2831.2	5.6052
395	12.692	2677.8	2893.5	5.7168	12.115	2665.1	2877.1	5.6829	11.565	2651.9	2860.1	5.6485
400	13.021	2696.9	2918.2	5.7536	12.447	2685.0	2902.9	5.7213	11.901	2672.8	2887.0	5.6887
405	13.336	2715.0	2941.7	5.7883	12.764	2703.9	2927.3	5.7574	12.221	2692.5	2912.4	5.7264
410	13.638	2732.3	2964.1	5.8213	13.068	2721.8	2950.5	5.7916	12.526	2711.1	2936.6	5.7618
415	13.930	2748.9	2985.7	5.8527	13.360	2739.0	2972.8	5.8241	12.818	2728.9	2959.6	5.7954
420	14.212	2764.9	3006.5	5.8828	13.641	2755.5	2994.2	5.8551	13.100	2745.9	2981.7	5.8274
425	14.485	2780.3	3026.6	5.9117	13.914	2771.4	3014.9	5.8848	13.373	2762.3	3003.0	5.8580
430	14.751	2795.3	3046.1	5.9396	14.179	2786.8	3034.9	5.9134	13.637	2778.1	3023.5	5.8873
435	15.011	2809.9	3065.0	5.9664	14.437	2801.7	3054.3	5.9409	13.893	2793.4	3043.4	5.9155
440	15.264	2824.0	3083.5	5.9925	14.688	2816.2	3073.2	5.9675	14.143	2808.2	3062.8	5.9428
445	15.511	2837.9	3101.6	6.0177	14.934	2830.4	3091.7	5.9933	14.387	2822.7	3081.7	5.9691
450	15.754	2851.5	3119.3	6.0422	15.174	2844.2	3109.7	6.0184	14.625	2836.8	3100.1	5.9947
460	16.224	2877.8	3153.6	6.0894	15.640	2871.0	3144.7	6.0664	15.087	2864.1	3135.7	6.0436
470	16.679	2903.2	3186.8	6.1343	16.089	2896.9	3178.4	6.1121	15.531	2890.5	3170.0	6.0901
480	17.120	2927.9	3219.0	6.1774	16.524	2922.0	3211.1	6.1558	15.961	2915.9	3203.2	6.1345
490	17.549	2952.0	3250.4	6.2188	16.947	2946.4	3243.0	6.1978	16.378	2940.7	3235.5	6.1771
500	17.967	2975.6	3281.1	6.2587	17.358	2970.3	3274.1	6.2383	16.784	2964.9	3267.0	6.2181
510	18.375	2998.8	3311.2	6.2974	17.760	2993.7	3304.5	6.2774	17.179	2988.6	3297.8	6.2577
520	18.775	3021.6	3340.7	6.3349	18.153	3016.7	3334.4	6.3153	17.566	3011.8	3328.0	6.2960
530	19.167	3044.0	3369.8	6.3714	18.538	3039.4	3363.8	6.3522	17.944	3034.7	3357.7	6.3333
540	19.552	3066.2	3398.5	6.4069	18.916	3061.8	3392.8	6.3880	18.316	3057.3	3387.0	6.3695
550	19.930	3088.1	3426.9	6.4416	19.288	3083.9	3421.4	6.4230	18.681	3079.6	3415.9	6.4048
560	20.303	3109.8	3454.9	6.4755	19.653	3105.7	3449.7	6.4571	19.039	3101.7	3444.4	6.4392
570	20.670	3131.3	3482.7	6.5086	20.013	3127.4	3477.7	6.4905	19.393	3123.5	3472.6	6.4728
580	21.032	3152.7	3510.2	6.5410	20.368	3148.9	3505.4	6.5232	19.741	3145.2	3500.5	6.5058
590	21.389	3173.9	3537.5	6.5728	20.718	3170.3	3532.9	6.5552	20.084	3166.7	3528.2	6.5380
600	21.74	3195.0	3564.6	6.6040	21.06	3191.5	3560.1	6.5866	20.42	3188.0	3555.6	6.5696
610	22.09	3215.9	3591.5	6.6347	21.41	3212.6	3587.2	6.6175	20.76	3209.3	3582.9	6.6007
620	22.44	3236.8	3618.2	6.6648	21.74	3233.6	3614.1	6.6478	21.09	3230.4	3610.0	6.6311
630	22.78	3257.6	3644.8	6.6944	22.08	3254.5	3640.8	6.6775	21.42	3251.4	3636.9	6.6611
640	23.12	3278.3	3671.3	6.7235	22.41	3275.3	3667.4	6.7068	21.74	3272.3	3663.6	6.6905
650	23.45	3298.9	3697.6	6.7522	22.74	3296.0	3693.9	6.7357	22.06	3293.1	3690.2	6.7195
660	23.79	3319.5	3723.8	6.7804	23.06	3316.7	3720.3	6.7640	22.38	3313.9	3716.7	6.7480
670	24.12	3340.0	3749.9	6.8082	23.38	3337.3	3746.5	6.7920	22.69	3334.5	3743.0	6.7761
680	24.44	3360.4	3775.9	6.8357	23.70	3357.8	3772.6	6.8196	23.01	3355.2	3769.3	6.8038
690	24.77	3380.8	3801.9	6.8627	24.02	3378.3	3798.7	6.8467	23.32	3375.7	3795.4	6.8311
700	25.09	3401.2	3827.7	6.8894	24.34	3398.7	3824.6	6.8736	23.62	3396.3	3821.5	6.8580
720	25.73	3441.8	3879.2	6.9418	24.96	3439.5	3876.3	6.9261	24.23	3437.2	3873.4	6.9108
740	26.36	3482.3	3930.5	6.9929	25.58	3480.2	3927.8	6.9774	24.83	3478.0	3925.0	6.9623
760	26.99	3522.8	3981.6	7.0429	26.18	3520.8	3979.0	7.0275	25.43	3518.7	3976.5	7.0126
780	27.60	3563.2	4032.5	7.0917	26.79	3561.3	4030.1	7.0765	26.02	3559.4	4027.7	7.0617
800	28.22	3603.7	4083.3	7.1395	27.38	3601.8	4081.1	7.1244	26.60	3600.0	4078.8	7.1098
850	29.73	3704.8	4210.1	7.2550	28.86	3703.2	4208.2	7.2402	28.04	3701.5	4206.2	7.2258
900	31.22	3806.1	4336.8	7.3653	30.31	3804.7	4335.1	7.3507	29.45	3803.2	4333.3	7.3365
950	32.69	3908.0	4463.6	7.4712	31.74	3906.7	4462.1	7.4568	30.84	3905.3	4460.5	7.4427
1000	34.14	4010.5	4590.9	7.5732	33.16	4009.3	4589.5	7.5589	32.23	4008.0	4588.1	7.5449
1100	37.03	4218.0	4847.6	7.7673	35.97	4216.9	4846.4	7.7531	34.96	4215.8	4845.1	7.7393
1200	39.90	4429.4	5107.7	7.9501	38.76	4428.3	5106.6	7.9360	37.68	4427.2	5105.5	7.9222
1300	42.76	4644.6	5371.6	8.1234	41.54	4643.5	5370.5	8.1093	40.39	4642.4	5369.4	8.0956

Definitions of symbols on page 1.

	18.5 (359.33)				19.0 (361.54)				19.5 (363.70)			p (t Sat.)
10^3 v	u	h	s	10^3 v	u	h	s	10^3 v	u	h	s	t
7.070	2357.0	2487.8	5.0649	6.657	2338.1	2464.5	5.0228	6.247	2317.0	2438.8	4.9772	Sat.
												350
												355
7.238	2370.3	2504.1	5.0908	*6.080*	*2289.0*	*2404.5*	*4.9281*					360
8.162	2440.5	2591.5	5.2282	7.440	2401.2	2542.6	5.1454	6.622	2349.2	2478.3	5.0393	365
8.829	2488.8	2652.2	5.3229	8.209	2460.1	2616.1	5.2602	7.573	2426.8	2574.5	5.1894	370
9.378	2527.3	2700.8	5.3982	8.806	2504.0	2671.3	5.3457	8.238	2478.2	2638.9	5.2891	375
9.855	2559.9	2742.3	5.4620	9.311	2540.1	2717.0	5.4159	8.780	2518.6	2689.8	5.3674	380
10.283	2588.7	2778.9	5.5179	9.757	2571.2	2756.6	5.4763	9.248	2552.6	2732.9	5.4332	385
10.675	2614.5	2812.0	5.5680	10.161	2598.8	2791.9	5.5298	9.666	2582.3	2770.8	5.4906	390
11.038	2638.2	2842.4	5.6137	10.534	2623.9	2824.1	5.5782	10.048	2609.0	2805.0	5.5419	395
11.380	2660.1	2870.7	5.6558	10.881	2647.0	2853.8	5.6224	10.402	2633.4	2836.3	5.5885	400
11.702	2680.7	2897.2	5.6951	11.207	2668.5	2881.5	5.6634	10.734	2656.0	2865.3	5.6315	405
12.010	2700.1	2922.2	5.7319	11.517	2688.7	2907.6	5.7018	11.047	2677.0	2892.4	5.6714	410
12.304	2718.5	2946.1	5.7667	11.813	2707.8	2932.3	5.7378	11.345	2696.9	2918.1	5.7088	415
12.586	2736.1	2968.9	5.7997	12.096	2726.0	2955.8	5.7719	11.629	2715.7	2942.5	5.7441	420
12.858	2752.9	2990.8	5.8312	12.369	2743.4	2978.4	5.8044	11.903	2733.7	2965.8	5.7776	425
13.122	2769.2	3011.9	5.8613	12.633	2760.1	3000.1	5.8354	12.166	2750.9	2988.1	5.8095	430
13.378	2784.9	3032.4	5.8903	12.888	2776.2	3021.1	5.8651	12.421	2767.4	3009.7	5.8400	435
13.626	2800.1	3052.2	5.9182	13.135	2791.9	3041.4	5.8937	12.668	2783.5	3030.5	5.8693	440
13.869	2814.9	3071.5	5.9451	13.376	2807.0	3061.2	5.9213	12.908	2799.0	3050.7	5.8975	445
14.105	2829.3	3090.3	5.9712	13.612	2821.7	3080.4	5.9479	13.142	2814.0	3070.3	5.9248	450
14.563	2857.2	3126.6	6.0211	14.066	2850.2	3117.4	5.9988	13.593	2843.0	3108.1	5.9767	460
15.003	2884.0	3161.5	6.0684	14.502	2877.4	3152.9	6.0469	14.025	2870.7	3144.2	6.0257	470
15.428	2909.8	3195.3	6.1135	14.922	2903.7	3187.2	6.0927	14.441	2897.4	3179.1	6.0722	480
15.839	2935.0	3228.0	6.1567	15.329	2929.2	3220.4	6.1365	14.844	2923.3	3212.8	6.1166	490
16.239	2959.5	3259.9	6.1982	15.724	2954.0	3252.7	6.1786	15.234	2948.4	3245.5	6.1593	500
16.629	2983.4	3291.1	6.2383	16.108	2978.2	3284.3	6.2192	15.613	2973.0	3277.4	6.2003	510
17.010	3006.9	3321.6	6.2771	16.483	3002.0	3315.2	6.2584	15.983	2997.0	3308.7	6.2400	520
17.383	3030.1	3351.6	6.3147	16.850	3025.3	3345.5	6.2964	16.345	3020.6	3339.3	6.2784	530
17.748	3052.8	3381.2	6.3512	17.210	3048.3	3375.3	6.3333	16.699	3043.8	3369.5	6.3156	540
18.106	3075.3	3410.3	6.3868	17.562	3071.0	3404.7	6.3692	17.046	3066.7	3399.1	6.3519	550
18.459	3097.6	3439.1	6.4216	17.909	3093.5	3433.7	6.4042	17.387	3089.3	3428.4	6.3872	560
18.805	3119.6	3467.5	6.4555	18.249	3115.7	3462.4	6.4384	17.721	3111.7	3457.3	6.4217	570
19.147	3141.4	3495.6	6.4887	18.585	3137.6	3490.7	6.4719	18.051	3133.8	3485.8	6.4554	580
19.484	3163.1	3523.5	6.5211	18.915	3159.4	3518.8	6.5046	18.376	3155.8	3514.1	6.4883	590
19.816	3184.6	3551.2	6.5530	19.241	3181.1	3546.6	6.5366	18.696	3177.5	3542.1	6.5206	600
20.144	3205.9	3578.6	6.5842	19.563	3202.5	3574.2	6.5680	19.012	3199.2	3569.9	6.5522	610
20.468	3227.1	3605.8	6.6149	19.881	3223.9	3601.6	6.5989	19.324	3220.6	3597.4	6.5832	620
20.789	3248.3	3632.9	6.6450	20.195	3245.1	3628.8	6.6292	19.632	3242.0	3624.8	6.6137	630
21.106	3269.3	3659.7	6.6746	20.506	3266.3	3655.9	6.6590	19.937	3263.2	3652.0	6.6436	640
21.42	3290.2	3686.5	6.7037	20.81	3287.3	3682.8	6.6882	20.24	3284.4	3679.0	6.6731	650
21.73	3311.1	3713.1	6.7324	21.12	3308.2	3709.5	6.7171	20.54	3305.4	3705.9	6.7020	660
22.04	3331.8	3739.6	6.7606	21.42	3329.1	3736.1	6.7454	20.83	3326.4	3732.6	6.7305	670
22.35	3352.6	3766.0	6.7884	21.72	3349.9	3762.6	6.7734	21.13	3347.3	3759.3	6.7586	680
22.65	3373.2	3792.2	6.8158	22.02	3370.7	3789.0	6.8009	21.42	3368.1	3785.8	6.7863	690
22.95	3393.8	3818.4	6.8429	22.31	3391.4	3815.3	6.8281	21.71	3388.9	3812.2	6.8135	700
23.55	3434.9	3870.5	6.8959	22.89	3432.6	3867.6	6.8813	22.28	3430.3	3864.7	6.8669	720
24.13	3475.8	3922.3	6.9475	23.47	3473.7	3919.6	6.9331	22.84	3471.5	3916.9	6.9190	740
24.71	3516.7	3973.9	6.9980	24.04	3514.6	3971.3	6.9837	23.39	3512.6	3968.8	6.9697	760
25.29	3557.5	4025.3	7.0472	24.60	3555.5	4022.9	7.0331	23.94	3553.6	4020.5	7.0193	780
25.86	3598.2	4076.5	7.0954	25.15	3596.3	4074.3	7.0814	24.49	3594.5	4072.0	7.0678	800
27.26	3699.9	4204.2	7.2117	26.52	3698.3	4202.3	7.1980	25.82	3696.7	4200.3	7.1846	850
28.64	3801.8	4331.6	7.3227	27.87	3800.3	4329.8	7.3091	27.14	3798.9	4328.1	7.2959	900
30.00	3904.0	4459.0	7.4290	29.20	3902.7	4457.4	7.4156	28.44	3901.4	4455.9	7.4026	950
31.35	4006.8	4586.7	7.5313	30.51	4005.6	4585.3	7.5181	29.72	4004.4	4583.9	7.5051	1000
34.01	4214.7	4843.9	7.7258	33.11	4213.5	4842.7	7.7127	32.26	4212.4	4841.5	7.6999	1100
36.66	4426.1	5104.3	7.9089	35.69	4425.0	5103.2	7.8958	34.78	4423.9	5102.1	7.8831	1200
39.30	4641.3	5368.3	8.0822	38.27	4640.2	5367.3	8.0692	37.29	4639.1	5366.2	8.0565	1300

Definitions of symbols on page 1.

Table 3. Vapor

p (t Sat.)	20 (365.81)				21 (369.89)				22 (373.80)			
t	$10^3 v$	u	h	s	$10^3 v$	u	n	s	$10^3 v$	u	h	s
Sat.	5.834	2293.0	2409.7	4.9269	4.952	2230.6	2334.6	4.8013	3.568	2087.1	2165.6	4.5327
365	*5.430*	*2255.1*	*2363.7*	*4.8548*								
370	6.898	2386.5	2524.4	5.1059	5.023	2238.1	2343.6	4.8152				
375	7.666	2449.2	2602.5	5.2269	6.455	2375.4	2511.0	5.0747	4.824	2239.0	2345.1	4.8101
380	8.255	2495.2	2660.3	5.3158	7.207	2440.7	2592.0	5.1993	6.101	2368.3	2502.6	5.0522
385	8.751	2532.7	2707.8	5.3881	7.782	2488.2	2651.6	5.2903	6.822	2434.5	2584.6	5.1774
390	9.186	2564.9	2748.7	5.4500	8.264	2526.8	2700.4	5.3641	7.377	2483.0	2645.3	5.2692
395	9.580	2593.5	2785.1	5.5047	8.688	2559.9	2742.3	5.4271	7.843	2522.3	2694.9	5.3437
400	9.942	2619.3	2818.1	5.5540	9.071	2589.1	2779.6	5.4826	8.253	2556.0	2737.6	5.4074
405	10.279	2643.0	2848.5	5.5990	9.422	2615.5	2813.3	5.5326	8.623	2585.7	2775.5	5.4635
410	10.596	2665.0	2876.9	5.6407	9.749	2639.7	2844.4	5.5783	8.964	2612.6	2809.8	5.5139
415	10.897	2685.6	2903.6	5.6796	10.057	2662.1	2873.3	5.6205	9.280	2637.2	2841.4	5.5600
420	11.183	2705.1	2928.8	5.7162	10.348	2683.2	2900.5	5.6598	9.578	2660.0	2870.8	5.6025
425	11.458	2723.7	2952.9	5.7507	10.625	2703.1	2926.2	5.6968	9.860	2681.4	2898.4	5.6422
430	11.722	2741.5	2975.9	5.7836	10.891	2722.0	2950.7	5.7317	10.128	2701.6	2924.5	5.6794
435	11.976	2758.5	2998.0	5.8150	11.146	2740.0	2974.1	5.7649	10.385	2720.8	2949.3	5.7146
440	12.223	2774.9	3019.4	5.8450	11.392	2757.4	2996.6	5.7965	10.632	2739.1	2973.0	5.7481
445	12.462	2790.8	3040.0	5.8739	11.631	2774.0	3018.3	5.8268	10.870	2756.7	2995.9	5.7799
450	12.695	2806.2	3060.1	5.9017	11.862	2790.2	3039.3	5.8560	11.101	2773.6	3017.9	5.8105
460	13.144	2835.8	3098.7	5.9547	12.306	2821.0	3079.5	5.9111	11.542	2805.9	3059.8	5.8681
470	13.572	2864.0	3135.5	6.0046	12.729	2850.3	3117.6	5.9629	11.960	2836.3	3099.4	5.9218
480	13.985	2891.2	3170.8	6.0518	13.135	2878.4	3154.2	6.0117	12.360	2865.3	3137.2	5.9723
490	14.383	2917.4	3205.0	6.0969	13.525	2905.4	3189.4	6.0582	12.744	2893.1	3173.5	6.0202
500	14.768	2942.9	3238.2	6.1401	13.903	2931.5	3223.5	6.1026	13.115	2920.0	3208.6	6.0658
510	15.143	2967.7	3270.6	6.1817	14.269	2957.0	3256.6	6.1452	13.473	2946.1	3242.6	6.1095
520	15.508	2992.0	3302.2	6.2218	14.625	2981.8	3289.0	6.1862	13.821	2971.6	3275.6	6.1515
530	15.865	3015.8	3333.1	6.2606	14.972	3006.2	3320.6	6.2258	14.160	2996.4	3307.9	6.1919
540	16.213	3039.3	3363.5	6.2982	15.312	3030.1	3351.6	6.2642	14.491	3020.8	3339.6	6.2311
550	16.555	3062.4	3393.5	6.3348	15.644	3053.6	3382.1	6.3015	14.815	3044.7	3370.6	6.2691
560	16.890	3085.2	3423.0	6.3705	15.969	3076.8	3412.1	6.3377	15.131	3068.3	3401.2	6.3059
570	17.220	3107.7	3452.1	6.4052	16.289	3099.7	3441.7	6.3730	15.442	3091.6	3431.3	6.3418
580	17.544	3130.0	3480.9	6.4391	16.603	3122.3	3471.0	6.4075	15.747	3114.5	3461.0	6.3768
590	17.863	3152.1	3509.4	6.4723	16.912	3144.7	3499.8	6.4412	16.046	3137.2	3490.3	6.4110
600	18.178	3174.0	3537.6	6.5048	17.216	3166.9	3528.4	6.4741	16.341	3159.7	3519.2	6.4444
610	18.488	3195.8	3565.5	6.5366	17.516	3188.9	3556.8	6.5064	16.632	3182.0	3547.9	6.4770
620	18.795	3217.4	3593.3	6.5679	17.812	3210.8	3584.8	6.5380	16.918	3204.2	3576.4	6.5090
630	19.097	3238.8	3620.8	6.5985	18.104	3232.5	3612.7	6.5690	17.201	3226.1	3604.5	6.5404
640	19.397	3260.2	3648.1	6.6286	18.393	3254.1	3640.3	6.5994	17.480	3247.9	3632.5	6.5712
650	19.693	3281.4	3675.3	6.6582	18.678	3275.5	3667.8	6.6293	17.756	3269.6	3660.2	6.6014
660	19.986	3302.6	3702.3	6.6873	18.960	3296.9	3695.1	6.6587	18.028	3291.2	3687.8	6.6311
670	20.276	3323.6	3729.2	6.7160	19.240	3318.2	3722.2	6.6876	18.298	3312.6	3715.2	6.6603
680	20.563	3344.6	3755.9	6.7442	19.517	3339.3	3749.2	6.7161	18.565	3334.0	3742.4	6.6890
690	20.848	3365.6	3782.5	6.7719	19.791	3360.4	3776.0	6.7441	18.830	3355.3	3769.5	6.7173
700	21.13	3386.4	3809.0	6.7993	20.06	3381.4	3802.8	6.7717	19.092	3376.4	3796.5	6.7451
720	21.69	3428.0	3861.7	6.8529	20.60	3423.3	3855.9	6.8258	19.609	3418.6	3850.0	6.7996
740	22.24	3469.3	3914.1	6.9052	21.13	3464.9	3908.6	6.8784	20.119	3460.5	3903.2	6.8526
760	22.78	3510.5	3966.2	6.9561	21.65	3506.4	3961.1	6.9296	20.621	3502.3	3955.9	6.9042
780	23.32	3551.6	4018.1	7.0058	22.17	3547.7	4013.2	6.9796	21.117	3543.8	4008.4	6.9545
800	23.85	3592.7	4069.7	7.0544	22.68	3589.0	4065.2	7.0285	21.61	3585.3	4060.6	7.0036
850	25.16	3695.1	4198.3	7.1715	23.93	3691.9	4194.4	7.1461	22.81	3688.6	4190.4	7.1218
900	26.45	3797.5	4326.4	7.2830	25.16	3794.6	4322.9	7.2581	23.99	3791.7	4319.5	7.2342
950	27.71	3900.1	4454.3	7.3898	26.37	3897.4	4451.2	7.3652	25.15	3894.8	4448.1	7.3416
1000	28.97	4003.1	4582.5	7.4925	27.57	4000.7	4579.6	7.4681	26.30	3998.2	4576.8	7.4448
1100	31.45	4211.3	4840.2	7.6874	29.94	4209.1	4837.8	7.6633	28.57	4206.8	4835.4	7.6402
1200	33.91	4422.8	5101.0	7.8707	32.29	4420.6	5098.7	7.8467	30.82	4418.5	5096.5	7.8238
1300	36.36	4638.0	5365.1	8.0442	34.63	4635.7	5363.0	8.0202	33.06	4633.5	5360.8	7.9974

Definitions of symbols on page 1.

10^3 v	u	h	s	10^3 v	u	h	s	10^3 v	u	h	s	t
1.8172	1716.9	1758.7	3.8991	1.7866	1706.1	1749.0	3.8811	1.7610	1696.7	1740.7	3.8654	365
1.9410	1772.8	1817.5	3.9909	1.8873	1756.1	1801.4	3.9629	1.8470	1742.8	1789.0	3.9407	370
2.211	1861.3	1912.2	4.1375	2.054	1821.6	1870.9	4.0705	1.9731	1798.7	1848.0	4.0320	375
4.721	2248.6	2357.2	4.8212	2.609	1962.2	2024.8	4.3068	2.2135	1879.1	1934.5	4.1649	380
5.820	2365.2	2499.0	5.0377	4.657	2261.5	2373.3	4.8386	3.1707	2075.7	2154.9	4.5009	385
6.499	2430.8	2580.3	5.1608	5.598	2365.4	2499.7	5.0301	4.6126	2275.7	2391.0	4.8585	390
7.029	2479.5	2641.2	5.2523	6.229	2429.4	2578.9	5.1491	5.4213	2368.4	2504.0	5.0282	395
7.478	2519.3	2691.2	5.3269	6.732	2477.8	2639.4	5.2393	6.004	2430.1	2580.2	5.1418	400
7.873	2553.3	2734.4	5.3907	7.160	2517.6	2689.4	5.3133	6.477	2477.7	2639.6	5.2299	405
8.230	2583.4	2772.7	5.4470	7.539	2551.8	2732.7	5.3770	6.885	2517.2	2689.3	5.3029	410
8.558	2610.6	2807.4	5.4977	7.883	2582.1	2771.3	5.4332	7.247	2551.3	2732.5	5.3659	415
8.864	2635.5	2839.4	5.5440	8.200	2609.5	2806.3	5.4839	7.577	2581.7	2771.1	5.4218	420
9.152	2658.7	2869.2	5.5868	8.495	2634.6	2838.5	5.5303	7.881	2609.2	2806.3	5.4723	425
9.425	2680.3	2897.1	5.6267	8.773	2658.0	2868.6	5.5731	8.166	2634.5	2838.7	5.5186	430
9.684	2700.8	2923.5	5.6641	9.036	2679.9	2896.8	5.6131	8.433	2658.0	2868.9	5.5614	435
9.933	2720.2	2948.7	5.6995	9.287	2700.6	2923.4	5.6506	8.687	2680.1	2897.3	5.6013	440
10.171	2738.8	2972.7	5.7331	9.527	2720.2	2948.8	5.6861	8.930	2700.9	2924.1	5.6389	445
10.402	2756.6	2995.8	5.7651	9.758	2738.9	2973.1	5.7198	9.162	2720.7	2949.7	5.6744	450
10.841	2790.3	3039.6	5.8253	10.196	2774.2	3018.9	5.7828	9.600	2757.7	2997.7	5.7403	460
11.256	2821.9	3080.8	5.8811	10.609	2807.2	3061.8	5.8408	10.011	2792.1	3042.4	5.8008	470
11.651	2852.0	3119.9	5.9334	11.000	2838.3	3102.3	5.8950	10.400	2824.4	3084.4	5.8570	480
12.030	2880.7	3157.4	5.9828	11.374	2868.0	3141.0	5.9459	10.769	2855.0	3124.3	5.9096	490
12.394	2908.3	3193.4	6.0297	11.733	2896.4	3178.0	5.9942	11.123	2884.3	3162.4	5.9592	500
12.746	2935.1	3228.3	6.0745	12.079	2923.9	3213.8	6.0402	11.464	2912.5	3199.1	6.0064	510
13.087	2961.1	3262.1	6.1175	12.414	2950.5	3248.5	6.0842	11.793	2939.8	3234.7	6.0515	520
13.419	2986.5	3295.2	6.1588	12.739	2976.5	3282.2	6.1265	12.112	2966.3	3269.2	6.0947	530
13.742	3011.4	3327.4	6.1988	13.055	3001.9	3315.2	6.1672	12.422	2992.2	3302.8	6.1363	540
14.057	3035.8	3359.1	6.2375	13.363	3026.7	3347.4	6.2066	12.724	3017.5	3335.6	6.1765	550
14.366	3059.7	3390.2	6.2750	13.664	3051.1	3379.0	6.2448	13.019	3042.4	3367.8	6.2154	560
14.668	3083.4	3420.7	6.3115	13.959	3075.1	3410.1	6.2819	13.307	3066.8	3399.4	6.2531	570
14.965	3106.7	3450.9	6.3470	14.248	3098.8	3440.7	6.3180	13.589	3090.8	3430.5	6.2897	580
15.256	3129.7	3480.6	6.3817	14.532	3122.1	3470.9	6.3532	13.866	3114.5	3461.1	6.3254	590
15.543	3152.5	3510.0	6.4155	14.811	3145.2	3500.7	6.3875	14.137	3137.9	3491.4	6.3602	600
15.825	3175.1	3539.1	6.4486	15.085	3168.1	3530.1	6.4210	14.404	3161.1	3521.2	6.3942	610
16.103	3197.5	3567.8	6.4810	15.355	3190.8	3559.3	6.4538	14.667	3184.0	3550.7	6.4274	620
16.377	3219.7	3596.4	6.5128	15.621	3213.2	3588.1	6.4860	14.926	3206.7	3579.9	6.4599	630
16.647	3241.7	3624.6	6.5439	15.883	3235.5	3616.7	6.5174	15.181	3229.3	3608.8	6.4917	640
16.914	3263.6	3652.7	6.5744	16.142	3257.6	3645.1	6.5483	15.433	3251.6	3637.4	6.5229	650
17.178	3285.4	3680.5	6.6044	16.398	3279.6	3673.2	6.5786	15.681	3273.8	3665.8	6.5536	660
17.439	3307.1	3708.2	6.6339	16.651	3301.5	3701.1	6.6084	15.926	3295.9	3694.0	6.5836	670
17.697	3328.6	3735.7	6.6629	16.901	3323.2	3728.9	6.6376	16.169	3317.8	3722.0	6.6131	680
17.953	3350.1	3763.0	6.6914	17.148	3344.9	3756.4	6.6664	16.409	3339.6	3749.9	6.6422	690
18.206	3371.4	3790.2	6.7195	17.393	3366.4	3783.8	6.6947	16.646	3361.3	3777.5	6.6707	700
18.705	3413.9	3844.1	6.7744	17.877	3409.2	3838.2	6.7501	17.114	3404.5	3832.3	6.7265	720
19.197	3456.1	3897.7	6.8278	18.351	3451.7	3892.1	6.8038	17.574	3447.3	3886.6	6.7806	740
19.681	3498.1	3950.8	6.8797	18.819	3494.0	3945.6	6.8561	18.027	3489.8	3940.5	6.8332	760
20.158	3539.9	4003.6	6.9303	19.280	3536.0	3998.7	6.9070	18.472	3532.1	3993.9	6.8845	780
20.63	3581.7	4056.2	6.9798	19.735	3578.0	4051.6	6.9567	18.912	3574.3	4047.1	6.9345	800
21.79	3685.4	4186.5	7.0985	20.852	3682.1	4182.6	7.0760	19.991	3678.9	4178.7	7.0543	850
22.92	3788.8	4316.0	7.2113	21.944	3785.9	4312.5	7.1892	21.045	3783.0	4309.1	7.1680	900
24.04	3892.2	4445.0	7.3190	23.018	3889.5	4442.0	7.2973	22.080	3886.9	4438.9	7.2763	950
25.14	3995.8	4574.0	7.4224	24.08	3993.4	4571.3	7.4009	23.10	3990.9	4568.5	7.3802	1000
27.32	4204.6	4833.0	7.6181	26.17	4202.4	4830.6	7.5969	25.12	4200.2	4828.2	7.5765	1100
29.48	4416.3	5094.3	7.8018	28.25	4414.2	5092.1	7.7807	27.11	4412.0	5089.9	7.7605	1200
31.62	4631.3	5358.7	7.9755	30.31	4629.1	5356.6	7.9544	29.10	4626.9	5354.4	7.9342	1300

Definitions of symbols on page 1.

Table 3. Vapor

p	26				27				28			
t	$10^3 v$	u	h	s	$10^3 v$	u	h	s	$10^3 v$	u	h	s
365	1.7389	1688.3	1733.6	3.8514	1.7195	1680.8	1727.2	3.8387	1.7021	1673.8	1721.4	3.8270
370	1.8146	1731.7	1778.8	3.9220	1.7875	1721.9	1770.2	3.9058	1.7641	1713.2	1762.6	3.8913
375	1.9180	1781.8	1831.7	4.0039	1.8760	1768.2	1818.9	3.9812	1.8422	1756.8	1808.3	3.9621
380	2.0813	1845.5	1899.6	4.1083	2.0028	1823.5	1877.6	4.0714	1.9472	1806.8	1861.3	4.0434
385	2.4446	1948.0	2011.5	4.2789	2.2202	1897.6	1957.6	4.1934	2.1054	1868.3	1927.3	4.1441
390	3.5144	2144.8	2236.2	4.6188	2.7317	2021.5	2095.3	4.4018	2.3944	1954.6	2021.7	4.2869
395	4.5781	2290.4	2409.5	4.8793	3.7090	2189.7	2289.8	4.6940	3.0007	2086d	2170.1	4.5099
400	5.280	2373.6	2510.9	5.0306	4.550	2305.3	2428.2	4.9003	3.831	2223.5	2330.7	4.7494
405	5.815	2432.5	2583.7	5.1383	5.165	2380.5	2520.0	5.0363	4.526	2320.1	2446.9	4.9213
410	6.258	2479.0	2641.7	5.2236	5.655	2436.5	2589.1	5.1379	5.070	2388.6	2530.6	5.0444
415	6.645	2518.0	2690.8	5.2952	6.070	2481.6	2645.5	5.2202	5.519	2441.7	2596.2	5.1401
420	6.990	2552.0	2733.7	5.3573	6.435	2520.0	2693.7	5.2899	5.907	2485.4	2650.8	5.2191
425	7.306	2582.3	2772.2	5.4127	6.764	2553.6	2736.2	5.3510	6.251	2522.9	2698.0	5.2869
430	7.598	2609.8	2807.4	5.4628	7.065	2583.7	2774.5	5.4056	6.563	2556.1	2739.9	5.3467
435	7.871	2635.1	2839.8	5.5088	7.345	2611.1	2809.5	5.4552	6.851	2585.9	2777.8	5.4004
440	8.129	2658.7	2870.1	5.5515	7.608	2636.5	2841.9	5.5009	7.120	2613.2	2812.6	5.4494
445	8.374	2680.9	2898.6	5.5913	7.856	2660.1	2872.2	5.5432	7.372	2638.5	2844.9	5.4946
450	8.608	2701.8	2925.6	5.6288	8.092	2682.3	2900.8	5.5829	7.610	2662.0	2875.1	5.5366
460	9.048	2740.8	2976.0	5.6980	8.534	2723.3	2953.7	5.6556	8.055	2705.3	2930.8	5.6131
470	9.458	2776.6	3022.5	5.7610	8.944	2760.8	3002.3	5.7214	8.465	2744.5	2981.6	5.6818
480	9.844	2810.2	3066.1	5.8193	9.328	2795.6	3047.5	5.7818	8.848	2780.8	3028.5	5.7446
490	10.210	2841.8	3107.3	5.8736	9.692	2828.4	3090.1	5.8380	9.210	2814.7	3072.6	5.8027
500	10.560	2872.0	3146.6	5.9248	10.038	2859.5	3130.5	5.8907	9.553	2846.8	3114.3	5.8570
510	10.897	2901.0	3184.3	5.9732	10.370	2889.3	3169.3	5.9405	9.881	2877.4	3154.1	5.9081
520	11.221	2929.0	3220.7	6.0194	10.690	2917.9	3206.6	5.9878	10.197	2906.8	3192.3	5.9566
530	11.534	2956.1	3256.0	6.0636	10.999	2945.7	3242.6	6.0330	10.501	2935.2	3229.2	6.0028
540	11.838	2982.5	3290.3	6.1061	11.298	2972.6	3277.7	6.0763	10.795	2962.7	3265.0	6.0471
550	12.134	3008.3	3323.8	6.1470	11.588	2998.9	3311.8	6.1181	11.081	2989.5	3299.8	6.0896
560	12.423	3033.5	3356.5	6.1866	11.871	3024.6	3345.2	6.1583	11.359	3015.7	3333.7	6.1307
570	12.705	3058.4	3388.7	6.2249	12.147	3049.9	3377.8	6.1973	11.630	3041.3	3366.9	6.1703
580	12.981	3082.8	3420.3	6.2621	12.417	3074.7	3409.9	6.2351	11.894	3066.5	3399.5	6.2087
590	13.251	3106.8	3451.3	6.2983	12.682	3099.0	3441.5	6.2719	12.153	3091.2	3431.5	6.2460
600	13.516	3130.5	3482.0	6.3336	12.941	3123.1	3472.5	6.3077	12.407	3115.6	3463.0	6.2823
610	13.776	3154.0	3512.2	6.3681	13.195	3146.9	3503.1	6.3425	12.655	3139.7	3494.1	6.3176
620	14.033	3177.2	3542.0	6.4017	13.445	3170.4	3533.4	6.3766	12.900	3163.5	3524.7	6.3521
630	14.285	3200.2	3571.6	6.4346	13.691	3193.6	3563.3	6.4099	13.140	3187.0	3554.9	6.3858
640	14.533	3223.0	3600.8	6.4668	13.933	3216.6	3592.8	6.4424	13.376	3210.3	3584.8	6.4187
650	14.778	3245.5	3629.8	6.4983	14.171	3239.5	3622.1	6.4743	13.609	3233.3	3614.4	6.4509
660	15.019	3268.0	3658.5	6.5292	14.407	3262.1	3651.1	6.5055	13.838	3256.2	3643.7	6.4825
670	15.258	3290.2	3686.9	6.5596	14.639	3284.6	3679.8	6.5362	14.064	3278.9	3672.7	6.5134
680	15.493	3312.4	3715.2	6.5894	14.868	3306.9	3708.4	6.5663	14.288	3301.4	3701.5	6.5438
690	15.727	3334.4	3743.3	6.6187	15.095	3329.1	3736.7	6.5958	14.508	3323.8	3730.1	6.5736
700	15.957	3356.3	3771.1	6.6475	15.319	3351.2	3764.8	6.6248	14.727	3346.1	3758.4	6.6029
720	16.411	3399.7	3826.4	6.7037	15.760	3395.0	3820.5	6.6815	15.156	3390.2	3814.5	6.6600
740	16.857	3442.8	3881.1	6.7582	16.193	3438.4	3875.6	6.7364	15.577	3433.9	3870.0	6.7153
760	17.295	3485.6	3935.3	6.8112	16.618	3481.4	3930.1	6.7897	15.990	3477.2	3924.9	6.7689
780	17.727	3528.1	3989.1	6.8627	17.037	3524.2	3984.2	6.8416	16.397	3520.2	3979.4	6.8211
800	18.153	3570.6	4042.5	6.9130	17.450	3566.8	4038.0	6.8922	16.797	3563.1	4033.4	6.8720
850	19.196	3675.6	4174.7	7.0334	18.460	3672.4	4170.8	7.0131	17.777	3669.1	4166.9	6.9935
900	20.214	3780.1	4305.7	7.1475	19.446	3777.2	4302.2	7.1276	18.732	3774.3	4298.8	7.1084
950	21.214	3884.3	4435.8	7.2561	20.412	3881.6	4432.8	7.2366	19.668	3879.0	4429.7	7.2177
1000	22.20	3988.5	4565.7	7.3602	21.37	3986.1	4562.9	7.3409	20.59	3983.6	4560.2	7.3223
1100	24.15	4198.0	4825.8	7.5568	23.24	4195.8	4823.4	7.5378	22.41	4193.6	4821.0	7.5195
1200	26.07	4409.9	5087.7	7.7409	25.10	4407.7	5085.5	7.7221	24.20	4405.6	5083.3	7.7039
1300	27.98	4624.7	5352.3	7.9147	26.95	4622.5	5350.2	7.8959	25.99	4620.3	5348.1	7.8778

Definitions of symbols on page 1.

	29				**30**				**31**			**p**
10^3 v	u	h	s	10^3 v	u	h	s	10^3 v	u	h	s	t
1.6864	1667.3	1716.2	3.8162	1.6720	1661.3	1711.5	3.8061	1.6588	1655.7	1707.1	3.7966	**365**
1.7435	1705.4	1755.9	3.8781	1.7252	1698.1	1749.9	3.8661	1.7086	1691.5	1744.4	3.8549	**370**
1.8137	1746.7	1799.3	3.9454	1.7892	1737.8	1791.5	3.9305	1.7676	1729.7	1784.5	3.9169	**375**
1.9042	1793.1	1848.3	4.0206	1.8691	1781.4	1837.5	4.0012	1.8395	1771.2	1828.2	3.9841	**380**
2.0302	1847.4	1906.3	4.1090	1.9749	1830.9	1890.2	4.0816	1.9313	1817.3	1877.1	4.0588	**385**
2.2287	1916.3	1980.9	4.2220	2.1270	1890.2	1954.0	4.1782	2.0554	1870.4	1934.1	4.1450	**390**
2.5927	2012.9	2088.1	4.3829	2.3720	1966.5	2037.6	4.3037	2.2379	1934.9	2004.2	4.2503	**395**
3.214	2137.9	2231.1	4.5962	2.790	2067.4	2151.1	4.4728	2.527	2016.6	2095.0	4.3856	**400**
3.913	2251.3	2364.8	4.7941	3.376	2179.7	2281.0	4.6651	2.966	2115.4	2207.4	4.5520	**405**
4.505	2334.9	2465.5	4.9421	3.972	2275.6	2394.8	4.8324	3.499	2214.4	2322.9	4.7216	**410**
4.991	2397.7	2542.4	5.0543	4.487	2349.4	2484.0	4.9625	4.016	2297.5	2422.0	4.8662	**415**
5.403	2448.0	2604.7	5.1444	4.924	2407.4	2555.1	5.0655	4.470	2363.7	2502.3	4.9825	**420**
5.765	2490.2	2657.4	5.2202	5.303	2455.1	2614.2	5.1504	4.866	2417.6	2568.5	5.0776	**425**
6.089	2526.8	2703.4	5.2859	5.641	2495.8	2665.0	5.2229	5.217	2462.9	2624.6	5.1578	**430**
6.386	2559.4	2744.6	5.3442	5.947	2531.5	2709.9	5.2865	5.532	2502.1	2673.6	5.2272	**435**
6.660	2588.9	2782.0	5.3969	6.228	2563.4	2750.3	5.3434	5.820	2536.8	2717.3	5.2886	**440**
6.917	2616.0	2816.5	5.4452	6.489	2592.5	2787.2	5.3950	6.087	2568.2	2756.9	5.3440	**445**
7.159	2641.1	2848.7	5.4898	6.735	2619.3	2821.4	5.4424	6.336	2596.8	2793.2	5.3944	**450**
7.607	2686.8	2907.4	5.5704	7.187	2667.7	2883.3	5.5274	6.793	2648.0	2858.6	5.4842	**460**
8.018	2727.9	2960.4	5.6422	7.600	2710.8	2938.8	5.6026	7.207	2693.3	2916.7	5.5630	**470**
8.401	2765.6	3009.2	5.7075	7.982	2750.1	2989.6	5.6705	7.590	2734.3	2969.6	5.6336	**480**
8.760	2800.7	3054.8	5.7676	8.340	2786.5	3036.7	5.7327	7.947	2772.1	3018.4	5.6981	**490**
9.101	2833.8	3097.8	5.8236	8.678	2820.7	3081.1	5.7905	8.283	2807.4	3064.1	5.7576	**500**
9.426	2865.3	3138.7	5.8761	9.001	2853.1	3123.1	5.8445	8.603	2840.7	3107.4	5.8132	**510**
9.738	2895.5	3177.9	5.9259	9.309	2884.0	3163.3	5.8955	8.908	2872.4	3148.6	5.8654	**520**
10.038	2924.5	3215.6	5.9732	9.605	2913.7	3201.9	5.9439	9.201	2902.9	3188.1	5.9149	**530**
10.328	2952.6	3252.1	6.0184	9.891	2942.5	3239.2	5.9900	9.483	2932.2	3226.2	5.9621	**540**
10.609	2980.0	3287.6	6.0617	10.168	2970.3	3275.4	6.0342	9.756	2960.6	3263.1	6.0072	**550**
10.882	3006.6	3322.2	6.1035	10.437	2997.5	3310.6	6.0768	10.021	2988.3	3298.9	6.0505	**560**
11.148	3032.7	3356.0	6.1438	10.698	3024.0	3344.9	6.1178	10.278	3015.2	3333.9	6.0922	**570**
11.407	3058.3	3389.1	6.1828	10.953	3050.0	3378.6	6.1574	10.528	3041.6	3368.0	6.1324	**580**
11.661	3083.4	3421.6	6.2207	11.202	3075.5	3411.5	6.1958	10.773	3067.5	3401.5	6.1714	**590**
11.910	3108.1	3453.5	6.2574	11.446	3100.5	3443.9	6.2331	11.012	3092.9	3434.3	6.2092	**600**
12.153	3132.5	3484.9	6.2932	11.684	3125.2	3475.8	6.2694	11.246	3117.9	3466.6	6.2460	**610**
12.392	3156.6	3515.9	6.3281	11.918	3149.6	3507.1	6.3047	11.475	3142.6	3498.3	6.2818	**620**
12.627	3180.3	3546.5	6.3622	12.148	3173.7	3538.1	6.3392	11.701	3167.0	3529.7	6.3166	**630**
12.858	3203.9	3576.7	6.3955	12.374	3197.5	3568.7	6.3729	11.922	3191.0	3560.6	6.3507	**640**
13.085	3227.2	3606.6	6.4281	12.596	3221.0	3598.9	6.4058	12.139	3214.8	3591.1	6.3840	**650**
13.309	3250.3	3636.2	6.4600	12.815	3244.3	3628.8	6.4380	12.353	3238.4	3621.3	6.4165	**660**
13.530	3273.2	3665.6	6.4912	13.031	3267.5	3658.4	6.4695	12.564	3261.7	3651.2	6.4483	**670**
13.747	3295.9	3694.6	6.5218	13.243	3290.4	3687.7	6.5004	12.772	3284.9	3680.8	6.4795	**680**
13.963	3318.5	3723.4	6.5519	13.453	3313.2	3716.8	6.5308	12.977	3307.8	3710.1	6.5102	**690**
14.175	3340.9	3752.0	6.5814	13.661	3335.8	3745.6	6.5606	13.180	3330.6	3739.2	6.5402	**700**
14.594	3385.4	3808.6	6.6390	14.069	3380.6	3802.6	6.6186	13.578	3375.8	3796.7	6.5987	**720**
15.003	3429.4	3864.5	6.6947	14.468	3424.9	3858.9	6.6747	13.967	3420.4	3853.4	6.6552	**740**
15.405	3473.0	3919.8	6.7487	14.859	3468.8	3914.6	6.7291	14.349	3464.6	3909.4	6.7099	**760**
15.801	3516.3	3974.5	6.8012	15.244	3512.3	3969.6	6.7819	14.724	3508.3	3964.8	6.7630	**780**
16.190	3559.3	4028.8	6.8523	15.623	3555.5	4024.2	6.8332	15.093	3551.8	4019.7	6.8147	**800**
17.141	3665.8	4162.9	6.9745	16.548	3662.6	4159.0	6.9560	15.993	3659.3	4155.1	6.9380	**850**
18.068	3771.4	4295.4	7.0898	17.448	3768.5	4291.9	7.0718	16.868	3765.6	4288.5	7.0542	**900**
18.975	3876.4	4426.7	7.1995	18.329	3873.8	4423.6	7.1817	17.725	3871.1	4420.6	7.1645	**950**
19.869	3981.2	4557.4	7.3042	19.196	3978.8	4554.7	7.2867	18.567	3976.4	4551.9	7.2697	**1000**
21.630	4191.4	4818.6	7.5017	20.903	4189.2	4816.3	7.4845	20.224	4187.0	4813.9	7.4678	**1100**
23.369	4403.4	5081.1	7.6862	22.589	4401.3	5079.0	7.6692	21.860	4399.2	5076.8	7.6526	**1200**
25.099	4618.2	5346.0	7.8602	24.266	4616.0	5344.0	7.8432	23.486	4613.8	5341.9	7.8267	**1300**

Definitions of symbols on page 1.

Table 3. Vapor

p		32				33				34		
t	$10^3\,v$	u	h	s	$10^3\,v$	u	h	s	$10^3\,v$	u	h	s
365	1.6465	1650.3	1703.0	3.7876	1.6350	1645.2	1699.2	3.7791	1.6243	1640.4	1695.6	3.7710
370	1.6934	1685.2	1739.4	3.8445	1.6795	1679.4	1734.8	3.8347	1.6665	1673.9	1730.5	3.8254
375	1.7483	1722.2	1778.2	3.9045	1.7308	1715.4	1772.5	3.8930	1.7149	1708.9	1767.2	3.8823
380	1.8139	1762.0	1820.1	3.9688	1.7914	1753.7	1812.8	3.9550	1.7712	1746.1	1806.3	3.9423
385	1.8953	1805.5	1866.2	4.0392	1.8648	1795.1	1856.7	4.0219	1.8383	1785.8	1848.3	4.0064
390	2.0009	1854.4	1918.4	4.1182	1.9571	1840.8	1905.4	4.0957	1.9206	1829.1	1894.4	4.0761
395	2.1465	1911.3	1980.0	4.2108	2.0786	1892.6	1961.2	4.1795	2.0252	1877.1	1945.9	4.1536
400	2.361	1980.4	2055.9	4.3239	2.247	1953.2	2027.3	4.2781	2.164	1931.7	2005.3	4.2421
405	2.683	2064.3	2150.2	4.4635	2.491	2025.3	2107.5	4.3967	2.356	1995.2	2075.3	4.3457
410	3.116	2157.0	2256.8	4.6200	2.828	2108.1	2201.4	4.5347	2.620	2068.4	2157.5	4.4663
415	3.594	2244.2	2359.3	4.7695	3.240	2193.2	2300.1	4.6786	2.959	2147.5	2248.1	4.5986
420	4.049	2317.6	2447.2	4.8969	3.669	2270.6	2391.7	4.8113	3.343	2225.0	2338.7	4.7297
425	4.455	2377.9	2520.4	5.0022	4.074	2336.5	2470.9	4.9251	3.730	2294.5	2421.3	4.8485
430	4.816	2428.2	2582.3	5.0905	4.441	2391.8	2538.4	5.0214	4.094	2354.3	2493.5	4.9515
435	5.141	2471.3	2635.7	5.1662	4.772	2439.0	2596.5	5.1038	4.428	2405.6	2556.1	5.0403
440	5.436	2509.0	2683.0	5.2327	5.074	2480.1	2647.5	5.1756	4.734	2450.0	2611.0	5.1175
445	5.708	2542.8	2725.5	5.2920	5.350	2516.5	2693.1	5.2392	5.014	2489.2	2659.7	5.1856
450	5.961	2573.5	2764.2	5.3458	5.607	2549.3	2734.4	5.2965	5.274	2524.4	2703.7	5.2466
460	6.423	2627.8	2833.3	5.4407	6.074	2607.0	2807.4	5.3969	5.745	2585.6	2780.9	5.3527
470	6.839	2675.4	2894.2	5.5232	6.492	2657.0	2871.2	5.4834	6.166	2638.2	2847.9	5.4434
480	7.221	2718.1	2949.2	5.5968	6.875	2701.6	2928.5	5.5600	6.550	2684.8	2907.5	5.5232
490	7.578	2757.3	2999.8	5.6635	7.231	2742.3	2981.0	5.6291	6.904	2727.1	2961.9	5.5948
500	7.912	2793.8	3047.0	5.7250	7.564	2780.0	3029.6	5.6925	7.236	2766.1	3012.1	5.6603
510	8.229	2828.1	3091.5	5.7821	7.879	2815.4	3075.4	5.7513	7.549	2802.5	3059.1	5.7207
520	8.532	2860.7	3133.7	5.8357	8.179	2848.8	3118.7	5.8063	7.846	2836.8	3103.6	5.7771
530	8.822	2891.9	3174.1	5.8864	8.465	2880.7	3160.1	5.8581	8.131	2869.5	3145.9	5.8302
540	9.100	2921.8	3213.1	5.9345	8.741	2911.4	3199.8	5.9073	8.403	2900.8	3186.5	5.8804
550	9.370	2950.8	3250.7	5.9805	9.007	2941.0	3238.2	5.9542	8.666	2931.0	3225.6	5.9282
560	9.631	2979.0	3287.2	6.0246	9.264	2969.6	3275.4	5.9991	8.920	2960.2	3263.5	5.9739
570	9.884	3006.4	3322.7	6.0670	9.514	2997.5	3311.5	6.0422	9.166	2988.6	3300.2	6.0178
580	10.130	3033.2	3357.4	6.1079	9.757	3024.8	3346.7	6.0837	9.405	3016.3	3336.0	6.0600
590	10.371	3059.5	3391.3	6.1475	9.993	3051.4	3381.2	6.1239	9.638	3043.3	3371.0	6.1007
600	10.606	3085.3	3424.6	6.1858	10.224	3077.6	3415.0	6.1628	9.865	3069.8	3405.2	6.1401
610	10.835	3110.6	3457.3	6.2230	10.450	3103.2	3448.1	6.2005	10.087	3095.8	3438.8	6.1784
620	11.060	3135.6	3489.5	6.2593	10.671	3128.5	3480.7	6.2372	10.304	3121.4	3471.8	6.2155
630	11.281	3160.2	3521.2	6.2946	10.887	3153.4	3512.7	6.2729	10.517	3146.6	3504.2	6.2516
640	11.498	3184.5	3552.5	6.3290	11.100	3178.0	3544.3	6.3077	10.726	3171.5	3536.2	6.2868
650	11.711	3208.6	3583.3	6.3626	11.309	3202.3	3575.5	6.3417	10.931	3196.1	3567.7	6.3212
660	11.921	3232.4	3613.8	6.3955	11.514	3226.4	3606.3	6.3749	11.132	3220.3	3598.8	6.3547
670	12.127	3255.9	3644.0	6.4276	11.717	3250.2	3636.8	6.4073	11.331	3244.3	3629.6	6.3875
680	12.331	3279.3	3673.9	6.4591	11.916	3273.7	3666.9	6.4391	11.526	3268.1	3660.0	6.4196
690	12.531	3302.5	3703.5	6.4900	12.112	3297.1	3696.8	6.4703	11.718	3291.7	3690.1	6.4510
700	12.729	3325.4	3732.8	6.5203	12.306	3320.2	3726.4	6.5008	11.908	3315.0	3719.9	6.4818
720	13.118	3370.9	3790.7	6.5792	12.687	3366.1	3784.7	6.5602	12.280	3361.2	3778.8	6.5416
740	13.498	3415.9	3847.8	6.6361	13.058	3411.3	3842.3	6.6175	12.644	3406.8	3836.7	6.5994
760	13.871	3460.3	3904.2	6.6912	13.422	3456.1	3899.0	6.6730	13.000	3451.8	3893.8	6.6552
780	14.237	3504.3	3959.9	6.7447	13.779	3500.4	3955.1	6.7268	13.349	3496.4	3950.2	6.7093
800	14.597	3548.0	4015.1	6.7966	14.130	3544.3	4010.6	6.7790	13.691	3540.5	4006.0	6.7618
850	15.474	3656.0	4151.2	6.9205	14.985	3652.7	4147.3	6.9035	14.526	3649.5	4143.4	6.8869
900	16.325	3762.7	4285.1	7.0372	15.815	3759.8	4281.7	7.0206	15.335	3756.9	4278.3	7.0044
950	17.158	3868.5	4417.6	7.1478	16.626	3865.9	4414.6	7.1315	16.125	3863.3	4411.5	7.1156
1000	17.977	3974.0	4549.2	7.2533	17.423	3971.5	4546.5	7.2372	16.901	3969.1	4543.8	7.2216
1100	19.587	4184.8	4811.6	7.4516	18.989	4182.6	4809.3	7.4359	18.426	4180.4	4806.9	7.4206
1200	21.176	4397.0	5074.7	7.6366	20.533	4394.9	5072.5	7.6210	19.929	4392.8	5070.4	7.6058
1300	22.756	4611.6	5339.8	7.8107	22.069	4609.4	5337.7	7.7951	21.424	4607.3	5335.7	7.7800

Definitions of symbols on page 1.

	35				36				37			p
10^3 v	u	h	s	10^3 v	u	h	s	10^3 v	u	h	s	t
1.6142	1635.8	1692.3	3.7632	1.6046	1631.4	1689.2	3.7558	1.5956	1627.2	1686.2	3.7486	365
1.6545	1668.6	1726.5	3.8166	1.6432	1663.7	1722.8	3.8083	1.6326	1658.9	1719.3	3.8003	370
1.7003	1702.9	1762.4	3.8722	1.6867	1697.2	1757.9	3.8627	1.6741	1691.8	1753.8	3.8537	375
1.7530	1739.0	1800.3	3.9305	1.7363	1732.4	1794.9	3.9195	1.7210	1726.2	1789.9	3.9092	380
1.8149	1777.3	1840.9	3.9923	1.7939	1769.6	1834.1	3.9793	1.7750	1762.3	1828.0	3.9673	385
1.8893	1818.6	1884.7	4.0587	1.8621	1809.2	1876.2	4.0430	1.8380	1800.6	1868.6	4.0288	390
1.9816	1863.7	1933.1	4.1313	1.9448	1852.0	1922.1	4.1119	1.9131	1841.6	1912.4	4.0945	395
2.100	1914.1	1987.6	4.2126	2.048	1899.1	1972.8	4.1875	2.005	1886.0	1960.2	4.1658	400
2.257	1971.2	2050.2	4.3053	2.180	1951.5	2030.0	4.2722	2.119	1934.8	2013.3	4.2443	405
2.468	2036.5	2122.8	4.4119	2.354	2010.5	2095.3	4.3681	2.266	1989.0	2072.8	4.3319	410
2.743	2108.6	2204.6	4.5312	2.579	2076.1	2168.9	4.4755	2.453	2048.9	2139.7	4.4293	415
3.073	2182.9	2290.5	4.6556	2.857	2145.7	2248.6	4.5908	2.687	2113.5	2212.9	4.5353	420
3.428	2253.4	2373.4	4.7747	3.172	2214.8	2329.0	4.7064	2.960	2179.7	2289.2	4.6450	425
3.779	2316.4	2448.7	4.8822	3.499	2279.2	2405.2	4.8152	3.257	2243.8	2364.3	4.7522	430
4.110	2371.3	2515.2	4.9764	3.820	2336.8	2474.3	4.9132	3.560	2302.9	2434.6	4.8519	435
4.416	2419.1	2573.7	5.0588	4.122	2387.6	2536.0	5.0000	3.854	2356.1	2498.6	4.9420	440
4.699	2461.2	2625.7	5.1314	4.405	2432.5	2591.1	5.0769	4.132	2403.4	2556.3	5.0226	445
4.961	2498.7	2672.4	5.1962	4.668	2472.4	2640.5	5.1455	4.394	2445.7	2608.3	5.0947	450
5.436	2563.7	2753.9	5.3083	5.145	2541.2	2726.5	5.2636	4.871	2518.4	2698.6	5.2187	460
5.859	2619.0	2824.1	5.4033	5.569	2599.4	2799.9	5.3631	5.296	2579.5	2775.4	5.3228	470
6.243	2667.7	2886.2	5.4864	5.953	2650.3	2864.6	5.4496	5.680	2632.6	2842.7	5.4128	480
6.597	2711.6	2942.5	5.5606	6.307	2695.9	2922.9	5.5265	6.033	2679.9	2903.2	5.4925	490
6.927	2751.9	2994.4	5.6282	6.636	2737.6	2976.4	5.5962	6.361	2723.0	2958.4	5.5644	500
7.238	2789.4	3042.7	5.6903	6.945	2776.2	3026.2	5.6602	6.668	2762.8	3009.5	5.6302	510
7.534	2824.7	3088.3	5.7482	7.238	2812.4	3073.0	5.7195	6.959	2800.0	3057.5	5.6911	520
7.815	2858.2	3131.7	5.8025	7.517	2846.7	3117.3	5.7751	7.236	2835.2	3102.9	5.7479	530
8.085	2890.2	3173.2	5.8538	7.785	2879.4	3159.7	5.8275	7.501	2868.6	3146.1	5.8014	540
8.345	2921.0	3213.0	5.9026	8.041	2910.9	3200.3	5.8772	7.755	2900.7	3187.6	5.8521	550
8.595	2950.7	3251.6	5.9491	8.289	2941.2	3239.6	5.9246	8.000	2931.5	3227.5	5.9003	560
8.838	2979.6	3288.9	5.9937	8.529	2970.5	3277.6	5.9699	8.237	2961.4	3266.2	5.9465	570
9.074	3007.7	3325.3	6.0366	8.762	2999.1	3314.5	6.0135	8.466	2990.4	3303.7	5.9907	580
9.303	3035.2	3360.8	6.0779	8.988	3027.0	3350.3	6.0554	8.689	3018.7	3340.2	6.0333	590
9.527	3062.0	3395.5	6.1179	9.208	3054.2	3385.7	6.0960	8.907	3046.4	3375.9	6.0744	600
9.745	3088.4	3429.5	6.1566	9.423	3080.9	3420.2	6.1352	9.118	3073.4	3410.8	6.1141	610
9.959	3114.3	3462.9	6.1942	9.633	3107.2	3454.0	6.1732	9.325	3100.0	3445.0	6.1526	620
10.168	3139.8	3495.7	6.2307	9.839	3133.0	3487.2	6.2102	9.528	3126.1	3478.6	6.1900	630
10.373	3165.0	3528.0	6.2663	10.040	3158.4	3519.8	6.2462	9.726	3151.8	3511.6	6.2264	640
10.575	3189.8	3559.9	6.3010	10.238	3183.5	3552.0	6.2813	9.920	3177.1	3544.2	6.2618	650
10.772	3214.3	3591.3	6.3349	10.432	3208.2	3583.8	6.3155	10.111	3202.1	3576.2	6.2964	660
10.967	3238.5	3622.4	6.3680	10.623	3232.7	3615.1	6.3489	10.299	3226.8	3607.9	6.3301	670
11.158	3262.5	3653.1	6.4004	10.811	3256.9	3646.1	6.3815	10.483	3251.2	3639.1	6.3631	680
11.347	3286.3	3683.4	6.4321	10.996	3280.9	3676.7	6.4135	10.665	3275.4	3670.0	6.3953	690
11.533	3309.8	3713.5	6.4631	11.179	3304.6	3707.0	6.4448	10.844	3299.4	3700.6	6.4269	700
11.898	3356.4	3772.8	6.5234	11.536	3351.5	3766.8	6.5056	11.195	3346.6	3760.8	6.4881	720
12.253	3402.2	3831.1	6.5816	11.885	3397.7	3825.5	6.5642	11.536	3393.1	3820.0	6.5471	740
12.601	3447.5	3888.6	6.6378	12.226	3443.3	3883.4	6.6207	11.870	3439.0	3878.2	6.6040	760
12.943	3492.4	3945.4	6.6922	12.560	3488.3	3940.5	6.6755	12.197	3484.3	3935.6	6.6591	780
13.278	3536.7	4001.5	6.7450	12.887	3533.0	3996.9	6.7285	12.518	3529.2	3992.4	6.7125	800
14.093	3646.2	4139.5	6.8706	13.684	3642.9	4135.5	6.8548	13.298	3639.6	4131.6	6.8393	850
14.883	3754.0	4274.9	6.9886	14.456	3751.1	4271.5	6.9732	14.052	3748.2	4268.1	6.9582	900
15.653	3860.7	4408.5	7.1002	15.208	3858.0	4405.5	7.0851	14.787	3855.4	4402.5	7.0704	950
16.410	3966.7	4541.1	7.2064	15.946	3964.3	4538.4	7.1916	15.507	3961.9	4535.7	7.1771	1000
17.895	4178.3	4804.6	7.4057	17.394	4176.1	4802.3	7.3911	16.921	4173.9	4800.0	7.3770	1100
19.360	4390.7	5068.3	7.5910	18.822	4388.6	5066.1	7.5766	18.313	4386.4	5064.0	7.5626	1200
20.815	4605.1	5333.6	7.7653	20.240	4602.9	5331.6	7.7509	19.696	4600.8	5329.5	7.7369	1300

Definitions of symbols on page 1.

Table 3. Vapor

p		38				39				40		
t	$10^3 v$	u	h	s	$10^3 v$	u	h	s	$10^3 v$	u	h	s
365	1.5870	1623.1	1683.4	3.7417	1.5788	1619.2	1680.7	3.7351	1.5710	1615.4	1678.2	3.7287
370	1.6226	1654.3	1716.0	3.7926	1.6131	1650.0	1712.9	3.7853	1.6041	1645.8	1709.9	3.7782
375	1.6623	1686.7	1749.9	3.8451	1.6511	1681.8	1746.2	3.8369	1.6407	1677.1	1742.8	3.8290
380	1.7069	1720.4	1785.2	3.8994	1.6937	1714.8	1780.9	3.8902	1.6813	1709.6	1776.8	3.8814
385	1.7577	1755.6	1822.4	3.9561	1.7417	1749.2	1817.2	3.9455	1.7270	1743.3	1812.3	3.9355
390	1.8163	1792.7	1861.7	4.0156	1.7967	1785.3	1855.4	4.0034	1.7788	1778.5	1849.6	3.9919
395	1.8853	1832.1	1903.8	4.0788	1.8605	1823.4	1896.0	4.0644	1.8383	1815.4	1889.0	4.0511
400	1.9679	1874.5	1949.2	4.1466	1.9359	1864.0	1939.5	4.1293	1.9077	1854.6	1930.9	4.1135
405	2.0691	1920.4	1999.0	4.2203	2.0266	1907.7	1986.7	4.1991	1.9899	1896.3	1975.9	4.1802
410	2.1953	1970.8	2054.2	4.3013	2.1375	1955.0	2038.3	4.2749	2.0888	1941.1	2024.7	4.2518
415	2.3543	2026.0	2115.5	4.3907	2.2746	2006.5	2095.2	4.3578	2.2091	1989.5	2077.8	4.3294
420	2.5517	2085.8	2182.8	4.4882	2.4436	2062.1	2157.4	4.4479	2.3557	2041.6	2135.8	4.4133
425	2.787	2148.4	2254.3	4.5910	2.647	2120.9	2224.2	4.5439	2.532	2096.9	2198.1	4.5029
430	3.052	2210.8	2326.8	4.6944	2.880	2180.9	2293.2	4.6425	2.737	2154.0	2263.5	4.5962
435	3.331	2270.3	2396.8	4.7937	3.134	2239.5	2361.7	4.7395	2.964	2211.1	2329.7	4.6900
440	3.611	2324.9	2462.1	4.8856	3.395	2294.8	2427.2	4.8316	3.206	2266.1	2394.3	4.7809
445	3.882	2374.3	2521.8	4.9690	3.655	2345.6	2488.1	4.9168	3.451	2317.6	2455.7	4.8666
450	4.140	2418.7	2576.0	5.0442	3.906	2391.7	2544.1	4.9944	3.693	2365.1	2512.8	4.9459
460	4.614	2495.1	2670.5	5.1739	4.375	2471.7	2642.3	5.1293	4.151	2448.1	2614.2	5.0852
470	5.038	2559.2	2750.7	5.2826	4.796	2538.7	2725.7	5.2424	4.569	2517.9	2700.7	5.2024
480	5.422	2614.6	2820.6	5.3761	5.179	2596.4	2798.4	5.3395	4.950	2578.0	2776.0	5.3030
490	5.774	2663.8	2883.2	5.4586	5.530	2647.4	2863.1	5.4248	5.299	2630.8	2842.8	5.3912
500	6.101	2708.3	2940.1	5.5328	5.855	2693.4	2921.8	5.5013	5.622	2678.4	2903.3	5.4700
510	6.407	2749.3	2992.7	5.6004	6.159	2735.6	2975.8	5.5708	5.925	2721.9	2958.9	5.5414
520	6.696	2787.5	3041.9	5.6628	6.446	2774.9	3026.3	5.6348	6.210	2762.2	3010.5	5.6070
530	6.970	2823.5	3088.4	5.7210	6.718	2811.8	3073.8	5.6943	6.480	2799.9	3059.1	5.6678
540	7.232	2857.7	3132.5	5.7757	6.978	2846.7	3118.9	5.7501	6.737	2835.7	3105.2	5.7248
550	7.484	2890.4	3174.8	5.8273	7.227	2880.1	3162.0	5.8028	6.984	2869.7	3149.1	5.7785
560	7.726	2921.9	3215.5	5.8764	7.467	2912.!	3203.3	5.8528	7.221	2902.3	3191.2	5.8293
570	7.960	2952.3	3254.8	5.9233	7.698	2943.1	3243.3	5.9004	7.450	2933.8	3231.8	5.8778
580	8.187	2981.8	3292.9	5.9682	7.922	2973.0	3282.0	5.9460	7.671	2964.2	3271.1	5.9241
590	8.407	3010.5	3329.9	6.0114	8.139	3002.1	3319.6	5.9899	7.886	2993.8	3309.2	5.9686
600	8.621	3038.5	3366.1	6.0531	8.351	3030.6	3356.2	6.0321	8.094	3022.6	3346.4	6.0114
610	8.830	3065.9	3401.4	6.0933	8.557	3058.3	3392.0	6.0728	8.297	3050.7	3382.6	6.0527
620	9.034	3092.8	3436.1	6.1323	8.757	3085.5	3427.1	6.1123	8.495	3078.3	3418.1	6.0926
630	9.233	3119.2	3470.0	6.1701	8.954	3112.3	3461.5	6.1506	8.689	3105.3	3452.9	6.1313
640	9.428	3145.2	3503.4	6.2069	9.146	3138.5	3495.2	6.1878	8.878	3131.9	3487.0	6.1689
650	9.619	3170.8	3536.3	6.2427	9.334	3164.4	3528.4	6.2239	9.063	3158.0	3520.6	6.2054
660	9.807	3196.0	3568.7	6.2776	9.519	3189.9	3561.1	6.2592	9.245	3183.8	3553.6	6.2410
670	9.992	3220.9	3600.6	6.3117	9.700	3215.1	3593.4	6.2935	9.424	3209.2	3586.1	6.2757
680	10.173	3245.6	3632.2	6.3449	9.879	3239.9	3625.2	6.3271	9.599	3234.3	3618.2	6.3096
690	10.351	3270.0	3663.3	6.3775	10.054	3264.5	3656.6	6.3599	9.772	3259.0	3649.9	6.3426
700	10.527	3294.1	3694.1	6.4093	10.227	3288.8	3687.7	6.3920	9.941	3283.6	3681.2	6.3750
720	10.871	3341.7	3754.8	6.4710	10.565	3336.8	3748.8	6.4542	10.274	3331.9	3742.9	6.4377
740	11.207	3388.6	3814.4	6.5304	10.894	3384.0	3808.8	6.5140	10.597	3379.4	3803.3	6.4979
760	11.534	3434.7	3873.0	6.5877	11.215	3430.4	3867.8	6.5717	10.912	3426.1	3862.6	6.5559
780	11.854	3480.3	3930.8	6.6431	11.529	3476.3	3925.9	6.6274	11.220	3472.3	3921.1	6.6120
800	12.169	3525.4	3987.8	6.6967	11.837	3521.6	3983.3	6.6813	11.523	3517.8	3978.7	6.6662
850	12.932	3636.3	4127.8	6.8242	12.585	3633.0	4123.9	6.8094	12.255	3629.7	4120.0	6.7948
900	13.670	3745.3	4264.7	6.9435	13.307	3742.3	4261.3	6.9291	12.962	3739.4	4257.9	6.9150
950	14.388	3852.8	4399.5	7.0560	14.009	3850.2	4396.5	7.0420	13.650	3847.5	4393.6	7.0282
1000	15.092	3959.5	4533.0	7.1630	14.698	3957.1	4530.3	7.1491	14.324	3954.6	4527.6	7.1356
1100	16.472	4171.7	4797.7	7.3631	16.047	4169.6	4795.4	7.3496	15.642	4167.4	4793.1	7.3364
1200	17.831	4384.3	5061.9	7.5489	17.374	4382.2	5059.8	7.5355	16.940	4380.1	5057.7	7.5224
1300	19.181	4598.6	5327.5	7.7233	18.693	4596.4	5325.5	7.7100	18.229	4594.3	5323.5	7.6969

Definitions of symbols on page 1.

42				44				46				p
10^3v	u	h	s	10^3v	u	h	s	10^3v	u	h	s	t
1.5563	1608.2	1673.5	3.7164	1.5427	1601.4	1669.2	3.7048	1.5301	1594.9	1665.3	3.6939	365
1.5873	1637.8	1704.5	3.7647	1.5720	1630.4	1699.6	3.7522	1.5578	1623.4	1695.1	3.7403	370
1.6213	1668.3	1736.4	3.8142	1.6038	1660.2	1730.7	3.8004	1.5878	1652.5	1725.6	3.7875	375
1.6588	1699.7	1769.4	3.8649	1.6386	1690.7	1762.8	3.8497	1.6204	1682.3	1756.9	3.8357	380
1.7004	1732.2	1803.6	3.9171	1.6769	1722.2	1796.0	3.9003	1.6560	1712.9	1789.1	3.8848	385
1.7469	1765.9	1839.3	3.9710	1.7194	1754.6	1830.3	3.9523	1.6950	1744.4	1822.3	3.9351	390
1.7996	1801.0	1876.6	4.0271	1.7667	1788.3	1866.0	4.0059	1.7382	1776.8	1856.8	3.9868	395
1.8597	1837.8	1915.9	4.0857	1.8200	1823.2	1903.3	4.0616	1.7862	1810.3	1892.5	4.0401	400
1.9293	1876.5	1957.6	4.1474	1.8806	1859.7	1942.5	4.1195	1.8401	1845.1	1929.7	4.0952	405
2.0108	1917.6	2002.0	4.2127	1.9501	1898.0	1983.8	4.1803	1.9009	1881.3	1968.7	4.1525	410
2.1071	1961.2	2049.7	4.2823	2.0305	1938.3	2027.7	4.2442	1.9700	1919.0	2009.7	4.2122	415
2.2218	2007.8	2101.1	4.3566	2.1242	1980.8	2074.3	4.3117	2.0492	1958.5	2052.8	4.2747	420
2.358	2057.1	2156.1	4.4357	2.234	2025.6	2123.9	4.3830	2.140	1999.9	2098.3	4.3401	425
2.517	2108.7	2214.4	4.5189	2.361	2072.5	2176.4	4.4579	2.245	2043.0	2146.3	4.4086	430
2.699	2161.6	2274.9	4.6047	2.506	2121.0	2231.3	4.5358	2.364	2087.8	2196.5	4.4798	435
2.898	2214.3	2336.0	4.6906	2.669	2170.4	2287.9	4.6153	2.498	2133.7	2248.6	4.5531	440
3.110	2265.4	2396.1	4.7745	2.847	2219.5	2344.8	4.6949	2.646	2180.1	2301.9	4.6275	445
3.326	2314.0	2453.7	4.8545	3.034	2267.4	2400.9	4.7728	2.806	2226.3	2355.3	4.7017	450
3.755	2401.4	2559.1	4.9993	3.422	2356.6	2507.1	4.9186	3.148	2314.8	2459.6	4.8448	460
4.158	2476.3	2650.9	5.1237	3.802	2435.0	2602.3	5.0476	3.498	2395.2	2556.1	4.9756	470
4.531	2540.7	2731.0	5.2308	4.162	2503.4	2686.6	5.1602	3.839	2466.6	2643.2	5.0921	480
4.875	2597.4	2802.1	5.3246	4.497	2563.6	2761.5	5.2590	4.163	2529.9	2721.4	5.1952	490
5.194	2648.0	2866.1	5.4079	4.811	2617.2	2828.9	5.3468	4.468	2586.4	2791.9	5.2870	500
5.492	2694.0	2924.7	5.4832	5.104	2665.8	2890.4	5.4258	4.755	2637.4	2856.2	5.3696	510
5.773	2736.4	2978.9	5.5520	5.380	2710.4	2947.1	5.4979	5.025	2684.2	2915.4	5.4447	520
6.039	2776.0	3029.7	5.6156	5.641	2751.9	3000.1	5.5642	5.281	2727.6	2970.5	5.5138	530
6.292	2813.3	3077.6	5.6749	5.889	2790.8	3049.9	5.6259	5.525	2768.1	3022.3	5.5779	540
6.534	2848.8	3123.2	5.7306	6.127	2827.6	3097.2	5.6837	5.758	2806.4	3071.2	5.6377	550
6.766	2882.6	3166.8	5.7833	6.354	2862.7	3142.3	5.7382	5.981	2842.7	3117.8	5.6940	560
6.990	2915.2	3208.7	5.8333	6.573	2896.4	3185.6	5.7898	6.195	2877.5	3162.5	5.7473	570
7.206	2946.6	3249.2	5.8811	6.785	2928.8	3227.3	5.8390	6.402	2910.9	3205.4	5.7979	580
7.415	2977.0	3288.5	5.9268	6.989	2960.1	3267.7	5.8860	6.602	2943.2	3246.9	5.8462	590
7.619	3006.6	3326.6	5.9707	7.188	2990.6	3306.8	5.9311	6.796	2974.4	3287.0	5.8925	600
7.817	3035.5	3363.8	6.0131	7.381	3020.2	3344.9	5.9745	6.985	3004.8	3326.1	5.9369	610
8.009	3063.7	3400.1	6.0540	7.569	3049.1	3382.1	6.0164	7.168	3034.4	3364.1	5.9798	620
8.198	3091.4	3435.7	6.0936	7.752	3077.4	3418.5	6.0569	7.347	3063.3	3401.3	6.0211	630
8.381	3118.5	3470.5	6.1320	7.931	3105.1	3454.1	6.0961	7.521	3091.6	3437.6	6.0611	640
8.561	3145.2	3504.8	6.1693	8.106	3132.3	3489.0	6.1341	7.692	3119.4	3473.2	6.0999	650
8.738	3171.4	3538.4	6.2055	8.278	3159.1	3523.3	6.1711	7.859	3146.7	3508.2	6.1376	660
8.911	3197.3	3571.6	6.2409	8.446	3185.5	3557.1	6.2071	8.022	3173.5	3542.6	6.1742	670
9.081	3222.9	3604.3	6.2753	8.611	3211.4	3590.3	6.2421	8.182	3200.0	3576.4	6.2099	680
9.248	3248.1	3636.5	6.3090	8.773	3237.1	3623.1	6.2763	8.340	3226.1	3609.7	6.2447	690
9.412	3273.0	3668.3	6.3418	8.932	3262.4	3655.4	6.3097	8.495	3251.8	3642.5	6.2786	700
9.733	3322.1	3730.9	6.4055	9.243	3312.2	3718.9	6.3743	8.797	3302.3	3707.0	6.3441	720
10.046	3370.2	3792.1	6.4665	9.546	3361.0	3781.0	6.4362	9.090	3351.8	3769.9	6.4069	740
10.350	3417.5	3852.3	6.5253	9.840	3408.9	3841.9	6.4957	9.375	3400.3	3831.5	6.4671	760
10.647	3464.2	3911.4	6.5820	10.127	3456.1	3901.7	6.5531	9.653	3448.0	3892.1	6.5251	780
10.939	3510.2	3969.7	6.6368	10.408	3502.6	3960.6	6.6085	9.925	3495.0	3951.6	6.5811	800
11.644	3623.2	4112.2	6.7666	11.089	3616.6	4104.5	6.7395	10.582	3609.9	4096.7	6.7133	850
12.323	3733.6	4251.2	6.8877	11.743	3727.8	4244.5	6.8615	11.213	3721.9	4237.8	6.8362	900
12.983	3842.3	4387.6	7.0016	12.378	3837.1	4381.7	6.9760	11.825	3831.8	4375.8	6.9514	950
13.629	3949.8	4522.2	7.1095	12.998	3945.0	4516.9	7.0844	12.423	3940.2	4511.6	7.0603	1000
14.892	4163.1	4788.5	7.3109	14.211	4158.7	4784.0	7.2864	13.589	4154.4	4779.5	7.2628	1100
16.134	4375.9	5053.6	7.4972	15.403	4371.7	5049.4	7.4730	14.735	4367.5	5045.3	7.4497	1200
17.368	4590.0	5319.4	7.6718	16.585	4585.7	5315.4	7.6476	15.871	4581.4	5311.4	7.6245	1300

Definitions of symbols on page 1.

Table 3. Vapor

p	48				50				52			
t	10^3 v	u	h	s	10^3 v	u	h	s	10^3 v	u	h	s
365	1.5183	1588.8	1661.7	3.6835	1.5073	1583.1	1658.4	3.6735	1.4969	1577.5	1655.4	3.6640
370	1.5447	1616.8	1691.0	3.7291	1.5324	1610.6	1687.2	3.7184	1.5209	1604.6	1683.7	3.7083
375	1.5730	1645.4	1720.9	3.7754	1.5594	1638.6	1716.6	3.7639	1.5466	1632.2	1712.6	3.7530
380	1.6037	1674.5	1751.5	3.8225	1.5884	1667.2	1746.6	3.8101	1.5741	1660.2	1742.1	3.7984
385	1.6370	1704.4	1782.9	3.8704	1.6197	1696.4	1777.3	3.8570	1.6037	1688.8	1772.2	3.8443
390	1.6733	1734.9	1815.3	3.9194	1.6536	1726.2	1808.9	3.9047	1.6356	1718.0	1803.1	3.8910
395	1.7130	1766.4	1848.6	3.9694	1.6905	1756.7	1841.3	3.9534	1.6702	1747.8	1834.7	3.9385
400	1.7569	1798.7	1883.0	4.0208	1.7309	1788.1	1874.6	4.0031	1.7077	1778.3	1867.1	3.9869
405	1.8054	1832.1	1918.7	4.0736	1.7753	1820.3	1909.1	4.0541	1.7486	1809.6	1900.5	4.0363
410	1.8596	1866.6	1955.9	4.1282	1.8243	1853.5	1944.7	4.1064	1.7934	1841.6	1934.9	4.0868
415	1.9204	1902.4	1994.6	4.1846	1.8786	1887.7	1981.6	4.1603	1.8427	1874.6	1970.4	4.1385
420	1.9890	1939.6	2035.1	4.2433	1.9393	1923.1	2020.0	4.2159	1.8971	1908.4	2007.1	4.1917
425	2.067	1978.3	2077.5	4.3042	2.007	1959.7	2060.0	4.2734	1.9574	1943.4	2045.1	4.2464
430	2.155	2018.5	2121.9	4.3676	2.083	1997.5	2101.7	4.3329	2.0244	1979.3	2084.6	4.3027
435	2.255	2060.1	2168.3	4.4334	2.169	2036.6	2145.1	4.3943	2.0989	2016.4	2125.5	4.3607
440	2.367	2103.0	2216.6	4.5013	2.264	2076.9	2190.1	4.4577	2.1816	2054.4	2167.8	4.4202
445	2.491	2146.6	2266.2	4.5707	2.370	2118.0	2236.5	4.5225	2.2730	2093.3	2211.5	4.4812
450	2.627	2190.5	2316.6	4.6406	2.486	2159.6	2284.0	4.5884	2.373	2132.8	2256.3	4.5434
460	2.925	2276.8	2417.2	4.7787	2.745	2242.8	2380.0	4.7203	2.599	2212.6	2347.7	4.6690
470	3.242	2357.6	2513.2	4.9088	3.028	2322.7	2474.1	4.8478	2.850	2290.8	2439.0	4.7927
480	3.560	2430.9	2601.8	5.0273	3.321	2396.8	2562.9	4.9665	3.117	2364.8	2526.9	4.9102
490	3.869	2496.6	2682.3	5.1335	3.611	2464.3	2644.9	5.0747	3.387	2433.3	2609.5	5.0191
500	4.163	2555.7	2755.5	5.2288	3.892	2525.5	2720.1	5.1726	3.653	2496.0	2686.0	5.1187
510	4.442	2609.1	2822.3	5.3146	4.161	2581.0	2789.1	5.2612	3.911	2553.4	2756.7	5.2096
520	4.706	2658.0	2883.9	5.3927	4.418	2631.8	2852.7	5.3420	4.159	2606.0	2822.2	5.2927
530	4.956	2703.2	2941.1	5.4644	4.662	2678.8	2911.9	5.4161	4.396	2654.6	2883.2	5.3691
540	5.195	2745.3	2994.7	5.5308	4.895	2722.5	2967.3	5.4847	4.623	2699.8	2940.2	5.4397
550	5.423	2785.0	3045.3	5.5926	5.118	2763.6	3019.5	5.5485	4.841	2742.3	2994.0	5.5054
560	5.641	2822.6	3093.4	5.6507	5.332	2802.5	3069.1	5.6084	5.049	2782.4	3045.0	5.5670
570	5.851	2858.5	3139.4	5.7056	5.537	2839.5	3116.4	5.6648	5.250	2820.5	3093.5	5.6249
580	6.054	2893.0	3183.5	5.7576	5.735	2875.0	3161.7	5.7183	5.444	2857.0	3140.1	5.6798
590	6.249	2926.1	3226.1	5.8072	5.927	2909.1	3205.4	5.7692	5.631	2892.0	3184.8	5.7319
600	6.439	2958.2	3267.3	5.8547	6.112	2942.0	3247.6	5.8178	5.813	2925.7	3228.0	5.7816
610	6.623	2989.3	3307.3	5.9002	6.292	2973.9	3288.5	5.8643	5.988	2958.4	3269.8	5.8293
620	6.802	3019.7	3346.2	5.9440	6.467	3004.9	3328.2	5.9091	6.159	2990.1	3310.4	5.8750
630	6.976	3049.2	3384.1	5.9863	6.637	3035.1	3367.0	5.9522	6.326	3021.0	3349.9	5.9190
640	7.147	3078.2	3421.2	6.0271	6.803	3064.6	3404.8	5.9939	6.488	3051.1	3388.5	5.9615
650	7.313	3106.5	3457.5	6.0666	6.966	3093.5	3441.8	6.0342	6.646	3080.6	3426.2	6.0025
660	7.475	3134.3	3493.1	6.1050	7.124	3121.8	3478.1	6.0732	6.801	3109.4	3463.1	6.0422
670	7.635	3161.6	3528.1	6.1423	7.279	3149.7	3513.6	6.1112	6.953	3137.7	3499.2	6.0808
680	7.791	3188.5	3562.5	6.1786	7.432	3177.0	3548.6	6.1480	7.101	3165.5	3534.8	6.1183
690	7.944	3215.0	3596.3	6.2139	7.581	3203.9	3583.0	6.1839	7.246	3192.9	3569.7	6.1547
700	8.094	3241.1	3629.7	6.2483	7.727	3230.5	3616.8	6.2189	7.389	3219.8	3604.1	6.1902
720	8.388	3292.4	3695.0	6.3148	8.013	3282.5	3683.1	6.2863	7.667	3272.6	3671.3	6.2586
740	8.673	3342.5	3758.8	6.3784	8.290	3333.3	3747.8	6.3508	7.937	3324.0	3736.7	6.3239
760	8.949	3391.7	3821.2	6.4394	8.558	3383.0	3810.9	6.4125	8.198	3374.3	3800.7	6.3863
780	9.219	3439.9	3882.4	6.4981	8.820	3431.8	3872.8	6.4718	8.453	3423.7	3863.2	6.4463
800	9.482	3487.4	3942.5	6.5546	9.076	3479.8	3933.6	6.5290	8.701	3472.1	3924.6	6.5041
850	10.118	3603.3	4089.0	6.6881	9.692	3596.7	4081.3	6.6636	9.300	3590.1	4073.7	6.6399
900	10.728	3716.1	4231.1	6.8118	10.283	3710.3	4224.4	6.7882	9.872	3704.4	4217.8	6.7654
950	11.319	3826.6	4369.9	6.9277	10.854	3821.3	4364.0	6.9048	10.426	3816.0	4358.2	6.8826
1000	11.896	3935.4	4506.3	7.0370	11.411	3930.5	4501.1	7.0146	10.964	3925.7	4495.9	6.9929
1100	13.020	4150.1	4775.0	7.2402	12.496	4145.7	4770.5	7.2184	12.014	4141.4	4766.1	7.1973
1200	14.123	4363.3	5041.2	7.4273	13.561	4359.1	5037.2	7.4058	13.042	4354.9	5033.1	7.3850
1300	15.217	4577.1	5307.5	7.6022	14.616	4572.8	5303.6	7.5808	14.061	4568.5	5299.7	7.5601

Definitions of symbols on page 1.

	54				56				58			p
$10^3 v$	u	h	s	$10^3 v$	u	h	s	$10^3 v$	u	h	s	t
1.4871	1572.2	1652.5	3.6549	1.4777	1567.1	1649.9	3.6461	1.4689	1562.3	1647.4	3.6377	**365**
1.5101	1598.9	1680.5	3.6986	1.4999	1593.5	1677.5	3.6893	1.4902	1588.3	1674.7	3.6803	**370**
1.5347	1626.1	1709.0	3.7427	1.5234	1620.3	1705.6	3.7327	1.5128	1614.7	1702.5	3.7232	**375**
1.5609	1653.7	1738.0	3.7873	1.5485	1647.4	1734.2	3.7767	1.5368	1641.5	1730.6	3.7665	**380**
1.5890	1681.8	1767.6	3.8324	1.5752	1675.1	1763.3	3.8211	1.5624	1668.7	1759.3	3.8103	**385**
1.6191	1710.4	1797.8	3.8781	1.6038	1703.1	1792.9	3.8660	1.5896	1696.3	1788.5	3.8544	**390**
1.6516	1739.5	1828.7	3.9246	1.6345	1731.7	1823.2	3.9115	1.6187	1724.3	1818.2	3.8991	**395**
1.6866	1769.3	1860.3	3.9718	1.6675	1760.8	1854.2	3.9576	1.6498	1752.9	1848.6	3.9444	**400**
1.7247	1799.7	1892.8	4.0198	1.7030	1790.5	1885.9	4.0045	1.6833	1781.9	1879.5	3.9902	**405**
1.7660	1830.8	1926.1	4.0688	1.7415	1820.8	1918.3	4.0522	1.7192	1811.5	1911.2	4.0367	**410**
1.8112	1862.6	1960.4	4.1188	1.7832	1851.7	1951.6	4.1007	1.7581	1841.6	1943.6	4.0839	**415**
1.8606	1895.3	1995.8	4.1700	1.8286	1883.3	1985.7	4.1502	1.8000	1872.4	1976.8	4.1320	**420**
1.9149	1928.8	2032.2	4.2224	1.8781	1915.7	2020.9	4.2006	1.8456	1903.7	2010.8	4.1808	**425**
1.9748	1963.2	2069.9	4.2761	1.9322	1948.8	2057.0	4.2523	1.8950	1935.7	2045.7	4.2307	**430**
2.0408	1998.6	2108.8	4.3312	1.9914	1982.7	2094.2	4.3050	1.9488	1968.4	2081.5	4.2814	**435**
2.1136	2034.8	2148.9	4.3877	2.0563	2017.4	2132.5	4.3589	2.0073	2001.8	2118.2	4.3332	**440**
2.1936	2071.8	2190.2	4.4454	2.1273	2052.8	2171.9	4.4139	2.0711	2035.8	2155.9	4.3858	**445**
2.281	2109.4	2232.6	4.5042	2.205	2088.8	2212.3	4.4699	2.140	2070.4	2194.6	4.4394	**450**
2.479	2185.9	2319.7	4.6239	2.379	2162.1	2295.4	4.5841	2.296	2141.0	2274.1	4.5487	**460**
2.702	2261.9	2407.8	4.7432	2.578	2235.8	2380.2	4.6989	2.474	2212.2	2355.7	4.6592	**470**
2.943	2335.1	2494.1	4.8585	2.796	2307.7	2464.3	4.8113	2.671	2282.5	2437.4	4.7684	**480**
3.193	2403.9	2576.4	4.9670	3.025	2376.3	2545.7	4.9187	2.880	2350.4	2517.5	4.8741	**490**
3.443	2467.7	2653.6	5.0676	3.259	2440.5	2623.0	5.0194	3.097	2414.8	2594.5	4.9742	**500**
3.689	2526.4	2725.6	5.1601	3.491	2500.3	2695.8	5.1129	3.316	2475.2	2667.5	5.0681	**510**
3.926	2580.5	2792.5	5.2451	3.718	2555.6	2763.8	5.1993	3.532	2531.5	2736.3	5.1555	**520**
4.156	2630.6	2855.0	5.3234	3.939	2607.0	2827.6	5.2792	3.743	2584.0	2801.1	5.2366	**530**
4.376	2677.3	2913.6	5.3959	4.152	2655.0	2887.5	5.3533	3.949	2633.1	2862.1	5.3121	**540**
4.588	2721.0	2968.8	5.4633	4.358	2700.0	2944.0	5.4224	4.148	2679.2	2919.8	5.3826	**550**
4.791	2762.3	3021.1	5.5265	4.556	2742.4	2997.6	5.4870	4.340	2722.7	2974.5	5.4486	**560**
4.988	2801.6	3070.9	5.5859	4.747	2782.7	3048.5	5.5479	4.526	2764.0	3026.5	5.5108	**570**
5.177	2839.0	3118.6	5.6421	4.932	2821.1	3097.3	5.6053	4.706	2803.3	3076.3	5.5694	**580**
5.360	2874.9	3164.3	5.6955	5.111	2857.9	3144.1	5.6599	4.881	2841.0	3124.1	5.6251	**590**
5.537	2909.5	3208.5	5.7463	5.284	2893.3	3189.2	5.7118	5.050	2877.2	3170.1	5.6781	**600**
5.709	2942.9	3251.2	5.7950	5.452	2927.5	3232.8	5.7615	5.214	2912.1	3214.5	5.7287	**610**
5.876	2975.3	3292.6	5.8416	5.615	2960.6	3275.0	5.8091	5.374	2945.9	3257.6	5.7772	**620**
6.039	3006.8	3333.0	5.8865	5.774	2992.7	3316.1	5.8548	5.529	2978.7	3299.4	5.8237	**630**
6.197	3037.6	3372.2	5.9298	5.929	3024.1	3356.1	5.8988	5.681	3010.6	3340.1	5.8686	**640**
6.352	3067.6	3410.6	5.9716	6.080	3054.6	3395.1	5.9413	5.829	3041.7	3379.8	5.9118	**650**
6.503	3097.0	3448.1	6.0120	6.228	3084.5	3433.3	5.9824	5.973	3072.1	3418.5	5.9536	**660**
6.651	3125.7	3484.9	6.0512	6.372	3113.8	3470.7	6.0223	6.114	3101.9	3456.5	5.9940	**670**
6.796	3154.0	3521.0	6.0893	6.514	3142.5	3507.3	6.0609	6.252	3131.0	3493.7	6.0333	**680**
6.938	3181.8	3556.5	6.1263	6.653	3170.7	3543.3	6.0985	6.388	3159.7	3530.2	6.0713	**690**
7.077	3209.1	3591.3	6.1623	6.789	3198.5	3578.6	6.1350	6.521	3187.8	3566.0	6.1084	**700**
7.348	3262.7	3659.5	6.2316	7.053	3252.7	3647.7	6.2053	6.779	3242.8	3636.0	6.1796	**720**
7.611	3314.8	3725.8	6.2977	7.309	3305.5	3714.8	6.2722	7.029	3296.2	3703.9	6.2473	**740**
7.866	3365.7	3790.4	6.3609	7.558	3357.0	3780.2	6.3361	7.271	3348.3	3770.1	6.3119	**760**
8.113	3415.5	3853.7	6.4215	7.799	3407.4	3844.1	6.3974	7.507	3399.3	3834.7	6.3739	**780**
8.355	3464.5	3915.6	6.4798	8.034	3456.8	3906.7	6.4563	7.736	3449.2	3897.9	6.4333	**800**
8.937	3583.5	4066.0	6.6168	8.600	3576.8	4058.4	6.5944	8.287	3570.2	4050.9	6.5727	**850**
9.493	3698.6	4211.2	6.7433	9.140	3692.7	4204.6	6.7218	8.813	3686.8	4198.0	6.7009	**900**
10.029	3810.8	4352.3	6.8611	9.661	3805.5	4346.6	6.8403	9.320	3800.2	4340.8	6.8200	**950**
10.551	3920.9	4490.7	6.9719	10.168	3916.0	4485.5	6.9516	9.812	3911.2	4480.3	6.9319	**1000**
11.567	4137.1	4761.7	7.1769	11.153	4132.7	4757.3	7.1572	10.768	4128.4	4753.0	7.1381	**1100**
12.562	4350.7	5029.1	7.3649	12.117	4346.6	5025.1	7.3454	11.703	4342.4	5021.1	7.3266	**1200**
13.548	4564.2	5295.8	7.5401	13.072	4559.9	5291.9	7.5207	12.629	4555.6	5288.1	7.5019	**1300**

Definitions of symbols on page 1.

Table 3. Vapor

p	60				62				64			
t	$10^3 v$	u	h	s	$10^3 v$	u	h	s	$10^3 v$	u	h	s
365	1.4604	1557.6	1645.2	3.6296	1.4523	1553.0	1643.1	3.6217	1.4446	1548.6	1641.1	3.6141
370	1.4810	1583.3	1672.2	3.6717	1.4722	1578.5	1669.8	3.6634	1.4638	1573.9	1667.5	3.6553
375	1.5028	1609.4	1699.5	3.7141	1.4932	1604.3	1696.8	3.7053	1.4841	1599.3	1694.3	3.6968
380	1.5258	1635.8	1727.4	3.7568	1.5154	1630.4	1724.3	3.7475	1.5055	1625.1	1721.5	3.7386
385	1.5503	1662.6	1755.6	3.7999	1.5389	1656.8	1752.2	3.7901	1.5281	1651.3	1749.0	3.7806
390	1.5763	1689.8	1784.4	3.8435	1.5638	1683.6	1780.5	3.8330	1.5520	1677.7	1777.0	3.8229
395	1.6040	1717.4	1813.6	3.8874	1.5902	1710.8	1809.3	3.8762	1.5773	1704.5	1805.4	3.8656
400	1.6335	1745.4	1843.4	3.9318	1.6183	1738.3	1838.6	3.9199	1.6041	1731.6	1834.2	3.9086
405	1.6651	1773.9	1873.8	3.9767	1.6483	1766.3	1868.4	3.9640	1.6327	1759.1	1863.5	3.9520
410	1.6989	1802.8	1904.7	4.0222	1.6803	1794.6	1898.8	4.0086	1.6630	1786.9	1893.4	3.9958
415	1.7353	1832.2	1936.3	4.0683	1.7145	1823.5	1929.7	4.0538	1.6953	1815.2	1923.7	4.0400
420	1.7744	1862.2	1968.7	4.1151	1.7511	1852.7	1961.3	4.0994	1.7298	1843.9	1954.6	4.0847
425	1.8165	1892.7	2001.7	4.1626	1.7904	1882.5	1993.5	4.1457	1.7667	1873.0	1986.1	4.1300
430	1.8621	1923.8	2035.5	4.2109	1.8327	1912.8	2026.4	4.1927	1.8061	1902.6	2018.1	4.1758
435	1.9114	1955.5	2070.1	4.2600	1.8782	1943.6	2060.0	4.2403	1.8484	1932.6	2050.9	4.2221
440	1.9647	1987.7	2105.6	4.3099	1.9272	1974.9	2094.3	4.2886	1.8938	1963.0	2084.2	4.2690
445	2.0225	2020.5	2141.9	4.3606	1.9801	2006.7	2129.4	4.3376	1.9425	1993.9	2118.2	4.3166
450	2.085	2053.9	2179.0	4.4121	2.037	2038.9	2165.2	4.3873	1.9947	2025.2	2152.9	4.3647
460	2.225	2121.9	2255.4	4.5170	2.164	2104.7	2238.9	4.4885	2.1107	2089.1	2224.1	4.4625
470	2.385	2190.9	2334.0	4.6235	2.309	2171.6	2314.7	4.5912	2.2425	2153.9	2297.5	4.5619
480	2.563	2259.5	2413.3	4.7295	2.470	2238.5	2391.6	4.6940	2.3900	2219.1	2372.1	4.6616
490	2.755	2326.5	2491.8	4.8329	2.646	2304.2	2468.3	4.7951	2.5510	2283.6	2446.9	4.7603
500	2.956	2390.6	2567.9	4.9321	2.832	2367.8	2543.4	4.8928	2.723	2346.5	2520.7	4.8564
510	3.160	2451.2	2640.8	5.0258	3.023	2428.4	2615.9	4.9860	2.901	2406.9	2592.6	4.9487
520	3.365	2508.2	2710.1	5.1137	3.216	2485.8	2685.2	5.0740	3.083	2464.4	2661.7	5.0365
530	3.567	2561.6	2775.6	5.1958	3.408	2539.9	2751.2	5.1567	3.265	2519.0	2727.9	5.1194
540	3.764	2611.7	2837.5	5.2724	3.597	2590.8	2813.8	5.2342	3.446	2570.5	2791.0	5.1975
550	3.956	2658.8	2896.2	5.3441	3.782	2638.8	2873.3	5.3069	3.624	2619.2	2851.1	5.2710
560	4.143	2703.3	2951.9	5.4113	3.963	2684.1	2929.8	5.3752	3.798	2665.4	2908.4	5.3402
570	4.324	2745.5	3004.9	5.4746	4.138	2727.2	2983.8	5.4395	3.968	2709.2	2963.2	5.4055
580	4.499	2785.7	3055.7	5.5345	4.309	2768.3	3035.4	5.5004	4.133	2751.1	3015.6	5.4673
590	4.669	2824.2	3104.3	5.5912	4.474	2807.5	3084.9	5.5581	4.294	2791.1	3065.9	5.5259
600	4.834	2861.1	3151.2	5.6452	4.635	2845.2	3132.6	5.6130	4.451	2829.5	3114.3	5.5817
610	4.995	2896.8	3196.5	5.6967	4.792	2881.5	3178.6	5.6655	4.604	2866.5	3161.1	5.6350
620	5.151	2931.2	3240.3	5.7461	4.944	2916.7	3223.2	5.7156	4.752	2902.2	3206.3	5.6859
630	5.303	2964.6	3282.8	5.7934	5.092	2950.7	3266.4	5.7638	4.897	2936.8	3250.2	5.7348
640	5.451	2997.1	3324.2	5.8390	5.237	2983.8	3308.5	5.8101	5.039	2970.4	3292.9	5.7818
650	5.595	3028.8	3364.5	5.8829	5.378	3015.9	3349.4	5.8547	5.177	3003.2	3334.5	5.8270
660	5.736	3059.7	3403.9	5.9253	5.516	3047.4	3389.4	5.8977	5.312	3035.1	3375.0	5.8707
670	5.874	3089.9	3442.4	5.9664	5.651	3078.1	3428.5	5.9394	5.444	3066.2	3414.7	5.9130
680	6.010	3119.6	3480.2	6.0062	5.784	3108.1	3466.7	5.9797	5.573	3096.8	3453.4	5.9539
690	6.142	3148.6	3517.2	6.0448	5.913	3137.6	3504.2	6.0189	5.700	3126.6	3491.4	5.9935
700	6.272	3177.2	3553.5	6.0824	6.040	3166.6	3541.1	6.0569	5.824	3156.0	3528.7	6.0320
720	6.524	3232.9	3624.4	6.1545	6.287	3223.0	3612.8	6.1299	6.065	3213.1	3601.3	6.1059
740	6.769	3287.0	3693.1	6.2230	6.526	3277.8	3682.4	6.1992	6.299	3268.5	3671.7	6.1760
760	7.005	3339.7	3760.0	6.2883	6.757	3331.0	3749.9	6.2653	6.525	3322.4	3740.0	6.2428
780	7.235	3391.1	3825.2	6.3509	6.981	3383.0	3815.9	6.3285	6.744	3374.9	3806.5	6.3066
800	7.459	3441.5	3889.1	6.4109	7.200	3433.9	3880.3	6.3891	6.958	3426.2	3871.5	6.3678
850	7.996	3563.5	4043.3	6.5515	7.724	3556.9	4035.8	6.5308	7.470	3550.3	4028.3	6.5106
900	8.508	3681.0	4191.5	6.6805	8.223	3675.1	4185.0	6.6607	7.957	3669.2	4178.5	6.6414
950	9.001	3795.0	4335.0	6.8004	8.704	3789.7	4329.3	6.7812	8.425	3784.4	4323.6	6.7626
1000	9.480	3906.4	4475.2	6.9127	9.170	3901.5	4470.1	6.8940	8.879	3896.7	4465.0	6.8758
1100	10.409	4124.1	4748.6	7.1195	10.074	4119.7	4744.3	7.1014	9.759	4115.4	4740.0	7.0839
1200	11.317	4338.2	5017.2	7.3083	10.956	4334.0	5013.3	7.2905	10.618	4329.8	5009.4	7.2732
1300	12.215	4551.4	5284.3	7.4837	11.829	4547.1	5280.5	7.4660	11.468	4542.8	5276.7	7.4488

Definitions of symbols on page 1.

66				68				70				p
10^3 v	u	h	s	10^3 v	u	h	s	10^3 v	u	h	s	t
1.4372	1544.4	1639.3	3.6067	1.4301	1540.3	1637.5	3.5995	1.4232	1536.3	1636.0	3.5925	365
1.4558	1569.4	1665.5	3.6476	1.4482	1565.0	1663.5	3.6400	1.4408	1560.9	1661.7	3.6327	370
1.4755	1594.6	1692.0	3.6886	1.4672	1590.0	1689.8	3.6807	1.4592	1585.6	1687.7	3.6730	375
1.4961	1620.1	1718.8	3.7299	1.4871	1615.3	1716.4	3.7216	1.4786	1610.6	1714.1	3.7135	380
1.5179	1645.9	1746.1	3.7715	1.5082	1640.8	1743.3	3.7627	1.4989	1635.8	1740.8	3.7542	385
1.5409	1672.0	1773.7	3.8133	1.5303	1666.6	1770.6	3.8040	1.5203	1661.4	1767.8	3.7951	390
1.5651	1698.4	1801.7	3.8554	1.5537	1692.7	1798.3	3.8456	1.5428	1687.1	1795.1	3.8362	395
1.5908	1725.2	1830.2	3.8978	1.5783	1719.0	1826.4	3.8874	1.5664	1713.2	1822.8	3.8775	400
1.6180	1752.2	1859.0	3.9405	1.6043	1745.7	1854.8	3.9295	1.5914	1739.5	1850.9	3.9191	405
1.6469	1779.6	1888.3	3.9836	1.6319	1772.7	1883.7	3.9720	1.6178	1766.1	1879.4	3.9609	410
1.6776	1807.4	1918.1	4.0270	1.6611	1800.0	1913.0	4.0147	1.6456	1793.0	1908.2	4.0030	415
1.7102	1835.6	1948.4	4.0709	1.6920	1827.7	1942.8	4.0578	1.6751	1820.3	1937.5	4.0454	420
1.7449	1864.1	1979.3	4.1152	1.7249	1855.7	1973.0	4.1013	1.7063	1847.8	1967.2	4.0881	425
1.7820	1893.0	2010.6	4.1599	1.7598	1884.1	2003.7	4.1451	1.7394	1875.6	1997.4	4.1311	430
1.8215	1922.3	2042.5	4.2052	1.7969	1912.8	2035.0	4.1894	1.7744	1903.8	2028.0	4.1745	435
1.8638	1952.0	2075.1	4.2509	1.8365	1941.8	2066.7	4.2341	1.8116	1932.3	2059.1	4.2182	440
1.9089	1982.2	2108.1	4.2972	1.8786	1971.2	2099.0	4.2792	1.8511	1961.1	2090.6	4.2623	445
1.9572	2012.7	2141.8	4.3439	1.9235	2001.0	2131.8	4.3247	1.8931	1990.2	2122.7	4.3068	450
2.0638	2074.7	2210.9	4.4388	2.0222	2061.5	2199.0	4.4170	1.9849	2049.2	2188.2	4.3968	460
2.1847	2137.8	2282.0	4.5351	2.1336	2123.0	2268.1	4.5105	2.0881	2109.3	2255.5	4.4879	470
2.3197	2201.4	2354.5	4.6320	2.2579	2185.0	2338.6	4.6048	2.2030	2169.9	2324.1	4.5797	480
2.4678	2264.6	2427.5	4.7282	2.3944	2246.9	2409.8	4.6987	2.3293	2230.5	2393.6	4.6713	490
2.626	2326.6	2499.9	4.8225	2.541	2307.9	2480.8	4.7911	2.466	2290.6	2463.2	4.7619	500
2.793	2386.5	2570.9	4.9138	2.696	2367.4	2550.7	4.8810	2.610	2349.4	2532.1	4.8504	510
2.964	2444.1	2639.7	5.0010	2.857	2424.7	2619.0	4.9676	2.761	2406.3	2599.6	4.9361	520
3.136	2498.8	2705.8	5.0840	3.020	2479.6	2685.0	5.0503	2.915	2461.2	2665.3	5.0184	530
3.309	2550.9	2769.2	5.1624	3.184	2531.9	2748.4	5.1288	3.071	2513.7	2728.7	5.0969	540
3.479	2600.2	2829.8	5.2364	3.347	2581.7	2809.3	5.2033	3.227	2563.9	2789.7	5.1715	550
3.647	2647.0	2887.7	5.3064	3.508	2629.1	2867.7	5.2738	3.382	2611.7	2848.4	5.2424	560
3.811	2691.6	2943.1	5.3725	3.667	2674.3	2923.7	5.3405	3.534	2657.5	2904.9	5.3097	570
3.971	2734.1	2996.2	5.4351	3.822	2717.5	2977.4	5.4039	3.684	2701.2	2959.1	5.3737	580
4.128	2774.8	3047.3	5.4946	3.974	2758.8	3029.1	5.4641	3.832	2743.1	3011.3	5.4345	590
4.280	2813.9	3096.4	5.5512	4.122	2798.5	3078.8	5.5215	3.976	2783.4	3061.7	5.4925	600
4.429	2851.5	3143.8	5.6052	4.267	2836.8	3126.9	5.5762	4.117	2822.2	3110.4	5.5480	610
4.574	2887.9	3189.8	5.6569	4.409	2873.7	3173.5	5.6286	4.255	2859.7	3157.5	5.6010	620
4.716	2923.1	3234.3	5.7065	4.547	2909.4	3218.6	5.6789	4.389	2895.9	3203.2	5.6519	630
4.854	2957.2	3277.6	5.7542	4.682	2944.1	3262.5	5.7272	4.521	2931.1	3247.6	5.7008	640
4.989	2990.5	3319.7	5.8001	4.814	2977.8	3305.2	5.7737	4.650	2965.3	3290.8	5.7479	650
5.121	3022.8	3360.8	5.8443	4.943	3010.7	3346.8	5.8185	4.776	2998.6	3332.9	5.7933	660
5.250	3054.5	3401.0	5.8871	5.069	3042.8	3387.4	5.8619	4.900	3031.1	3374.1	5.8371	670
5.376	3085.4	3440.3	5.9286	5.193	3074.1	3427.2	5.9038	5.021	3062.9	3414.4	5.8796	680
5.500	3115.7	3478.7	5.9687	5.314	3104.8	3466.2	5.9445	5.139	3094.0	3453.8	5.9207	690
5.622	3145.4	3516.5	6.0077	5.433	3134.9	3504.4	5.9839	5.256	3124.5	3492.4	5.9606	700
5.850	3203.3	3589.9	6.0824	5.664	3193.3	3578.7	6.0593	5.482	3183.7	3567.3	6.0370	720
6.086	3259.3	3661.0	6.1533	5.887	3250.2	3650.5	6.1311	5.701	3241.0	3640.1	6.1094	740
6.307	3313.8	3730.1	6.2208	6.104	3305.2	3720.2	6.1993	5.913	3296.5	3710.5	6.1782	760
6.522	3366.8	3797.3	6.2852	6.314	3358.7	3788.1	6.2643	6.118	3350.6	3778.9	6.2438	780
6.731	3418.6	3862.9	6.3469	6.518	3411.0	3854.2	6.3265	6.318	3403.4	3845.7	6.3066	800
7.231	3543.6	4020.9	6.4909	7.007	3537.0	4013.5	6.4716	6.797	3530.3	4006.1	6.4528	850
7.707	3663.4	4172.0	6.6226	7.472	3657.5	4165.6	6.6042	7.252	3651.6	4159.2	6.5862	900
8.164	3779.1	4317.9	6.7444	7.919	3773.8	4312.3	6.7266	7.688	3768.5	4306.7	6.7093	950
8.607	3891.8	4459.9	6.8581	8.351	3887.0	4454.9	6.8408	8.110	3882.1	4449.8	6.8240	1000
9.465	4111.1	4735.7	7.0667	9.188	4106.7	4731.5	7.0501	8.927	4102.4	4727.3	7.0338	1100
10.301	4325.6	5005.5	7.2564	10.003	4321.5	5001.7	7.2400	9.722	4317.3	4997.8	7.2240	1200
11.128	4538.6	5273.0	7.4321	10.809	4534.3	5269.3	7.4158	10.509	4530.0	5265.6	7.3999	1300

Definitions of symbols on page 1.

Table 3. Vapor

t	$10^3 v$	u	h	s	$10^3 v$	u	h	s	$10^3 v$	u	h	s
		72				**74**				**76**		
365	1.4166	1532.5	1634.5	3.5858	1.4102	1528.7	1633.1	3.5792	1.4041	1525.1	1631.8	3.5727
370	1.4337	1556.8	1660.0	3.6256	1.4269	1552.9	1658.4	3.6187	1.4203	1549.0	1657.0	3.6120
375	1.4516	1581.3	1685.8	3.6656	1.4443	1577.2	1684.0	3.6584	1.4372	1573.2	1682.4	3.6514
380	1.4704	1606.1	1711.9	3.7057	1.4625	1601.7	1709.9	3.6982	1.4549	1597.5	1708.0	3.6908
385	1.4900	1631.1	1738.4	3.7460	1.4816	1626.5	1736.1	3.7381	1.4734	1622.0	1734.0	3.7304
390	1.5107	1656.3	1765.1	3.7865	1.5016	1651.5	1762.6	3.7782	1.4928	1646.8	1760.2	3.7701
395	1.5324	1681.8	1792.1	3.8271	1.5225	1676.7	1789.3	3.8184	1.5131	1671.7	1786.7	3.8099
400	1.5552	1707.5	1819.5	3.8679	1.5445	1702.1	1816.4	3.8588	1.5344	1696.9	1813.5	3.8499
405	1.5792	1733.6	1847.2	3.9090	1.5677	1727.8	1843.8	3.8993	1.5567	1722.4	1840.7	3.8900
410	1.6045	1759.8	1875.3	3.9503	1.5920	1753.8	1871.6	3.9401	1.5801	1748.0	1868.1	3.9303
415	1.6312	1786.4	1903.8	3.9918	1.6176	1780.0	1899.7	3.9811	1.6047	1773.9	1895.8	3.9708
420	1.6593	1813.2	1932.6	4.0335	1.6445	1806.4	1928.1	4.0222	1.6305	1800.0	1923.9	4.0114
425	1.6890	1840.3	1961.9	4.0756	1.6728	1833.1	1956.9	4.0636	1.6577	1826.3	1952.3	4.0522
430	1.7204	1867.7	1991.5	4.1179	1.7028	1860.1	1986.1	4.1053	1.6862	1852.9	1981.0	4.0933
435	1.7536	1895.3	2021.6	4.1604	1.7343	1887.3	2015.6	4.1471	1.7163	1879.7	2010.1	4.1345
440	1.7887	1923.3	2052.1	4.2033	1.7676	1914.8	2045.6	4.1893	1.7480	1906.7	2039.6	4.1759
445	1.8260	1951.5	2083.0	4.2466	1.8028	1942.5	2075.9	4.2317	1.7813	1934.0	2069.4	4.2176
450	1.8654	1980.0	2114.3	4.2901	1.8399	1970.5	2106.7	4.2743	1.8165	1961.5	2099.6	4.2595
460	1.9513	2037.8	2178.3	4.3779	1.9206	2027.2	2169.3	4.3604	1.8926	2017.2	2161.0	4.3438
470	2.0473	2096.6	2244.0	4.4669	2.0105	2084.7	2233.5	4.4473	1.9770	2073.6	2223.8	4.4290
480	2.1540	2155.9	2310.9	4.5564	2.1100	2142.8	2298.9	4.5348	2.0701	2130.6	2287.9	4.5146
490	2.2713	2215.3	2378.8	4.6459	2.2192	2201.1	2365.3	4.6223	2.1722	2187.8	2352.9	4.6003
500	2.398	2274.3	2447.0	4.7347	2.338	2259.1	2432.1	4.7093	2.283	2244.9	2418.4	4.6856
510	2.533	2332.4	2514.8	4.8218	2.464	2316.4	2498.8	4.7949	2.401	2301.4	2483.9	4.7698
520	2.675	2388.9	2581.5	4.9065	2.597	2372.5	2564.6	4.8785	2.526	2356.9	2548.9	4.8522
530	2.820	2443.6	2646.7	4.9881	2.734	2426.9	2629.3	4.9595	2.656	2411.0	2612.9	4.9324
540	2.968	2496.2	2709.9	5.0664	2.875	2479.5	2692.2	5.0374	2.790	2463.4	2675.4	5.0098
550	3.117	2546.6	2771.1	5.1411	3.017	2530.0	2753.2	5.1120	2.925	2514.0	2736.3	5.0842
560	3.266	2594.9	2830.0	5.2122	3.159	2578.5	2812.3	5.1833	3.061	2562.7	2795.3	5.1555
570	3.412	2641.0	2886.7	5.2799	3.300	2625.0	2869.3	5.2513	3.197	2609.5	2852.5	5.2237
580	3.557	2685.3	2941.4	5.3444	3.440	2669.7	2924.3	5.3161	3.332	2654.5	2907.7	5.2888
590	3.700	2727.7	2994.1	5.4058	3.578	2712.6	2977.4	5.3780	3.465	2697.8	2961.2	5.3511
600	3.840	2768.5	3045.0	5.4644	3.714	2753.9	3028.7	5.4371	3.596	2739.6	3012.9	5.4107
610	3.977	2807.8	3094.2	5.5205	3.847	2793.7	3078.4	5.4937	3.726	2779.8	3063.0	5.4677
620	4.111	2845.8	3141.8	5.5741	3.977	2832.2	3126.5	5.5479	3.853	2818.7	3111.5	5.5224
630	4.242	2882.6	3188.0	5.6256	4.105	2869.4	3173.2	5.5999	3.977	2856.4	3158.7	5.5749
640	4.371	2918.2	3233.0	5.6750	4.231	2905.5	3218.6	5.6499	4.100	2893.0	3204.5	5.6254
650	4.497	2952.9	3276.7	5.7227	4.354	2940.6	3262.8	5.6981	4.220	2928.5	3249.2	5.6740
660	4.620	2986.7	3319.3	5.7686	4.474	2974.8	3305.9	5.7445	4.337	2963.1	3292.7	5.7209
670	4.741	3019.6	3360.9	5.8130	4.592	3008.1	3348.0	5.7893	4.453	2996.8	3335.2	5.7662
680	4.859	3051.8	3401.6	5.8559	4.708	3040.7	3389.1	5.8327	4.566	3029.7	3376.8	5.8100
690	4.975	3083.2	3441.5	5.8975	4.822	3072.5	3429.4	5.8747	4.677	3061.9	3417.4	5.8524
700	5.089	3114.1	3480.5	5.9378	4.933	3103.7	3468.8	5.9154	4.786	3093.5	3457.2	5.8936
720	5.311	3174.0	3556.4	6.0150	5.150	3164.3	3545.5	5.9934	4.999	3154.7	3534.6	5.9723
740	5.525	3231.9	3629.7	6.0881	5.360	3222.8	3619.5	6.0672	5.205	3213.8	3609.3	6.0468
760	5.733	3288.0	3700.8	6.1575	5.563	3279.5	3691.2	6.1373	5.404	3271.0	3681.7	6.1175
780	5.934	3342.6	3769.8	6.2238	5.761	3334.6	3760.9	6.2041	5.597	3326.6	3751.9	6.1849
800	6.130	3395.8	3837.1	6.2871	5.952	3388.2	3828.7	6.2679	5.785	3380.6	3820.3	6.2492
850	6.599	3523.7	3998.8	6.4344	6.412	3517.1	3991.6	6.4163	6.235	3510.5	3984.4	6.3986
900	7.044	3645.7	4152.9	6.5686	6.847	3639.9	4146.6	6.5514	6.662	3634.0	4140.3	6.5345
950	7.470	3763.2	4301.1	6.6923	7.265	3757.9	4295.6	6.6757	7.071	3752.6	4290.0	6.6595
1000	7.883	3877.3	4444.8	6.8075	7.669	3872.4	4439.9	6.7914	7.466	3867.5	4435.0	6.7756
1100	8.681	4098.0	4723.1	7.0179	8.449	4093.7	4718.9	7.0024	8.229	4089.3	4714.8	6.9873
1200	9.458	4313.1	4994.0	7.2084	9.208	4308.9	4990.3	7.1932	8.971	4304.7	4986.5	7.1783
1300	10.225	4525.8	5262.0	7.3844	9.957	4521.5	5258.3	7.3693	9.704	4517.3	5254.7	7.3545

Definitions of symbols on page 1.

	78				80				82			p
$10^3 v$	u	h	s	$10^3 v$	u	h	s	$10^3 v$	u	h	s	t
1.3981	1521.6	1630.6	3.5665	1.3924	1518.1	1629.5	3.5604	1.3868	1514.8	1628.5	3.5544	365
1.4139	1545.3	1655.6	3.6055	1.4078	1541.7	1654.3	3.5991	1.4018	1538.2	1653.1	3.5929	370
1.4304	1569.3	1680.8	3.6445	1.4239	1565.5	1679.4	3.6379	1.4175	1561.8	1678.0	3.6314	375
1.4476	1593.4	1706.3	3.6837	1.4406	1589.4	1704.6	3.6767	1.4338	1585.5	1703.1	3.6700	380
1.4656	1617.7	1732.0	3.7229	1.4581	1613.5	1730.2	3.7157	1.4509	1609.5	1728.4	3.7086	385
1.4844	1642.2	1758.0	3.7623	1.4764	1637.9	1756.0	3.7547	1.4686	1633.6	1754.0	3.7473	390
1.5041	1667.0	1784.3	3.8017	1.4955	1662.4	1782.0	3.7938	1.4872	1657.9	1779.8	3.7861	395
1.5247	1691.9	1810.8	3.8413	1.5154	1687.1	1808.3	3.8330	1.5065	1682.4	1805.9	3.8250	400
1.5462	1717.1	1837.7	3.8810	1.5362	1712.0	1834.9	3.8724	1.5267	1707.1	1832.2	3.8640	405
1.5688	1742.4	1864.8	3.9209	1.5581	1737.1	1861.7	3.9118	1.5478	1731.9	1858.8	3.9030	410
1.5925	1768.0	1892.2	3.9609	1.5809	1762.4	1888.8	3.9514	1.5699	1756.9	1885.7	3.9422	415
1.6173	1793.8	1919.9	4.0010	1.6048	1787.9	1916.2	3.9911	1.5929	1782.2	1912.8	3.9814	420
1.6434	1819.8	1948.0	4.0413	1.6299	1813.6	1943.9	4.0309	1.6171	1807.6	1940.2	4.0208	425
1.6707	1846.0	1976.3	4.0818	1.6561	1839.5	1971.9	4.0708	1.6423	1833.2	1967.8	4.0603	430
1.6995	1872.5	2005.0	4.1224	1.6836	1865.5	2000.2	4.1109	1.6687	1858.9	1995.8	4.0999	435
1.7297	1899.1	2034.0	4.1633	1.7125	1891.8	2028.8	4.1512	1.6964	1884.9	2024.0	4.1396	440
1.7614	1926.0	2063.4	4.2042	1.7428	1918.3	2057.7	4.1915	1.7254	1911.0	2052.5	4.1794	445
1.7948	1953.0	2093.0	4.2454	1.7746	1945.0	2086.9	4.2321	1.7557	1937.3	2081.3	4.2193	450
1.8668	2007.7	2153.4	4.3283	1.8429	1998.8	2146.3	4.3135	1.8208	1990.4	2139.7	4.2996	460
1.9463	2063.2	2215.0	4.4118	1.9182	2053.3	2206.8	4.3955	1.8921	2044.0	2199.2	4.3802	470
2.0339	2119.1	2277.8	4.4956	2.0007	2108.3	2268.4	4.4779	1.9702	2098.2	2259.7	4.4611	480
2.1296	2175.3	2341.4	4.5797	2.0908	2163.6	2330.9	4.5603	2.0553	2152.6	2321.1	4.5421	490
2.233	2231.5	2405.7	4.6633	2.188	2218.9	2394.0	4.6425	2.147	2207.0	2383.1	4.6228	500
2.345	2287.2	2470.1	4.7461	2.293	2273.9	2457.3	4.7239	2.246	2261.3	2445.4	4.7029	510
2.462	2342.1	2534.2	4.8274	2.404	2328.2	2520.5	4.8040	2.351	2314.9	2507.7	4.7819	520
2.585	2395.8	2597.5	4.9067	2.520	2381.4	2583.1	4.8824	2.461	2367.7	2569.5	4.8593	530
2.712	2448.1	2659.6	4.9833	2.640	2433.4	2644.6	4.9586	2.575	2419.4	2630.6	4.9349	540
2.841	2498.6	2720.2	5.0576	2.763	2483.9	2704.9	5.0323	2.692	2469.7	2690.5	5.0081	550
2.971	2547.4	2779.2	5.1288	2.888	2532.7	2763.7	5.1033	2.811	2518.5	2749.1	5.0788	560
3.101	2594.5	2836.4	5.1971	3.013	2579.9	2821.0	5.1716	2.931	2565.8	2806.2	5.1470	570
3.231	2639.8	2891.8	5.2625	3.138	2625.4	2876.5	5.2371	3.052	2611.5	2861.8	5.2125	580
3.360	2683.4	2945.5	5.3250	3.262	2669.4	2930.4	5.2999	3.172	2655.7	2915.8	5.2755	590
3.487	2725.5	2997.5	5.3850	3.386	2711.8	2982.7	5.3601	3.291	2698.4	2968.3	5.3360	600
3.613	2766.2	3048.0	5.4424	3.507	2752.8	3033.4	5.4179	3.409	2739.8	3019.3	5.3941	610
3.736	2805.5	3096.9	5.4975	3.627	2792.5	3082.7	5.4734	3.526	2779.8	3068.9	5.4499	620
3.858	2843.6	3144.5	5.5505	3.745	2831.0	3130.6	5.5267	3.640	2818.6	3117.1	5.5036	630
3.977	2880.6	3190.8	5.6014	3.862	2868.3	3177.3	5.5781	3.754	2856.3	3164.1	5.5553	640
4.094	2916.5	3235.8	5.6505	3.976	2904.6	3222.7	5.6276	3.865	2893.0	3209.9	5.6052	650
4.209	2951.4	3279.7	5.6978	4.088	2940.0	3267.0	5.6753	3.975	2928.6	3254.6	5.6533	660
4.322	2985.6	3322.7	5.7436	4.198	2974.4	3310.3	5.7215	4.082	2963.4	3298.2	5.6999	670
4.433	3018.9	3364.6	5.7878	4.307	3008.1	3352.6	5.7661	4.188	2997.4	3340.9	5.7449	680
4.541	3051.4	3405.6	5.8307	4.413	3041.0	3394.0	5.8093	4.292	3030.6	3382.6	5.7884	690
4.648	3083.3	3445.8	5.8722	4.518	3073.2	3434.6	5.8512	4.395	3063.1	3423.5	5.8307	700
4.856	3145.1	3523.9	5.9516	4.722	3135.6	3513.4	5.9314	4.595	3126.2	3503.0	5.9115	720
5.058	3204.8	3599.3	6.0268	4.919	3195.9	3589.4	6.0072	4.788	3187.0	3579.6	5.9879	740
5.253	3262.5	3672.3	6.0981	5.111	3254.1	3663.0	6.0791	4.976	3245.7	3653.8	6.0604	760
5.443	3318.6	3743.1	6.1660	5.296	3310.6	3734.3	6.1475	5.158	3302.7	3725.7	6.1293	780
5.627	3373.1	3812.0	6.2308	5.477	3365.6	3803.8	6.2128	5.336	3358.1	3795.6	6.1951	800
6.068	3503.9	3977.2	6.3813	5.910	3497.3	3970.1	6.3643	5.760	3490.7	3963.1	6.3477	850
6.487	3628.1	4134.1	6.5180	6.320	3622.3	4127.9	6.5018	6.163	3616.4	4121.8	6.4859	900
6.887	3747.4	4284.6	6.6436	6.713	3742.1	4279.1	6.6280	6.548	3736.8	4273.7	6.6128	950
7.274	3862.7	4430.1	6.7602	7.093	3857.8	4425.2	6.7451	6.920	3852.9	4420.3	6.7303	1000
8.021	4085.0	4710.6	6.9724	7.824	4080.6	4706.6	6.9579	7.637	4076.3	4702.5	6.9437	1100
8.747	4300.5	4982.8	7.1638	8.534	4296.4	4979.1	7.1495	8.332	4292.2	4975.5	7.1356	1200
9.464	4513.0	5251.2	7.3400	9.236	4508.8	5247.6	7.3259	9.019	4504.5	5244.1	7.3120	1300

Definitions of symbols on page 1.

Table 3.   Vapor

p		84				86				88		
t	$10^3 v$	u	h	s	$10^3 v$	u	h	s	$10^3 v$	u	h	s
365	1.3814	1511.5	1627.5	3.5486	1.3761	1508.3	1626.6	3.5429	1.3710	1505.2	1625.8	3.5373
370	1.3961	1534.8	1652.0	3.5868	1.3905	1531.4	1651.0	3.5809	1.3851	1528.2	1650.0	3.5751
375	1.4114	1558.2	1676.7	3.6251	1.4054	1554.7	1675.6	3.6189	1.3997	1551.3	1674.5	3.6129
380	1.4273	1581.8	1701.7	3.6634	1.4210	1578.1	1700.3	3.6570	1.4149	1574.6	1699.1	3.6507
385	1.4439	1605.5	1726.8	3.7018	1.4372	1601.7	1725.3	3.6951	1.4306	1598.0	1723.9	3.6886
390	1.4612	1629.5	1752.2	3.7402	1.4540	1625.5	1750.5	3.7332	1.4470	1621.6	1748.9	3.7264
395	1.4792	1653.6	1777.8	3.7787	1.4715	1649.4	1775.9	3.7714	1.4641	1645.3	1774.1	3.7644
400	1.4980	1677.9	1803.7	3.8172	1.4898	1673.5	1801.6	3.8097	1.4819	1669.2	1799.6	3.8023
405	1.5176	1702.3	1829.8	3.8558	1.5088	1697.7	1827.4	3.8480	1.5004	1693.2	1825.3	3.8403
410	1.5380	1726.9	1856.1	3.8945	1.5286	1722.1	1853.6	3.8863	1.5196	1717.4	1851.2	3.8783
415	1.5594	1751.7	1882.7	3.9333	1.5493	1746.7	1879.9	3.9247	1.5397	1741.8	1877.3	3.9164
420	1.5816	1776.7	1909.5	3.9722	1.5709	1771.4	1906.5	3.9632	1.5605	1766.3	1903.6	3.9546
425	1.6049	1801.8	1936.6	4.0111	1.5933	1796.3	1933.3	4.0018	1.5823	1790.9	1930.2	3.9927
430	1.6292	1827.1	1964.0	4.0502	1.6168	1821.3	1960.4	4.0404	1.6050	1815.7	1957.0	4.0310
435	1.6546	1852.6	1991.6	4.0893	1.6413	1846.5	1987.7	4.0791	1.6286	1840.7	1984.0	4.0693
440	1.6812	1878.2	2019.5	4.1285	1.6668	1871.9	2015.2	4.1178	1.6532	1865.7	2011.2	4.1076
445	1.7090	1904.0	2047.6	4.1678	1.6935	1897.4	2043.0	4.1567	1.6789	1891.0	2038.7	4.1460
450	1.7380	1930.0	2076.0	4.2072	1.7214	1923.0	2071.0	4.1956	1.7056	1916.3	2066.4	4.1844
460	1.8001	1982.3	2133.6	4.2863	1.7808	1974.7	2127.8	4.2736	1.7626	1967.4	2122.5	4.2614
470	1.8680	2035.2	2192.1	4.3656	1.8455	2026.9	2185.6	4.3518	1.8245	2018.9	2179.4	4.3386
480	1.9421	2088.6	2251.7	4.4452	1.9160	2079.4	2244.2	4.4302	1.8916	2070.8	2237.2	4.4158
490	2.0226	2142.2	2312.0	4.5248	1.9923	2132.3	2303.6	4.5085	1.9643	2122.9	2295.8	4.4930
500	2.110	2195.8	2373.0	4.6042	2.075	2185.2	2363.6	4.5867	2.043	2175.1	2354.9	4.5700
510	2.203	2249.3	2434.4	4.6831	2.163	2238.0	2424.1	4.6643	2.127	2227.3	2414.4	4.6465
520	2.302	2302.4	2495.8	4.7610	2.257	2290.5	2484.6	4.7411	2.216	2279.1	2474.1	4.7223
530	2.406	2354.7	2556.8	4.8375	2.356	2342.3	2544.9	4.8167	2.310	2330.5	2533.7	4.7970
540	2.515	2406.0	2617.3	4.9123	2.459	2393.2	2604.7	4.8907	2.408	2381.0	2592.9	4.8702
550	2.626	2456.1	2676.8	4.9850	2.566	2443.1	2663.8	4.9629	2.510	2430.6	2651.5	4.9418
560	2.740	2504.9	2735.1	5.0554	2.675	2491.7	2721.8	5.0329	2.614	2479.1	2709.1	5.0114
570	2.856	2552.2	2792.1	5.1234	2.785	2539.0	2778.6	5.1007	2.720	2526.3	2765.7	5.0789
580	2.971	2598.0	2847.6	5.1889	2.897	2584.9	2834.1	5.1661	2.827	2572.3	2821.1	5.1442
590	3.087	2642.4	2901.8	5.2520	3.009	2629.5	2888.2	5.2292	2.935	2616.9	2875.2	5.2073
600	3.203	2685.4	2954.4	5.3126	3.120	2672.6	2941.0	5.2900	3.043	2660.2	2928.0	5.2681
610	3.317	2727.0	3005.6	5.3709	3.231	2714.5	2992.4	5.3485	3.150	2702.3	2979.5	5.3268
620	3.430	2767.3	3055.5	5.4271	3.341	2755.1	3042.4	5.4049	3.257	2743.1	3029.8	5.3833
630	3.542	2806.5	3104.0	5.4811	3.450	2794.5	3091.2	5.4592	3.363	2782.8	3078.7	5.4379
640	3.652	2844.5	3151.3	5.5332	3.557	2832.8	3138.7	5.5116	3.467	2821.4	3126.5	5.4905
650	3.761	2881.5	3197.4	5.5834	3.663	2870.1	3185.1	5.5621	3.571	2859.0	3173.2	5.5413
660	3.868	2917.5	3242.4	5.6319	3.767	2906.5	3230.4	5.6109	3.672	2895.6	3218.8	5.5904
670	3.973	2952.6	3286.3	5.6787	3.870	2941.9	3274.7	5.6581	3.773	2931.3	3263.3	5.6379
680	4.077	2986.9	3329.3	5.7241	3.971	2976.5	3318.0	5.7038	3.872	2966.2	3306.9	5.6839
690	4.179	3020.4	3371.4	5.7680	4.071	3010.3	3360.4	5.7480	3.969	3000.3	3349.6	5.7284
700	4.279	3053.2	3412.6	5.8106	4.169	3043.4	3401.9	5.7909	4.065	3033.7	3391.4	5.7716
720	4.475	3116.9	3492.7	5.8921	4.361	3107.6	3482.6	5.8730	4.253	3098.4	3472.7	5.8543
740	4.664	3178.2	3570.0	5.9690	4.547	3169.4	3560.4	5.9505	4.435	3160.7	3551.0	5.9324
760	4.848	3237.4	3644.7	6.0421	4.727	3229.1	3635.7	6.0241	4.613	3220.9	3626.8	6.0065
780	5.027	3294.8	3717.1	6.1115	4.903	3287.0	3708.6	6.0941	4.785	3279.2	3700.3	6.0769
800	5.201	3350.7	3787.6	6.1778	5.074	3343.2	3779.6	6.1608	4.952	3335.8	3771.7	6.1441
850	5.618	3484.2	3956.1	6.3313	5.483	3477.6	3949.1	6.3152	5.354	3471.1	3942.3	6.2995
900	6.013	3610.6	4115.7	6.4704	5.870	3604.8	4109.6	6.4551	5.735	3598.9	4103.6	6.4401
950	6.391	3731.5	4268.3	6.5978	6.241	3726.2	4263.0	6.5831	6.099	3720.9	4257.7	6.5687
1000	6.756	3848.1	4415.5	6.7158	6.600	3843.2	4410.8	6.7015	6.451	3838.3	4406.0	6.6875
1100	7.459	4071.9	4698.4	6.9297	7.289	4067.6	4694.4	6.9161	7.128	4063.2	4690.4	6.9027
1200	8.140	4288.0	4971.8	7.1219	7.958	4283.8	4968.2	7.1086	7.783	4279.7	4964.6	7.0954
1300	8.813	4500.3	5240.6	7.2985	8.617	4496.1	5237.1	7.2852	8.430	4491.8	5233.7	7.2722

Definitions of symbols on page 1.

	90				95				100			p
10^3 v	u	h	s	10^3 v	u	h	s	10^3 v	u	h	s	t
1.3661	1502.1	1625.1	3.5318	1.3543	1494.8	1623.5	3.5187	1.3432	1487.9	1622.2	3.5062	365
1.3798	1525.0	1649.2	3.5694	1.3673	1517.4	1647.3	3.5558	1.3556	1510.2	1645.7	3.5428	370
1.3941	1548.0	1673.4	3.6070	1.3808	1540.0	1671.2	3.5929	1.3685	1532.5	1669.4	3.5794	375
1.4089	1571.1	1697.9	3.6446	1.3949	1562.8	1695.3	3.6299	1.3818	1555.0	1693.2	3.6160	380
1.4243	1594.4	1722.6	3.6822	1.4094	1585.7	1719.6	3.6670	1.3956	1577.6	1717.1	3.6526	385
1.4403	1617.8	1747.4	3.7198	1.4245	1608.8	1744.1	3.7040	1.4098	1600.3	1741.3	3.6891	390
1.4570	1641.3	1772.5	3.7575	1.4402	1631.9	1768.7	3.7410	1.4246	1623.1	1765.5	3.7256	395
1.4743	1665.1	1797.7	3.7952	1.4564	1655.2	1793.6	3.7781	1.4399	1646.0	1790.0	3.7620	400
1.4923	1688.9	1823.2	3.8329	1.4732	1678.6	1818.6	3.8151	1.4558	1669.1	1814.6	3.7985	405
1.5110	1712.9	1848.9	3.8706	1.4907	1702.2	1843.8	3.8522	1.4722	1692.2	1839.4	3.8349	410
1.5304	1737.1	1874.8	3.9084	1.5089	1725.9	1869.2	3.8892	1.4892	1715.5	1864.4	3.8713	415
1.5507	1761.3	1900.9	3.9462	1.5277	1749.7	1894.8	3.9263	1.5068	1738.9	1889.5	3.9077	420
1.5717	1785.8	1927.2	3.9840	1.5473	1773.6	1920.6	3.9633	1.5251	1762.3	1914.8	3.9441	425
1.5937	1810.3	1953.8	4.0219	1.5676	1797.6	1946.5	4.0004	1.5440	1785.9	1940.3	3.9804	430
1.6165	1835.0	1980.5	4.0598	1.5886	1821.8	1972.7	4.0374	1.5636	1809.6	1965.9	4.0168	435
1.6403	1859.9	2007.5	4.0977	1.6105	1846.0	1999.0	4.0745	1.5838	1833.4	1991.7	4.0531	440
1.6650	1884.8	2034.7	4.1357	1.6332	1870.4	2025.5	4.1116	1.6048	1857.2	2017.7	4.0893	445
1.6908	1909.9	2062.0	4.1737	1.6568	1894.9	2052.3	4.1486	1.6266	1881.1	2043.8	4.1256	450
1.7455	1960.4	2117.5	4.2498	1.7066	1944.1	2106.2	4.2227	1.6724	1929.3	2096.5	4.1979	460
1.8048	2011.3	2173.7	4.3260	1.7603	1993.6	2160.8	4.2967	1.7215	1977.6	2149.8	4.2701	470
1.8689	2062.5	2230.7	4.4022	1.8181	2043.4	2216.1	4.3706	1.7741	2026.2	2203.6	4.3421	480
1.9382	2114.0	2288.4	4.4783	1.8802	2093.4	2272.0	4.4444	1.8303	2075.0	2258.0	4.4138	490
2.013	2165.6	2346.7	4.5542	1.9466	2143.5	2328.5	4.5178	1.8903	2123.8	2312.8	4.4852	500
2.093	2217.1	2405.4	4.6296	2.0176	2193.6	2385.3	4.5908	1.9541	2172.6	2368.1	4.5562	510
2.177	2268.4	2464.3	4.7044	2.0930	2243.6	2442.4	4.6633	2.0218	2221.4	2423.5	4.6266	520
2.267	2319.2	2523.2	4.7782	2.1726	2293.2	2499.6	4.7349	2.0932	2269.9	2479.2	4.6963	530
2.361	2369.3	2581.8	4.8507	2.2561	2342.3	2556.6	4.8055	2.1681	2318.0	2534.8	4.7651	540
2.458	2418.6	2639.8	4.9216	2.343	2390.8	2613.4	4.8749	2.246	2365.7	2590.3	4.8329	550
2.558	2466.9	2697.1	4.9908	2.433	2438.5	2669.6	4.9428	2.327	2412.7	2645.4	4.8995	560
2.659	2514.1	2753.4	5.0580	2.525	2485.3	2725.2	5.0091	2.411	2459.0	2700.1	4.9647	570
2.763	2560.0	2808.7	5.1231	2.619	2531.1	2779.9	5.0736	2.496	2504.5	2754.1	5.0284	580
2.867	2604.7	2862.7	5.1861	2.714	2575.8	2833.6	5.1362	2.583	2549.0	2807.4	5.0905	590
2.971	2648.2	2915.6	5.2469	2.810	2619.4	2886.3	5.1969	2.671	2592.7	2859.8	5.1509	600
3.075	2690.4	2967.1	5.3057	2.906	2661.9	2938.0	5.2557	2.760	2635.3	2911.3	5.2095	610
3.178	2731.4	3017.5	5.3624	3.002	2703.4	2988.5	5.3127	2.849	2677.0	2961.9	5.2665	620
3.281	2771.4	3066.7	5.4171	3.097	2743.8	3038.0	5.3677	2.938	2717.7	3011.5	5.3217	630
3.383	2810.2	3114.7	5.4700	3.192	2783.1	3086.4	5.4210	3.027	2757.4	3060.1	5.3752	640
3.483	2848.0	3161.6	5.5211	3.287	2821.5	3133.7	5.4726	3.115	2796.2	3107.7	5.4271	650
3.583	2884.9	3207.4	5.5704	3.380	2859.0	3180.1	5.5225	3.203	2834.1	3154.4	5.4774	660
3.681	2920.9	3252.2	5.6182	3.472	2895.5	3225.4	5.5709	3.290	2871.2	3200.2	5.5262	670
3.778	2956.1	3296.0	5.6645	3.564	2931.3	3269.9	5.6178	3.376	2907.5	3245.1	5.5736	680
3.873	2990.4	3339.0	5.7093	3.654	2966.3	3313.4	5.6632	3.462	2943.0	3289.1	5.6196	690
3.967	3024.0	3381.1	5.7528	3.743	3000.5	3356.1	5.7073	3.546	2977.7	3332.3	5.6642	700
4.151	3089.3	3462.9	5.8360	3.918	3066.9	3439.1	5.7918	3.712	3045.2	3416.4	5.7497	720
4.330	3152.1	3541.8	5.9146	4.088	3130.9	3519.2	5.8716	3.875	3110.1	3497.6	5.8306	740
4.504	3212.7	3618.1	5.9892	4.254	3192.6	3596.7	5.9473	4.033	3172.8	3576.1	5.9074	760
4.673	3271.5	3692.0	6.0601	4.415	3252.3	3671.8	6.0193	4.187	3233.5	3652.2	5.9803	780
4.837	3328.5	3763.8	6.1277	4.573	3310.3	3744.7	6.0879	4.338	3292.3	3726.1	6.0499	800
5.232	3464.6	3935.5	6.2840	4.950	3448.5	3918.7	6.2465	4.700	3432.5	3902.5	6.2105	850
5.606	3593.1	4097.7	6.4253	5.309	3578.6	4083.0	6.3896	5.044	3564.3	4068.6	6.3553	900
5.964	3715.6	4252.4	6.5545	5.652	3702.5	4239.4	6.5201	5.373	3689.4	4226.7	6.4872	950
6.309	3833.5	4401.3	6.6738	5.982	3821.3	4389.6	6.6405	5.690	3809.2	4378.2	6.6087	1000
6.973	4058.9	4686.5	6.8895	6.618	4048.0	4676.7	6.8577	6.299	4037.1	4667.1	6.8272	1100
7.617	4275.5	4961.0	7.0826	7.233	4265.1	4952.2	7.0514	6.889	4254.7	4943.6	7.0216	1200
8.252	4487.6	5230.3	7.2594	7.839	4477.1	5221.8	7.2285	7.469	4466.6	5213.6	7.1989	1300

Definitions of symbols on page 1.

Table 4. Liquid

p (t Sat.)	0				2.5 (223.99)				5.0 (263.99)			
t	$10^3 v$	u	h	s	$10^3 v$	u	h	s	$10^3 v$	u	h	s
Sat.					1.1973	959.1	962.1	2.5546	1.2859	1147.8	1154.2	2.9202
0	1.0002	-.03	-.03	-.0001	.9990	-.00	2.50	-.0000	.9977	.04	5.04	.0001
20	1.0018	83.95	83.95	.2966	1.0006	83.80	86.30	.2961	.9995	83.65	88.65	.2956
40	1.0078	167.56	167.56	.5725	1.0067	167.25	169.77	.5715	1.0056	166.95	171.97	.5705
60	1.0172	251.12	251.12	.8312	1.0160	250.67	253.21	.8298	1.0149	250.23	255.30	.8285
80	1.0291	334.87	334.87	1.0753	1.0280	334.29	336.86	1.0737	1.0268	333.72	338.85	1.0720
100	1.0436	418.96	418.96	1.3069	1.0423	418.24	420.85	1.3050	1.0410	417.52	422.72	1.3030
120	1.0604	503.57	503.57	1.5278	1.0590	502.68	505.33	1.5255	1.0576	501.80	507.09	1.5233
140	1.0800	588.89	588.89	1.7395	1.0784	587.82	590.52	1.7369	1.0768	586.76	592.15	1.7343
160	1.1024	675.19	675.19	1.9434	1.1006	673.90	676.65	1.9404	1.0988	672.62	678.12	1.9375
180	1.1283	762.72	762.72	2.1410	1.1261	761.16	763.97	2.1375	1.1240	759.63	765.25	2.1341
200	1.1581	851.8	851.8	2.3334	1.1555	849.9	852.8	2.3294	1.1530	848.1	853.9	2.3255
210	1.1749	897.1	897.1	2.4281	1.1720	895.0	898.0	2.4238	1.1691	893.0	898.8	2.4195
220	1.1930	943.0	943.0	2.5221	1.1898	940.7	943.7	2.5174	1.1866	938.4	944.4	2.5128
230	1.2129	989.6	989.6	2.6157	1.2092	987.0	990.1	2.6105	1.2056	984.5	990.6	2.6055
240	1.2347	1037.1	1037.1	2.7091	1.2305	1034.2	1037.2	2.7034	1.2264	1031.4	1037.5	2.6979
250	1.2590	1085.6	1085.6	2.8027	1.2540	1082.3	1085.4	2.7964	1.2493	1079.1	1085.3	2.7902
260	1.2862	1135.4	1135.4	2.8970	1.2804	1131.6	1134.8	2.8898	1.2749	1127.9	1134.3	2.8830
270	1.3173	1186.8	1186.8	2.9926	1.3102	1182.4	1185.7	2.9844	1.3036	1178.2	1184.7	2.9766
280	1.3535	1240.4	1240.4	3.0904	1.3447	1235.1	1238.5	3.0808	1.3365	1230.2	1236.8	3.0717
290	1.3971	1297.0	1297.0	3.1918	1.3855	1290.5	1294.0	3.1801	1.3750	1284.4	1291.3	3.1693
300	1.4520	1358.1	1358.1	3.2992	1.4357	1349.6	1353.2	3.2843	1.4214	1341.9	1349.0	3.2708
310									1.4803	1404.1	1411.5	3.3789

p (t Sat.)	7.5 (290.59)				10.0 (311.06)				12.5 (327.89)			
t	$10^3 v$	u	h	s	$10^3 v$	u	h	s	$10^3 v$	u	h	s
Sat.	1.3677	1282.0	1292.2	3.1649	1.4524	1393.0	1407.6	3.3596	1.5466	1492.2	1511.5	3.5286
0	.9965	.06	7.54	.0002	.9952	.09	10.04	.0002	.9940	.13	12.56	.0003
20	.9984	83.50	90.99	.2950	.9972	83.36	93.33	.2945	.9961	83.21	95.67	.2940
40	1.0045	166.64	174.18	.5696	1.0034	166.35	176.38	.5686	1.0024	166.05	178.58	.5676
60	1.0138	249.79	257.40	.8272	1.0127	249.36	259.49	.8258	1.0116	248.94	261.58	.8245
80	1.0256	333.15	340.84	1.0704	1.0245	332.59	342.83	1.0688	1.0233	332.03	344.82	1.0672
100	1.0397	416.81	424.62	1.3011	1.0385	416.12	426.50	1.2992	1.0373	415.42	428.39	1.2973
120	1.0562	500.94	508.86	1.5211	1.0549	500.08	510.64	1.5189	1.0535	499.24	512.41	1.5167
140	1.0752	585.72	593.78	1.7317	1.0737	584.68	595.42	1.7292	1.0722	583.66	597.07	1.7267
160	1.0970	671.37	679.59	1.9346	1.0953	670.13	681.08	1.9317	1.0935	668.91	682.58	1.9288
180	1.1219	758.13	766.55	2.1308	1.1199	756.65	767.84	2.1275	1.1178	755.19	769.16	2.1242
200	1.1505	846.3	854.9	2.3216	1.1480	844.5	856.0	2.3178	1.1456	842.8	857.1	2.3141
210	1.1664	891.0	899.8	2.4154	1.1637	889.1	900.7	2.4113	1.1610	887.1	901.6	2.4073
220	1.1835	936.2	945.1	2.5083	1.1805	934.1	945.9	2.5039	1.1776	932.0	946.7	2.4995
230	1.2022	982.1	991.1	2.6006	1.1988	979.7	991.7	2.5958	1.1955	977.4	992.3	2.5911
240	1.2225	1028.6	1037.8	2.6925	1.2187	1026.0	1038.1	2.6872	1.2150	1023.4	1038.5	2.6821
250	1.2448	1076.0	1085.3	2.7843	1.2405	1073.0	1085.4	2.7785	1.2363	1070.1	1085.6	2.7729
260	1.2696	1124.4	1134.0	2.8763	1.2645	1121.1	1133.7	2.8699	1.2597	1117.8	1133.5	2.8637
270	1.2973	1174.1	1183.9	2.9691	1.2913	1170.3	1183.2	2.9618	1.2857	1166.5	1182.6	2.9548
280	1.3288	1225.4	1235.4	3.0631	1.3216	1220.9	1234.1	3.0548	1.3148	1216.6	1233.0	3.0469
290	1.3653	1278.8	1289.0	3.1591	1.3564	1273.4	1287.0	3.1495	1.3481	1268.4	1285.2	3.1404
300	1.4087	1334.9	1345.4	3.2584	1.3972	1328.4	1342.3	3.2469	1.3867	1322.3	1339.6	3.2361
310	1.4622	1394.9	1405.8	3.3629	1.4465	1386.6	1401.1	3.3485	1.4326	1379.1	1397.0	3.3353
320	1.5330	1461.1	1472.6	3.4764	1.5093	1449.8	1464.9	3.4569	1.4895	1439.9	1458.5	3.4399
330					1.5972	1521.5	1537.5	3.5783	1.5646	1507.1	1526.6	3.5538
340									1.6795	1586.9	1607.9	3.6874

Definitions of symbols on page 1.

Table 4. Liquid

p (t Sat.)	15.0 (342.24)				17.5 (354.75)				20.0 (365.81)			
t	$10^3 v$	u	h	s	$10^3 v$	u	h	s	$10^3 v$	u	h	s
Sat.	1.6581	1585.6	1610.5	3.6848	1.8035	1679.4	1710.9	3.8394	2.036	1785.6	1826.3	4.0139
0	.9928	.15	15.05	.0004	.9916	.17	17.53	.0004	.9904	.19	20.01	.0004
20	.9950	83.06	97.99	.2934	.9939	82.91	100.31	.2928	.9928	82.77	102.62	.2923
40	1.0013	165.76	180.78	.5666	1.0002	165.46	182.96	.5656	.9992	165.17	185.16	.5646
60	1.0105	248.51	263.67	.8232	1.0095	248.09	265.76	.8219	1.0084	247.68	267.85	.8206
80	1.0222	331.48	346.81	1.0656	1.0211	330.94	348.80	1.0640	1.0199	330.40	350.80	1.0624
100	1.0361	414.74	430.28	1.2955	1.0349	414.06	432.17	1.2936	1.0337	413.39	434.06	1.2917
120	1.0522	498.40	514.19	1.5145	1.0509	497.58	515.97	1.5123	1.0496	496.76	517.76	1.5102
140	1.0707	582.66	598.72	1.7242	1.0692	581.67	600.38	1.7217	1.0678	580.69	602.04	1.7193
160	1.0918	667.71	684.09	1.9260	1.0901	666.52	685.60	1.9232	1.0885	665.35	687.12	1.9204
180	1.1159	753.76	770.50	2.1210	1.1139	752.35	771.84	2.1178	1.1120	750.95	773.20	2.1147
200	1.1433	841.0	858.2	2.3104	1.1410	839.4	859.3	2.3067	1.1388	837.7	860.5	2.3031
220	1.1748	929.9	947.5	2.4953	1.1720	927.9	948.4	2.4911	1.1693	925.9	949.3	2.4870
240	1.2114	1020.8	1039.0	2.6771	1.2080	1018.4	1039.5	2.6722	1.2046	1016.0	1040.0	2.6674
260	1.2550	1114.6	1133.4	2.8576	1.2505	1111.5	1133.4	2.8517	1.2462	1108.6	1133.5	2.8459
280	1.3084	1212.5	1232.1	3.0393	1.3023	1208.5	1231.3	3.0319	1.2965	1204.7	1230.6	3.0248
300	1.3770	1316.6	1337.3	3.2260	1.3680	1311.2	1335.2	3.2163	1.3596	1306.1	1333.3	3.2071
310	1.4201	1372.1	1393.4	3.3231	1.4087	1365.7	1390.3	3.3117	1.3982	1359.6	1387.6	3.3010
320	1.4724	1431.1	1453.2	3.4247	1.4573	1423.1	1448.6	3.4107	1.4437	1415.7	1444.6	3.3979
330	1.5388	1495.0	1518.1	3.5332	1.5172	1484.5	1511.1	3.5153	1.4987	1475.2	1505.2	3.4992
340	1.6311	1567.5	1591.9	3.6546	1.5961	1552.3	1580.3	3.6290	1.5684	1539.7	1571.0	3.6075
350	*1.7958*	*1661.2*	*1688.2*	*3.8102*	1.7139	1632.1	1662.1	3.7613	1.6640	1612.3	1645.6	3.7281
360									1.8226	1702.8	1739.3	3.8772

p (t Sat.)	22.09 (374.14)				25				30			
t	$10^3 v$	u	h	s	$10^3 v$	u	h	s	$10^3 v$	u	h	s
Sat.	3.155	2029.6	2099.3	4.4298								
0	.9894	.22	22.08	.0004	.9880	.24	24.95	.0003	.9856	.25	29.82	.0001
20	.9919	82.64	104.56	.2918	.9907	82.47	107.24	.2911	.9886	82.17	111.84	.2899
40	.9983	164.93	186.98	.5638	.9971	164.60	189.52	.5626	.9951	164.04	193.89	.5607
60	1.0075	247.33	269.59	.8195	1.0063	246.86	272.02	.8180	1.0042	246.06	276.19	.8154
80	1.0190	329.95	352.46	1.0611	1.0178	329.34	354.78	1.0592	1.0156	328.30	358.77	1.0561
100	1.0327	412.84	435.65	1.2902	1.0313	412.08	437.85	1.2881	1.0290	410.78	441.66	1.2844
120	1.0485	496.09	519.25	1.5084	1.0470	495.16	521.33	1.5060	1.0445	493.59	524.93	1.5018
140	1.0666	579.88	603.44	1.7173	1.0649	578.76	605.39	1.7145	1.0621	576.88	608.75	1.7098
160	1.0871	664.39	688.40	1.9181	1.0853	663.06	690.19	1.9149	1.0821	660.82	693.28	1.9096
180	1.1105	749.81	774.33	2.1120	1.1083	748.23	775.94	2.1085	1.1047	745.59	778.73	2.1024
200	1.1369	836.3	861.5	2.3002	1.1344	834.5	862.8	2.2961	1.1302	831.4	865.3	2.2893
220	1.1671	924.3	950.1	2.4836	1.1641	922.0	951.1	2.4789	1.1590	918.3	953.1	2.4711
240	1.2019	1014.0	1040.56	2.6634	1.1982	1011.3	1041.3	2.6580	1.1920	1006.9	1042.6	2.6490
260	1.2427	1106.1	1133.6	2.8412	1.2380	1102.8	1133.8	2.8349	1.2303	1097.4	1134.3	2.8243
280	1.2918	1201.6	1230.2	3.0191	1.2856	1197.5	1229.6	3.0113	1.2755	1190.7	1229.0	2.9986
300	1.3530	1302.1	1331.9	3.1998	1.3442	1296.6	1330.2	3.1900	1.3304	1287.9	1327.8	3.1741
310	1.3901	1354.8	1385.5	3.2925	1.3794	1348.5	1383.0	3.2812	1.3629	1338.5	1379.3	3.2632
320	1.4333	1409.9	1441.6	3.3878	1.4200	1402.4	1437.9	3.3746	1.3997	1390.7	1432.7	3.3539
330	1.4849	1468.1	1500.9	3.4869	1.4677	1458.9	1495.6	3.4711	1.4422	1444.9	1488.2	3.4467
340	1.5488	1530.4	1564.6	3.5917	1.5253	1518.8	1557.0	3.5719	1.4920	1501.7	1546.5	3.5426
350	1.6326	1599.1	1635.2	3.7058	1.5977	1583.6	1623.5	3.6796	1.5518	1561.9	1608.5	3.6428
360	1.7561	1679.4	1718.2	3.8379	1.6955	1655.9	1698.3	3.7986	1.6265	1626.6	1675.4	3.7494
370	2.0138	1793.9	1838.4	4.0261	1.8470	1742.8	1789.0	3.9407	1.7252	1698.1	1749.9	3.8661
380					2.2135	1879.1	1934.5	4.1649	1.8691	1781.4	1837.5	4.0012

Definitions of symbols on page 1.

Table 4. Liquid

p	35				40				50			
t	$10^3 v$	u	h	s	$10^3 v$	u	h	s	$10^3 v$	u	h	s
0	.9833	.27	34.70	-.0001	.9811	.26	39.51	-.0004	.9766	.20	49.03	-.0014
20	.9865	81.89	116.41	.2887	.9845	81.59	120.97	.2874	.9804	81.00	130.02	.2848
40	.9931	163.48	198.24	.5587	.9911	162.93	202.58	.5567	.9872	161.86	211.21	.5527
60	1.0022	245.27	280.34	.8128	1.0002	244.49	284.49	.8103	.9962	242.98	292.79	.8052
80	1.0135	327.28	362.75	1.0531	1.0114	326.28	366.74	1.0500	1.0073	324.34	374.70	1.0440
100	1.0267	409.52	445.46	1.2809	1.0245	408.28	449.26	1.2773	1.0201	405.88	456.89	1.2703
120	1.0420	492.06	528.52	1.4977	1.0396	490.56	532.14	1.4937	1.0348	487.65	539.39	1.4857
140	1.0594	575.04	612.12	1.7051	1.0567	573.25	615.52	1.7005	1.0515	569.77	622.35	1.6915
160	1.0790	658.64	696.41	1.9043	1.0760	656.51	699.56	1.8991	1.0703	652.41	705.92	1.8891
180	1.1012	743.02	781.55	2.0965	1.0978	740.51	784.42	2.0907	1.0912	735.69	790.25	2.0794
200	1.1261	828.3	867.7	2.2826	1.1222	825.4	870.3	2.2760	1.1146	819.7	875.5	2.2634
220	1.1542	914.8	955.1	2.4635	1.1496	911.3	957.3	2.4561	1.1408	904.7	961.7	2.4419
240	1.1862	1002.6	1044.1	2.6403	1.1806	998.5	1045.7	2.6319	1.1702	990.7	1049.2	2.6158
260	1.2230	1092.2	1135.0	2.8141	1.2161	1087.3	1136.0	2.8044	1.2034	1078.1	1138.2	2.7860
280	1.2662	1184.4	1228.7	2.9865	1.2574	1178.3	1228.6	2.9751	1.2415	1167.2	1229.3	2.9537
300	1.3179	1279.9	1326.0	3.1593	1.3064	1272.3	1324.6	3.1455	1.2860	1258.7	1323.0	3.1200
310	1.3481	1329.3	1376.5	3.2466	1.3347	1320.8	1374.2	3.2312	1.3112	1305.5	1371.1	3.2032
320	1.3820	1380.1	1428.5	3.3350	1.3662	1370.4	1425.1	3.3178	1.3388	1353.3	1420.2	3.2868
330	1.4204	1432.6	1482.3	3.4250	1.4015	1421.5	1477.5	3.4054	1.3693	1402.0	1470.5	3.3708
340	1.4647	1487.1	1538.3	3.5172	1.4414	1474.1	1531.8	3.4947	1.4032	1452.0	1522.1	3.4557
350	1.5164	1544.1	1597.1	3.6123	1.4873	1528.8	1588.3	3.5860	1.4411	1503.2	1575.3	3.5417
360	1.5782	1604.2	1659.5	3.7116	1.5407	1585.8	1647.4	3.6802	1.4838	1556.0	1630.2	3.6291
370	1.6545	1668.6	1726.5	3.8166	1.6041	1645.8	1709.9	3.7782	1.5324	1610.6	1687.2	3.7184
380	1.7530	1739.0	1800.3	3.9305	1.6813	1709.6	1776.8	3.8813	1.5884	1667.2	1746.6	3.8101

p	60				70				80			
t	$10^3 v$	u	h	s	$10^3 v$	u	h	s	$10^3 v$	u	h	s
0	.9723	.10	58.45	-.0026	.9682	-.06	67.71	-.0042	.9642	-.25	76.89	-.0060
20	.9765	80.41	139.00	.2820	.9727	79.83	147.92	.2792	.9690	79.25	156.77	.2763
40	.9834	160.81	219.81	.5487	.9796	159.80	228.38	.5447	.9760	158.82	236.90	.5407
60	.9924	241.52	301.06	.8002	.9886	240.11	309.31	.7952	.9849	238.74	317.54	.7903
80	1.0033	322.47	382.66	1.0380	.9994	320.66	390.61	1.0322	.9955	318.91	398.56	1.0265
100	1.0159	403.57	464.53	1.2635	1.0118	401.34	472.16	1.2568	1.0078	399.19	479.82	1.2503
120	1.0303	484.85	546.67	1.4780	1.0259	482.17	553.98	1.4704	1.0216	479.57	561.30	1.4630
140	1.0465	566.44	629.23	1.6828	1.0417	563.25	636.17	1.6743	1.0371	560.17	643.14	1.6660
160	1.0647	648.48	712.37	1.8793	1.0595	644.72	718.89	1.8698	1.0544	641.12	725.46	1.8606
180	1.0850	731.10	796.20	2.0685	1.0791	726.72	802.26	2.0580	1.0735	722.53	808.41	2.0478
200	1.1076	814.4	880.9	2.2513	1.1009	809.3	886.4	2.2397	1.0945	804.5	892.0	2.2284
220	1.1326	898.5	966.4	2.4284	1.1249	892.6	971.4	2.4155	1.1177	887.1	976.5	2.4032
240	1.1605	983.5	1053.1	2.6007	1.1515	976.7	1057.3	2.5863	1.1431	970.3	1061.7	2.5726
260	1.1918	1069.6	1141.1	2.7689	1.1811	1061.7	1144.3	2.7527	1.1713	1054.3	1148.0	2.7375
280	1.2273	1157.1	1230.7	2.9340	1.2144	1147.8	1232.8	2.9157	1.2026	1139.2	1235.4	2.8985
300	1.2681	1246.5	1322.6	3.0970	1.2522	1235.5	1323.1	3.0760	1.2379	1225.4	1324.4	3.0566
310	1.2909	1292.1	1369.5	3.1782	1.2731	1280.0	1369.1	3.1555	1.2572	1269.0	1369.6	3.1347
320	1.3157	1338.3	1417.2	3.2594	1.2956	1325.0	1415.7	3.2348	1.2779	1313.1	1415.3	3.2125
330	1.3427	1385.3	1465.9	3.3407	1.3200	1370.7	1463.1	3.3140	1.3001	1357.6	1461.6	3.2899
340	1.3723	1433.3	1515.6	3.4224	1.3464	1417.1	1511.3	3.3933	1.3240	1402.7	1508.6	3.3672
350	1.4049	1482.1	1566.4	3.5047	1.3751	1464.1	1560.4	3.4727	1.3497	1448.4	1556.4	3.4444
360	1.4409	1532.1	1618.6	3.5877	1.4064	1512.1	1610.5	3.5525	1.3776	1494.7	1604.9	3.5217
370	1.4810	1583.3	1672.2	3.6717	1.4408	1560.9	1661.7	3.6327	1.4078	1541.7	1654.3	3.5991
380	1.5258	1635.8	1727.4	3.7568	1.4786	1610.6	1714.1	3.7135	1.4406	1589.4	1704.6	3.6767

Definitions of symbols on page 1.

Table 4.   Liquid

p		90				100				110		
t	$10^3 v$	u	h	s	$10^3 v$	u	h	s	$10^3 v$	u	h	s
0	.9603	- .45	85.97	- .0079	.9566	- .69	94.97	- .0101	.9530	- .93	103.90	- .0123
20	.9653	78.69	165.56	.2733	.9617	78.14	174.31	.2703	.9582	77.61	183.02	.2672
40	.9724	157.87	245.37	.5367	.9688	156.94	253.82	.5326	.9654	156.05	262.24	.5286
60	.9813	237.42	325.74	.7854	.9777	236.14	333.91	.7805	.9742	234.90	342.06	.7757
80	.9918	317.22	406.49	1.0208	.9882	315.59	414.41	1.0152	.9846	314.01	422.31	1.0096
100	1.0039	397.12	487.46	1.2438	1.0000	395.11	495.11	1.2375	.9963	393.16	502.8	1.2312
120	1.0174	477.08	568.65	1.4557	1.0134	474.66	576.00	1.4486	1.0095	472.32	583.4	1.4416
140	1.0326	557.21	650.14	1.6579	1.0283	554.35	657.18	1.6500	1.0241	551.58	664.2	1.6422
160	1.0495	637.65	732.10	1.8516	1.0448	634.30	738.78	1.8429	1.0402	631.08	745.5	1.8343
180	1.0681	718.51	814.63	2.0379	1.0629	714.65	820.93	2.0283	1.0579	710.93	827.3	2.0190
200	1.0885	799.9	897.8	2.2176	1.0827	795.4	903.7	2.2071	1.0772	791.2	909.7	2.1969
220	1.1108	881.8	981.7	2.3913	1.1043	876.7	987.2	2.3798	1.0982	871.9	992.7	2.3687
240	1.1353	964.2	1066.4	2.5595	1.1279	958.5	1071.3	2.5470	1.1209	953.0	1076.3	2.5349
260	1.1622	1047.3	1151.9	2.7231	1.1536	1040.8	1156.2	2.7093	1.1456	1034.6	1160.6	2.6961
280	1.1918	1131.2	1238.5	2.8825	1.1818	1123.8	1242.0	2.8672	1.1725	1116.8	1245.7	2.8528
300	1.2249	1216.1	1326.4	3.0385	1.2130	1207.5	1328.8	3.0215	1.2021	1199.5	1331.8	3.0056
310	1.2429	1259.0	1370.9	3.1154	1.2299	1249.8	1372.8	3.0975	1.2180	1241.2	1375.2	3.0807
320	1.2621	1302.2	1415.8	3.1919	1.2478	1292.3	1417.1	3.1728	1.2347	1283.1	1418.9	3.1551
330	1.2825	1345.9	1461.3	3.2679	1.2667	1335.1	1461.8	3.2476	1.2524	1325.3	1463.0	3.2288
340	1.3043	1389.9	1507.3	3.3435	1.2868	1378.3	1507.0	3.3219	1.2711	1367.7	1507.5	3.3019
350	1.3277	1434.4	1553.9	3.4190	1.3083	1421.8	1552.7	3.3958	1.2909	1410.4	1552.4	3.3746
360	1.3528	1479.4	1601.2	3.4942	1.3312	1465.8	1598.9	3.4694	1.3120	1453.5	1597.8	3.4468
370	1.3798	1525.0	1649.2	3.5694	1.3556	1510.2	1645.7	3.5428	1.3343	1496.9	1643.6	3.5187
380	1.4089	1571.1	1697.9	3.6446	1.3818	1555.0	1693.2	3.6160	1.3581	1540.6	1690.0	3.5902

p		120				130				140		
t	$10^3 v$	u	h	s	$10^3 v$	u	h	s	$10^3 v$	u	h	s
0	.9495	- 1.17	112.77	- .0147	.9462	- 1.37	121.64	- .0169	.9430	- 1.56	130.46	.0192
20	.9548	77.11	191.68	.2641	.9514	76.62	200.31	.2611	.9481	76.17	208.91	.2580
40	.9619	155.18	270.61	.5246	.9586	154.34	278.95	.5206	.9552	153.54	287.26	.5165
60	.9708	233.69	350.19	.7709	.9674	232.52	358.28	.7661	.9640	231.38	366.34	.7613
80	.9811	312.47	430.20	1.0041	.9776	310.98	438.07	.9987	.9742	309.52	445.91	.9933
100	.9927	391.28	510.4	1.2250	.9891	389.45	518.0	1.2189	.9856	387.67	525.6	1.2129
120	1.0056	470.06	590.7	1.4348	1.0019	467.86	598.1	1.4280	.9983	465.72	605.5	1.4213
140	1.0200	548.90	671.3	1.6346	1.0160	546.31	678.4	1.6272	1.0122	543.79	685.5	1.6198
160	1.0358	627.96	752.3	1.8260	1.0316	624.94	759.0	1.8178	1.0275	622.01	765.9	1.8097
180	1.0531	707.34	833.7	2.0098	1.0485	703.87	840.2	2.0009	1.0441	700.51	846.7	1.9922
200	1.0719	787.1	915.7	2.1869	1.0669	783.2	921.9	2.1773	1.0620	779.4	928.0	2.1679
220	1.0923	867.3	998.3	2.3579	1.0867	862.8	1004.1	2.3475	1.0814	858.6	1009.9	2.3374
240	1.1143	947.8	1081.5	2.5233	1.1080	942.8	1086.9	2.5121	1.1021	938.1	1092.3	2.5012
260	1.1381	1028.8	1165.3	2.6835	1.1310	1023.2	1170.2	2.6714	1.1243	1017.9	1175.3	2.6597
280	1.1639	1110.2	1249.8	2.8390	1.1557	1103.9	1254.1	2.8259	1.1481	1097.9	1258.7	2.8133
300	1.1920	1192.1	1335.1	2.9905	1.1826	1185.0	1338.7	2.9761	1.1738	1178.4	1342.7	2.9625
310	1.2070	1233.2	1378.1	3.0648	1.1968	1225.8	1381.3	3.0498	1.1874	1218.7	1385.0	3.0356
320	1.2228	1274.6	1421.3	3.1384	1.2118	1266.6	1424.2	3.1227	1.2016	1259.2	1427.4	3.1078
330	1.2394	1316.2	1464.9	3.2112	1.2274	1307.7	1467.3	3.1947	1.2164	1299.8	1470.1	3.1791
340	1.2568	1357.9	1508.8	3.2833	1.2438	1348.9	1510.6	3.2660	1.2319	1340.5	1513.0	3.2497
350	1.2753	1400.0	1553.0	3.3549	1.2611	1390.3	1554.3	3.3366	1.2481	1381.4	1556.2	3.3195
360	1.2948	1442.3	1597.6	3.4260	1.2792	1432.0	1598.3	3.4067	1.2651	1422.5	1599.6	3.3887
370	1.3154	1484.8	1642.7	3.4965	1.2983	1473.8	1642.6	3.4761	1.2829	1463.7	1643.3	3.4572
380	1.3372	1527.7	1688.2	3.5667	1.3185	1515.9	1687.3	3.5451	1.3017	1505.2	1687.4	3.5252

Definitions of symbols on page 1.

Table 5. Critical Region

10^3v (t Sat.)	1.70 (346.45)				1.75 (350.80)				1.80 (354.51)			
t	p	u	h	s	p	u	h	s	p	u	h	s
Sat.	15.806	1615.5	1642.3	3.7342	16.676	1647.9	1677.1	3.7877	17.450	1677.4	1708.8	3.8362
350	18.106	1626.7	1657.5	3.7523	*16 195*	*1645.4*	*1673.7*	*3.7836*	*14.912*	*1662.6*	*1689.5*	*3.8125*
352	19.415	1633.0	1666.0	3.7623	17.404	1651.8	1682.3	3.7939	*16.030*	*1669.2*	*1698.1*	*3.8231*
354	20.733	1639.2	1674.5	3.7723	18.622	1658.2	1690.8	3.8041	*17.159*	*1675.8*	*1706.6*	*3.8335*
356	22.060	1645.4	1682.9	3.7822	19.850	1664.5	1699.3	3.8142	18.298	1682.3	1715.2	3.8439
358	23.395	1651.6	1691.4	3.7920	21.086	1670.8	1707.7	3.8242	19.447	1688.7	1723.7	3.8541
360	24.74	1657.7	1699.8	3.8017	22.33	1677.1	1716.2	3.8341	20.60	1695.1	1732.2	3.8642
362	26.09	1663.8	1708.2	3.8113	23.58	1683.3	1724.6	3.8439	21.77	1701.5	1740.6	3.8742
364	27.44	1669.9	1716.6	3.8208	24.85	1689.5	1733.0	3.8536	22.95	1707.8	1749.1	3.8842
366	28.81	1676.0	1724.9	3.8303	26.11	1695.7	1741.4	3.8633	24.13	1714.0	1757.5	3.8940
368	30.18	1682.0	1733.3	3.8397	27.39	1701.8	1749.7	3.8728	25.32	1720.3	1765.8	3.9037
370	31.55	1688.0	1741.6	3.8490	28.67	1707.9	1758.0	3.8823	26.52	1726.5	1774.2	3.9134
372	32.93	1693.9	1749.9	3.8583	29.96	1713.9	1766.3	3.8917	27.72	1732.6	1782.5	3.9229
374	34.32	1699.8	1758.2	3.8675	31.26	1719.9	1774.6	3.9010	28.93	1738.7	1790.8	3.9324
376	35.71	1705.8	1766.5	3.8766	32.56	1725.9	1782.9	3.9102	30.15	1744.8	1799.1	3.9418
378	37.11	1711.6	1774.7	3.8856	33.86	1731.9	1791.1	3.9194	31.37	1750.9	1807.3	3.9511
380	38.51	1717.5	1783.0	3.8946	35.17	1737.8	1799.4	3.9285	32.60	1756.9	1815.6	3.9603
382	39.91	1723.3	1791.2	3.9035	36.49	1743.7	1807.6	3.9376	33.84	1762.9	1823.8	3.9695
384	41.32	1729.2	1799.4	3.9124	37.81	1749.6	1815.8	3.9465	35.08	1768.9	1832.0	3.9786
386	42.74	1735.0	1807.6	3.9212	39.13	1755.5	1824.0	3.9554	36.32	1774.8	1840.2	3.9876
388	44.15	1740.7	1815.8	3.9300	40.46	1761.3	1832.1	3.9643	37.57	1780.7	1848.3	3.9965
390	45.57	1746.5	1824.0	3.9387	41.79	1767.1	1840.3	3.9731	38.82	1786.6	1856.5	4.0054
392	46.99	1752.2	1832.1	3.9473	43.13	1772.9	1848.4	3.9818	40.08	1792.4	1864.6	4.0142
394	48.42	1758.0	1840.3	3.9559	44.47	1778.7	1856.5	3.9905	41.34	1798.3	1872.7	4.0230
396	49.84	1763.7	1848.4	3.9645	45.81	1784.5	1864.6	3.9991	42.61	1804.1	1880.8	4.0317
398	51.27	1769.4	1856.5	3.9730	47.16	1790.2	1872.7	4.0076	43.88	1809.9	1888.8	4.0403
400	52.70	1775.0	1864.6	3.9814	48.51	1795.9	1880.8	4.0161	45.15	1815.6	1896.9	4.0489

10^3v (t Sat.)	1.85 (357.68)				1.90 (360.38)				1.95 (362.67)			
t	p	u	h	s	p	u	h	s	p	u	h	s
Sat.	18.134	1704.3	1737.8	3.8802	18.737	1728.8	1764.4	3.9205	19.261	1751.3	1788.8	3.9574
350	*14.088*	*1678.6*	*1704.7*	*3.8393*	*13.592*	*1693.5*	*1719.3*	*3.8643*	*13.332*	*1707.3*	*1733.3*	*3.8876*
352	*15.127*	*1685.4*	*1713.4*	*3.8502*	*14.561*	*1700.4*	*1728.1*	*3.8754*	*14.239*	*1714.4*	*1742.2*	*3.8990*
354	*16.177*	*1692.1*	*1722.0*	*3.8609*	*15.541*	*1707.3*	*1736.8*	*3.8864*	*15.158*	*1721.5*	*1751.0*	*3.9102*
356	*17.237*	*1698.7*	*1730.6*	*3.8715*	*16.531*	*1714.1*	*1745.5*	*3.8972*	*16.087*	*1728.4*	*1759.8*	*3.9213*
358	18.307	1705.3	1739.2	3.8819	*17.532*	*1720.8*	*1754.1*	*3.9079*	*17.027*	*1735.4*	*1768.6*	*3.9323*
360	19.387	1711.9	1747.7	3.8923	*18.543*	*1727.5*	*1762.8*	*3.9185*	*17.977*	*1742.2*	*1777.3*	*3.9431*
362	20.476	1718.4	1756.2	3.9025	19.563	1734.2	1771.3	3.9290	*18 936*	*1749.0*	*1785.9*	*3.9538*
364	21.574	1724.8	1764.7	3.9126	20.592	1740.8	1779.9	3.9393	19.904	1755.7	1794.5	3.9644
366	22.680	1731.2	1773.2	3.9227	21.630	1747.3	1788.4	3.9496	20.882	1762.4	1803.1	3.9749
368	23.794	1737.6	1781.6	3.9326	22.676	1753.8	1796.9	3.9597	21.868	1769.0	1811.6	3.9852
370	24.92	1743.9	1790.0	3.9424	23.73	1760.2	1805.3	3.9697	22.86	1775.6	1820.1	3.9954
372	26.05	1750.1	1798.3	3.9522	24.79	1766.6	1813.7	3.9796	23.86	1782.1	1828.6	4.0055
374	27.18	1756.4	1806.6	3.9618	25.86	1772.9	1822.1	3.9894	24.87	1788.5	1837.0	4.0155
376	28.32	1762.6	1815.0	3.9713	26.94	1779.3	1830.4	3.9992	25.89	1794.9	1845.4	4.0254
378	29.47	1768.7	1823.2	3.9808	28.02	1785.5	1838.7	4.0088	26.91	1801.3	1853.8	4.0352
380	30.63	1774.8	1831.5	3.9902	29.11	1791.7	1847.0	4.0183	27.94	1807.6	1862.1	4.0449
382	31.79	1780.9	1839.7	3.9995	30.20	1797.9	1855.3	4.0278	28.98	1813.9	1870.4	4.0545
384	32 96	1787.0	1847.9	4.0087	31.31	1804.0	1863.5	4.0371	30.02	1820.1	1878.7	4.0640
386	34.13	1793.0	1856.1	4.0178	32.41	1810.1	1871.7	4.0464	31.07	1826.3	1886.9	4.0734
388	35.31	1799.0	1864.3	4.0269	33.52	1816.2	1879.9	4.0556	32.12	1832.5	1895.1	4.0827
390	36.49	1804.9	1872.4	4.0359	34.64	1822.2	1888.0	4.0647	33.18	1838.6	1903.3	4.0919
392	37.67	1810.8	1880.5	4.0448	35.76	1828.2	1896.2	4.0737	34.24	1844.7	1911.5	4.1011
394	38.86	1816.7	1888.6	4.0537	36.89	1834.2	1904.3	4.0827	35.31	1850.7	1919.6	4.1102
396	40.06	1822.6	1896.7	4.0625	38.02	1840.1	1912.4	4.0916	36.38	1856.7	1927.7	4.1192
398	41.26	1828.5	1904.8	4.0712	39.15	1846.1	1920.4	4.1004	37.46	1862.7	1935.8	4.1281
400	42.46	1834.3	1912.8	4.0799	40.29	1851.9	1928.5	4.1092	38.54	1868.7	1943.8	4.1369

Definitions of symbols on page 1.

Table 5. Critical Region

$10^3 v$ (t Sat.)

t	2.0 (364.61)				2.1 (367.65)				2.2 (369.77)			
	p	u	h	s	p	u	h	s	p	u	h	s
Sat.	19.715	1771.9	1811.3	3.9913	20.44	1808.6	1851.5	4.0518	20.97	1840.2	1886.3	4.1043
350	13.235	1720.3	1746.7	3.9094	13.333	1743.9	1771.9	3.9495	13.622	1765.2	1795.2	3.9859
352	14.089	1727.6	1755.7	3.9211	14.100	1751.6	1781.2	3.9618	14.323	1773.2	1804.7	3.9987
354	14.954	1734.8	1764.7	3.9326	14.877	1759.1	1790.4	3.9739	15.035	1781.1	1814.2	4.0113
356	15.829	1741.9	1773.6	3.9440	15.665	1766.6	1799.5	3.9858	15.757	1788.9	1823.6	4.0237
358	16.716	1749.0	1782.4	3.9552	16.464	1774.0	1808.6	3.9975	16.488	1796.6	1832.9	4.0359
360	17.612	1756.0	1791.2	3.9663	17.272	1781.3	1817.6	4.0091	17.229	1804.2	1842.1	4.0479
362	18.518	1762.9	1800.0	3.9772	18.089	1788.5	1826.5	4.0205	17.978	1811.7	1851.3	4.0598
364	19.433	1769.8	1808.7	3.9880	18.916	1795.7	1835.4	4.0317	18.736	1819.2	1860.4	4.0715
366	20.356	1776.6	1817.3	3.9987	19.751	1802.8	1844.3	4.0428	19.503	1826.5	1869.4	4.0830
368	21.289	1783.3	1825.9	4.0092	20.594	1809.8	1853.0	4.0538	20.277	1833.8	1878.4	4.0943
370	22.23	1790.0	1834.5	4.0197	21.45	1816.7	1861.8	4.0646	21.06	1841.0	1887.3	4.1055
372	23.18	1796.7	1843.0	4.0300	22.31	1823.6	1870.5	4.0753	21.85	1848.1	1896.2	4.1166
374	24.13	1803.2	1851.5	4.0401	23.17	1830.4	1879.1	4.0858	22.65	1855.1	1905.0	4.1275
376	25.10	1809.8	1860.0	4.0502	24.05	1837.2	1887.7	4.0962	23.45	1862.1	1913.7	4.1383
378	26.07	1816.2	1868.4	4.0602	24.93	1843.9	1896.2	4.1065	24.26	1869.0	1922.4	4.1489
380	27.04	1822.7	1876.8	4.0700	25.81	1850.5	1904.7	4.1167	25.07	1875.9	1931.0	4.1594
382	28.03	1829.0	1885.1	4.0798	26.71	1857.1	1913.2	4.1268	25.90	1882.6	1939.6	4.1697
384	29.02	1835.4	1893.4	4.0894	27.61	1863.6	1921.6	4.1367	26.73	1889.4	1948.2	4.1800
386	30.01	1841.7	1901.7	4.0990	28.51	1870.1	1930.0	4.1465	27.56	1896.0	1956.6	4.1901
388	31.01	1847.9	1909.9	4.1084	29.42	1876.5	1938.3	4.1563	28.40	1902.6	1965.1	4.2001
390	32.02	1854.1	1918.1	4.1178	30.34	1882.9	1946.6	4.1659	29.24	1909.2	1973.5	4.2100
392	33.03	1860.3	1926.3	4.1271	31.26	1889.2	1954.9	4.1754	30.09	1915.7	1981.9	4.2198
394	34.05	1866.4	1934.5	4.1363	32.19	1895.5	1963.1	4.1849	30.95	1922.1	1990.2	4.2294
396	35.07	1872.5	1942.6	4.1454	33.12	1901.8	1971.3	4.1942	31.80	1928.5	1998.5	4.2390
398	36.09	1878.5	1950.7	4.1544	34.06	1908.0	1979.5	4.2035	32.67	1934.8	2006.7	4.2485
400	37.12	1884.6	1958.8	4.1634	35.00	1914.1	1987.6	4.2126	33.53	1941.1	2014.9	4.2579

$10^3 v$ (t Sat.)

t	2.3 (371.26)				2.4 (372.27)				2.6 (373.40)			
	p	u	h	s	p	u	h	s	p	u	h	s
Sat.	21.35	1867.9	1917.0	4.1506	21.60	1892.5	1944.4	4.1921	21.90	1935.2	1992.1	4.2650
350	13.964	1784.8	1816.9	4.0194	14.299	1802.9	1837.3	4.0509	14.862	1836.5	1875.2	4.1095
352	14.617	1793.1	1826.7	4.0327	14.915	1811.5	1847.3	4.0646	15.430	1845.6	1885.7	4.1240
354	15.279	1801.3	1836.4	4.0458	15.540	1820.0	1857.3	4.0782	16.005	1854.6	1896.2	4.1383
356	15.950	1809.3	1846.0	4.0587	16.173	1828.4	1867.2	4.0915	16.587	1863.4	1906.5	4.1524
358	16.631	1817.3	1855.6	4.0714	16.815	1836.6	1877.0	4.1046	17.175	1872.1	1916.8	4.1662
360	17.320	1825.2	1865.0	4.0838	17.464	1844.8	1886.7	4.1174	17.769	1880.7	1926.9	4.1798
362	18.017	1833.0	1874.4	4.0961	18.121	1852.8	1896.3	4.1301	18.370	1889.1	1936.9	4.1931
364	18.722	1840.7	1883.7	4.1082	18.785	1860.7	1905.8	4.1426	18.976	1897.5	1946.8	4.2062
366	19.435	1848.3	1893.0	4.1201	19.457	1868.5	1915.2	4.1549	19.587	1905.7	1956.6	4.2191
368	20.156	1855.8	1902.2	4.1318	20.135	1876.3	1924.6	4.1669	20.205	1913.8	1966.4	4.2318
370	20.88	1863.2	1911.3	4.1434	20.82	1883.9	1933.9	4.1788	20.83	1921.8	1976.0	4.2443
372	21.62	1870.6	1920.3	4.1548	21.51	1891.5	1943.1	4.1906	21.45	1929.7	1985.5	4.2565
374	22.36	1877.8	1929.3	4.1661	22.21	1898.9	1952.2	4.2021	22.09	1937.6	1995.0	4.2686
376	23.11	1885.0	1938.2	4.1771	22.91	1906.3	1961.3	4.2135	22.72	1945.3	2004.4	4.2805
378	23.86	1892.1	1947.0	4.1881	23.62	1913.6	1970.3	4.2247	23.36	1952.9	2013.6	4.2923
380	24.62	1899.2	1955.8	4.1989	24.33	1920.8	1979.2	4.2358	24.01	1960.4	2022.9	4.3038
382	25.38	1906.1	1964.5	4.2095	25.05	1928.0	1988.1	4.2467	24.66	1967.9	2032.0	4.3152
384	26.16	1913.0	1973.2	4.2200	25.77	1935.1	1996.9	4.2575	25.31	1975.2	2041.0	4.3264
386	26.93	1919.9	1981.8	4.2304	26.50	1942.1	2005.7	4.2681	25.97	1982.5	2050.0	4.3375
388	27.71	1926.7	1990.4	4.2407	27.24	1949.0	2014.4	4.2786	26.63	1989.7	2059.0	4.3484
390	28.50	1933.4	1998.9	4.2508	27.98	1955.8	2023.0	4.2890	27.30	1996.8	2067.8	4.3591
392	29.29	1940.0	2007.4	4.2608	28.72	1962.6	2031.6	4.2992	27.97	2003.9	2076.6	4.3697
394	30.08	1946.6	2015.8	4.2707	29.46	1969.4	2040.1	4.3093	28.64	2010.9	2085.3	4.3802
396	30.88	1953.1	2024.2	4.2805	30.21	1976.0	2048.6	4.3193	29.31	2017.8	2094.0	4.3905
398	31.69	1959.6	2032.5	4.2902	30.97	1982.7	2057.0	4.3292	29.99	2024.6	2102.6	4.4007
400	32.49	1966.1	2040.8	4.2998	31.73	1989.2	2065.4	4.3389	30.67	2031.4	2111.1	4.4108

Table 5. Critical Region

10^3v (t Sat.)	2.8 (373.86)				3.0 (374.02)				3.1547 (374.14)			
t	p	u	h	s	p	u	h	s	p	u	h	s
Sat.	22.02	1972.1	2033.7	4.3288	22.06	2005.3	2071.5	4.3870	22.09	2029.6	2099.3	4.4298
350	15.281	1867.7	1910.5	4.1644	15.598	1897.5	1944.3	4.2171	15.799	1919.7	1969.6	4.2567
352	15.820	1877.2	1921.5	4.1796	16.118	1907.2	1955.6	4.2327	16.307	1929.6	1981.1	4.2725
354	16.364	1886.6	1932.4	4.1945	16.641	1916.8	1966.8	4.2480	16.817	1939.4	1992.4	4.2881
356	16.914	1895.7	1943.1	4.2091	17.168	1926.3	1977.8	4.2631	17.331	1949.0	2003.7	4.3034
358	17.468	1904.8	1953.7	4.2235	17.699	1935.6	1988.7	4.2778	17.846	1958.4	2014.7	4.3184
360	18.027	1913.7	1964.2	4.2376	18.233	1944.8	1999.5	4.2923	18.364	1967.7	2025.7	4.3331
362	18.590	1922.5	1974.5	4.2514	18.770	1953.8	2010.1	4.3066	18.885	1976.9	2036.5	4.3476
364	19.158	1931.1	1984.8	4.2650	19.310	1962.7	2020.6	4.3205	19.407	1985.9	2047.1	4.3617
366	19.730	1939.7	1994.9	4.2784	19.853	1971.4	2031.0	4.3343	19.932	1994.8	2057.7	4.3756
368	20.307	1948.1	2005.0	4.2915	20.399	1980.1	2041.3	4.3478	20.459	2003.6	2068.1	4.3893
370	20.89	1956.4	2014.9	4.3045	20.95	1988.6	2051.4	4.3610	20.99	2012.2	2078.4	4.4027
372	21.47	1964.6	2024.7	4.3172	21.50	1997.0	2061.5	4.3740	21.52	2020.7	2088.6	4.4160
374	22.06	1972.7	2034.4	4.3297	22.05	2005.2	2071.4	4.3868	22.05	2029.1	2098.6	4.4289
376	22.65	1980.6	2044.1	4.3420	22.61	2013.4	2081.2	4.3994	22.59	2037.3	2108.6	4.4417
378	23.24	1988.5	2053.6	4.3541	23.17	2021.5	2091.0	4.4118	23.12	2045.5	2118.4	4.4542
380	23.84	1996.3	2063.0	4.3660	23.73	2029.4	2100.6	4.4240	23.66	2053.5	2128.2	4.4666
382	24.44	2004.0	2072.4	4.3778	24.29	2037.3	2110.1	4.4360	24.20	2061.5	2137.8	4.4787
384	25.05	2011.5	2081.7	4.3893	24.86	2045.0	2119.6	4.4478	24.74	2069.3	2147.4	4.4907
386	25.65	2019.0	2090.9	4.4007	25.43	2052.7	2128.9	4.4595	25.28	2077.1	2156.8	4.5024
388	26.26	2026.5	2100.0	4.4119	26.00	2060.2	2138.2	4.4709	25.83	2084.7	2166.2	4.5140
390	26.87	2033.8	2109.0	4.4230	26.57	2067.7	2147.4	4.4822	26.37	2092.3	2175.4	4.5254
392	27.49	2041.0	2118.0	4.4339	27.14	2075.1	2156.5	4.4933	26.92	2099.7	2184.6	4.5366
394	28 11	2048.2	2126.9	4.4446	27.71	2082.4	2165.5	4.5043	27.47	2107.1	2193.7	4.5477
396	28 72	2055.3	2135.7	4.4552	28.29	2089.6	2174.5	4.5151	28.01	2114.4	2202.8	4.5586
398	29.35	2062.3	2144.4	4.4657	28.87	2096.7	2183.3	4.5257	28.56	2121.6	2211.7	4.5694
400	29.97	2069.2	2153.1	4.4760	29.45	2103.8	2192.1	4.5362	29.11	2128.7	2220.6	4.5800

10^3v (t Sat.)	3.5 (373.86)				4.0 (373.06)				4.5 (371.60)			
t	p	u	h	s	p	u	h	s	p	u	h	s
Sat.	22.02	2078.2	2155.2	4.5166	21.81	2139.5	2226.8	4.6284	21.43	2191.0	2287.5	4.7249
350	16.181	1967.5	2024.1	4.3422	16.634	2032.3	2098.8	4.4593	16.990	2091.7	2168.2	4.5682
352	16.665	1977.6	2035.9	4.3584	17.086	2042.4	2110.7	4.4755	17.408	2101.7	2180.0	4.5842
354	17.151	1987.5	2047.6	4.3742	17.538	2052.3	2122.5	4.4914	17.825	2111.4	2191.7	4.5998
356	17.638	1997.3	2059.0	4.3898	17.989	2062.1	2134.1	4.5070	18.240	2121.1	2203.1	4.6151
358	18.125	2006.9	2070.4	4.4051	18.439	2071.7	2145.5	4.5222	18.654	2130.5	2214.5	4.6301
360	18.613	2016.4	2081.5	4.4200	18.888	2081.2	2156.8	4.5372	19.067	2139.8	2225.6	4.6448
362	19.102	2025.7	2092.6	4.4347	19.337	2090.5	2167.9	4.5519	19.478	2149.0	2236.6	4.6592
364	19.592	2034.9	2103.5	4.4492	19.785	2099.7	2178.8	4.5663	19.887	2158.0	2247.5	4.6734
366	20.083	2043.9	2114.2	4.4633	20.233	2108.7	2189.7	4.5805	20.296	2166.9	2258.2	4.6873
368	20.574	2052.8	2124.8	4.4772	20.680	2117.6	2200.3	4.5944	20.703	2175.6	2268.8	4.7009
370	21.07	2061.6	2135.3	4.4909	21.13	2126.4	2210.9	4.6080	21.11	2184.2	2279.2	4.7144
372	21.56	2070.2	2145.7	4.5043	21.57	2135.0	2221.3	4.6214	21.51	2192.7	2289.5	4.7275
374	22.05	2078.7	2155.9	4.5175	22.02	2143.5	2231.6	4.6346	21.92	2201.1	2299.7	4.7405
376	22.54	2087.1	2166.0	4.5304	22.46	2151.9	2241.8	4.6475	22.32	2209.3	2309.7	4.7532
378	23.04	2095.4	2176.0	4.5431	22.91	2160.2	2251.8	4.6602	22.72	2217.4	2319.7	4.7657
380	23.53	2103.6	2185.9	4.5557	23.35	2168.3	2261.7	4.6727	23.12	2225.4	2329.5	4.7780
382	24.03	2111.6	2195.7	4.5680	23.79	2176.4	2271.5	4.6850	23.52	2233.4	2339.2	4.7901
384	24.52	2119.6	2205.4	4.5801	24.24	2184.3	2281.2	4.6971	23.92	2241.2	2348.8	4.8020
386	25.02	2127.4	2215.0	4.5920	24.68	2192.1	2290.8	4.7090	24.31	2248.9	2358.3	4.8137
388	25.52	2135.2	2224.5	4.6037	25.12	2199.9	2300.3	4.7207	24.71	2256.5	2367.7	4.8252
390	26.01	2142.8	2233.8	4.6153	25.56	2207.5	2309.7	4.7322	25.11	2264.0	2376.9	4.8365
392	26.51	2150.3	2243.1	4.6266	26.00	2215.0	2319.0	4.7436	25.50	2271.4	2386.1	4.8477
394	27.01	2157.8	2252.3	4.6378	26.44	2222.4	2328.2	4.7547	25.89	2278.7	2395.2	4.8587
396	27.51	2165.2	2261.4	4.6489	26.88	2229.8	2337.3	4.7657	26.29	2285.9	2404.2	4.8695
398	28.00	2172.5	2270.5	4.6597	27.32	2237.1	2346.3	4.7766	26.68	2293.1	2413.1	4.8802
400	28.50	2179.7	2279.4	4.6704	27.76	2244.2	2355.3	4.7873	27.07	2300.1	2421.9	4.8907

Definitions of symbols on page 1.

Table 5. Critical Region

<table>
<tr><td>10³v (t Sat.)</td><td colspan="4">5 (369.69)</td><td colspan="4">6 (364.97)</td><td colspan="4">7 (359.70)</td></tr>
<tr><td>t</td><td>p</td><td>u</td><td>h</td><td>s</td><td>p</td><td>u</td><td>h</td><td>s</td><td>p</td><td>u</td><td>h</td><td>s</td></tr>
<tr><td>Sat.</td><td>20.95</td><td>2234.5</td><td>2339.3</td><td>4.8089</td><td>19.800</td><td>2303.0</td><td>2421.8</td><td>4.9477</td><td>18.584</td><td>2354.0</td><td>2484.1</td><td>5.0581</td></tr>
<tr><td>350</td><td>17.240</td><td>2145.9</td><td>2232.1</td><td>4.6688</td><td>17.426</td><td>2238.9</td><td>2343.5</td><td>4.8461</td><td>17.277</td><td>2314.6</td><td>2435.5</td><td>4.9953</td></tr>
<tr><td>352</td><td>17.625</td><td>2155.5</td><td>2243.6</td><td>4.6843</td><td>17.750</td><td>2247.9</td><td>2354.4</td><td>4.8605</td><td>17.550</td><td>2322.9</td><td>2445.8</td><td>5.0087</td></tr>
<tr><td>354</td><td>18.008</td><td>2165.0</td><td>2255.1</td><td>4.6995</td><td>18.071</td><td>2256.8</td><td>2365.2</td><td>4.8746</td><td>17.821</td><td>2331.2</td><td>2455.9</td><td>5.0218</td></tr>
<tr><td>356</td><td>18.389</td><td>2174.4</td><td>2266.3</td><td>4.7143</td><td>18.391</td><td>2265.5</td><td>2375.9</td><td>4.8885</td><td>18.090</td><td>2339.3</td><td>2465.9</td><td>5.0347</td></tr>
<tr><td>358</td><td>18.768</td><td>2183.5</td><td>2277.4</td><td>4.7289</td><td>18.708</td><td>2274.1</td><td>2386.3</td><td>4.9021</td><td>18.358</td><td>2347.2</td><td>2475.8</td><td>5.0474</td></tr>
<tr><td>360</td><td>19.145</td><td>2192.6</td><td>2288.3</td><td>4.7432</td><td>19.024</td><td>2282.5</td><td>2396.7</td><td>4.9155</td><td>18.623</td><td>2355.1</td><td>2485.5</td><td>5.0599</td></tr>
<tr><td>362</td><td>19.521</td><td>2201.5</td><td>2299.1</td><td>4.7573</td><td>19.337</td><td>2290.9</td><td>2406.9</td><td>4.9286</td><td>18.887</td><td>2362.9</td><td>2495.1</td><td>5.0721</td></tr>
<tr><td>364</td><td>19.894</td><td>2210.3</td><td>2309.7</td><td>4.7711</td><td>19.649</td><td>2299.1</td><td>2417.0</td><td>4.9415</td><td>19.150</td><td>2370.6</td><td>2504.6</td><td>5.0842</td></tr>
<tr><td>366</td><td>20.267</td><td>2218.9</td><td>2320.3</td><td>4.7846</td><td>19.959</td><td>2307.2</td><td>2426.9</td><td>4.9542</td><td>19.411</td><td>2378.1</td><td>2514.0</td><td>5.0960</td></tr>
<tr><td>368</td><td>20.637</td><td>2227.4</td><td>2330.6</td><td>4.7979</td><td>20.268</td><td>2315.2</td><td>2436.8</td><td>4.9667</td><td>19.670</td><td>2385.6</td><td>2523.3</td><td>5.1077</td></tr>
<tr><td>370</td><td>21.01</td><td>2235.8</td><td>2340.9</td><td>4.8110</td><td>20.57</td><td>2323.0</td><td>2446.5</td><td>4.9789</td><td>19.927</td><td>2392.9</td><td>2532.4</td><td>5.1192</td></tr>
<tr><td>372</td><td>21.37</td><td>2244.1</td><td>2351.0</td><td>4.8238</td><td>20.88</td><td>2330.8</td><td>2456.1</td><td>4.9910</td><td>20.184</td><td>2400.2</td><td>2541.5</td><td>5.1304</td></tr>
<tr><td>374</td><td>21.74</td><td>2252.2</td><td>2360.9</td><td>4.8364</td><td>21.18</td><td>2338.5</td><td>2465.6</td><td>5.0029</td><td>20.438</td><td>2407.4</td><td>2550.5</td><td>5.1416</td></tr>
<tr><td>376</td><td>22.11</td><td>2260.3</td><td>2370.8</td><td>4.8488</td><td>21.49</td><td>2346.0</td><td>2474.9</td><td>5.0145</td><td>20.692</td><td>2414.5</td><td>2559.3</td><td>5.1525</td></tr>
<tr><td>378</td><td>22.47</td><td>2268.2</td><td>2380.5</td><td>4.8610</td><td>21.79</td><td>2353.5</td><td>2484.2</td><td>5.0260</td><td>20.944</td><td>2421.5</td><td>2568.1</td><td>5.1633</td></tr>
<tr><td>380</td><td>22.83</td><td>2276.0</td><td>2390.2</td><td>4.8730</td><td>22.08</td><td>2360.8</td><td>2493.3</td><td>5.0373</td><td>21.19</td><td>2428.4</td><td>2576.8</td><td>5.1739</td></tr>
<tr><td>382</td><td>23.19</td><td>2283.7</td><td>2399.7</td><td>4.8848</td><td>22.38</td><td>2368.1</td><td>2502.4</td><td>5.0484</td><td>21.44</td><td>2435.2</td><td>2585.3</td><td>5.1843</td></tr>
<tr><td>384</td><td>23.55</td><td>2291.3</td><td>2409.1</td><td>4.8964</td><td>22.68</td><td>2375.3</td><td>2511.4</td><td>5.0593</td><td>21.69</td><td>2442.0</td><td>2593.8</td><td>5.1946</td></tr>
<tr><td>386</td><td>23.91</td><td>2298.9</td><td>2418.4</td><td>4.9078</td><td>22.97</td><td>2382.4</td><td>2520.2</td><td>5.0701</td><td>21.94</td><td>2448.7</td><td>2602.3</td><td>5.2048</td></tr>
<tr><td>388</td><td>24.27</td><td>2306.3</td><td>2427.6</td><td>4.9191</td><td>23.27</td><td>2389.4</td><td>2529.0</td><td>5.0807</td><td>22.19</td><td>2455.3</td><td>2610.6</td><td>5.2148</td></tr>
<tr><td>390</td><td>24.62</td><td>2313.6</td><td>2436.7</td><td>4.9301</td><td>23.56</td><td>2396.3</td><td>2537.7</td><td>5.0912</td><td>22.43</td><td>2461.8</td><td>2618.8</td><td>5.2246</td></tr>
<tr><td>392</td><td>24.98</td><td>2320.9</td><td>2445.7</td><td>4.9410</td><td>23.85</td><td>2403.2</td><td>2546.3</td><td>5.1015</td><td>22.67</td><td>2468.3</td><td>2627.0</td><td>5.2344</td></tr>
<tr><td>394</td><td>25.33</td><td>2328.0</td><td>2454.7</td><td>4.9518</td><td>24.14</td><td>2409.9</td><td>2554.8</td><td>5.1116</td><td>22.92</td><td>2474.7</td><td>2635.1</td><td>5.2440</td></tr>
<tr><td>396</td><td>25.68</td><td>2335.1</td><td>2463.5</td><td>4.9623</td><td>24.43</td><td>2416.6</td><td>2563.2</td><td>5.1217</td><td>23.16</td><td>2481.0</td><td>2643.1</td><td>5.2534</td></tr>
<tr><td>398</td><td>26.03</td><td>2342.1</td><td>2472.2</td><td>4.9728</td><td>24.72</td><td>2423.2</td><td>2571.5</td><td>5.1315</td><td>23.40</td><td>2487.3</td><td>2651.1</td><td>5.2628</td></tr>
<tr><td>400</td><td>26.38</td><td>2349.0</td><td>2480.9</td><td>4.9830</td><td>25.01</td><td>2429.8</td><td>2579.8</td><td>5.1413</td><td>23.64</td><td>2493.5</td><td>2658.9</td><td>5.2720</td></tr>
</table>

<table>
<tr><td>10³v (t Sat.)</td><td colspan="4">8 (354.32)</td><td colspan="4">9 (349.02)</td><td colspan="4">10 (343.91)</td></tr>
<tr><td>t</td><td>p</td><td>u</td><td>h</td><td>s</td><td>p</td><td>u</td><td>h</td><td>s</td><td>p</td><td>u</td><td>h</td><td>s</td></tr>
<tr><td>Sat.</td><td>17.409</td><td>2393.0</td><td>2532.3</td><td>5.1485</td><td>16.316</td><td>2423.6</td><td>2570.5</td><td>5.2246</td><td>15.316</td><td>2448.2</td><td>2601.4</td><td>5.2898</td></tr>
<tr><td>350</td><td>16.909</td><td>2376.3</td><td>2511.6</td><td>5.1219</td><td>16.415</td><td>2427.2</td><td>2575.0</td><td>5.2303</td><td>15.857</td><td>2469.7</td><td>2628.3</td><td>5.3243</td></tr>
<tr><td>352</td><td>17.141</td><td>2384.1</td><td>2521.2</td><td>5.1343</td><td>16.615</td><td>2434.5</td><td>2584.1</td><td>5.2420</td><td>16.032</td><td>2476.6</td><td>2636.9</td><td>5.3353</td></tr>
<tr><td>354</td><td>17.372</td><td>2391.8</td><td>2530.8</td><td>5.1466</td><td>16.815</td><td>2441.7</td><td>2593.0</td><td>5.2535</td><td>16.206</td><td>2483.3</td><td>2645.4</td><td>5.3462</td></tr>
<tr><td>356</td><td>17.602</td><td>2399.3</td><td>2540.2</td><td>5.1586</td><td>17.013</td><td>2448.8</td><td>2601.9</td><td>5.2648</td><td>16.378</td><td>2490.0</td><td>2653.8</td><td>5.3568</td></tr>
<tr><td>358</td><td>17.830</td><td>2406.8</td><td>2549.4</td><td>5.1705</td><td>17.209</td><td>2455.8</td><td>2610.7</td><td>5.2759</td><td>16.550</td><td>2496.7</td><td>2662.2</td><td>5.3674</td></tr>
<tr><td>360</td><td>18.056</td><td>2414.2</td><td>2558.6</td><td>5.1821</td><td>17.404</td><td>2462.8</td><td>2619.4</td><td>5.2869</td><td>16.720</td><td>2503.2</td><td>2670.4</td><td>5.3777</td></tr>
<tr><td>362</td><td>18.281</td><td>2421.4</td><td>2567.7</td><td>5.1936</td><td>17.598</td><td>2469.6</td><td>2628.0</td><td>5.2976</td><td>16.889</td><td>2509.7</td><td>2678.6</td><td>5.3879</td></tr>
<tr><td>364</td><td>18.504</td><td>2428.6</td><td>2576.7</td><td>5.2049</td><td>17.791</td><td>2476.4</td><td>2636.5</td><td>5.3083</td><td>17.057</td><td>2516.1</td><td>2686.7</td><td>5.3980</td></tr>
<tr><td>366</td><td>18.726</td><td>2435.7</td><td>2585.5</td><td>5.2160</td><td>17.982</td><td>2483.0</td><td>2644.9</td><td>5.3187</td><td>17.225</td><td>2522.4</td><td>2694.7</td><td>5.4079</td></tr>
<tr><td>368</td><td>18.947</td><td>2442.7</td><td>2594.3</td><td>5.2269</td><td>18.173</td><td>2489.6</td><td>2653.2</td><td>5.3290</td><td>17.391</td><td>2528.7</td><td>2702.6</td><td>5.4177</td></tr>
<tr><td>370</td><td>19.166</td><td>2449.6</td><td>2602.9</td><td>5.2377</td><td>18.362</td><td>2496.2</td><td>2661.4</td><td>5.3392</td><td>17.556</td><td>2534.9</td><td>2710.4</td><td>5.4273</td></tr>
<tr><td>372</td><td>19.385</td><td>2456.4</td><td>2611.5</td><td>5.2483</td><td>18.550</td><td>2502.6</td><td>2669.6</td><td>5.3492</td><td>17.720</td><td>2541.0</td><td>2718.2</td><td>5.4369</td></tr>
<tr><td>374</td><td>19.601</td><td>2463.2</td><td>2620.0</td><td>5.2587</td><td>18.737</td><td>2509.0</td><td>2677.6</td><td>5.3591</td><td>17.884</td><td>2547.1</td><td>2725.9</td><td>5.4463</td></tr>
<tr><td>376</td><td>19.817</td><td>2469.9</td><td>2628.4</td><td>5.2690</td><td>18.923</td><td>2515.3</td><td>2685.6</td><td>5.3689</td><td>18.046</td><td>2553.1</td><td>2733.6</td><td>5.4555</td></tr>
<tr><td>378</td><td>20.031</td><td>2476.5</td><td>2636.7</td><td>5.2792</td><td>19.108</td><td>2521.6</td><td>2693.5</td><td>5.3785</td><td>18.208</td><td>2559.1</td><td>2741.1</td><td>5.4647</td></tr>
<tr><td>380</td><td>20.24</td><td>2483.0</td><td>2645.0</td><td>5.2892</td><td>19.292</td><td>2527.7</td><td>2701.4</td><td>5.3880</td><td>18.369</td><td>2564.9</td><td>2748.6</td><td>5.4737</td></tr>
<tr><td>382</td><td>20.46</td><td>2489.4</td><td>2653.1</td><td>5.2990</td><td>19.475</td><td>2533.9</td><td>2709.1</td><td>5.3973</td><td>18.529</td><td>2570.8</td><td>2756.1</td><td>5.4827</td></tr>
<tr><td>384</td><td>20.67</td><td>2495.8</td><td>2661.2</td><td>5.3088</td><td>19.657</td><td>2539.9</td><td>2716.8</td><td>5.4066</td><td>18.688</td><td>2576.6</td><td>2763.4</td><td>5.4915</td></tr>
<tr><td>386</td><td>20.88</td><td>2502.1</td><td>2669.2</td><td>5.3184</td><td>19.839</td><td>2545.9</td><td>2724.5</td><td>5.4157</td><td>18.847</td><td>2582.3</td><td>2770.8</td><td>5.5002</td></tr>
<tr><td>388</td><td>21.09</td><td>2508.4</td><td>2677.1</td><td>5.3278</td><td>20.019</td><td>2551.9</td><td>2732.0</td><td>5.4247</td><td>19.004</td><td>2588.0</td><td>2778.0</td><td>5.5088</td></tr>
<tr><td>390</td><td>21.30</td><td>2514.6</td><td>2684.9</td><td>5.3372</td><td>20.20</td><td>2557.8</td><td>2739.6</td><td>5.4336</td><td>19.161</td><td>2593.6</td><td>2785.2</td><td>5.5173</td></tr>
<tr><td>392</td><td>21.50</td><td>2520.7</td><td>2692.7</td><td>5.3464</td><td>20.38</td><td>2563.6</td><td>2747.0</td><td>5.4424</td><td>19.318</td><td>2599.2</td><td>2792.4</td><td>5.5257</td></tr>
<tr><td>394</td><td>21.71</td><td>2526.8</td><td>2700.4</td><td>5.3555</td><td>20.56</td><td>2569.4</td><td>2754.4</td><td>5.4510</td><td>19.473</td><td>2604.7</td><td>2799.5</td><td>5.5340</td></tr>
<tr><td>396</td><td>21.91</td><td>2532.8</td><td>2708.1</td><td>5.3645</td><td>20.73</td><td>2575.1</td><td>2761.7</td><td>5.4596</td><td>19.628</td><td>2610.2</td><td>2806.5</td><td>5.5423</td></tr>
<tr><td>398</td><td>22.12</td><td>2538.7</td><td>2715.7</td><td>5.3734</td><td>20.91</td><td>2580.8</td><td>2769.0</td><td>5.4681</td><td>19.782</td><td>2615.7</td><td>2813.5</td><td>5.5504</td></tr>
<tr><td>400</td><td>22.32</td><td>2544.6</td><td>2723.2</td><td>5.3822</td><td>21.08</td><td>2586.4</td><td>2776.2</td><td>5.4765</td><td>19.936</td><td>2621.1</td><td>2820.5</td><td>5.5584</td></tr>
</table>

Definitions of symbols on page 1.

Table 6. Saturation: Solid-Vapor

		Specific Volume		Internal Energy			Enthalpy			Entropy		
Temp. °C	Press. kPa	Sat. Solid	Sat. Vapor	Sat. Solid	Subl.	Sat. Vapor	Sat. Solid	Subl.	Sat. Vapor	Sat. Solid	Subl.	Sat. Vapor
t	p	$10^3 v_i$	v_g	u_i	u_{ig}	u_g	h_i	h_{ig}	h_g	s_i	s_{ig}	s_g
.01	.6113	1.0908	206.1	−333.40	2708.7	2375.3	−333.40	2834.8	2501.4	−1.221	10.378	9.156
0	.6108	1.0908	206.3	−333.43	2708.8	2375.3	−333.43	2834.8	2501.3	−1.221	10.378	9.157
−2	.5176	1.0904	241.7	−337.62	2710.2	2372.6	−337.62	2835.3	2497.7	−1.237	10.456	9.219
−4	.4375	1.0901	283.8	−341.78	2711.6	2369.8	−341.78	2835.7	2494.0	−1.253	10.536	9.283
−6	.3689	1.0898	334.2	−345.91	2712.9	2367.0	−345.91	2836.2	2490.3	−1.268	10.616	9.348
−8	.3102	1.0894	394.4	−350.02	2714.2	2364.2	−350.02	2836.6	2486.6	−1.284	10.698	9.414
−10	.2602	1.0891	466.7	−354.09	2715.5	2361.4	−354.09	2837.0	2482.9	−1.299	10.781	9.481
−12	.2176	1.0888	553.7	−358.14	2716.8	2358.7	−358.14	2837.3	2479.2	−1.315	10.865	9.550
−14	.1815	1.0884	658.8	−362.15	2718.0	2355.9	−362.15	2837.6	2475.5	−1.331	10.950	9.619
−16	.1510	1.0881	786.0	−366.14	2719.2	2353.1	−366.14	2837.9	2471.8	−1.346	11.036	9.690
−18	.1252	1.0878	940.5	−370.10	2720.4	2350.3	−370.10	2838.2	2468.1	−1.362	11.123	9.762
−20	.1035	1.0874	1128.6	−374.03	2721.6	2347.5	−374.03	2838.4	2464.3	−1.377	11.212	9.835
−22	.0853	1.0871	1358.4	−377.93	2722.7	2344.7	−377.93	2838.6	2460.6	−1.393	11.302	9.909
−24	.0701	1.0868	1640.1	−381.80	2723.7	2342.0	−381.80	2838.7	2456.9	−1.408	11.394	9.985
−26	.0574	1.0864	1986.4	−385.64	2724.8	2339.2	−385.64	2838.9	2453.2	−1.424	11.486	10.062
−28	.0469	1.0861	2413.7	−389.45	2725.8	2336.4	−389.45	2839.0	2449.5	−1.439	11.580	10.141
−30	.0381	1.0858	2943.	−393.23	2726.8	2333.6	−393.23	2839.0	2445.8	−1.455	11.676	10.221
−32	.0309	1.0854	3600.	−396.98	2727.8	2330.8	−396.98	2839.1	2442.1	−1.471	11.773	10.303
−34	.0250	1.0851	4419.	−400.71	2728.7	2328.0	−400.71	2839.1	2438.4	−1.486	11.872	10.386
−36	.0201	1.0848	5444.	−404.40	2729.6	2325.2	−404.40	2839.1	2434.7	−1.501	11.972	10.470
−38	.0161	1.0844	6731.	−408.06	2730.5	2322.4	−408.06	2839.0	2430.9	−1.517	12.073	10.556
−40	.0129	1.0841	8354.	−411.70	2731.3·	2319.6	−411.70	2838.9	2427.2	−1.532	12.176	10.644

Definitions of symbols on page 1.

Table 7. Dynamic Viscosity (micropoise)

Pressure MPa	lbf/in²	Temp. °C: 0 / °F: 32	50 / 122	100 / 212	150 / 302	200 / 392	250 / 482	300 / 572	350 / 662	375 / 707
.1	14.504	17500	5440	121.1	141.5	161.8	182.2	202.5	223	233
.5	72.52	17500	5440	2790	1810	160.2	181.4	202.3		234
1.0	145.04	17500	5440	2790	1810	158.5	180.6	202.2		234
2.5	362.6	17500	5440	2800	1820	1340	177.8	201.6		236
5.0	725.2	17500	5450	2800	1820	1350	1070	200.6		240
7.5	1087.8	17500	5450	2800	1830	1350	1080	199.2		244
10.0	1450.4	17500	5450	2810	1830	1360	1080	905		249
12.5	1813.0	17500	5460	2810	1840	1360	1090	911		254
15.0	2176	17400	5460	2820	1840	1370	1100	917		262
17.5	2538	17400	5460	2820	1850	1380	1100	924		273
20.0	2901	17400	5460	2830	1860	1380	1110	930	735	291
22.5	3263	17400	5460	2830	1860	1390	1120	936	747	491
25.0	3626	17400	5470	2840	1870	1390	1120	943	760	597
27.5	3989	17400	5470	2840	1870	1400	1130	949	772	633
30	4351	17400	5470	2850	1880	1400	1130	955	785	657
35	5076	17300	5480	2860	1890	1420	1150	968	805	693
40	5802	17300	5480	2870	1900	1430	1160	981	825	721
45	6527	17300	5490	2880	910	1440	1170	993	837	743
50	7252	17200	5490	2890	1920	1450	1180	1010	850	762
55	7977	17200	5500	2900	1930	1460	1200	1020	860	780
60	8702	17200	5500	2910	1940	1480	1210	1030	870	795
65	9427	17200	5510	2920	1960	1490	1220	1040	882	809
70	10153	17100	5510	2930	1970	1500	1230	1060	895	822
75	10878	17100	5520	2940	1980	1510	1240	1070	905	835
80	11603	17100	5520	2950	1990	1520	1260	1080	915	846

Pressure MPa	lbf/in²	Temp. °C: 400 / °F: 752	425 / 797	450 / 842	475 / 887	500 / 932	550 / 1022	600 / 1112	650 / 1202	700 / 1292
.1	14.504	243	253	264	274	284	304	325	345	365
.5	72.52	244	254	264	274	284	305	325	345	366
1.0	145.04	244	255	265	275	285	305	326	346	366
2.5	362.6	246	256	266	276	287	307	327	347	367
5.0	725.2	250	259	269	279	289	309	329	349	369
7.5	1087.8	253	263	273	282	292	312	332	352	372
10.0	1450.4	258	267	276	286	295	315	334	354	374
12.5	1813.0	263	271	280	289	299	318	337	357	376
15.0	2176	269	276	285	294	302	321	340	359	379
17.5	2538	276	282	290	298	307	324	343	362	381
20.0	2901	286	289	296	303	311	328	346	365	384
22.5	3263	299	298	302	309	316	332	350	368	386
25.0	3626	321	309	310	315	321	336	353	371	389
27.5	3989	367	324	320	322	327	341	357	374	392
30	4351	458	345	331	330	334	346	361	377	395
35	5076	573	416	363	351	349	357	369	385	401
40	5802	628	503	411	379	369	369	379	392	408
45	6527	664	565	468	415	393	383	389	401	415
50	7252	693	609	521	456	421	400	401	410	423
55	7977	716	643	564	497	453	418	414	420	431
60	8702	736	670	600	534	485	439	428	430	439
65	9427	754	693	629	567	516	460	442	441	448
70	10153	770	713	654	596	545	482	458	453	458
75	10878	784	732	676	621	572	504	474	466	468
80	11603	798	748	695	644	596	526	491	478	478

Conversion Factors for Viscosity

10^6 micropoise = 0.0020885 lbf × sec/ft²   = 241.91 lb/(hr × ft)   = 0.1 kg/(m·s)

= 0.067197 lb/(ft × sec)   = 0.58015 × 10^{-6} lbf × hr/ft²   = 0.1 N·s/m²

(113)

Table 8.   Kinematic Viscosity $\times 10^7$ (m²/s)

Pressure MPa	lbf/in²	Temp. °C °F	0 32	50 122	100 212	150 302	200 392	250 482	300 572	350 662	375 707
.1	14.504		17.5	5.51	205.4	273.9	351.4	438.3	534.4	640.2	695.9
.5	72.52		17.5	5.51	2.91	1.98	68.1	86.1	105.7		138.9
1.0	145.04		17.5	5.50	2.91	1.98	32.7	42.0	52.2		68.9
2.5	362.6		17.5	5.50	2.91	1.98	1.55	15.47	19.94		27.1
5.0	725.2		17.5	5.50	2.91	1.98	1.55	1.34	9.09		13.2
7.5	1087.8		17.4	5.50	2.92	1.99	1.56	1.34	5.32		8.5
10.0	1450.4		17.4	5.49	2.92	1.99	1.56	1.34	1.26		6.11
12.5	1813.0		17.4	5.49	2.92	1.99	1.56	1.35	1.26		4.64
15.0	2176		17.3	5.49	2.92	2.00	1.57	1.35	1.26		3.64
17.5	2538		17.3	5.48	2.92	2.00	1.57	1.36	1.27		2.88
20.0	2901		17.2	5.48	2.93	2.00	1.58	1.36	1.27	1.22	2.24
22.5	3263		17.2	5.48	2.93	2.00	1.58	1.36	1.27	1.22	1.23
25.0	3626		17.2	5.47	2.93	2.01	1.58	1.37	1.27	1.22	1.18
27.5	3989		17.1	5.47	2.93	2.01	1.59	1.37	1.27	1.22	1.18
30	4351		17.1	5.47	2.93	2.01	1.59	1.37	1.27	1.22	1.18
35	5076		17.0	5.46	2.94	2.02	1.60	1.38	1.28	1.22	1.18
40	5802		17.0	5.45	2.94	2.03	1.61	1.39	1.28	1.22	1.19
45	6527		16.9	5.45	2.94	2.03	1.61	1.40	1.29	1.23	1.19
50	7252		16.8	5.44	2.95	2.04	1.62	1.40	1.30	1.23	1.19

Pressure MPa	lbf/in²	Temp. °C	400 752	425 797	450 842	475 887	500 932	550 1022	600 1112	650 1202	700 1292
.1	14.504		754.0	814.4	880	945	1013	1154	1309	1469	1639
.5	72.52		150.6	162.8	175	188	202	231	261	293	328
1.0	145.04		74.8	81.2	88	94	101	115	131	147	164
2.5	362.6		29.5	32.0	34.6	37.3	40.2	46.0	52.1	58.6	65.4
5.0	725.2		14.5	15.7	17.0	18.4	19.8	22.8	25.9	29.2	32.7
7.5	1087.8		9.3	10.3	11.2	12.1	13.1	15.1	17.2	19.4	21.8
10.0	1450.4		6.82	7.51	8.21	8.95	9.67	11.23	12.82	14.52	16.30
12.5	1813.0		5.26	5.84	6.44	7.03	7.66	8.91	10.21	11.60	13.01
15.0	2176		4.21	4.73	5.26	5.78	6.28	7.36	8.47	9.62	10.85
17.5	2538		3.44	3.93	4.40	4.86	5.33	6.25	7.23	8.23	9.27
20.0	2901		2.85	3.31	3.76	4.18	4.59	5.43	6.29	7.19	8.12
22.5	3263		2.35	2.83	3.25	3.65	4.03	4.79	5.58	6.38	7.20
25.0	3626		1.93	2.44	2.84	3.22	3.58	4.28	4.99	5.73	6.48
27.5	3989		1.54	2.11	2.51	2.87	3.20	3.86	4.52	5.19	5.89
30	4351		1.28	1.83	2.23	2.57	2.90	3.52	4.13	4.75	5.40
35	5076		1.21	1.43	1.80	2.13	2.42	2.98	3.52	4.07	4.63
40	5802		1.20	1.28	1.52	1.81	2.08	2.58	3.07	3.55	4.06
45	6527		1.20	1.24	1.37	1.59	1.82	2.27	2.72	3.17	3.62
50	7252		1.20	1.22	1.30	1.45	1.64	2.05	2.45	2.86	3.27

Conversion Factors for Kinematic Viscosity

$$10^{-7}\,\text{m}^2/\text{s} = 1.076 \times 10^{-6}\,\text{ft}^2/\text{s}$$
$$= 3.875 \times 10^{-3}\,\text{ft}^2/\text{hr}$$
$$= 3.6 \times 10^{-4}\quad \text{m}^2/\text{hr}$$
$$= 10^{-3}\qquad \text{cm}^2/\text{s}$$

Table 9. Thermal Conductivity (mW / m · K)

Pressure MPa	lbf/in.²	Temp. °C 0 °F 32	50 122	100 212	150 302	200 392	250 482	300 572	350 662	375 707
.1	14.504	569	643	24.8	28.7	33.2	38.2	43.4	49.0	51.9
.5	72.52	569	644	681	687	33.8	38.6	43.8	49.4	52.3
1.0	145.04	570	644	681	687	35.1	39.3	44.4	49.9	52.8
2.5	362.6	571	645	682	688	665	42.9	46.5	51.6	54.3
5.0	725.2	573	647	684	690	668	618	52.5	55.4	57.6
7.5	1087.8	575	649	686	691	670	622	63.7	60.8	62.0
10.0	1450.4	577	651	688	693	672	625	545	68.8	67.9
12.5	1813.0	579	653	689	695	674	629	552	81.3	75.9
15.0	2176	581	655	691	696	676	633	559	104.0	87.5
17.5	2538	583	657	693	698	679	636	565	442	106.0
20.0	2901	585	659	695	700	681	639	571	454	126
22.5	3263	587	661	696	701	683	642	577	465	297
25.0	3626	589	662	698	703	685	646	582	476	376
27.5	3989	591	664	699	705	687	649	588	486	402
30	4351	592	666	701	706	689	652	592	496	419
35	5076	596	669	704	710	693	657	601	514	444
40	5802	599	672	707	713	697	662	609	529	468
45	6527	603	675	710	716	701	667	616	541	486
50	7252	606	678	713	720	704	671	622	552	501

Pressure MPa	lbf/in.²	Temp. °C 400 °F 752	425 797	450 842	475 887	500 932	550 1022	600 1112	650 1202	700 1292
.1	14.504	54.9	58.0	61.1	64.2	67.4	73.9	80.6	87.4	94.3
.5	72.52	55.3	58.3	61.4	64.5	67.7	74.3	80.9	87.7	94.6
1.0	145.04	55.7	58.8	61.8	65.0	68.2	74.7	81.4	88.2	95.0
2.5	362.6	57.2	60.2	63.3	66.4	69.6	76.1	82.7	89.5	96.3
5.0	725.2	60.2	63.0	65.9	68.9	72.0	78.4	85.0	91.7	98.6
7.5	1087.8	63.9	66.3	68.9	71.7	74.7	80.9	87.4	94.0	101.0
10.0	1450.4	68.6	70.2	72.4	74.9	77.6	83.5	89.8	96	103
12.5	1813.0	74.5	74.9	76.4	78.4	80.8	86.3	92.4	99	105
15.0	2176	82.2	80.7	81.0	82.4	84.3	89.3	95.1	101	108
17.5	2538	92.6	87.9	86.5	86.9	88.3	92.5	98.0	104	110
20.0	2901	107	97	93	92	93	96	101	107	113
22.5	3263	130	109	101	98	97	100	104	110	115
25.0	3626	157	125	111	105	103	104	107	112	118
27.5	3989	200	147	123	113	109	108	111	115	121
30	4351	264	171	138	122	116	112	114	118	124
35	5076	351	239	182	147	132	122	122	125	129
40	5802	390	296	220	177	153	134	130	132	135
45	6527	416	338	264	210	180	148	139	139	142
50	7252	436	370	301	246	206	163	149	147	148

Conversion Factors for Thermal Conductivity

$$1000 \text{ mW}/(\text{m} \cdot \text{K}) = 0.57779 \quad \text{Btu}/(\text{hr} \times \text{ft} \times \text{°F})$$
$$= 449.62 \quad \text{lbf}/(\text{hr} \times \text{°F})$$
$$= 0.16933 \quad \text{W}/(\text{ft} \times \text{°F})$$
$$= 367.10 \quad \text{kgf}/(\text{hr} \times \text{K})$$
$$= 0.0023885 \text{ cal}/(\text{s} \times \text{cm} \times \text{K})$$
$$= 0.85985 \quad \text{kcal}/(\text{hr} \times \text{m} \times \text{K})$$

Table 10.   Conversion Factors
Basic Equivalents

Length: One inch = 2.54000 cm. = 0.025400 m
Mass: One pound (lb) = 0.45359237 kg
Force: One kilogram (kgf) = 9.80665 N
            One pound (lbf) = 4.448222 N
            One pascal = 1 N/m^2
            One bar = 10^5 N/m^2
Pressure: One bar = 10^6 dynes/cm^2 = 0.1 MPa
            One atmosphere = 1.01325 bars = 0.101325 MPa
Energy: One calorie (Int. Table) = 4.1868 joules
            One joule = one watt-second = 1 $N \cdot m$
            = 10^7 dyne-cm = 10^7 ergs

Specific Entropy and Specific Heat Capacity:
            One Btu/lb deg Rankine = 1 kcal/(kg·K) = 4.1868 kJ/(kg·K)
Temperature: On the Kelvin scale the temperature of the
            triple point of water is 273.16.
            T (Rankine) = 1.8 T (Kelvin)
            t (Celsius) = T (Kelvin) − 273.15
            t (Fahrenheit) = 32 + 1.8 t (Celsius)
            The Kelvin, Rankine, Celsius, and
            Fahrenheit scales used here are all
            thermodynamic scales of temperature.

Specific Volume

$$1 ft^3/lb = 0.062428\ m^3/kg \qquad 1\ m^3/kg = 16.0185\ ft^3/lb$$

Pressure[1]

	atm	kgf/cm^2	lbf/in^2	bar	mm Hg	in Hg	MPa
1 atm	1*	1.03323	14.6959	1.01325*	760.000	29.9213	.101325
1 kgf/cm^2	0.967841	1*	14.2233	0.980665*	735.559	28.9590	.0980665
10 lbf/in^2	0.680460	0.703070	10*	0.689476	517.149	20.3602	.0689476
1 bar	0.986923	1.01972	14.5038	1*	750.062	29.5300	.1*
10^3 mm Hg	1.31579	1.35951*	19.3368	1.33322	1000*	39.3701	.133322
10 in Hg	0.334211	0.345316	4.91154	0.338639	254*	10*	.0338639
1 MPa	9.86923	10.1972	145.038	10*	7500.62	295.300	1*

The units mm Hg and in Hg are conventional barometric and manometric units of pressure. The conventional "mercury" is, by definition, fluid having exactly an invariable density of 13.5951 g/cm^3. Gravity is 980.665 cm/s^2.

Energy[1]

	joule	kgf m	ft lbf	W h	cal	Btu	(lbf/in^2) $\times ft^3$	atm cm^3	hp h
10^4 joules	10000*	1019.72	7375.62	2.77778	2388.46	9.47817	51.2196	98692.3	0.00372506
10^4 kgf m	98066.5*	10000*	72330.1	27.2407	23422.8	92.9491	502.293	967841	0.0365304
10^4 ft lbf	13558.2	1382.55	10000*	3.76616	3238.32	12.8507	69.4444	133809	0.00505051
10^{-3} kW h	3600*	367.098	2655.22	1*	859.845	3.41214	18.4391	35529.2	0.00134102
10^4 cal	41868*	4269.35	30880.3	11.63*	10000*	39.6832	214.446	413205	0.0155961
10 Btu	10550.6	1075.86	7781.69	2.93071	2519.96	10*	54.0395	104126	0.00393015
10^2(lbf/in^2) $\times ft^3$	19523.8	1990.87	14400*	5.42327	4663.17	18.5050	100*	192685	0.00727273
10^5 atm $\times cm^3$	10132.5*	1033.23	7473.35	2.81458	2420.11	9.60376	51.8983	100000*	0.00377442
10^{-3} hp h	2684.52	273.745	1980*	0.745700	641.186	2.54443	13.75*	26494.1	0.001*
10^{23} el-volts	16020.7	1633.66	11816.26	4.45020	3826.48	15.18469	82.0574	158112.0	0.00596781

The joule is the unit formerly called the "Absolute" joule. The "International" joule was abandoned on 31st December, 1947.

The calorie used in this table is the "International Table Calorie (1956)".

Specific Energy[1]

	$\dfrac{kJ}{kg}$	$\dfrac{kgf\ m}{g}$	$\dfrac{ft\ lbf}{lb}$	$\dfrac{W\ h}{g}$	$\dfrac{cal}{g}$	$\dfrac{Btu}{lb}$	$\dfrac{(lbf/in^2)}{(lb/ft^3)}$	$\dfrac{atm\ cm^3}{g}$	$\dfrac{hp\ h}{lb}$
$10^3\dfrac{kJ}{kg}$	1000*	101.972	334553	0.277778	238.846	429.923	2323.28	9869.23	0.168966
$10\dfrac{kgf\ m}{g}$	98.0665*	10*	32808.4	0.0272407	23.4228	42.1610	227.836	967.841	0.0165699
$10^6\dfrac{ft\ lbf}{lb}$	2989.07	304.8*	1000000*	0.830296	713.926	1285.07	6944.44	29499.8	0.505051
$10^{-5}\dfrac{kW\ h}{g}$	36*	3.67098	12043.9	0.01*	8.59845	15.4772	83.6381	355.292	0.00608277
$10\dfrac{cal}{g}$	41.868*	4.26935	14007.0	0.01163*	10*	18*	97.2712	413.205	0.00707427
$10^2\dfrac{Btu}{lb}$	232.6*	23.7186	77816.9	0.0646111	55.5556	100*	540.395	2295.58	0.0393015
$10^4\dfrac{(lbf/in^2)}{(lb/ft^3)}$	4304.26	438.912	1440000*	1.19563	1028.05	1850.50	10000*	42479.7	0.727273
$10^4\dfrac{atm\ cm^3}{g}$	1013.25*	103.323	338985	0.281458	242.011	435.619	2354.07	10000*	0.171205
$1\dfrac{hp\ h}{lb}$	5918.35	603.504	1980000*	1.64399	1413.57	2544.43	13750*	58409.6	1*

* Exact value by definition

[1] Lefevre, E. J., "Conversion Factors for Specific Energy, Energy, and Pressure" Heat Division Paper, DSIR, Aug., 1956.

Table 10 (Cont.)

Temperature Conversion. Celsius (C) to Fahrenheit (F)

°C	°F	°C	°F	°C	°F	°C	°F	°C	°F	°C	°F	°C	°F
−50	−58	150	302	350	662	550	1022	750	1382	950	1742	1150	2102
−45	−49	155	311	355	671	555	1031	755	1391	955	1751	1155	2111
−40	−40	160	320	360	680	560	1040	760	1400	960	1760	1160	2120
−35	−31	165	329	365	689	565	1049	765	1409	965	1769	1165	2129
−30	−22	170	338	370	698	570	1058	770	1418	970	1778	1170	2138
−25	−13	175	347	375	707	575	1067	775	1427	975	1787	1175	2147
−20	−4	180	356	380	716	580	1076	780	1436	980	1796	1180	2156
−15	5	185	365	385	725	585	1085	785	1445	985	1805	1185	2165
−10	14	190	374	390	734	590	1094	790	1454	990	1814	1190	2174
−5	23	195	383	395	743	595	1103	795	1463	995	1823	1195	2183
0	32	200	392	400	752	600	1112	800	1472	1000	1832	1200	2192
5	41	205	401	405	761	605	1121	805	1481	1005	1841	1205	2201
10	50	210	410	410	770	610	1130	810	1490	1010	1850	1210	2210
15	59	215	419	415	779	615	1139	815	1499	1015	1859	1215	2219
20	68	220	428	420	788	620	1148	820	1508	1020	1868	1220	2228
25	77	225	437	425	797	625	1157	825	1517	1025	1877	1225	2237
30	86	230	446	430	806	630	1166	830	1526	1030	1886	1230	2246
35	95	235	455	435	815	635	1175	835	1535	1035	1895	1235	2255
40	104	240	464	440	824	640	1184	840	1544	1040	1904	1240	2264
45	113	245	473	445	833	645	1193	845	1553	1045	1913	1245	2273
50	122	250	482	450	842	650	1202	850	1562	1050	1922	1250	2282
55	131	255	491	455	851	655	1211	855	1571	1055	1931	1255	2291
60	140	260	500	460	860	660	1220	860	1580	1060	1940	1260	2300
65	149	265	509	465	869	665	1229	865	1589	1065	1949	1265	2309
70	158	270	518	470	878	670	1238	870	1598	1070	1958	1270	2318
75	167	275	527	475	887	675	1247	875	1607	1075	1967	1275	2327
80	176	280	536	480	896	680	1256	880	1616	1080	1976	1280	2336
85	185	285	545	485	905	685	1265	885	1625	1085	1985	1285	2345
90	194	290	554	490	914	690	1274	890	1634	1090	1994	1290	2354
95	203	295	563	495	923	695	1283	895	1643	1095	2003	1295	2363
100	212	300	572	500	932	700	1292	900	1652	1100	2012	1300	2372
105	221	305	581	505	941	705	1301	905	1661	1105	2021	1305	2381
110	230	310	590	510	950	710	1310	910	1670	1110	2030	1310	2390
115	239	315	599	515	959	715	1319	915	1679	1115	2039	1315	2399
120	248	320	608	520	968	720	1328	920	1688	1120	2048	1320	2408
125	257	325	617	525	977	725	1337	925	1697	1125	2057	1325	2417
130	266	330	626	530	986	730	1346	930	1706	1130	2066	1330	2426
135	275	335	635	535	995	735	1355	935	1715	1135	2075	1335	2435
140	284	340	644	540	1004	740	1364	940	1724	1140	2084	1340	2444
145	293	345	653	545	1013	745	1373	945	1733	1145	2093	1345	2453

TABLE OF VALUES FOR INTERPOLATION IN ABOVE

1°C = 1.8°F	4°C = 7.2°F	7°C = 12.6°F	1°F = 0.55°C	4°F = 2.22°C	7°F = 3.88°C
2 = 3.6	5 = 9.0	8 = 14.4	2 = 1.11	5 = 2.77	8 = 4.44
3 = 5.4	6 = 10.8	9 = 16.2	3 = 1.66	6 = 3.33	9 = 5.00

All decimals are exact.   All decimals are repeating decimals.

Temperature Conversion. International Practical Scale (int)* to Thermodynamic (th) Celsius Scale

t_{int}	$t_{th}-t_{int}$	t_{int}	$t_{th}-t_{int}$	t_{int}	$t_{th}-t_{int}$	t_{int}	$t_{th}-t_{int}$	t_{int}	$t_{th}-t_{int}$	t_{int}	$t_{th}-t_{int}$	t_{int}	$t_{th}-t_{int}$	t_{int}	$t_{th}-t_{int}$
0	0	100	−.006	200	.041	300	.096	400	.115	500	.081	700	.175	900	.525
5	−.002	105	−.005	205	.044	305	.098	405	.114	510	.080	710	.186	910	.549
10	−.004	110	−.003	210	.047	310	.100	410	.114	520	.079	720	.198	920	.574
15	−.006	115	−.001	215	.050	315	.102	415	.112	530	.079	730	.211	930	.600
20	−.008	120	.001	220	.052	320	.104	420	.111	540	.079	740	.224	940	.626
25	−.009	125	.003	225	.055	325	.106	425	.109	550	.080	750	.238	950	.653
30	−.010	130	.005	230	.058	330	.107	430	.108	560	.082	760	.253	960	.680
35	−.011	135	.007	235	.061	335	.109	435	.105	570	.085	770	.268	970	.708
40	−.012	140	.009	240	.064	340	.110	440	.103	580	.088	780	.284	980	.737
45	−.012	145	.011	245	.067	345	.112	445	.100	590	.092	790	.301	990	.766
50	−.012	150	.014	250	.070	350	.113	450	.098	600	.096	800	.318	1000	.796
55	−.013	155	.016	255	.073	355	.114	455	.095	610	.101	810	.336	1050	.956
60	−.012	160	.019	260	.076	360	.115	460	.093	620	.107	820	.354	1100	1.131
65	−.012	165	.021	265	.078	365	.115	465	.091	630	.113	830	.373	1150	1.323
70	−.012	170	.024	270	.081	370	.116	470	.089	640	.120	840	.393	1200	1.531
75	−.011	175	.027	275	.084	375	.116	475	.087	650	.128	850	.414	1250	1.754
80	−.010	180	.029	280	.086	380	.116	480	.086	660	.136	860	.435	1300	1.994
85	−.010	185	.032	285	.089	385	.116	485	.084	670	.144	870	.456		
90	−.009	190	.035	290	.091	390	.116	490	.083	680	.154	880	.479		
95	−.007	195	.038	295	.094	395	.116	495	.082	690	.164	890	.501		

*t_{int} in this table refers to the International Practical Scale of 1948. Since the completion of these tables the International Practical Scale of 1968 has been adopted and published. The differences between t_{th} used here and the IPTS of 1968 are considerably smaller than $t_{th}-t_{int}$ used in this table. For virtually all purposes they may be assumed to be zero.

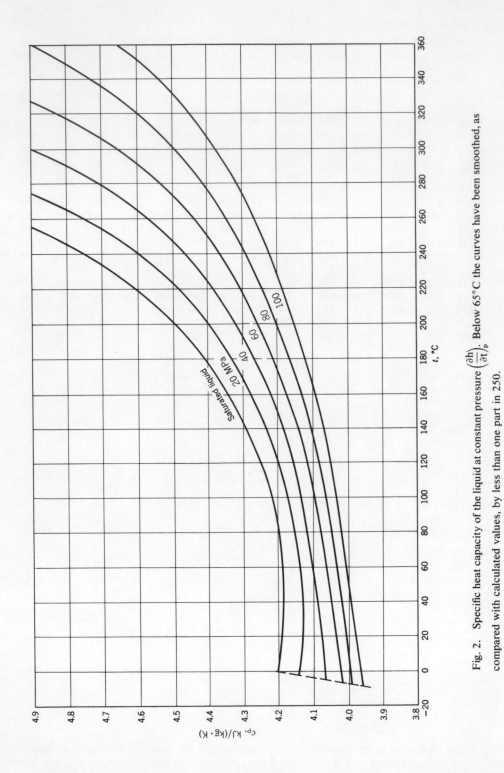

Fig. 2. Specific heat capacity of the liquid at constant pressure $\left(\frac{\partial h}{\partial t}\right)_p$. Below 65°C the curves have been smoothed, as compared with calculated values, by less than one part in 250.

(120)

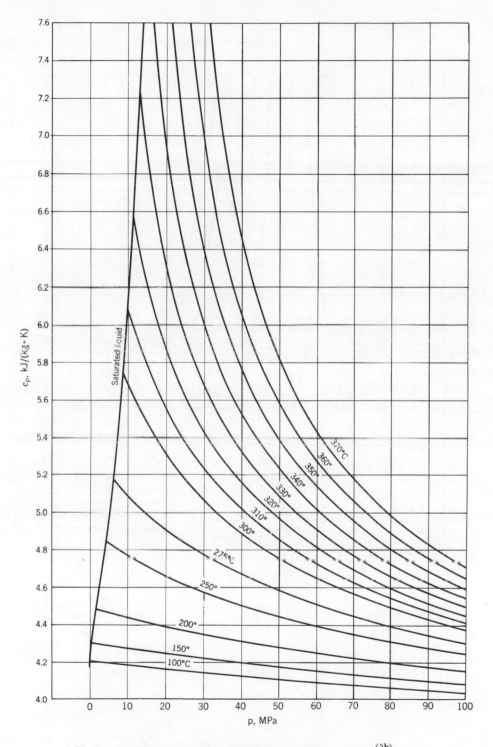

Fig. 3.   Specific heat capacity of the liquid at constant pressure $\left(\dfrac{\partial h}{\partial t}\right)_p$;

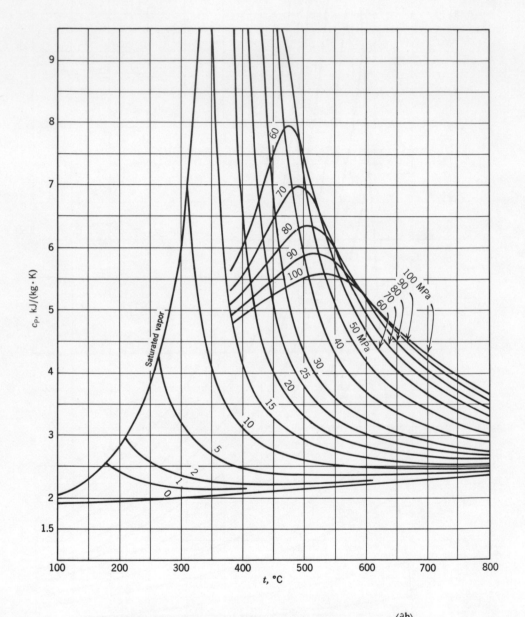

Fig. 4. Specific heat capacity of the vapor at constant pressure $\left(\dfrac{\partial h}{\partial t}\right)_p$.

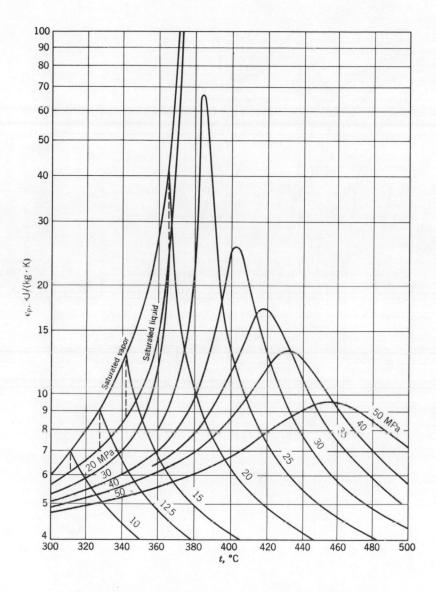

Fig. 5. Specific heat capacity at constant pressure near the critical point $\left(\dfrac{\partial h}{\partial t}\right)_p$.

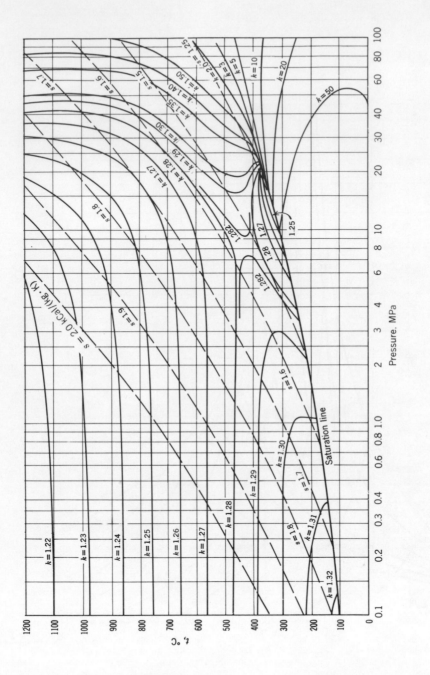

Fig. 6. Isentropic expansion exponents

$$k = -\left[\frac{\partial(\log p)}{\partial(\log v)}\right]_s = -\frac{v}{p}\left(\frac{\partial p}{\partial v}\right)_s = -\frac{v}{p}\left(\frac{\partial p}{\partial v}\right)_T \frac{c_p}{c_v}.$$

For small changes in pressure (or volume) along an isentrope, pv^k = constant.

1 kcal/(kg·K) = 4.1868 kJ/(kg·K)

(124)

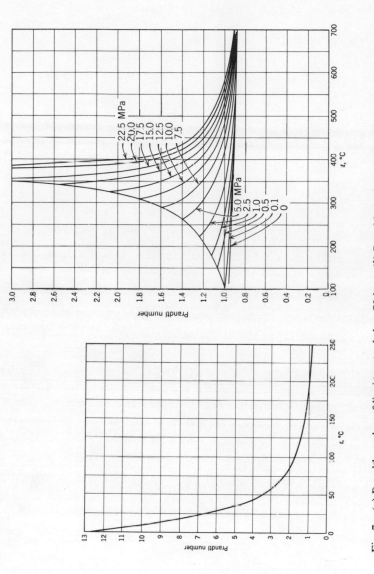

Fig. 7. (a) Prandtl number of liquid water below 70 bars; (b) Prandtl number of superheated water vapor.

Appendix

Appendix

THE FUNDAMENTAL EQUATION

All values that make up the present tables of thermodynamic properties of liquid and vapor water (i.e., exclusive of values for the solid and for transport properties) were obtained from the following fundamental equation which expresses the characteristic function ψ, called the Helmholtz free energy, in terms of the independent variables density ρ and temperature on the Kelvin scale T:

$$\psi = \psi_0(T) + RT \left[\ln \rho + \rho Q(\rho, \tau)\right], \tag{1}$$

where

$$\psi_0 = \sum_{i=1}^{6} C_i/\tau^{i-1} + C_7 \ln T + C_8 \ln T/\tau \tag{2}$$

and

$$Q = (\tau - \tau_c) \sum_{j=1}^{7} (\tau - \tau_{aj})^{j-2} \left[\sum_{i=1}^{8} A_{ij} (\rho - \rho_{aj})^{i-1} + e^{-E\rho} \sum_{i=9}^{10} A_{ij} \rho^{i-9}\right]. \tag{3}$$

In (1), (2), and (3) T denotes temperature on the Kelvin scale, τ denotes $1000/T$, ρ denotes density in g/cm³, $R = 4.6151$ bar cm³/g°K or 0.46151 J/g°K, $\tau_c \equiv 1000/T_{crit} = 1.544912$, $E = 4.8$, and

$$\tau_{aj} = \tau_c (j=1) \qquad \rho_{aj} = 0.634 (j=1)$$
$$= 2.5 (j > 1), \qquad = 1.0 (j > 1).$$

The coefficients for ψ_0 in joules per gram are given as follows:

$$C_1 = 1857.065 \qquad C_4 = 36.6649 \qquad C_7 = 46.$$
$$C_2 = 3229.12 \qquad C_5 = -20.5516 \qquad C_8 = -1011.249$$
$$C_3 = -419.465 \qquad C_6 = 4.85233$$

the coefficients A_{ij} are listed in Table I.

THERMODYNAMIC PROPERTIES

The advantage of using a fundamental equation is that all thermodynamic properties can be obtained through derivatives of the characteristic function. Because differentiation, unlike integration, results in no undetermined functions

Table I. The Coefficients A_{ij} in Equation 3

i \ j	1	2	3	4	5	6	7
1	29.492937	-5.1985860	6.8335354	-0.1564104	-6.3972405	-3.9661401	-0.69048554
2	-132.13917	7.7779182	-26.149751	-0.72546108	26.409282	15.453061	2.7407416
3	274.64632	-33.301902	65.326396	-9.2734289	-47.740374	-29.142470	-5.1028070
4	-360.93828	-16.254622	-26.181978	4.3125840	56.323130	29.568796	3.9636085
5	342.18431	-177.31074	0	0	0	0	0
6	-244.50042	127.48742	0	0	0	0	0
7	155.18535	137.46153	0	0	0	0	0
8	5.9728487	155.97836	0	0	0	0	0
9	-410.30848	337.31180	-137.46618	6.7874983	136.87317	79.847970	13.041253
10	-416.05860	-209.88866	-733.96848	10.401717	645.81880	399.17570	71.531353

or constants, the information yielded is complete and unambiguous.

The basic relations [1] for determining values of pressure, specific internal energy, and specific entropy are as follows:

$$p = \rho^2 \left(\frac{\partial \psi}{\partial \rho}\right)_\tau, \tag{4}$$

$$u = \left[\frac{\partial(\psi\tau)}{\partial\tau}\right]_\rho, \tag{5}$$

$$s = -\left(\frac{\partial\psi}{\partial T}\right)_\rho. \tag{6}$$

Values for specific enthalpy and specific Gibbs free energy are found in turn from their definitions in these forms:

$$h \equiv u + pv, \tag{7}$$

and

$$\zeta \equiv \psi + pv. \tag{8}$$

Specific heat capacities at constant volume and constant pressure are found by further differentiation:

$$c_v = \left(\frac{\partial u}{\partial T}\right)_\rho \tag{9}$$

and

$$c_p = \left(\frac{\partial h}{\partial T}\right)_p = \left(\frac{\partial h}{\partial T}\right)_\rho - \frac{(\partial h/\partial\rho)_T (\partial p/\partial T)_\rho}{(\partial p/\partial\rho)_T}. \tag{10}$$

Substitution from (1) into (4) to (6) yields

$$p = \rho RT \left[1 + \rho Q + \rho^2 \left(\frac{\partial Q}{\partial\rho}\right)_\tau \right], \tag{11}$$

$$u = RT \rho\tau \left(\frac{\partial Q}{\partial\tau}\right)_\rho + \frac{d(\psi_0\tau)}{d\tau}, \tag{12}$$

and

$$s = -R\left[\ln\rho + \rho Q - \rho\tau\left(\frac{\partial Q}{\partial\tau}\right)_\rho \right] - \frac{d\psi_0}{dT}. \tag{13}$$

It follows from (7), (11), and (12) that

$$h = RT \left[\rho\tau\left(\frac{\partial Q}{\partial\tau}\right)_\rho + 1 + \rho Q + \rho^2\left(\frac{\partial Q}{\partial\rho}\right)_\tau \right] + \frac{d(\psi_0\tau)}{d\tau}. \tag{14}$$

Other properties of thermodynamic interest may likewise be expressed in terms of the function Q; for example, the second virial coefficient B is given by

$$B \equiv \left\{\left[\frac{\partial(p/\rho RT)}{\partial\rho}\right]_T\right\}_{\rho=0} = Q_{\rho=0}. \tag{15}$$

(130)

TEMPERATURE SCALES

Experimental observations on the properties of water are almost invariably reported in terms of the International Practical Temperature Scale (the I. P. scale) [2]. In order, however, to relate one kind of observation to another, it is necessary to associate these observed quantities with a thermodynamic scale of temperature. For this purpose the curve of differences between temperatures on the I. P. scale and the Celsius thermodynamic scale shown in Fig. A-1 was devised. Between 0°C and the sulfur boiling point of 444.6°C the differences in Fig. A-1 conform to the observations of James A. Beattie [3] and his colleagues. At the gold point the Celsius thermodynamic temperature was taken to be in excess of the International Practical temperature by one degree. A curve continuing the curve of Beattie and passing through +1 at the gold-point temperature was drawn as shown in Figure A-1. At the temperatures of the silver and gold points this curve falls within the range of spread of experimental determinations [4] of thermodynamic temperature.

Thermodynamic temperatures for all experimental observations used in determining the coefficients A_{ij} or in comparisons with values calculated from (1) were obtained by adding to the reported observed temperatures the corrections given by Figure A-1. Thus a measured temperature of 600°C (I.P.) is taken to be 600.096°C (thermodynamic) or 873.246°K.

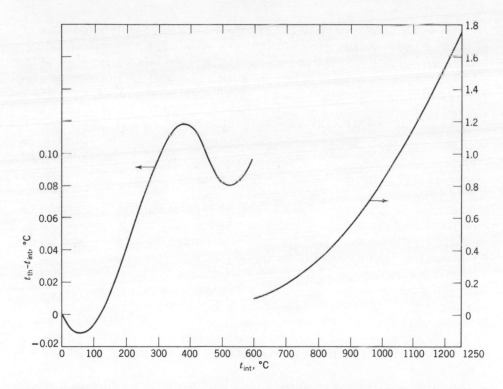

Fig. A-1. Difference between thermodynamic Celsius scale and International Practical Scale of Temperature.

Because the I.P. scale will doubtless be revised to bring it into better accord with the thermodynamic Celsius scale, it was decided to use thermodynamic Celsius temperatures. For the present, therefore, any temperature measured on an instrument calibrated in the usual way by a national calibrating laboratory should be corrected in accordance with Figure A-1 before it is used for calculating properties from (1) or for finding values from the tables. Whether or not this correction is made will in most engineering applications be a matter of indifference. Moreover, on any subsequent revision of the I.P. scale, the correspondingly revised correction should shrink to a quantity with an effect of a smaller order of magnitude than many other uncertainties in the available experimental data on water vapor.

DETERMINING COEFFICIENTS FROM EXPERIMENTAL DATA

The coefficients of the fundamental equation (1) appear in the expressions for pressure (4), energy (5), entropy (6), enthalpy (7), second virial coefficient (15), and heat capacity, (9) and (10). To determine these coefficients a number of fixed conditions smaller than the number of coefficients were established, and a large number of selected observed values were weighted and introduced into a least-squares procedure. The procedure was to minimize a quantity of the form

$$\Sigma \left\{ \frac{P_{obs} - P_{calc}}{\Delta} \right\}^2 + \lambda_1 G_1 + \lambda_2 G_2 + \cdots \tag{16}$$

with respect to all desired A_{ij} and λ_n, where P_{obs} denotes the observed value of a property P (either pressure or enthalpy), P_{calc} is the value of P corresponding to (1), Δ is an assigned measure of precision which is small for observations of high precision and large for those of low precision, $\lambda_1, \lambda_2, \cdots$ are LaGrangian multipliers, and $G_1 = 0$, $G_2 = 0$, $\cdots$ are the constraints corresponding to fixed conditions.

The selection of fixed constraints and observed values used in minimizing (16) is explained in the following paragraphs.

FIXED CONSTRAINTS

The fixed constraints satisfied by (1), which has a total of 59 constants, are outlined in Table II and shown graphically in Figure A-2. The purpose of item 1 in Table II is to conform to a convention established by the Fifth International Conference on the Properties of Steam.

Item 2 is in recognition of the detailed observations and study of the modes of behavior of the water molecule which have been carried out over the last three or four decades. The equation of item 2 represents the values of Friedman and Haar faithfully to 1300°C (2400°F). Because at 1300°C water dissociates appreciably at a pressure of one atmosphere, the values given here, which are for undissociated water vapor at this high temperature, are of limited usefulness near zero pressure.

The conditions imposed by items 1 and 2 in Table II determined the first eight coefficients in (1) as they appear in (2). These conditions therefore did not determine LaGrangian multipliers and were omitted from the minimization procedure.

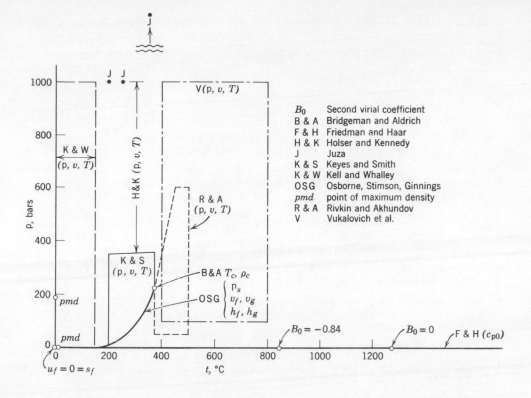

Fig. A-2. Fixed constraints and observed values selected for determining coefficients of equation (1).

A detailed and painstaking re-examination of the measurements of saturation properties by the U.S. National Bureau of Standards was made by O. C. Bridgeman and E. W. Aldrich [6]. Among their conclusions were values for temperature, density, and pressure at the critical state. Item 3 adopts these values for temperature and density. The corresponding calculated value for the pressure is 220.88 bars, compared with 220.91 bars from Bridgeman and Aldrich.

Item 4 was introduced to control the characteristics of the fundamental equation at temperatures above the highest at which observed values of pressure and density are available. It is known from kinetic theory [7] that the second virial coefficient changes sign at a temperature (the Boyle temperature) well in excess of the critical temperature. The value of that temperature chosen here is consistent with an extrapolation of second virial coefficients found from the data of Vukalovich, discussed below, and, in terms of the critical temperature, is consistent with that found by F. G. Keyes for other vapors [8].

The observed pressure-volume-temperature data for the liquid region, by their influence on the coefficients in the least-squares procedure, resulted in a line of maximum density such as liquid water is known to exhibit. The location of this line, however, was not at first in accurate accord with the observed course of the line [9, 10], in pressure and temperature. By means of item 5 the accord was made as good as the available data seem to warrant.

Table II. Fixed Values for Determining Coefficients

1. For the saturated liquid at the triple point, $t_{tp} = 0.01°C$, $p_{tp} = 0.006113$ bar, $u_f = 0 = s_f$.

2. Heat capacity at zero pressure and density is given by the following equation which represents values obtained from spectroscopic observations:

$$c_{po} = 0.046\tau + 1.47276 + 0.83893/\tau$$
$$-0.219989/\tau^2 + 0.246619/\tau^3$$
$$-0.0970466/\tau^4, \text{ in joules/gram.}$$

3. The critical point is fixed at the values of temperature (thermodynamic) and density given by O. C. Bridgeman and E. W. Aldrich [11]: for $t_c = 374.136°C$ and $\rho_c = 0.317$ g/cm³

$$\left(\frac{\partial p}{\partial \rho}\right)_T = 0 = \left(\frac{\partial^2 p}{\partial \rho^2}\right)_T.$$

   The value of the critical pressure was not fixed.

4. Two values of the second virial coefficient B_0 are fixed as follows: for $\tau = 0.9$ ($t \cong 840°C$), $B_0 = -0.84$ cm³/g; for $\tau = 0.65$ ($t \cong 1265.5°C$), $B_0 = 0$.

5. Two points on the line of maximum density in the liquid region are fixed for $\rho = 1$ g/cm³, $t = 4°C$ ($p \cong 2.5$ bars); for $\rho = 1.0089$, $t = 0°C$ ($p \cong 182$ bars)

$$\left(\frac{\partial \rho}{\partial T}\right)_p = 0.$$

   (The equivalent condition $(\partial p/\partial T)_\rho = 0$ was used instead as a matter of convenience.)

OBSERVED VALUES FOR LEAST-SQUARES METHOD

The observed values selected for use in the least-squares procedure are indicated in Figure A-2. The total number of observed values of all kinds was about 600.

Along the line of saturated states the specific volumes and enthalpies were taken from Osborne, Stimson, and Ginnings (O.S.G.) [11], except at temperatures below 50°C, where the liquid specific volumes were based on the data of Kell and Whalley [12]. The O.S.G. data were corrected for the temperature derivative of the difference between the I.P. temperature scale, as used by O.S.G., and the thermodynamic temperature scale.*

For the compressed liquid region the precise measurements of Kell and Whalley were used from 0 to 150°C. From 200 to 370°C to a maximum pressure of 360 bars the coefficients were controlled by the older measurements of Keyes and Smith [13]. The measurements of Holser and Kennedy [14] at 300°C to 1000 bars and those of Juza [15] at 200 and 250°C at 1000 bars and at 350°C at 1500 bars provided control at higher pressures.

In the critical region the excellent measurements of S. L. Rivkin and T. C. Akhundov [16] provided an important control which was modified, as explained

---

*At 0°C, however, the uncorrected value for vapor volume is in better accord with the calculated one. In view of the slight departure here from ideality, either the measured value of O.S.G., the effect of the temperature correction, or both must be in doubt by one part in one or two thousand.

below, by the critical point of Bridgeman and Aldrich. Extension of knowledge of properties to higher pressures and temperatures than before was represented by the extensive and high quality observations of M. P. Vukalovich and his colleagues [17]. The importance of these measurements is reflected in the large area covered by them in Figure A-2.

It will be observed that no measured values except those along the saturation line were used in the vapor region at temperatures below the critical temperature. Built into the fundamental equation (1), of course, is the condition at zero pressure that pv/RT is unity. This along with saturation values and the Rivkin data at the critical temperature provided triangular boundary conditions that by all subsequent tests proved to be sufficient to define the subcritical vapor region.

SATURATION STATES

It is characteristic of a fluid consisting of a single molecular species that two states can coexist in equilibrium over a certain range of temperature. The lower limit of this range is the triple-point temperature below which no stable equilibrium high-density fluid state exists in equilibrium with a low-density fluid state. The upper limit is the critical-point temperature above which no reversible constant-pressure, constant-temperature transition between high- and low-density states can occur. The states within this range that can coexist are called saturated-liquid and saturated-vapor states.

The conjugate saturation states could be found from (1) by identifying at each temperature those two states for which the Gibbs free energy and the pressure are identical. Alternatively, pairs of states may be found for which the pressure is the observed vapor pressure, that is, the pressure of equilibrium of coexisting states. Because, in determining the coefficients of the fundamental equation, the observed values of properties of saturation states used were consistent with equality of the Gibbs free energy at the corresponding vapor pressures, the two methods should yield substantially the same result.

A pair of saturation states at any temperature was taken to be the states of highest and lowest density corresponding to the vapor pressure given by Bridgeman and Aldrich [18] for that temperature. As a matter of convenience and for exact consistency with the critical point of (1) their values of vapor pressure were represented by the following equation:

$$\frac{p_s}{p_c} = \exp\left[\tau \, 10^{-5} \, (t_c - t) \sum_{i=1}^{8} F_i \, (0.65 - 0.01t)^{i-1}\right], \tag{17}$$

where p_s denotes vapor pressure, t saturation temperature (°C$_{th}$), p_c critical pressure (220.88 bars), t_c critical temperature (374.136°C$_{th}$), $\tau = 1000/T$°K, and F_i is given by the following table:

F_1	F_2	F_3	F_4	F_5	F_6	F_7	F_8
−741.9242	−29.72100	−11.55286	−0.8685635	0.1094098	0.439993	0.2520658	0.05218684

The standard deviation of the values of Bridgeman and Aldrich from (17) is about one part in 12,000; the maximum deviation is about one part in 6000.

METASTABLE STATES

The fundamental equation (1) is a representation of a continuum of equilibrium states which includes both liquid and vapor regions. Liquid and vapor states that can coexist in equilibrium at any one temperature may be identified, as explained above, by finding a pair of states for which pressure and Gibbs free energy are identical. Such a pair is made up of a saturated liquid state and a saturated vapor state. Single-phase states corresponding to (1) which are intermediate between saturated liquid and vapor states are, as in the Van der Waals equation, either metastable or unstable.

At temperatures of 300°C to the critical the curve of nonstable states between saturation states has, like the Van der Waals equation, a single maximum and a single minimum. At lower temperatures two maxima and two minima may be found. No significance is attached to this number of extrema. It is doubtless a natural consequence of having a large region of states, such as that between saturation states at low temperatures, without observed values to control the characteristics of an equation as complicated as (1). Because no measurements of properties of nonstable states are available, or likely to become available in the foreseeable future, no control in this region was considered necessary.

Nevertheless, metastable states are of sufficient engineering and scientific interest to justify offering corresponding calculated values of properties as the best values available. Properties of superheated liquid and supersaturated vapor are therefore given in italics as extensions of the tables for compressed liquid and superheated vapor. These extensions are stopped short of the first extremum at each temperature.

COMPARISON WITH OBSERVED VOLUMES

Between 0 and 220°C the calculated specific volume of the saturated liquid is generally within one part in 10,000 of the observed values of Osborne, Stimson, and Ginnings [11], as shown in Figure A-3. To a temperature of 360°C the agreement is within one part in 2000 on volume; near the critical point the agreement on pressure is within one part in 2000.

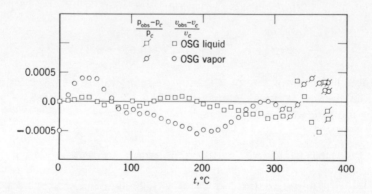

Fig. A-3. Specific volumes of saturated liquid and vapor. Comparison of observed and calculated values.

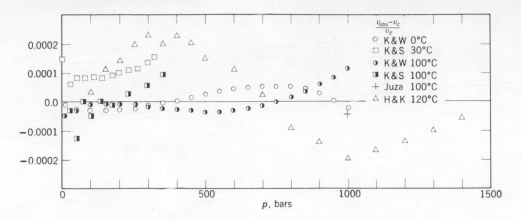

Fig. A-4. Specific volumes of compressed liquid, 0° to 120°C. Comparison of observed and calculated values.

Because the precision of measurement of properties of the vapor is less at low temperatures than for properties of the liquid, the difference between O.S.G. observations of saturated vapor volumes and calculated values is somewhat greater than for liquid. Below 320°C the volume differences are generally less than one part in 2000. Above that temperature the differences in pressure are within the same limit (Figure A-3).

In the compressed-liquid region for temperatures of 150°C and below Kell and Whalley [12] have made extremely precise measurements of pressure, volume, and temperature. Equation 1 represents these measurements within one part in 10,000 on volume (Figure A-4). Although better agreement could doubtless have been attained, further refinement of the coefficients of (1) seemed unwarranted for the present purpose.

The Keyes and Smith p-v-T measurements of nearly 40 years ago [13], which extend to a temperature of 374°C and a pressure of 350 bars, still meet high standards of quality. The agreement between these measurements and the calculated values is with few exceptions within one part in 1500 (Figure A-4, A-5, A-6, A-8).

Juza [15] made measurements at rather wide intervals at 1000 bars and higher and at 100 to 350°C by 50° intervals which were invaluable in providing some control of the coefficients of (1), particularly at high pressures above 150°C. Between 350 and 1000 bars and between 150 and 400°C the measurements by Holser and Kennedy [14] were the sole recourse. Their measurements at 300°C were used in determining the coefficients. Equation 1 represents the measurements of Juza and of Holser and Kennedy in the liquid region to within one part in 1000 to 1000 bars (Figures A-4, A-5, A-6). Although (1) was not intended for use above 1000 bars, it appears that to 350°C it may be useful to 2000 bars.

SUPERHEATED VAPOR

The p-v-T measurements of Keyes, Smith, and Gerry (K.S.G.) [19] for subcritical pressures and temperatures were not used in determining the coefficients

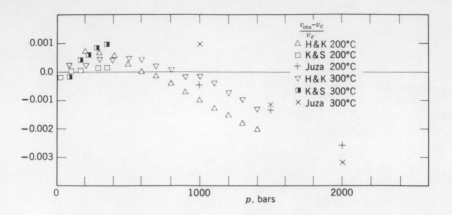

Fig. A-5. Specific volumes of compressed liquid at 200° and 300°C. Comparison of observed and calculated values.

in (1). At the lower temperatures and at specific volumes greater than about 25 cm³/g these measurements are doubtless affected by unknown amounts of adsorption of water on the inner wall of the container. The agreement between measured and calculated values is shown in Figure A-7.

NEAR THE CRITICAL TEMPERATURE

The careful measurements by Rivkin and Akhundov [16] have for the first time yielded detailed information concerning isotherms near the critical point. The differences from calculated values are shown in Figures A-6, A-8, and A-9 to be generally within one part in 1000 except for the immediate neighborhood of the critical point. There a definite *N*-shaped pattern appears in several isotherms. That this pattern is a result of a basic incompatibility between the Rivkin data and the Bridgeman and Aldrich critical point is indicated in Figure A-10. The ordinate of this figure is the slope of an isotherm on a pressure-density plot, that

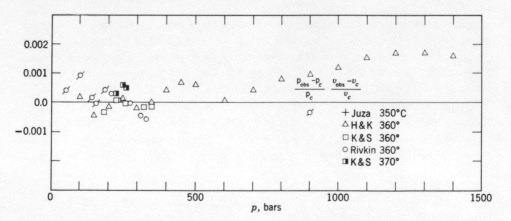

Fig. A-6. Specific volumes of compressed liquid, 350° to 370°C. Comparison of observed and calculated values.

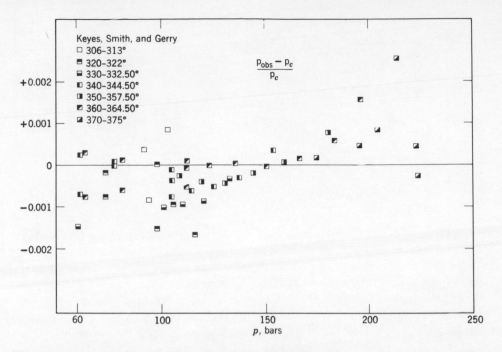

Fig. A-7. Keyes, Smith, and Gerry observations of specific volumes of vapor, 306° to 375°C. Comparison of observed and calculated values.

is, $(\partial p/\partial \rho)_T$, at the critical density. The line represents values calculated from (1). It passes through zero value of the slope at the critical temperature, 374.14°C thermodynamic (374.02°C I.P.), given by Bridgeman and Aldrich. The circles represent values obtained graphically from the Rivkin measurements at temperatures of 374.15, 375, and 380°C (I.P.). It is clear from Figure A-10 that the Rivkin isotherm at 374.15° is incompatible with the Bridgeman and Aldrich critical state and with any probable value of the critical temperature.

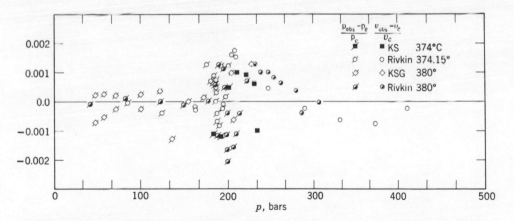

Fig. A-8. Specific volumes near the critical temperature. Comparison of observed and calculated values.

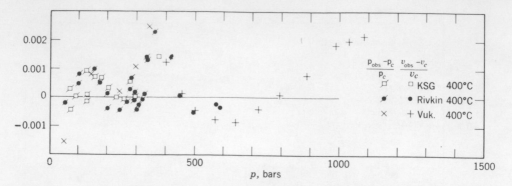

FIG. A-9. Specific volumes at 400°C. Comparison of observed and calculated values.

Approach to equilibrium is so slow in the neighborhood of the critical point that attainment of equilibrium is open to question in all but the most carefully controlled measurements. In view of the nature of the O.S.G. observations, it seems likely that they are less affected by departures from equilibrium than any other measurements available. It was therefore decided that when experimental results were in conflict with the critical state deduced from O.S.G. data by Bridgeman and Aldrich the latter would always be given greater weight.

SUPERCRITICAL TEMPERATURES

At temperatures above the critical temperature the excellent and extensive series of measurements by M. P. Vukalovich and his colleagues [17] are the principal resource. They are supplemented by those of Keyes, Smith, and Gerry [19] to 460°C, by those of Rivkin and Akhundov [16] to 500°C, and by the very extensive but somewhat less precise measurements of Holser and Kennedy [14].

For all temperatures of 400 to 800°C the agreement with Vukalovich is generally within one part in 1000 (Figure A-9, A-12). A slightly larger discrepancy occurs at 400°C and 1000 bars, probably because of a paucity of data at high pressures just below 400°C. The discrepancies, which are progressively larger at 850 and 900°C, attain a maximum of one part in 300 at 900°C and 800 bars

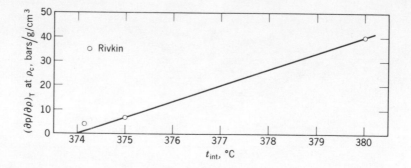

Fig. A-10. Values of $\left(\dfrac{\partial p}{\partial \rho}\right)_T$ at the critical density. Comparison of observed and calculated values.

(140)

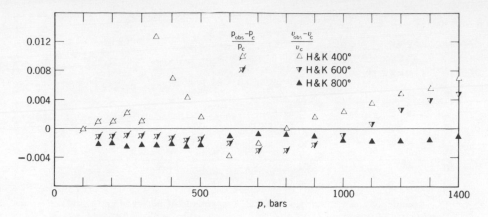

Fig. A-11. Holser and Kennedy observations of specific volumes at 400°, 600°, and 800°C. Comparison of observed and calculated values.

(Figure A-13). As explained above, the observed values at 900°C were not used in determining coefficients. At this temperature, according to the experimenters [17], the container experiences plastic deformation so that the magnitude of the enclosed volume is in doubt.

Comparisons with the data of Holser and Kennedy between 400 and 1000°C are shown in Figures A-11 and A-13.

SECOND VIRIAL COEFFICIENT

The second virial coefficient is B in the virial expansion

$$\frac{pv}{RT} = 1 + B\rho + C\rho^2 + \cdots.$$

For equations of state in general, all of which may be expanded into the virial form,

$$B \equiv \left\{ \left[\frac{\partial(pv/RT)}{\partial \rho} \right]_T \right\}_{\rho=0}$$

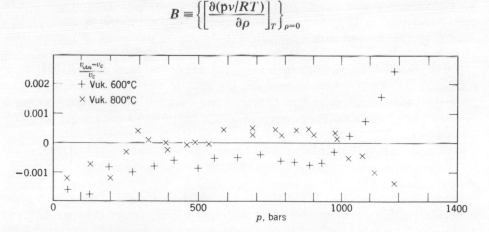

Fig. A-12. Vukalovich observations of specific volumes at 600° and 800°C. Comparison of observed and calculated values.

(141)

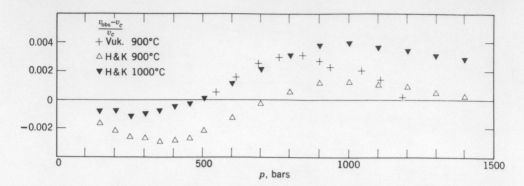

Fig. A-13. Specific volumes at 900° and 1000°C. Comparison of observed and calculated values.

Second virial coefficients are usually deduced from p-v-T measurements at low pressure. Because of adsorption effects on measurements of specific volume, this method is open to question at temperatures below the critical temperature.

In Figure A-14 calculated values of second virial coefficients are compared with those deduced from p-v-T measurements by Keyes, Smith, and Gerry [19] and by Kell, McLaurin, and Whalley [20]. The calculated values are in closer accord with the former. In 1949 Keyes [21] published an equation that gave second virial coefficients, but later high-temperature data prompted a revision that led to the equation

$$B = 2.0624 - 2.61204\tau \, 10^{(0.1008\tau^2/(1 + 0.0349\tau^2))}.$$

The values from this revised equation, shown in Figure A-14, are in close accord with the constant-temperature coefficient $(\partial h/\partial p)_T$ measured by Collins and Keyes [22] from 38 to 125°C.

The Collins and Keyes measurements, being flow measurements, should be free from the effects of adsorption. For a different reason the saturated vapor volumes, determined by O.S.G., are also free of these effects. The observations were calorimetric under circumstances in which adsorption could not be significant. Moreover, at low temperatures the saturated vapor volume must closely determine the second virial coefficient because of the small deviation from ideality.

If it is assumed that an isotherm is a straight line on the chart of pv/RT versus ρ, the saturated vapor volume determines the second virial coefficient. In Figure A-14 the values determined from O.S.G. vapor volumes are plotted for temperatures of 150°C and less. They converge on the calculated values as the temperature decreases.

Goff and Gratch [23] deduced from the O.S.G. vapor volumes and the Collins and Keyes flow measurements values of second virial coefficients at low temperatures (Figure A-14).

Stockmayer's analysis [24] of the second virial coefficient of water as a polar molecule yielded excellent agreement with the K.S.G. values between 127 and 477°C with assumed molecular constants which are in good agreement with other evidence. An equally good agreement could doubtless be found with the present calculated values.

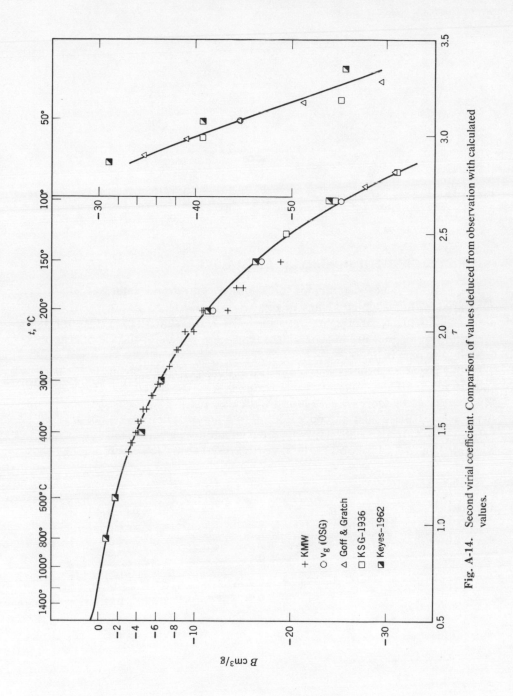

Fig. A-14. Second virial coefficient. Comparison of values deduced from observation with calculated values.

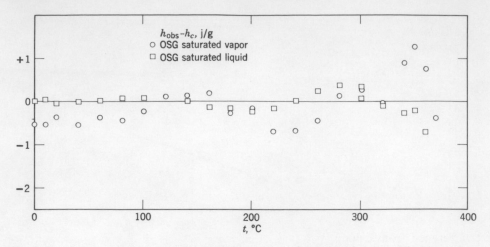

Fig. A-15. Osborne, Stimson, and Ginnings observations of enthalpy of saturated liquid and vapor. Comparison of observed and calculated values.

COMPARISONS WITH OBSERVED ENTHALPIES

The O.S.G. values of enthalpy of saturated liquid and saturated vapor are compared with calculated values in Figure A-15. Between 0 and 150°C the saturated-liquid discrepancies are less than 0.1 J/g. From 0 to 370°C for both saturation states they are in general less than 1 J/g, which for the vapor is approximately one part in 2500.

In the major experimental program on the properties of steam in the 1920's and 1930's enthalpy of superheated vapor was measured by Havlicek and Miskowsky [25] and by Egerton and Callendar [26]. The methods, which consisted of measuring heat on condensation to a liquid state, were similar.

In Figure A-16 these observations are compared with calculated values for 1 to 25 bars. The two sets of observations are in accord with each other, but they show

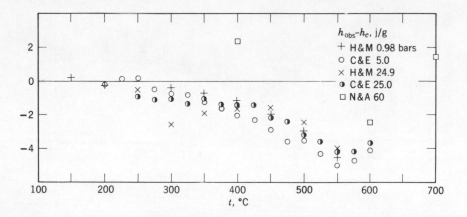

Fig. A-16. Enthalpy of vapor, 0.98 to 60 bars. Comparison of observed and calculated values.

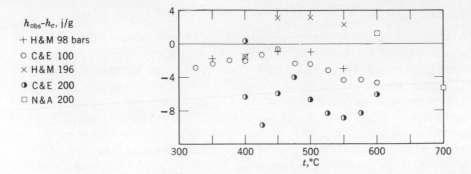

Fig. A-17. Enthalpy of vapor, 98 to 200 bars. Comparison of observed and calculated values.

a difference from calculated values that increases with increasing temperature to a maximum of about 5 J/g at 550°C. It is clear from the constancy of this difference as the pressure decreases toward zero that the observed values are discordant with the calculated values of enthalpy at zero pressure. This discord, which is an incompatibility between the enthalpy observations and the spectroscopically observed heat capacity at zero pressure, was pointed out by F. G. Keyes in 1949 [21].

The comparisons are extended to higher pressures in Figures A-17 and A-18. Also shown are comparisons with the recent excellent measurements by Newitt and Angus (N and Λ) [27]. At the lowest N and A observed pressure of 60 bars the differences indicate no conflict with the spectroscopic observations (Figure A-16). At 200 to 800 bars the N and A differences are of both signs and of reasonable magnitude. The observed difference in enthalpy between 800 and 1000 bars at 400°C is inordinately large in view of the relatively moderate values of derivatives in this region. This difference is in conflict with the observed p-v-t values of Vukalovich.

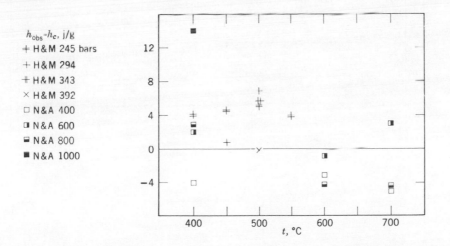

Fig. A-18. Enthalpy of vapor, 245 to 1000 bars. Comparison of observed and calculated values.

COMPARISONS WITH THROTTLING EXPERIMENTS

No observed values of the Joule-Thomson coefficient were used to determine the coefficients of (1). In Figure A-19 the calculated values are compared with the measurements of Davis and Kleinschmidt (Davis and Keenan [28]) of more than 40 years ago. The agreement is excellent except for low pressures at which the observed values decline with pressure as zero pressure is approached. At temperatures of 225°C and above the observed values indicate nearly straight isotherms at all pressures above 3 bars. Below this pressure a downward trend develops as if some new influence were coming into play with decreasing density. Because the behavior of vapors generally simplifies as the intermolecular dis-

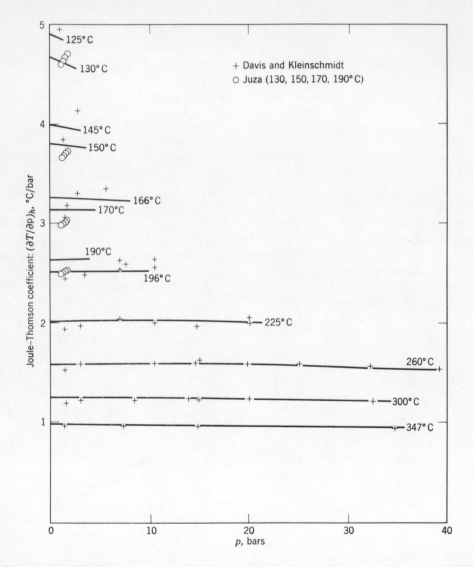

Fig. A-19. Joule-Thomson coefficients, 125° to 347°C. Comparison of observed and calculated values.

tances increase, this trend is open to a suspicion of some kind of experimental error.

The suspicion is strengthened by the exaggeration of this trend in observed values at lower temperatures at which Juza et al. have recently made some measurements [29]. These measurements are in good accord with those of Davis and Kleinschmidt. Although the calculated values for each isotherm are on the average compatible with the observations, the trend of the calculated isotherm with pressure is slightly less than zero, whereas the observed is substantially greater. It can be shown that adoption of an approximately linear isotherm passing through the observed values would not correspond to the O.S.G. enthalpies of the saturated vapor. Only by adopting a strongly curved inverted-U-shaped isotherm could observed values of the Joule-Thomson coefficient and O.S.G. saturated-vapor enthalpies be brought into accord. Taken all together, the evidence seems to cast doubt on the observed slopes of all isotherms in Figure A-19 at low pressures.

For temperatures of 38 to 125°C Collins and Keyes [22] measured the isothermal coefficient $(\partial h/\partial p)_T$ in a throttling experiment and extrapolated to zero pressure. From 60 to 125°C the observed values confirm the calculated ones (Figure A-20). Because the Joule-Thomson coefficient is simply the negative of the quotient of the isothermal coefficient and the heat capacity, this confirmation extends to the calculated Joule-Thomson coefficient. The small discrepancy between observed and calculated values at 38°C is of little significance in view of the small range of pressure for vapor states at that temperature.

COMPARISONS WITH OBSERVED HEAT CAPACITY

Over several decades measurements of heat capacity initiated by Professor Oscar Knoblauch [30, 31] were continued in Germany and were extended to 200 atm by Koch [32]. More recently an extensive series of measurements has been carried out by Sirota and Maltsev [33]. None of these were used in determining the coefficients of (1).

At low pressures the calculated values represent only a slight modification of the zero-pressure spectroscopic measurements and should therefore be highly reliable. The measurements of Knoblauch and Winkhaus [30] at pressures of less than 2 bars are in good accord.

At pressures approaching 200 atm the calculated values rise a little above the the Koch observations near saturation (Figure A-21) but they represent accurately the more recent values of Sirota and Maltsev at 200 kg/cm² (Figure A-22). In fact the accord with the Sirota and Maltsev values at high pressures is extremely good, as shown in Figure A-22. Even near the critical point where the heat capacity rises to infinity the agreement is good. As shown in Figure A-23, a displacement of a fraction of a degree in temperature is all that is needed to bring perfect agreement.

It has been pointed out how difficult it is to attain an equilibrium state in measurements near the critical state, even in static measurements such as p-v-T observations. Because heat-capacity measurements are observations on a flowing fluid, it seems unlikely that equilibrium will prevail for states near the critical

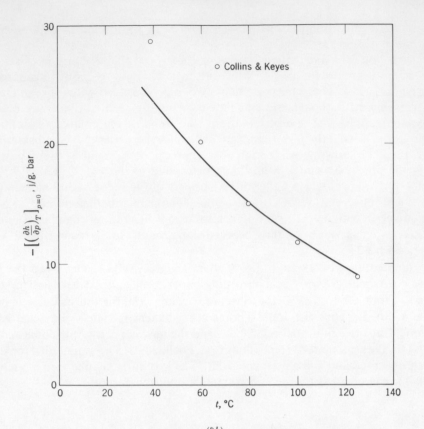

Fig. A-20. Collins and Keyes observations of $\left(\dfrac{\partial h}{\partial p}\right)_T$ at zero pressure. Comparison of observed and calculated values.

point. It is quite possible that a formulation based heavily in this region on the nonflow O.S.G. measurements and reflecting the p-v-T measurements beyond a short radius from the critical point would be at least as reliable as any flow measurement in this difficult region.

COMPARISON WITH SKELETON TABLES OF 1963

The present tables are generally in good accord with the Skeleton Tables of the Sixth International Conference on the Properties of Steam of 1963 [34], but because the fundamental equation (1) is the product of an entirely independent study of the available experimental data, which was completed subsequent to the Sixth Conference, some areas of small disagreement should be expected to appear.

All states for which the calculated specific volume differs from the corresponding value in the Skeleton Tables by more than the Skeleton Table tolerance are given in Table III. Complete tables with deviations appear in [35]. Here the calculated value of specific volume is given for each tabulated state; immediately below that is the Skeleton Table value minus the calculated value, followed by

Table III. Comparisons with International Skeleton Tables of 1963

Liquid and vapor saturation points outside the tolerance

Temp deg C	Pressure Bars	Enthalpy J/g liquid
30.		125.75 / −.09 dev / .08 tol / −.05 osg
40	0.073791 / −.000041 dev / .000038 tol / −.000013 b,a	
50	0.123419 / −.00007 dev / .00006 tol / −.00002 b,a	
374.02 / +.13 dev / .10 tol / 0.0 b,a	220.89 / −.3 dev / .1 tol / +.03 b,a	

Specific volume cm³/g points outside the tolerance

T deg C / P bars	300	350	375	400	425
500	1.28620 / +.002 dev / .001 tol / −.0005 h,k				
600	1.26828 / +.002 dev / .001 tol / +.0005 h,k				
700	1.25237 / −.002 dev / .001 tol / +.0002 h,k	1.37541 / +.005 dev / .003 tol / +.0007 h,k			
750		1.36224 / +.005 dev / .003 tol / −.001 h,k*	1.44112 / +.005 dev / .004 tol / −.001 h,k*		
800		1.35003 / +.005 dev / .003 tol / +.001 h,k*	1.42423 / +.006 dev / .004 tol / −.001 h,k*		
850		1.33864 / +.006 dev / .004 tol / +.001 h,k*	1.40874 / +.006 dev / .004 tol / +.0003 hk*	1.49427 / +.004 dev / .003 tol / .0006 v*	
900		1.32798 / +.006 dev / .004 tol / +.001 h,k*	1.39444 / +.007 dev / .004 tol / +.001 h,k*	1.47468 / +.005 dev / .003 tol / +.001 v*	1.57222 / +.004 dev / .003 tol / +.0005 v*
950		1.31797 / +.006 dev / .004 tol / +.001 h,k*	1.38116 / +.007 dev / .004 tol / +.002 h,k*	1.45677 / +.006 dev / .003 tol / +.002 v*	1.54769 / +.004 dev / .003 tol / +.001 v*
1000		1.30853 / +.005 dev / .004 tol / +.001 h,k*	1.36879 / −.007 dev / .004 tol / −.002 h,k*	1.44029 / +.007 dev / .003 tol / +.003 v*	1.52549 / +.005 dev / .003 tol / +.001 v*

Specific volume cm³/g points outside the tolerance

T deg C / P bars	425	450	475	500	550	600
25	125.17 / −.2 dev / .1 tol / −.03 ksg*	130.16 / −.2 dev / .1 tol / −.03 ksg*				
400	2.53609 / +.01 dev / .009 tol / +.004 r*					
450			3.82495 / −.011 dev / .010 tol / +.002 r*			
500				3.89456 / −.011 dev / .008 tol / +.0003 v*		
550				3.34982 / −.008 dev / .007 tol / +.0002 v*		
600				2.95734 / −.007 dev / .006 tol / −.0003 v*		
550					3.55104 / −.008 dev / .007 tol / −.002 v*	
700					3.22824 / −.007 dev / .006 tol / −.003 v*	
850						3.16178 / −.007 dev / .006 tol / −.002 v*

Specific volume cm³/g points outside the tolerance

T deg C / P bars	150	200	250
50	1.08744 / +.0004 dev / .0003 tol / +.00002 k,w		
100	1.08414 / +.0005 dev / .0004 tol / +.00005 k,w		
125	1.08252 / +.0005 dev / .0004 tol / +.00007 k,w		
300			1.21054 / +.0006 dev / .0005 tol / +.0007 k,s*
900		1.08852 / −.0006 dev / .0005 tol / −.0008 h,k	
950		1.08561 / −.0008 dev / .0005 tol / −.0010 h,k*	
1000		1.08276 / .0010 dev / .0005 tol / −.0006 j / −.0012 h,k	

Values are from Equation 1

dev Sixth International Conference value minus Equation 1
tol Sixth International Conference tolerance

Observed deviations given are observed minus calculated

b,a	Bridgeman and Aldrich
h,k	Holser and Kennedy
j	Juza
k,s	Keyes and Smith
ksg	Keyes, Smith and Gerry
k,w	Kell and Whalley
osg	Osborne, Stimson and Ginnings
r	Rivkin
v	Vukalovich
*	Interpolated

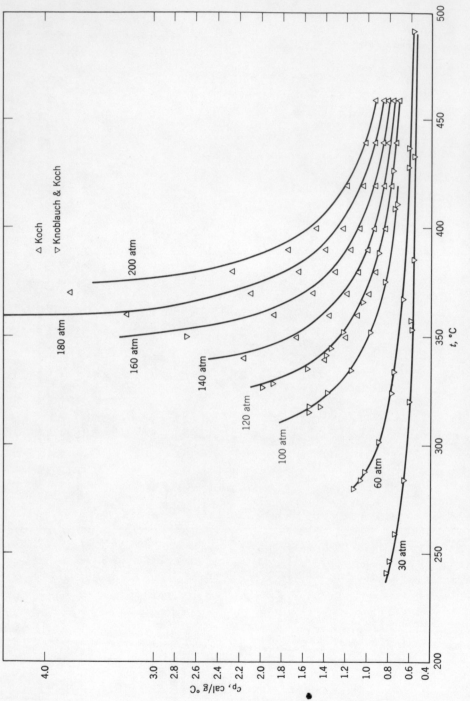

Fig. A-21. Specific heat capacity, 30 to 200 bars. Comparison of observed and calculated values.

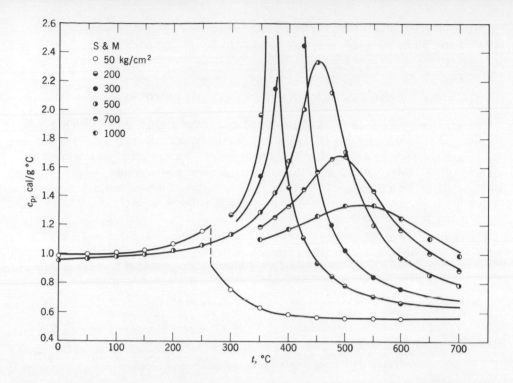

Fig. A-22. Sirota and Maltscv observations of specific heat capacity, 50 to 1000 kg/cm². Comparison of observed and calculated values.

the value of the tolerance in the Skeleton Tables. Finally there is an observed value for that state minus the calculated value.

From Table III it appears that the observed values are in substantially better agreement with (1) than with the Skeleton Tables. At 200°C and 900 to 1000 bars, since Table III shows that the spread of observed values from different experimenters exceeds the stated tolerances, an increase in tolerance seems called for. Because the volume at 250°C and 300 bars lies in a region in which the observations scatter by one part in 1000 on either side of the calculated values, an increase in tolerance again seems called for.

All calculated values of enthalpy but one saturation value differ from the Skeleton Table values by an amount usually less than but occasionally equal to the Skeleton Table tolerance. The largest actual differences appear at the highest pressures and temperatures, where they are less than one part in 300. Two calculated values of vapor pressure and one of saturated liquid enthalpy fall outside the tolerances. These values are given in Table III which shows that in each instance the observed value is in better agreement with the calculated value than with that from the Skeleton Table. The calculated critical temperature, which here is that of Bridgeman and Aldrich, is also outside the tolerance.

Equation 1 provides a critical test of the values of the International Skeleton Tables, as would any independent correlation of the available experimental data. In general it confirms the values of the Skeleton Tables of 1963, although it suggests that small modifications of Skeleton Table specific-volume values should be made. It also suggests that improvements could be made by small modifications of the Skeleton Table enthalpy values at high pressures and temperatures. In some parts of these tables an increase in tolerances seems to be required.

COMPARISON WITH KEENAN AND KEYES TABLES OF 1936

The amounts by which the Keenan and Keyes tables [36] differ from (1) are shown in [35]. Volume differences are generally small fractions of 1 percent. They become as large as 1 percent only above 4000 psi and are at a maximum at about 1300°F, being smaller at both lower and higher temperatures. A somewhat similar characteristic appears in enthalpy differences. They attain a maximum above 4000 psi in excess of 10 Btu/lb at about 1000°F, where the enthalpy is about 1400 Btu/lb.

The upper limit of observed specific volumes available to Keenan and Keyes in 1936 was at 460°C, or 860°F. The extrapolation above this temperature by means

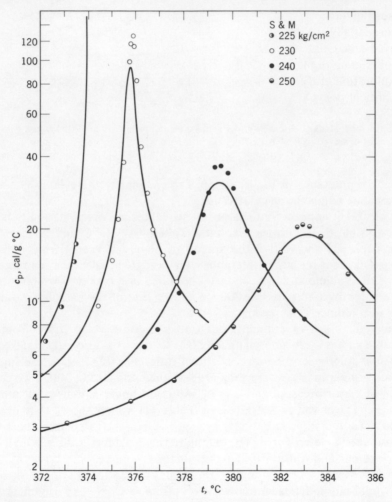

Fig. A-23. Sirota and Maltsev observations of specific heat capacity near the critical point. Comparison of observed and calculated values.

of (13) and (15) in [36] resulted in an increasing error in volume to about 1200°F and a decreasing error from 1200 to 1600°F. At the highest temperature of extrapolation the error is less than ½ percent for all pressures below 3500 psi.

Absence of the "stretching correction" in the Gordon values of heat capacity at zero pressure resulted in slightly lower values of enthalpy at zero pressure for the Keenan and Keyes tables. Errors in the p-v-T relation had the fortunate effect of compensating for the zero-pressure errors over most of the vapor region. The line of zero error in enthalpy stretches diagonally across the vapor region and results in excellent agreement between the old and the new tables except for the immediate neighborhood of the upper bound of pressure in the older tables. Agreement at the critical point in both enthalpy and volume is remarkably good.

TRANSPORT PROPERTIES

For reasons given in the Preface, values for viscosity and thermal conductivity are taken essentially from the Skeleton Tables prepared by a committee of the Sixth International Conference on the Properties of Steam [37]. The thermal conductivity of steam requires further investigation and correlation along with detailed exploration of the critical region at least to 50 Celsius degrees above and below the critical point.

References

[1] G. N. Hatsopoulos and J. H. Keenan, *Principles of General Thermodynamics,* Wiley, New York, 1965, p. 256

[2] H. F. Stimson, *J. Res. Nat. Bur. Stand.*—A. Physics and Chemistry, 65A, No. 3, 139—145 (May-June 1961).

[3] J. A. Beattie, *Proc. Am. Acad. Arts Sci.,* 77, 255—335 (1949).

[4] H. F. Stimson, *J. Nat. Bur. Stand.,* 131 (May-June 1961).

[5] A. S. Friedman and L. Haar, *J. Chem. Phys.,* 22, 2051—2058 (1954).

[6] O. C. Bridgeman and E. W. Aldrich, *Trans. ASME,* 266—274 (May 1965).

[7] J. O. Hirschfelder, C. F. Curtiss, and R. B. Bird, *Molecular Theory of Gases and Liquids,* Wiley, New York, 1954, p. 164.

[8] F. G. Keyes, *Temperature, Its Measurement and Control in Science and Industry,* Reinhold, New York, 1941, p. 59.

[9] N. E. Dorsey, *Properties of Ordinary Water Substance,* Reinhold, New York, 1940, pp. 275—277.

[10] R. J. Zaworski and J. H. Keenan, *Trans. ASME; J. Appl. Mech.,* 34, Series E, 478—483 (1967).

[11] N. S. Osborne, H. F. Stimson and D. C. Ginnings, Thermal Properties of Saturated Water and Steam, *Nat. Bur. Stand., Res. Paper RP1229,* 1939.

[12] G. S. Kell and E. Whalley, *Phil. Trans. Roy. Soc. (London),* 258, 565—617 (1965).

[13] F. G. Keyes and L. B. Smith, *Proc. Am. Acad. Arts Sci.,* 69, 285—314 (1934).

[14] W. T. Holser and G. C. Kennedy, *Am. J. Sci.,* 256, 744—753 (1958), and 257, 71—77 (1959).

[15] J. Juza, V. Kmonicek and O. Sifner, Appendix to J. Juza. An Equation of State for Water and Steam, *Acad., Naklad. Cesk. Acad. Ved,* Prague, 131—142 (1966).

[16] S. L. Rivkin and T. C. Akhundov, *Teploenerg.,* No. 1, 57—65 (1962), and No. 9, 66—68 (1963).

[17] M. P. Vukalovich, V. N. Zubarev, and A. A. Alexandrov, *Teploenerg.,* No. 10, 79—85 (1961), and No. 1, 49—51 (1962).

[18] O. C. Bridgeman and E. W. Aldrich, *Trans. ASME, J. Heat Transfer Series C-D,* 80, 279—286 (1964).

[19] F. G. Keyes, L. B. Smith and H. T. Gerry, *Proc. Am. Acad. Arts Sci.,* 70, 319—364 (1936).

[20] G. S. Kell, G. E. McLaurin and E. Whalley, Report No. 8316, Division of Applied Chemistry, National Research Council, Canada, 1966.

[21] F. G. Keyes, *J. Chem. Phys.,* 17, 923—934 (1949).

[22] S. C. Collins and F. G. Keyes, *Proc. Am. Acad. Arts Sci.,* 72, 283—299 (1938).

[23] J. Goff and S. Gratch, *Trans. A.S.H.V.E.,* 51, 125—158 (1945).

[24] W. H. Stockmayer, *J. Chem. Phys.,* 9, 398—402 (1941).

[25] J. Havlicek and L. Miskowsky, *Hel. Phys. Acta.,* 9, 161—207 (1936).

[26] A. Egerton and G. S. Callendar, *Phil. Trans. Roy. Soc. (London),* 252, 133—164 (1960).

[27] S. Angus and D. M. Newitt, *Phil. Trans. Roy. Soc. (London),* 259, 107—132 (1966).

[28] Reported in H. N. Davis and J. H. Keenan, *World. Eng. Cong. Rept.* No. 455, Tokyo, 1929.

[29] J. Juza, V. Kmonicek and K. Schovanec, document of Mechanical Engineering Research Institute of the Czechoslovak Academy of Sciences, Prague, 1963.

[30] O. Knoblauch and A. Winkhaus, *Mitt. Forsch. Gebiete. Ingenieurw.*, **195**, 1−20 (1917), and *Z. Ver. deutsch. Ingenieure,* **59**, 1915, pp. 376−379, 400−405.

[31] O. Knoblauch and W. Koch, *Z. Ver. deutsch. Ingenieure,* **72,** 1733−1739 (1928).

[32] W. Koch. *Forsch. Gebiete Ingenieurw,* **3,** 1−10, 189 (1932).

[33] A. M. Sirota and B. K. Maltsev. Teploenerg., No. 1, 52−57 (1962), No. 5, 64 (1963) and No. 8, 61 (1966).

[34] Proceedings of the Sixth International Conference on the Properties of Steam, October 1963, issued by ASME, New York.

[35] F. G. Keyes, J. H. Keenan, P. G. Hill and J. G. Moore, "A Fundamental Equation for Liquid and Vapor Water." Paper presented at the Seventh International Conference on the Properties of Steam, Tokyo, Japan, 1968.

[36] J. H. Keenan and F. G. Keyes, *Thermodynamic Properties of Steam,* Wiley, New York, 1936.

[37] Proceedings of the Sixth International Conference on the Properties of Steam, Supplementary Release on Transport Properties, November 1964.